Fundamentals of the Analysis and Design of Shell Structures

Prentice-Hall International Series
in Civil Engineering and Engineering Mechanics

William J. Hall, Editor

Fundamentals of the Analysis and Design of Shell Structures

VASANT S. KELKAR
Consulting Associate Professor of Civil Engineering
Stanford University
Chairman, Dr. Kelkar Consultants Pvt. Ltd.
Bombay, India

ROBERT T. SEWELL
Consulting Engineer

PRENTICE-HALL, INC., *Englewood Cliffs, New Jersey 07632*

Library of Congress Cataloging-in-Publication Data

KELKAR, VASANT S.

Fundamentals of the analysis and design of shell structures.

(Prentice-Hall international series in civil engineering and engineering mechanics)

Includes bibliographies and index.

1. Shells (Engineering) I. Sewell, Robert Thomas. II. Title. III. Series.

TA660.S5K45 1987 624.1'7762 86-25558

ISBN 0-13-334327-8

Editorial/production supervision and interior design: Elena Le Pera
Cover design: Lundgren Graphics, Ltd.
Manufacturing buyer: Rhett Conklin

Printed in the United States of America

10 9 8 7 6 5 4 3 2 1

ISBN 0-13-334327-8 025

Prentice-Hall International (UK) Limited, *London*
Prentice-Hall of Australia Pty. Limited, *Sydney*
Prentice-Hall Canada Inc., *Toronto*
Prentice-Hall Hispanoamericana, S. A., *Mexico*
Prentice-Hall of India Private Limited, *New Delhi*
Prentice-Hall of Japan, Inc., *Tokyo*
Prentice-Hall of Southeast Asia Pte. Ltd., *Singapore*
Editora Prentice-Hall do Brasil, Ltda., *Rio de Janeiro*

TO OUR PARENTS

Contents

Preface

The analysis and design of shell structures is a topic of interest in a variety of engineering disciplines. The civil engineer is concerned with the design of large-span roofs, liquid storage facilities, silos and many other structures. The mechanical engineer is interested in the design of pressure vessels, including nuclear reactor containment vessels and pipes. The aeronautical engineer is involved in the structural design of aircrafts, rockets and aerospace vehicles. All of these structures require the analysis and design of shells in one form or another.

Although many engineers and engineering students need an understanding of the analysis and design of shell structures, they may not be able to devote the time necessary to study in detail or become a specialist in the mathematical theory of shells. The goal of this text is to explain, in a simple and concise manner, the various aspects of shell analysis and design. One cannot, of course, avoid mathematical equations when studying the analysis of shells. However, an effort has been made to explain in clear detail the derivations and solutions of such equations, and to give at every stage their physical interpretation. In this way the reader can gain a clear insight into shell behavior, rather than simply being in awe of the complex mathematical equations. The analytical portions of the text involve concepts covered in courses on strength of materials and ordinary differential equations, which are normally taught in any undergraduate curriculum. The approach taken is consistent and shows the reader that the derivation of equations for plates or shells is only an extension of the derivation of equations for bars or beams, based on equilibrium, kinematics, and Hooke's law.

In design practice, when shell analysis is performed using an available computer program, the designer must have sufficient knowledge of shell behavior to be able to verify the accuracy of the results and interpret them correctly. Using the understanding of shell behavior gained from this text, including the approximate methods and the ready tables for quick solutions given, a reader will be able to judge the computer results, before designing a shell structure. The text also discusses some methods of computer analysis for different shell problems, without delving into the particulars of numerical analysis, such as in the finite element method.

The text begins by reviewing the theories for analysis of bars and beams. The reader can thus easily see the extension of these theories and methods of solutions to the more complicated problem of shell analysis. Theory of circular rings, concepts of stress resultants and middle surface are also introduced in the first chapter. A chapter on the bending of circular plates is included, as the derivations covered are useful for solving subsequent problems. Circular plate theory also forms an intermediate step for a gradual introduction or transition to shell analysis from the analysis of bars and beams.

Chapter 3 gives a basic discussion of surface characterization and classification, as well as a discussion of the differences in the membrane and bending analyses of thin shells. This serves as an introduction to the more detailed treatment on membrane and bending analyses of different types of shells of revolution of single or double curvature under axisymmetric loading, discussed in Chapters 4, 5, 6 and 7.

In Chapter 8 equations of the membrane theory of general shells of double curvature are derived, and these are then applied in Chapter 9 to the solutions of various problems of hyperbolic paraboloid shell roofs with straight-line boundaries. Concepts of Fourier series necessary for solving the equations for other types of shells with double curvature, for nonaxisymmetric load conditions on shells of revolution or for open cylindrical barrel shells, are discussed in detail in Chapter 10. These concepts are then used in Chapter 11 to solve problems of elliptic paraboloids, and hyperbolic paraboloids with generating parabolas as boundaries, as well as for problems involving nonaxisymmetric loads on shells of revolution. Fourier series concepts are also useful for the analysis of cylindrical barrel shell roofs, considered in Chapters 12 and 13. Approximate methods and the use of tables for the analyses of barrel shells and folded plates are discussed in Chapter 14; their design details are considered separately in Chapter 15. It should be emphasized that design examples for various shells (drawn from 15 years of shell design experience of the first author), and general code regulations are presented regularly throughout the text.

The text was developed from a set of notes used for instructing a course given by the first author at Stanford University and also earlier at the Indian Institute of Technology and V.J.T. Institute of Bombay. The aim of this course is to provide advanced undergraduate and first-year graduate students in civil, mechanical, or aeronautical engineering with a working knowledge of shell analysis and design, either as a foundation for further study or as an adequate background for practical design. The text covers shell analysis in a manner useful to all engineering disciplines; however, the design examples emphasize problems normally encountered in civil engineering.

The first author wishes to express his gratitude to Professor W. Flügge for his excellent teaching and advice when the first author was a doctoral student at Stanford University. Inevitably, much of the invaluable knowledge in shell analysis gained from Professor Flügge, is reflected in this text. Both authors wish to thank Professor H. C. Shah of Stanford University for his encouragement during the course of this work. The thoughtful reviews and suggestions by Professors W. A. Bradley, W. C. Schnobrich, and C. R. Steele are sincerely appreciated. Thanks are also due to Pat Simmons and Ditter Peschcke-Koedt for their excellence in typing the manuscript. Finally, the authors wish to thank Prentice-Hall and its staff, and in particular, Dan Joraanstad, Doug Humphrey, and Elena Le Pera, for their extraordinary dedication in seeing this text through its various stages to publication.

Vasant S. Kelkar
Bombay, India

Robert T. Sewell
Stanford, California

1987

1

Introduction

1.1 GENERAL

In designing any structure, whether it be a building, pressure vessel, or an aerospace structure, designers always aim at achieving economy by minimizing costs within the constraints of functional and aesthetic requirements. They thus endeavor to find newer materials of construction which are cheaper and stronger, or with the available materials, try to evolve new forms that resist the loads more efficiently than when the structure is designed in a conventional form.

Thus, for covering a given area by a roof, conventionally one designs a slab and beam structure, with slabs spanning on secondary beams which themselves span between the main beams supported on columns. As column spacing becomes larger and larger—a requirement of most clients—the sizes of the beams increase, thereby making the structure uneconomical and aesthetically unpleasing. Alternatively, to cover the same area, we can conceive a curved surface that carries the loads mainly in direct compression or tension, rather than in bending and in shear as done by the slab and beam structure. Even with a relatively small thickness, such curved surfaces can sustain large loads over large column-free areas with a minimum of deflection. The behavior of such surface structures can be compared to that of a soap-film membrane which covers a large area with extremely small thickness, or to that of an eggshell, which resists considerable pressure in spite of being very thin. Such curved surfaces, which have thicknesses that are small compared to their other two spatial dimensions, are called shells.

Even for carrying internal gas pressure in a pressure vessel or water load in a water tank, a cylindrical, conical, or spherical form that resists loads in direct tension is much more economical than a rectangular form that develops bending moments and shears. The thin curved surfaces of such pressure vessels and water tanks, or of cooling towers, fuselages of airplanes, or hulls of submarines would all fall under the category of shell structures.

Historically, in civil engineering construction, shells in the form of dome roofs have been used from as early as the Pantheon to the domes of such monuments as St. Peter's Church in Rome and the Taj Mahal near New Delhi. The igloo houses used by Eskimos, and the inverted "basket-like" huts made from tree branches and grass reeds, are more primitive forms of dome structures which were used for centuries [1].* The actual study of the mathematical theory of shells, however, was taken up in Germany in the 1920s and 1930s for the analysis and design of planetarium domes and cylindrical barrel shell roofs over factories and fair halls constructed during those years. Theoretical studies then made by Dischinger, Finsterwalder, Flügge [2], and others have spurred, over the past 60 years, the design and construction of a large number of remarkable shell structures throughout the world, most notable among them being the works of such engineers as Candela [3], Torroja [4], Nervi [5], and Tedesko [6]. The Sports Palace in Rome, the M.I.T. Auditorium in Cambridge, Massachusetts, the King Dome in Seattle, a restaurant at Xochimilco in Mexico, and the water towers of Kuwait are just a few of the outstanding civil engineering structures constructed in recent decades which demonstrate the enormous aesthetic appeal of shell structures.

For analyzing a shell structure subjected to a given loading, we could obtain the state of stress by solving the general equations of the three-dimensional *theory of elasticity* [7]. These equations are very complex and are amenable to solution for only a few idealized cases. Hence, for solving practical problems, such methods are not suitable. We therefore try to develop different simplified theories for different categories of problems encountered in practice, so that such problems can readily be solved.

Thus, in undergraduate years, in *strength of materials* [8], we studied theories for solving problems of bars, beams, shafts, cables, etc.—structures that could be represented by straight or curved lines. These linear structures had dimensions in one direction that were large compared to those in the other two directions. By making a few simplifying assumptions, we developed simplified theories for solving problems of bars—structures that resist loads in axial tension or compression; beams—structures that resist loading in bending and shear; shafts—structures that resist loading in torsion; and cables—curved structures that resist loading in axial tension. The equations obtained from the simplified theories were sufficient for solving practical problems, and gave results that were in agreement with experimental observations and more advanced theoretical calculations.

These theories of "line" structures are, however, not applicable for solving problems of plate or shell structures which have two of their dimensions large compared to the third, i.e., the thickness. Hence these structures have to be rep-

*Numbers in brackets refer to references at the end of each chapter.

resented by a surface which may be plane or curved. Simplified theories have been developed for solving such problems. Thus for structures that can be idealized by plane surfaces and have loadings perpendicular to these planes, we have the simplified *theory of plates*.* Also, for shell structures that can be idealized by curved surfaces, we have the simplified *theory of shells*.

For beams, which resist loads mainly in bending and shear, we obtained, in strength of materials, a fourth-order ordinary differential equation in the deflection w and axial coordinate x. Similarly, for plates, which also resist loads in bending and shear, we get a fourth-order partial differential equation in terms of w and coordinates x and y. Plates that have loadings in their planes, as in plane stress problems, resist the loadings by development of in-plane stresses without bending or shear. For such problems, we also get a fourth-order partial differential equation for in-plane deformations in terms of x and y. Shells have loadings both normal and tangential to their surfaces. They resist such loading by development of in-plane stresses, as well as by bending and shear stresses. For shell problems, therefore, we have, in general, an eighth-order partial differential equation in terms of deformations and coordinates x and y. For specific categories of shell problems, e.g., a circular cylindrical shell under axisymmetric loading, this equation can be simplified from a partial to an ordinary differential equation and from an eighth-order to a fourth-order equation.

As we shall see later, the derivations of governing equations for plates or shells follow the same steps as those used for deriving equations for bars or beams in strength of materials. We shall, therefore, first briefly review the derivations of these simplified theories in this chapter.

1.2 THEORY OF BARS

Consider a rectangular bar of length L whose cross-sectional dimensions—width b and thickness t—are small compared to L. For simplicity, we consider b as unity. The bar can be represented by an axis passing through the centroids of all cross sections. This centroidal axis is assumed to be a straight line. The bar is subjected to loading $p_x(x)$ along its axis (Fig. 1.1). Let u be the displacement of the bar along x. We consider now the equilibrium of an element of the bar bounded by sections $x = constant$ and $x + dx = constant$ (Fig. 1.2). N_x is the axial stress resultant on the bar cross section.

Equilibrium

By considering equilibrium in the x-direction, we get

*When the loading is in the plane of the plate, we have the simplified *plane stress* solution for such two-dimensional solids, as obtained from the theory of elasticity [7].

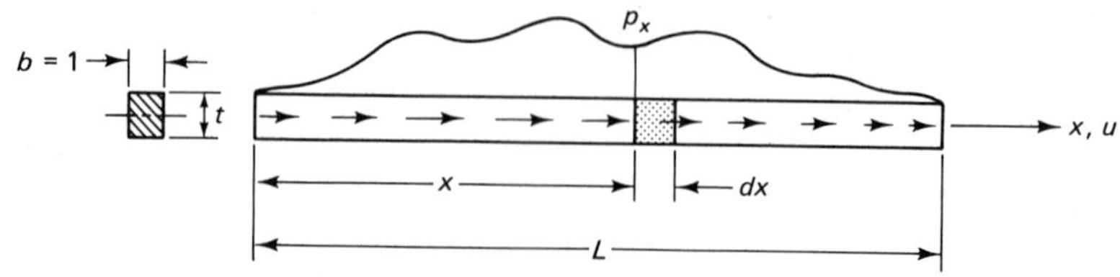

Figure 1.1 Bar subjected to an axial load intensity $p_x(x)$.

$$N_x + \left(\frac{dN_x}{dx}\right) dx - N_x + p_x dx = 0$$

Therefore,

$$N_x' = -p_x \tag{1.1}$$

where $(\cdot)'$ denotes differentiation with respect to x.

This equation contains only one unknown, N_x, which can be obtained for a given loading p_x by simple integration. The constant of integration can be determined from a boundary condition where N_x would be prescribed. Thus this problem could be solved by considering equilibrium equations alone; i.e., it is *statically determinate*. However, for solving general problems wherein not only forces but displacements are also prescribed as boundary conditions, we must derive a more general governing equation. This is done by considering the equations of *kinematics* (strain-displacement relations) and *Hooke's law* (stress-strain relations).

Kinematics

Figure 1.3 shows the deformation of an element of the bar bounded by sections x and $x + dx$. If u is the axial displacement at section x, then $u + (du/dx)\, dx$ will be the displacement at the neighboring location $x + dx$. The axial strain, ε_x, is then given by:

$$\varepsilon_x = \frac{du}{dx} \quad \text{or} \quad \varepsilon_x = u' \tag{1.2}$$

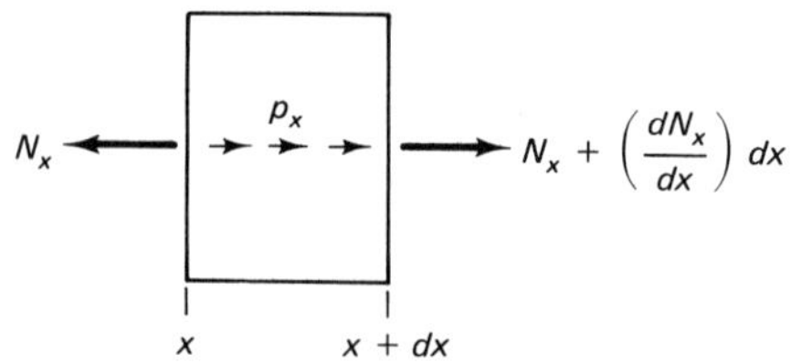

Figure 1.2 Forces acting on a bar element.

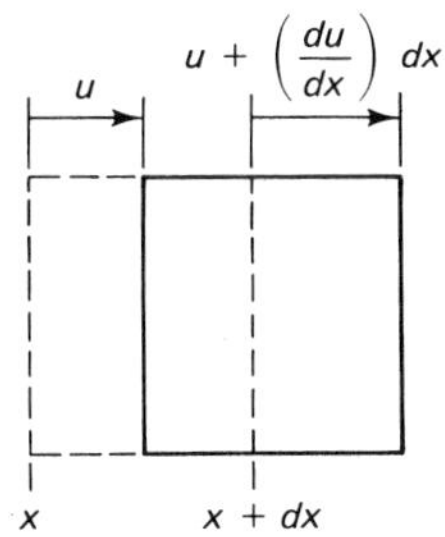

Figure 1.3 Deformation of a bar element.

Hooke's Law

We assume that the material of the bar is homogeneous, isotropic, and linearly elastic, following Hooke's law. Hence we can write the following expression for the axial stress σ_x:

$$\sigma_x = E\varepsilon_x$$

where E is the elastic proportionality constant or *Young's modulus*. Now stress resultant N_x is given by $N_x = 1 \cdot t \cdot \sigma_x$, which gives us

$$N_x = Et\varepsilon_x \qquad \text{or} \qquad N_x = D\varepsilon_x \tag{1.3}$$

where $D = Et$ is called the *extensional rigidity* of the bar. We can now summarize Eqs. (1.1) to (1.3) as follows:

$$\text{Equilibrium:} \quad N_x' = -p_x \tag{1.1}$$

$$\text{Kinematics:} \quad \varepsilon_x = u' \tag{1.2}$$

$$\text{Hooke's law:} \quad N_x = D\varepsilon_x \tag{1.3}$$

These three equations contain three unknowns: N_x, ε_x, and u. From Eqs. (1.2) and (1.3) we have

$$N_x = Du' \tag{1.4}$$

Substituting this in Eq. (1.1), we get

$$u'' = -\frac{p_x}{D} \tag{1.5}$$

This is a second-order ordinary differential equation which governs the displacement u under a given loading p_x. Its general solution will be

$$u = C_1 + C_2 x + u_p \tag{1.6}$$

where C_1 and C_2 are constants to be determined from the boundary conditions, and u_p is a *particular solution* of the equation for a given loading. Using this solution, many practical problems can be solved.

Example 1.1

For the bar shown in Fig. 1.4, fixed at one end and free at the other, with a linearly varying axial load, we can write

$$p_x(x) = p_o \frac{x}{L}$$

Equation (1.5) becomes

$$u'' = -\frac{p_o x}{DL}$$

We can assume that $u_p = Ax^3$, where A is a constant to be determined. Substituting this in the equation above, we get

$$A = -\frac{p_o}{6DL} \quad \text{and} \quad u_p = -\frac{p_o x^3}{6DL}$$

and hence

$$u = C_1 + C_2 x - \frac{p_o x^3}{6DL}$$

We can substitute into the solution above the boundary conditions at $x = 0$, $u = 0$ and at $x = L$, $N_x = 0$ (i.e., $u' = 0$). This gives $C_1 = 0$ and $C_2 = p_o L/2D$. Therefore, the solution becomes

$$u = \frac{p_o L^2}{6D}\left(\frac{3x}{L} - \frac{x^3}{L^3}\right)$$

Expressions for N_x and σ_x can be obtained from this solution.

1.3 THEORY OF BEAMS

We now consider a beam of length L and cross-sectional dimensions, width $b = 1$ and thickness t, which are assumed to be small compared to L. The beam is subjected to a load $p_z(x)$ in a direction normal to its centroidal axis (Fig. 1.5). We also make the following simplifying assumptions:

1. The material of the beam is homogeneous, isotropic, and linearly elastic, following Hooke's law.

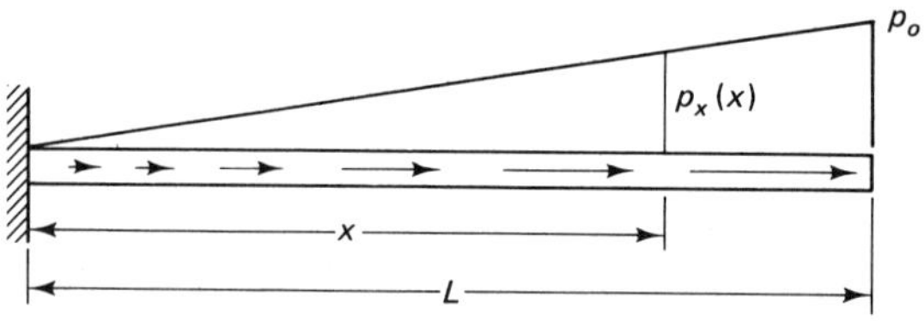

Figure 1.4 Fixed-free bar subjected to a linearly varying axial load.

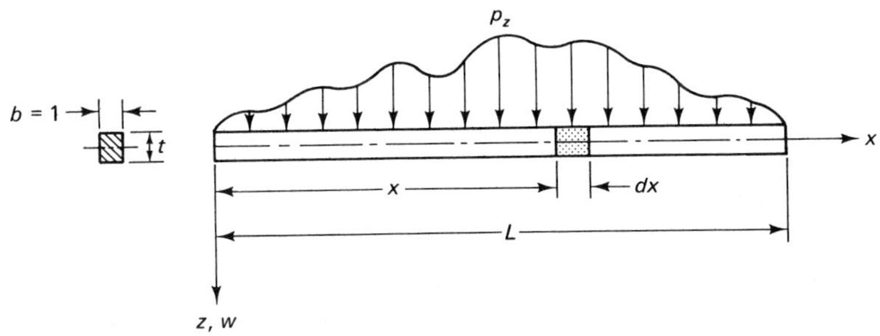

Figure 1.5 Beam subjected to a transverse load $p_z(x)$.

2. The deflection w is small compared to L, so that the slope $w' = dw/dx$ at any point can be considered to be small compared to unity.
3. Plane sections remain plane after bending; i.e., deformations due to shear are neglected.

Proceeding in the same manner as for the bar, we can derive equations of (a) equilibrium, (b) kinematics, and (c) Hooke's law.

Equilibrium

First, considering equilibrium of an element between sections x and $x + dx$ (Fig. 1.6), we can easily write the following equilibrium equations:

$$Q_x + \left(\frac{dQ_x}{dx}\right) dx - Q_x + p_z\, dx = 0$$

$$M_x + \left(\frac{dM_x}{dx}\right) dx - M_x - Q_x\, dx = 0^*$$

Upon simplification, these equations take the form

$$Q'_x = -p_z \tag{1.7}$$

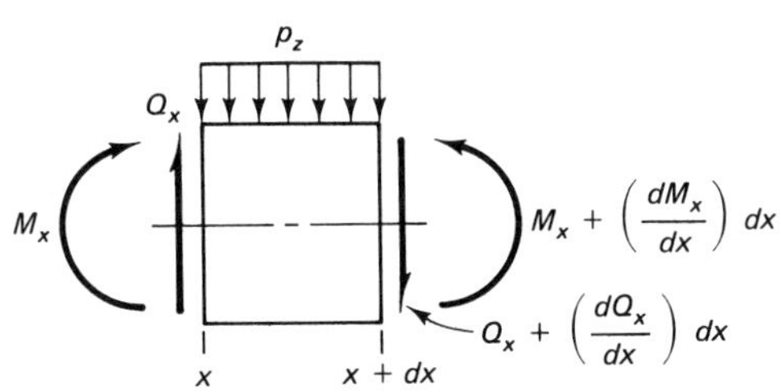

Figure 1.6 Forces acting on a beam element.

*Here we neglect a term $(p_z\, dx)(dx/2)$ which is small in comparison with the other terms.

and

$$M_x' = Q_x \tag{1.8}$$

where $(\cdot)'$ denotes differentiation with respect to x.

Kinematics

Since deformations due to shear are neglected, we consider only the change of curvature* which is produced by the bending moments, and which we shall call κ_x. Referring to Fig. 1.7, we call χ (clockwise positive) the slope at section x. We can then write

$$\chi \approx \tan \chi = \frac{dw}{dx} = w'$$

By definition, we have

$$|\kappa_x| = \frac{\text{change of angle between tangents}}{\text{arc length}}$$

Therefore,

$$|\kappa_x| = \frac{(d\chi/dx)\,dx}{dx} = \frac{d\chi}{dx} \tag{a}$$

Now we assume that κ_x is positive in the sense of positive M_x (Fig. 1.6), i.e., positive for a "sagging" curve. It can easily be seen from Fig. 1.7 that as the slope χ increases from A to B, i.e., when $d\chi/dx$ is positive, we have a "hogging" curve (negative κ_x), and hence we must write

$$\kappa_x = -\frac{d\chi}{dx} \tag{1.9}$$

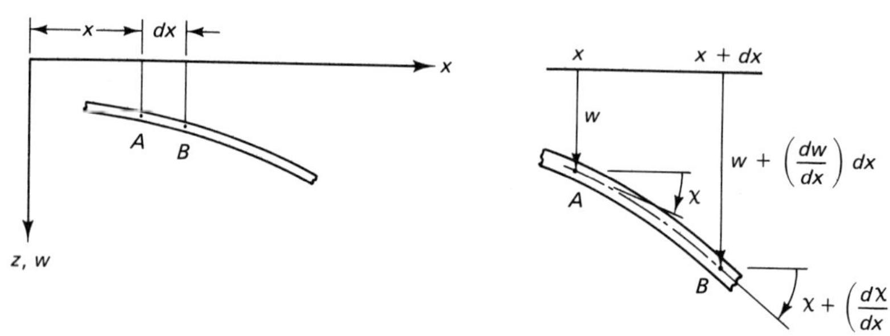

Figure 1.7 Deformation of a beam element.

*Here change of curvature is the curvature itself, since initially the beam axis is straight, and hence the initial curvature is zero.

and substituting for χ, we get*

$$\kappa_x = -w'' \tag{1.10}$$

Hooke's Law

Consider the deformation of the element bounded by sections x and $x + dx$ (Fig. 1.8). Assuming that section $x + dx$ undergoes a positive rotation $(d\chi/dx)\,dx$ relative to the section x, we find that the elongation Δ of the fiber AB at a distance z from the centroidal axis is

$$\Delta = -z\left(\frac{d\chi}{dx}\right) dx = -z\,d\chi$$

Hence the strain along AB is

$$\varepsilon_x(z) = \frac{\Delta}{dx} = -z\,\frac{d\chi}{dx}$$

Substituting from Eq. (1.9), we get

$$\varepsilon_x(z) = z\kappa_x \tag{1.11}$$

and hence stress $\sigma_x(z) = E \cdot \varepsilon_x(z)$ becomes

$$\sigma_x(z) = Ez\kappa_x$$

Now taking moments about the centroidal axis, we get

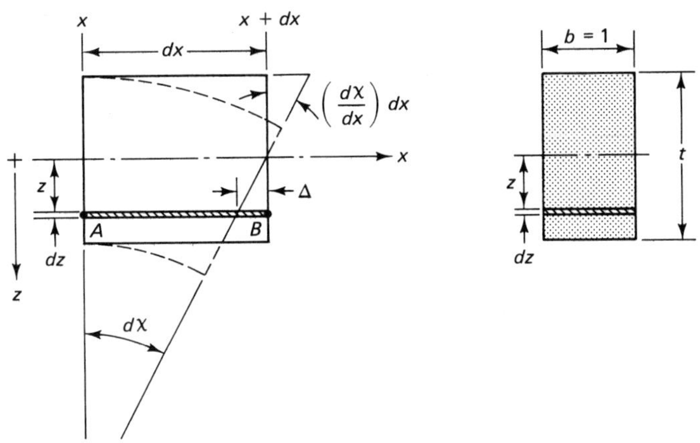

Figure 1.8 Bending deformation of a beam element.

*Note: If ds is used as the arc length instead of the approximation dx in Eq. (a), we get an exact expression for the curvature κ_x. In this case we have $ds = dx\sqrt{1 + (dw/dx)^2}$. Hence by using the approximate equation (1.10), we have, in effect, neglected slope w' compared to unity, which is consistent with our assumption 2.

$$M_x = \int_{-t/2}^{+t/2} \sigma_x(z) \cdot 1 \cdot dz \cdot z$$

$$= E\kappa_x \int_{-t/2}^{+t/2} z^2\, dz$$

$$= \frac{Et^3}{12}\kappa_x$$

or

$$M_x = K\kappa_x \tag{1.12}$$

where $K = Et^3/12$ is called the *flexural rigidity* of the beam.

We can now summarize our equations for the beam as follows:

$$\text{Equilibrium:} \quad Q_x' = -p_z \tag{1.7}$$

$$M_x' = Q_x \tag{1.8}$$

$$\text{Kinematics:} \quad \kappa_x = -w'' \tag{1.10}$$

$$\text{Hooke's law:} \quad M_x = K\kappa_x \tag{1.12}$$

Eliminating all unknowns but w from the equations above, we get the following differential equation for the deflection of the beam:

$$(Kw'')'' = p_z \tag{1.13}$$

or if K does not vary with x, i.e., if t and E are constant along x, we get

$$w'''' = \frac{p_z}{K} \tag{1.14}$$

Thus, for beams, the governing differential equation is of fourth order, as compared to the second-order equation obtained for bars. The general solution of this equation can be written as

$$w = C_1 + C_2x + C_3x^2 + C_4x^3 + w_p \tag{1.15}$$

where w_p is a particular solution of Eq. (1.14) for a given loading. Constants C_1, C_2, C_3, and C_4 are to be determined from the boundary conditions. Several problems can be solved using Eqs. (1.14) and (1.15).

Example 1.2

Consider a beam of length L, fixed at $x = 0$ and simply supported at $x = L$, carrying a uniformly distributed load p_o (Fig. 1.9). Taking $p_z(x) = p_o$ in Eq. (1.14), by integration we can obtain

$$w_p = \frac{p_o x^4}{24K}$$

Substituting this expression in Eq. (1.15) and using boundary conditions $w = 0$, $w' = 0$

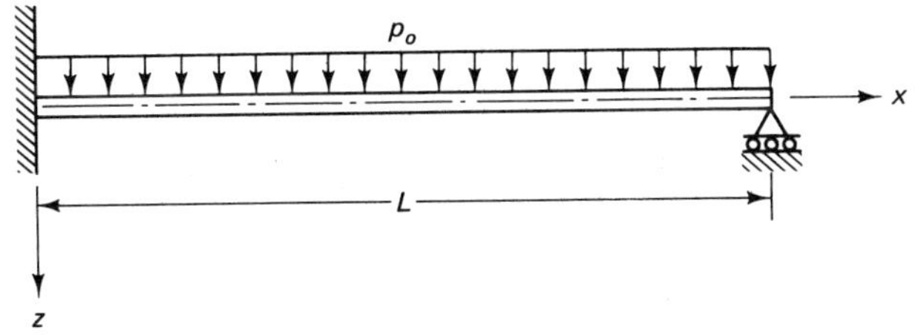

Figure 1.9 Propped cantilever beam subjected to a uniform load p_o.

at $x = 0$ and $w = 0$, $w'' = 0$* at $x = L$, we can readily obtain the following solution for w:

$$w = \frac{p_o L^4}{48K}\left(3\frac{x^2}{L^2} - 5\frac{x^3}{L^3} + 2\frac{x^4}{L^4}\right)$$

Although many problems such as the one above can easily be solved, problems such as those shown in Fig. 1.10 become much more complicated to solve by the general equation derived here. There are alternative simpler methods for solving such problems. However, because for corresponding similar problems of plates or shells there are no such alternative simpler methods of solution, we shall consider here the procedure for solving the problem of Fig. 1.10(a) by using the general equation derived above.

Example 1.3

We have to consider the beam in two parts, AC and CB, and apply the conditions of continuity at their junction. We take x_1 as coordinate for AC, and x_2 as coordinate for the region CB. Since there is no load on either AC or CB, we can write the solutions for the deflections w_1 and w_2 in AC and CB, respectively, as

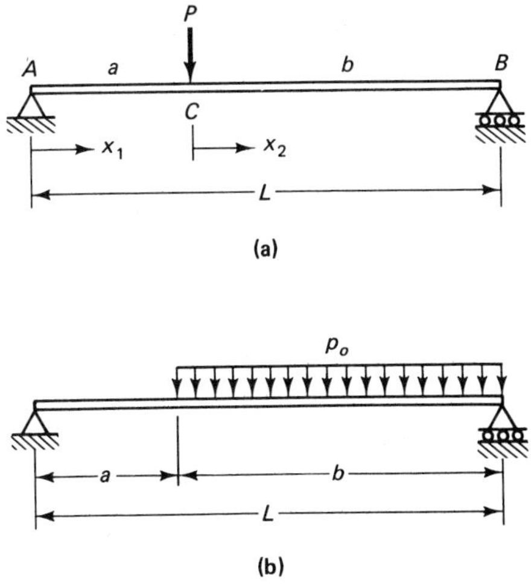

Figure 1.10 Simply supported beams with more complex loadings. (a) Concentrated load P. (b) Partial uniform load p_o.

*At $x = L$, we have $M_x = 0$, which gives $w'' = 0$ by virtue of Eqs. (1.12) and (1.10).

$$w_1 = C_1 + C_2x_1 + C_3x_1^2 + C_4x_1^3$$

$$w_2 = C_5 + C_6x_2 + C_7x_2^2 + C_8x_2^3$$

where $C_1, \ldots, C_8$ are constants to be determined. The boundary conditions are:

$$\text{at } x_1 = 0: \quad w_1 = 0 \quad \text{and} \quad w_1'' = 0$$

$$\text{at } x_2 = b: \quad w_2 = 0 \quad \text{and} \quad w_2'' = 0$$

In addition, the continuity conditions at B can be written as

$$w_1(x_1 = a) = w_2(x_2 = 0)$$

$$w_1'(x_1 = a) = w_2'(x_2 = 0)$$

$$M_{x_1}(x_1 = a) = M_{x_2}(x_2 = 0)$$

As a fourth continuity condition, we consider equilibrium of an element just below the load P (Fig. 1.11). Considering equilibrium of vertical forces, we can write the final equation as

$$Q_{x_1}(x_1 = a) - Q_{x_2}(x_2 = 0) - P = 0$$

Using these four continuity equations, together with the four boundary conditions, we get a total of eight equations to solve for constants C_1 to C_8. We can then obtain the solution for w. The solution of the problem becomes much simpler if P is applied at midspan of the beam (i.e., if $a = b = L/2$). By using principles of symmetry [9], we consider only half of the beam, and write two boundary conditions as follows:

$$\text{at } x_1 = a = \frac{L}{2}: \quad w_1' = 0 \quad \text{and} \quad Q_{x_1} = \frac{P}{2}$$

Using these conditions with the two boundary conditions at $x_1 = 0$, we can solve directly for the constants C_1, C_2, C_3, and C_4.

The problem shown in Fig. 1.10(b) can be solved in a manner similar to that described above. In this case the fourth continuity condition would be merely $Q_{x_1}(x_1 = a) = Q_{x_2}(x_2 = 0)$.

The reader should become familiar with the procedures followed for deriving and solving the differential equations for bars and beams, since we shall be following

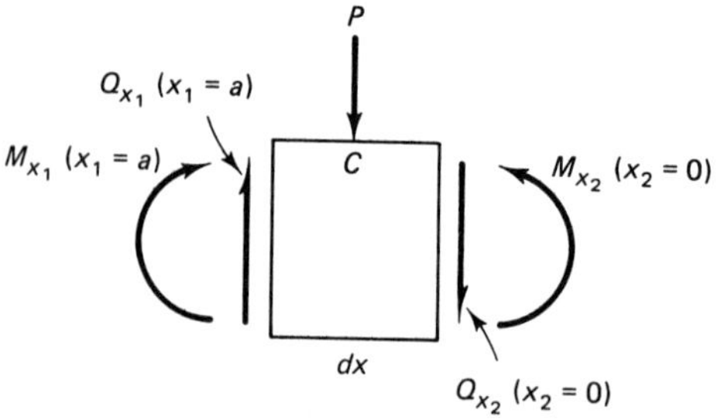

Figure 1.11 Freebody diagram of beam element below concentrated load of Fig. 1.10(a).

very similar steps for deriving and solving the more complex equations of plates and shells.

1.4 CIRCULAR RINGS

It will be appropriate to consider briefly here the theory of circular rings, since we shall be requiring this theory again and again in our later work. We shall presently consider circular rings subjected only to axisymmetric loadings in their own plane. Figure 1.12(a) shows such a ring of radius a, subjected to an internal load intensity p_r per unit length of its circumference. We make the following assumptions:

1. The material of the ring is homogeneous, isotropic, and linearly elastic, following Hooke's law.
2. The ring is "thin"; i.e., its cross-sectional dimensions are small compared to its radius a.
3. The ring is represented by a circular axis of radius a passing through the centroids of all cross sections.
4. One principal axis of the cross sections lies in the plane of the ring.

Equilibrium

We consider equilibrium of an element bounded by sections θ and $\theta + d\theta$ [Fig. 1.12(b)]. Due to axisymmetry, shears Q_θ will be zero, while N_θ and M_θ will be constant in θ.

As seen from the triangle of forces in Fig. 1.12(c), the resultant of the N_θ forces, acting on the end faces of the element, is a radially inward force ($N_\theta\, d\theta$).

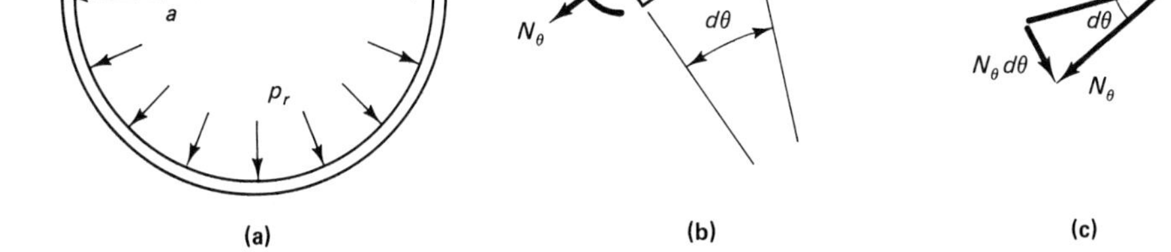

Figure 1.12 Circular ring subjected to a uniform radial load intensity p_r. (a) Ring and associated deformations. (b) Freebody diagram of ring element. (c) Radially inward resultant force.

Hence, from equilibrium of forces in the radial direction, we get

$$N_\theta \, d\theta = p_r a \, d\theta$$

or

$$N_\theta = p_r a \tag{1.16}$$

Since the equation of equilibrium of moments will yield simply $M_\theta = M_\theta$, moment M_θ is statically indeterminate, and we must therefore consider kinematics and Hooke's law relations in order to obtain its solution.

Kinematics

Due to axisymmetry the tangential displacement v of the ring will be zero, and the radial displacement w will be constant in θ. We can write

$$\varepsilon_\theta = \frac{(a + w)\, d\theta - a\, d\theta}{a\, d\theta}$$

or

$$\varepsilon_\theta = \frac{w}{a} \tag{1.17}$$

Now the change of curvature, κ_θ, is given by

$$\begin{aligned} |\kappa_\theta| &= |\text{new curvature} - \text{old curvature}| \\ &= \left| \frac{d\theta}{(a + w)\, d\theta} - \frac{d\theta}{a\, d\theta} \right| = \left| \frac{1}{a + w} - \frac{1}{a} \right| \end{aligned}$$

Neglecting w with respect to a, we approximately take*

$$\kappa_\theta = 0 \tag{1.18}$$

Hooke's Law

Because the ring is assumed to be thin, strain $\varepsilon_\theta = w/a$ is constant over the ring thickness, and hence $\sigma_\theta = E\varepsilon_\theta$ is also constant over the thickness. Thus, if the width of the cross section of the ring is taken as unity and the thickness is t, we get

$$\begin{aligned} N_\theta &= 1 \cdot t \cdot \sigma_\theta \\ &= Et\varepsilon_\theta \end{aligned}$$

or

*Actually, there is a small change of curvature; $|\kappa_\theta| = \{[1/(a + w)] - (1/a)\} \approx -(w/a^2)$, which, being a small quantity, is neglected. (Refer also to the discussion of κ_θ in Chapter 5.)

$$N_\theta = D\varepsilon_\theta \tag{1.19a}$$

where the extensional rigidity of the ring is $D = Et$. Also, as for beams, we can easily derive

$$M_\theta = K\kappa_\theta \tag{1.19b}$$

where $K = Et^3/12$.

Equations (1.16) to (1.19) form the basic equations governing the deformation w of a ring under axisymmetric loading. From these equations we obtain

$$M_\theta = 0$$

and

$$p_r a = N_\theta = D\varepsilon_\theta = \frac{Dw}{a}$$

Hence,

$$w = p_r \frac{a^2}{D}$$

which gives the relation between deflection w and the loading p_r. Thus, in this case, instead of a differential equation, we have an algebraic equation for w which can readily be solved for a known loading. Knowing w, we can obtain the solution for N_θ, while M_θ is zero throughout.

1.5 MIDDLE SURFACE

We represented bars and beams by a straight line passing through the centroids of all cross sections. Loads were assumed to be applied on this centroidal axis. Similarly, a plate or a shell is represented by its *middle surface,* a surface that passes midway between its two faces [Fig. 1.13(a)]. We assume that all loads are applied on this middle surface.

For corrugated or folded structures, as in Fig. 1.13(b) and (c), a surface passing midway between the two faces could be taken as the middle surface. However, it is more convenient to take the middle surfaces as shown, and treat the structure as an orthotropic shell with equivalent thicknesses (rigidities) about the middle surfaces in the two directions.

If we can define the shape of the middle surface and the thickness of the shell/plate at all points, we know the geometry of the entire structure. For plates, the middle surface is a plane that can easily be defined. The middle surface of a shell is a three-dimensional surface, and to define it mathematically a knowledge of *differential geometry* or *theory of surfaces* would be advantageous. We discuss the basics of such concepts briefly in Chapter 3.

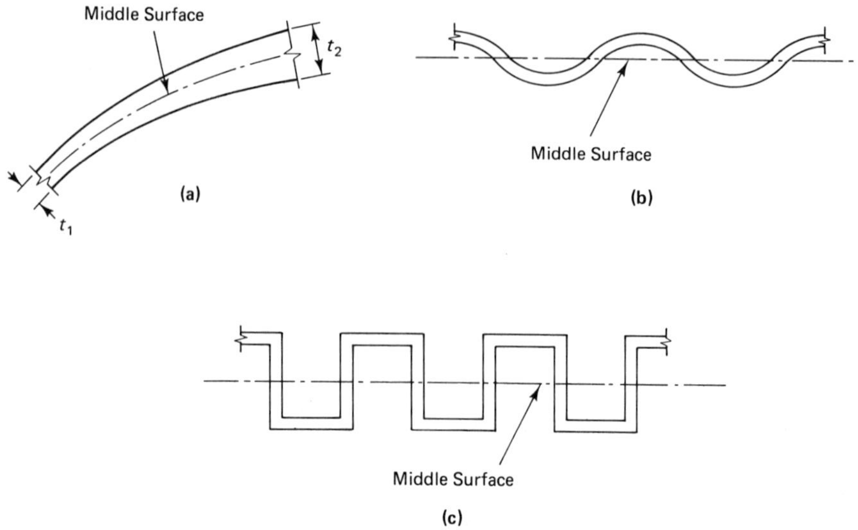

Figure 1.13 Middle surface of various shell configurations. (a) Shell of changing thickness. (b) Corrugated section. (c) Folded plate.

1.6 INTERNAL STRESS RESULTANTS

The general state of stress on the cross section of a bar/beam member carrying axial as well as normal loads will consist of stresses σ_x, τ_{xy}, and τ_{xz}, as shown in Fig. 1.14(a). Since we consider centroidal loads acting only in the x- and z-directions, τ_{xy} will be zero. Also, by our basic assumptions (see Section 1.3), σ_x and τ_{xz} will be constant in y, but they will vary with z, and of course, x. Instead of working with stresses σ_x and τ_{xz}, which vary with x and z, for deriving our equations, we prefer to work with the statically equivalent quantities N_x, Q_x, and M_x [Fig. 1.14(b)]. These are called the *internal stress resultants,* and they vary with x only. We can write the following relations for a bar/beam of unit width:

$$N_x = \int_{-t/2}^{+t/2} \sigma_x \, dz \tag{1.20a}$$

$$M_x = \int_{-t/2}^{+t/2} \sigma_x \, z \, dz \tag{1.20b}$$

$$Q_x = \int_{-t/2}^{+t/2} \tau_{xz} dz \tag{1.20c}$$

From our assumption that plane sections remain plane after bending, we can conclude that axial strains ε_x, and hence axial stresses σ_x, vary linearly in z. Thus we assume that $\sigma_x = \alpha + \beta z$, where α and β are some unknown constants. Substituting this expression for σ_x in Eqs. (1.20a,b) and integrating, we get

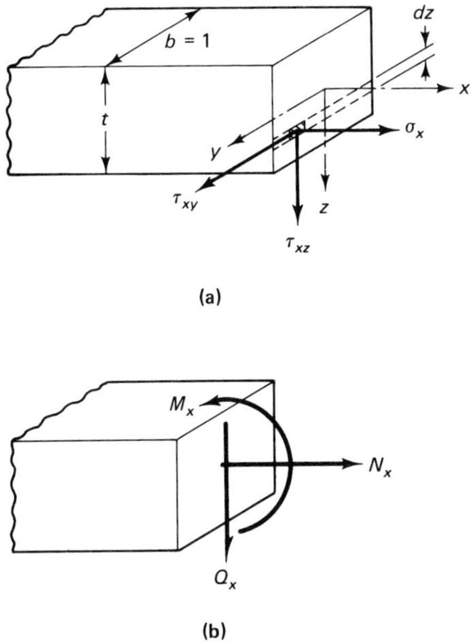

Figure 1.14 Beam/bar member. (a) Stresses acting on a section $x = constant$. (b) Internal stress resultants N_x, Q_x, and M_x.

$$N_x = \alpha t \quad \text{and} \quad M_x = \beta \frac{t^3}{12}$$

or

$$\alpha = \frac{N_x}{t} \quad \text{and} \quad \beta = \frac{M_x}{t^3/12}$$

and hence

$$\sigma_x = \frac{N_x}{t} + \frac{M_x z}{t^3/12} \tag{1.21}$$

Thus, knowing N_x and M_x, we can always calculate stress σ_x over the depth of the cross section by using the relation above. Also, knowing Q_x, we can calculate τ_{xz}.*

Similarly, for shells or plates also, it is easier to work with internal stress resultants rather than working with the stresses themselves. Figure 1.15(a) shows a general curved element of a shell, obtained by taking sections x, $x + dx$, y, and $y + dy = constant$. On the section $x = constant$ of this element, instead of working with stresses σ_x, τ_{xy}, and τ_{xz}, we use for deriving the governing equations the statically equivalent internal stress resultants N_x, N_{xy}, M_x, Q_x, and M_{xy}, which are defined per unit length along the y-axis. Similarly, on the section $y = constant$, we use N_y, N_{yx}, M_y, Q_y, and M_{yx} as the other five internal stress resultants. Thus

*The equation for τ_{xz} is not derived here, but may be found in any textbook on strength of materials [8] as $\tau_{xz} = [(Q_x/2t^3) \cdot (3t^2 - 12z^2)]$ for a rectangular cross section of unit width.

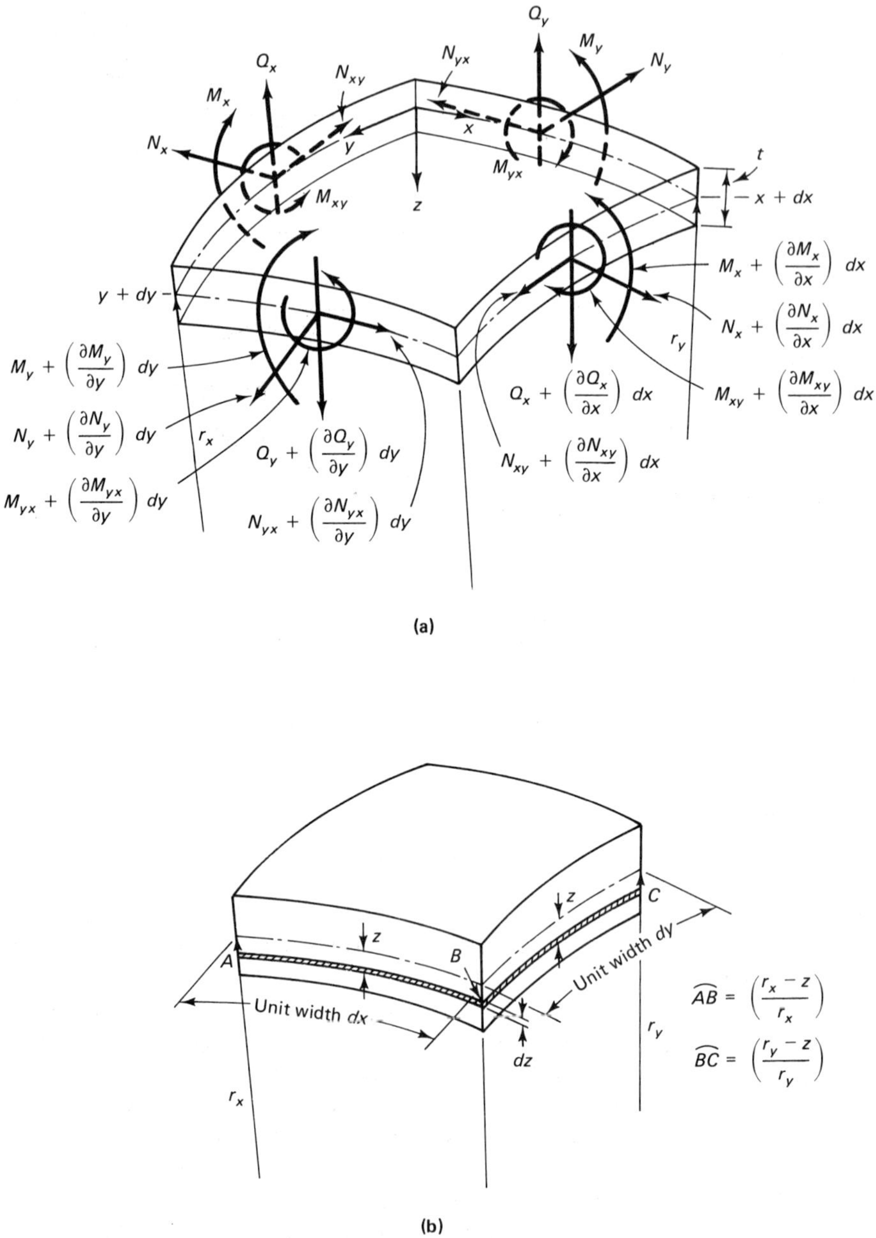

Figure 1.15 General shell element. (a) Internal stress resultants. (b) Variation of cross section width over thickness.

we have a total of 10 internal stress resultants for a general shell element. For plates subjected to loads normal to their planes, this number would reduce to six, since the in-plane resultants N_x, N_y, N_{xy}, and N_{yx} would be absent (assuming small deflections).

Considering the variation in width of the cross section over the thickness (this width being unity at $z = 0$), we can write from Fig. 1.15(b),

$$N_x = \int_{-t/2}^{+t/2} \sigma_x \cdot 1 \cdot \frac{r_y - z}{r_y}\, dz$$

or

$$N_x = \int_{-t/2}^{+t/2} \sigma_x \left(1 - \frac{z}{r_y}\right) dz$$

where r_y is the radius of curvature of the middle surface ($z = 0$) in the y-direction. For thin shells, z/r_y is a small quantity and is usually neglected compared to unity [2].* Hence we get an equation that is identical to Eq. (1.20a) obtained earlier for beams, namely,

$$N_x = \int_{-t/2}^{+t/2} \sigma_x\, dz$$

Similar expressions can be obtained for the other stress resultants as well. As for beams, for plates and shells also, we assume that plane sections remain plane after bending. Hence we can derive equations similar to Eq. (1.21) for obtaining the normal stresses once the stress resultants are known. Also, due to the small thicknesses of plates and shells, shear stresses are obtained simply as N_{xy}/t or Q_x/t, etc., once stress resultants N_{xy}, Q_x, etc., are known.

For solving plate and shell problems, our aim will thus be to obtain the variation of internal stress resultants that satisfies the appropriate boundary conditions, under a given loading. Once the stress resultants are known, stresses can be obtained by simple formulas as discussed above.

REFERENCES

[1] Martienssen, H., *The Shapes of Structure,* Oxford University Press, Oxford, 1976.

[2] Flügge, W., *Stresses in Shells,* 2nd ed., Springer-Verlag, New York, 1973.

[3] Colin, F., *CANDELA/The Shell Builder,* Reinhold Publishing Corporation, New York, 1963.

[4] Torroja Miret, E., *The Structures of Eduardo Torroja—An Autobiography of Engineering Accomplishment,* F. W. Dodge Corporation, New York, 1958.

[5] Nervi, P. L., *New Structures,* The Architectural Press, Ltd., London, 1963.

*This approximation is equivalent to taking the "trapezoidal" shape of the element cross section [Fig. 1.15(b)] as rectangular.

[6] Billington, D. P., "Anton Tedesko: Thin Shells and Aesthetics," *ASCE J. Struct. Div.*, Nov. 1982.

[7] Timoshenko, S. P., and Goodier, J. N., *Theory of Elasticity,* 3rd ed., McGraw-Hill Book Company, New York, 1970.

[8] Timoshenko, S. P., *Strength of Materials,* 3rd ed., Pt I, D. Van Nostrand Company, Princeton, N.J., 1955.

[9] Glockner, P. G., "Symmetry in Structural Mechanics," *ASCE J. Struct. Div.*, Jan. 1973, pp. 71–89.

PROBLEMS

1.1. For the bars shown in Fig. P1.1(a) and (b), subjected to axial loads along their centroidal axes, find expressions for u and N_x. The bars are free at their ends and carry self-equilibrating loadings. The loadings applied in Fig. P1.1(a) and (b) are the first harmonics ($n = 1$) of a general Fourier series. Any general loading can be expanded in a Fourier series that contains a summation over several harmonics—terms of the type $p_n \cos(n\pi x/L)$ and $q_n \sin(n\pi x/L)$. The solution for such a loading can be obtained by summation of the solutions for each harmonic. Such a procedure can be used to solve the case of a beam to which a prestress load F is applied at the ends. Load F is considered to be distributed over a small length c, and this loading [Fig. P1.1(c)] is then expanded in a Fourier series. Fourier series solutions are discussed in more detail in Chapter 10. Presently, without the use of Fourier series, solve the problem in Fig. P1.1(c) by considering half of the bar (due to symmetry) and applying boundary conditions at $x = 0$ and $x = L/2$ and applying continuity conditions at $x = c$.

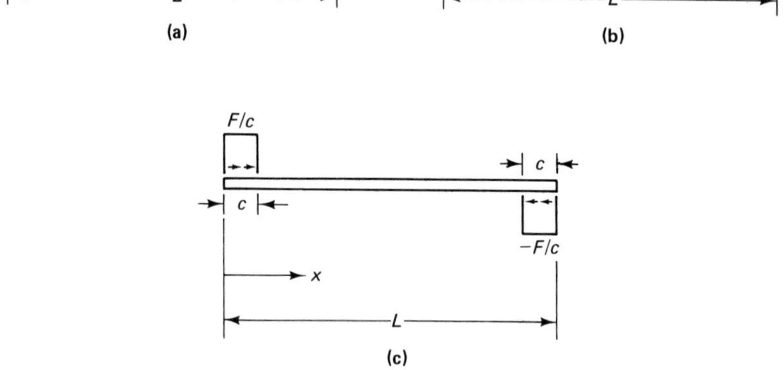

Figure P1.1 Bars subjected to self-equilibrating axial loads. (a) Member in compression. (b) Member in tension. (c) Member subjected to prestress compression at ends.

1.2. For solving the problem of a beam carrying a concentrated moment M_o (Fig. P1.2), write the boundary conditions when $a \neq b$. Also, write the simplified boundary conditions when $a = b$.

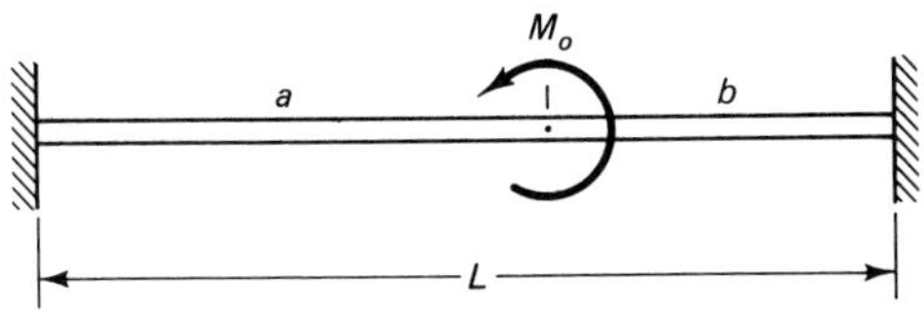

Figure P1.2 Fixed-end beam loaded by a concentrated moment M_o.

1.3. Solve the problem of the beam carrying a uniformly distributed load (UDL) equal to p_o. The beam is fixed at one end and supported by a spring at the other (Fig. P1.3). (This problem should be solved by releasing the spring support, assuming an unknown reaction X_1 between the beam and the spring, and then obtaining its value by writing an equation of compatibility of vertical displacements of the beam and the spring at their junction.)

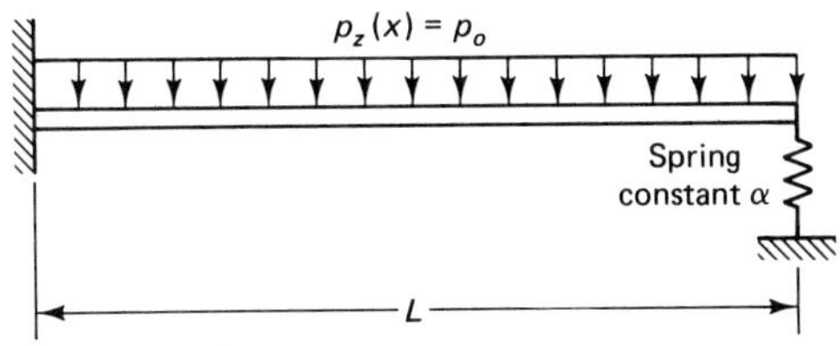

Figure P1.3 Beam under UDL supported by a spring at one end and fixed at the other.

1.4. The edge beams of cylindrical barrel shells are supported at their ends by diaphragms, which are rigid for deformations in their own planes, but are flexible for deformations outside their planes [Fig. P1.4(a)]. Figure P1.4(b) and (c) illustrate two such edge beams, which are subjected to the loadings shown, due to their interaction with the

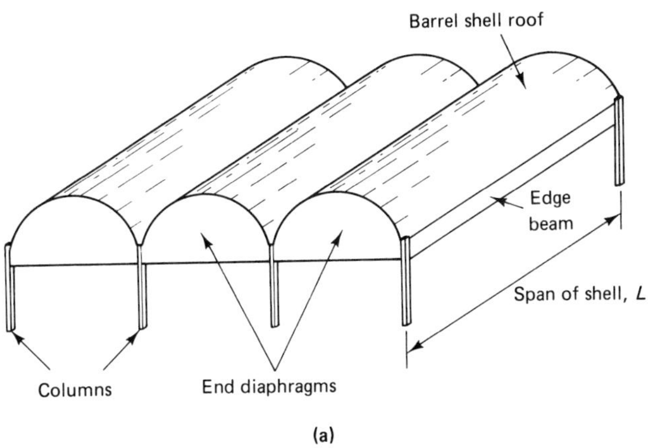

Figure P1.4 Barrel shell roof. (a) Various elements of the barrel shell roof. (*continued*)

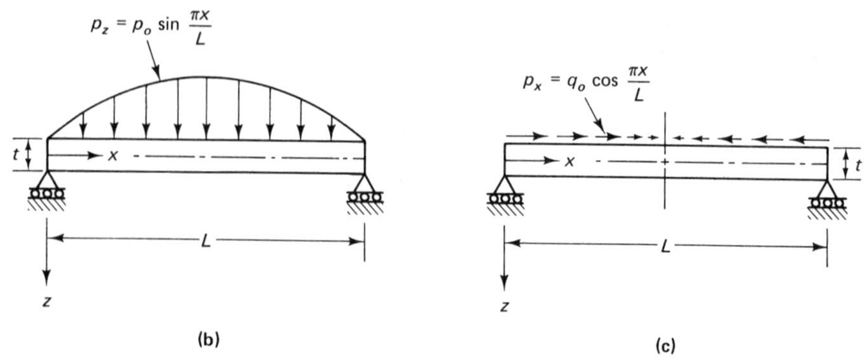

Figure P1.4 (*continued*) (b) Transverse loading of edge beam. (c) Axial loading of edge beam.

shells. Obtain the deflection w of the beam for the loadings in Fig. P1.4(b) and (c). For Fig. P1.4(b), find an expression for the horizontal deflection u at the top of the beam ($z = -t/2$) using the horizontal deflection at $x = L/2$ as the reference datum. Note that for both cases, the loads are applied at the top of the beam; they are not applied through the centroidal axis.

1.5. For beam-column problems, the effect of both axial and bending forces is considered in writing the equilibrium equations, and hence Eq. (1.8) should be modified. We now consider equilibrium of forces on an element of the deflected beam-column, as shown in Fig. P1.5. From this figure, Eq. (1.8) can be rewritten as

$$M_x' = Q_x - (N_x w')$$

Proceeding from this equation, obtain the differential equation for deflection w of a beam-column. Also, write down the equation for obtaining the buckling load of a column subjected to an axial compressive load P. Solve this equation for the lowest critical load of the column under simply supported end conditions.

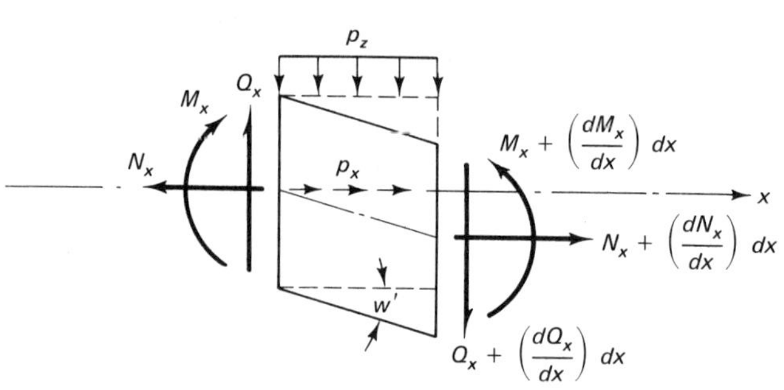

Figure P1.5 Freebody diagram of an element of a beam-column member.

2

Bending of Circular Plates

2.1 INTRODUCTION AND ASSUMPTIONS

Although we are interested primarily in studying the analysis of shell structures, for problems such as cylindrical water tanks, we may have to consider interaction forces between the shell walls and the circular slabs at the top and bottom of the shell (i.e., roof and raft slabs). It is therefore desirable that we study the theory of circular plates at this stage.

We consider plates that are subjected to axisymmetric loadings normal to their middle surfaces. Several such problems arise in practice, and the theory thus developed will be useful for practical design. Because of axisymmetry, all quantities—loads, deflections, and internal stress resultants—will be independent of θ. Thus the derivation of governing equations becomes much simplified.

We represent the plate by its middle surface and use the cylindrical coordinates r, θ, and z [Fig. 2.1(a)]. The deflection w is measured along the z-axis. We also make the following assumptions to simplify our equations:

1. The material of the plate is homogeneous, isotropic, and linearly elastic, following Hooke's law.
2. The thickness t of the plate is small compared to its radius a.
3. Deflection w is small compared to the radius a. From this assumption we can conclude that the slopes are small compared to unity. Thus we can use the approximate expression for curvature, as we did in deriving equations for beams (Chapter 1).

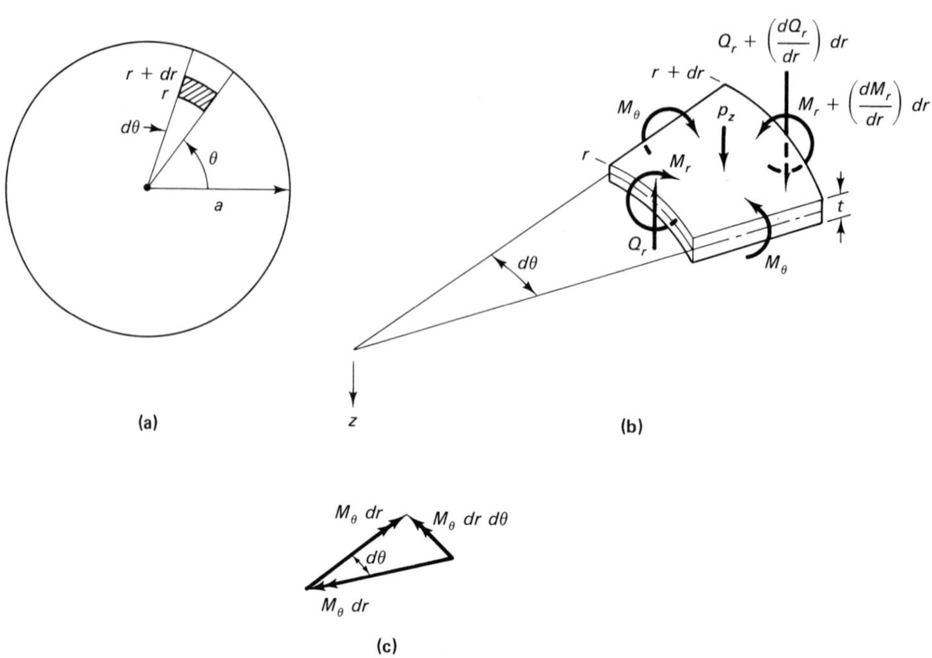

Figure 2.1 Circular plate. (a) Coordinate system. (b) Freebody diagram of hatched element. (c) Resultant of moment vectors $M_\theta \cdot dr$.

4. Deflections are also small compared to the thickness t. Under loading, the plate is deformed into a nondevelopable surface, which causes stretching of its middle surface. This stretching induces in-plane forces (N_r, N_θ, $N_{r\theta}$) in the plate. Timoshenko [1,2] has shown that if $w > 0.4t$, say, the in-plane stresses can be as much as 20% of the bending stresses. Since we shall not consider interaction of in-plane stresses with bending stresses in our derivations, we have to restrict the application of our equations to problems where the in-plane stresses are not significant, i.e., where w is small compared to t.*
5. Plane sections remain plane after bending. Thus, as for beams, we neglect shear deformations for plates also.

2.2 DERIVATION OF EQUATIONS

As was our approach in Chapter 1 for the derivations of governing equations of bars, beams, and circular rings, we shall now apply equations of (a) equilibrium, (b) kinematics, and (c) Hooke's law to the case of the circular plate with axisym-

*For statically determinate beams subjected to bending, there is no stretching of the centroidal axis of the beam, since at least one end of the beam is always free to move horizontally. Hence no axial stresses exist in such a beam subjected to vertical loads only. Therefore, we did not make this assumption (i.e., that w/t is small) for beams in Section 1.3.

metric loading. We shall then reduce this set of equations to a single fourth-order ordinary differential equation for deflection w of the plate.

Equilibrium Equations

Figure 2.1(a) shows a circular plate in which we consider equilibrium of an element (shown hatched), bounded by sections r, $r + dr$, and θ, $\theta + d\theta$. An enlarged view of this element, with the internal stress resultants and distributed loading p_z acting on it, is shown in Fig. 2.1(b). Because of axisymmetry, Q_θ, $M_{r\theta}$, and $M_{\theta r}$ are zero, and also M_r, Q_r, M_θ (internal stress resultants per unit length along the middle surface), and p_z are constant in θ. Considering equilibrium of forces in the z-direction, we can write

$$\left(Q_r + \frac{dQ_r}{dr}\,dr\right)(r + dr)\,d\theta - Q_r\,r\,d\theta + p_z \cdot r\,dr\,d\theta = 0 \qquad \text{(a)}$$

On expanding the terms in parentheses, and canceling two terms ($Q_r r d\theta$ and $-Q_r r d\theta$), we obtain

$$Q_r\,dr\,d\theta + r\left(\frac{dQ_r}{dr}\right)dr\,d\theta + \left(\frac{dQ_r}{dr}\right)(dr)^2\,d\theta + p_z \cdot r\,dr\,d\theta = 0$$

The third term in the equation above contains $(dr)^2 \cdot d\theta$, which is smaller (i.e., of higher order) than the other terms which contain only $(dr)(d\theta)$. Hence neglecting this smaller term and canceling $(dr)(d\theta)$ throughout, we have

$$Q_r + r\left(\frac{dQ_r}{dr}\right) + p_z \cdot r = 0$$

which can be written as

$$(Q_r \cdot r)' = -p_z \cdot r \qquad \text{(2.1a)}$$

where $(\cdot)'$ denotes a derivative with respect to r.

Instead of writing the first two terms separately in Eq. (a), we could have directly written the increment of the total force ($Q_r \cdot r d\theta$), from the edge $r = \textit{constant}$ to the edge $r + dr = \textit{constant}$, as

$$\frac{d}{dr}(Q_r \cdot r\,d\theta)\,dr$$

Hence the equilibrium equation would be written as

$$\frac{d}{dr}(Q_r \cdot r)\,d\theta\,dr + p_z \cdot r\,d\theta\,dr = 0$$

which gives us Eq. (2.1a) directly. We shall use this procedure for deriving such equilibrium equations in this chapter as well as in future chapters.

Now taking equilibrium of radial moments, we get

$$(M_r \cdot r\,d\theta)'\,dr - M_\theta\,dr\,d\theta - (Q_r \cdot r)\,d\theta\,dr = 0$$

or

$$(M_r \cdot r)' - M_\theta = (Q_r \cdot r) \tag{2.1b}$$

The contribution of M_θ to this equation can easily be seen by representing the moments M_θ on the two faces in vectorial form, and by taking the resultant of the two vectors [Fig. 2.1(c)].

Equations 2.1(a,b) are the two equations of equilibrium for the plate element. In the theories of bars and beams, we could obtain bending moments and shears for some problems by considering the equilibrium equations alone. The present plate problem, however, is not statically determinate since we have only two equations of equilibrium which contain three unknowns (M_r, M_θ, and Q_r). We must therefore necessarily consider kinematics and Hooke's law relations to solve any circular plate problem.

Kinematics

We refer to Fig. 2.2 and consider the deflections of two neighboring points at r and $r + dr$. As done earlier for beams, we can write the change of curvature κ_r (which is the same as the total curvature here, since the plate is initially flat) as

$$\kappa_r = -\frac{d\chi}{dr} \tag{2.2}$$

where χ is the angle of rotation of the tangent, measured clockwise from the horizontal (when considering the portion of the plate to the right of the z-axis). The negative sign above is required because κ_r is considered positive corresponding to positive M_r; i.e., a "sagging" curvature is positive. However, as seen from Fig. 2.2, a positive $d\chi/dr$ produces a "hogging" curvature.

We can also see from Fig. 2.2 that

$$\chi \approx \tan \chi = \frac{dw}{dr}$$

and hence we get

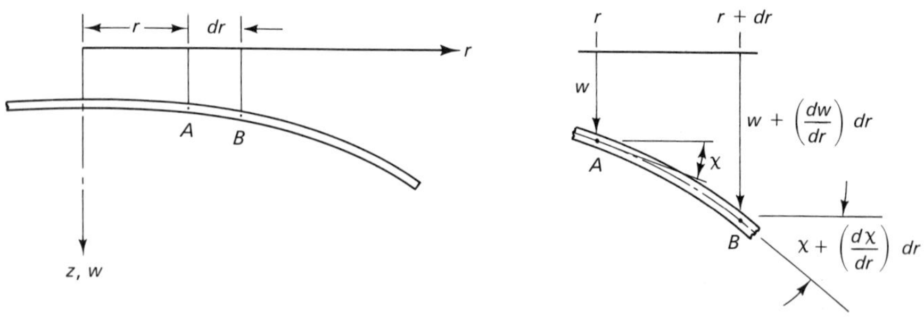

Figure 2.2 Axisymmetric deformation of a circular plate element.

$$\kappa_r = -\frac{d^2w}{dr^2} = -w'' \tag{2.3a}$$

We shall refer to κ_θ as the change of curvature caused by M_θ, taken as positive when sagging. To obtain an expression for κ_θ, we see in Fig. 2.3 that due to axisymmetry, the circular plate becomes a surface of revolution after deformation.* For such surfaces it is well known that if $r_1 = \overline{OB}$ is the radius of curvature of the meridian (i.e., the curve formed by intersection of the surface with a plane containing the axis of revolution) at a point B, then $\overline{AB}$, the distance between point B and the intersection of r_1 with the axis of revolution, is the radius of curvature of the principal curve in the θ-direction. This radius of curvature in the θ-direction is called r_2. Hence we have

$$\kappa_\theta = -\frac{1}{r_2} = -\frac{\sin\chi}{r} \approx -\frac{\chi}{r}$$

We insert the negative sign here because as $1/r_2$ increases, κ_θ decreases from zero to a negative (i.e., hogging) curvature. Substituting for χ, we get

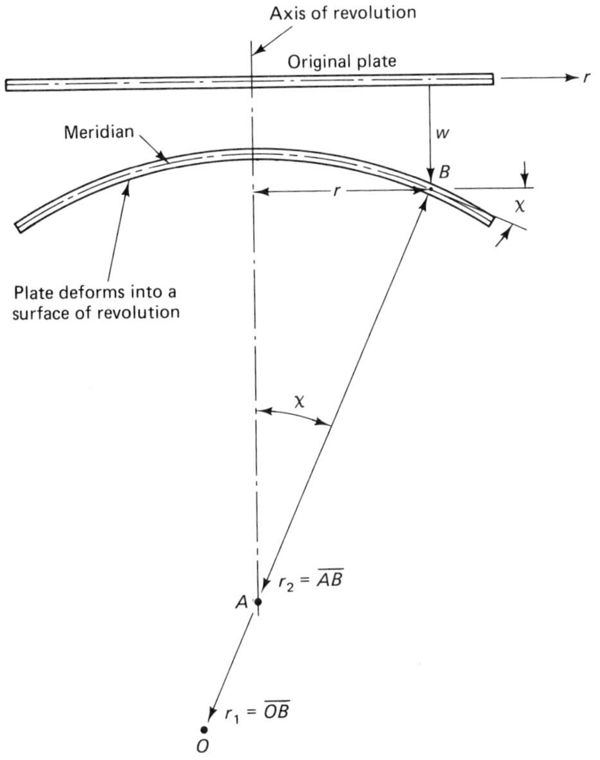

Figure 2.3 Deformation of a circular plate's midsurface into a surface of revolution.

*For the definition of a surface of revolution and its curvatures, see Chapter 3.

$$\kappa_\theta = -\frac{w'}{r} \tag{2.3b}$$

Again, the negative sign indicates that a positive w' produces a hogging (negative) κ_θ.

Hooke's Law

As for beams, we consider the strain in a fiber CD at a distance z from the middle surface (Fig. 2.4). Let the section at $r + dr$ undergo a positive incremental rotation $(d\chi/dr)\,dr$ with respect to the section at r. Then the elongation of fiber CD is

$$\Delta = -z\left(\frac{d\chi}{dr}\right) dr = -z\,d\chi$$

Hence the strain along CD is given by

$$\varepsilon_r(z) = \frac{\Delta}{dr} = -z\left(\frac{d\chi}{dr}\right)$$

Substituting from Eq. (2.2), we get

$$\varepsilon_r(z) = z\kappa_r \tag{2.4a}$$

Similarly, we can derive

$$\varepsilon_\theta(z) = z\kappa_\theta \tag{2.4b}$$

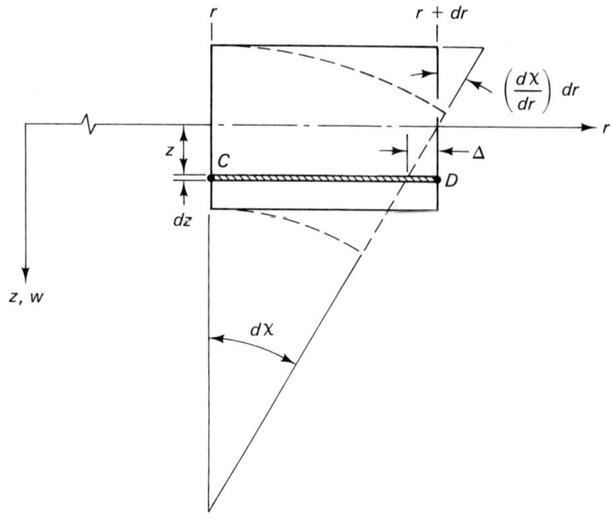

Figure 2.4 Deformation of plate element due to radial moment.

Now if we consider a thin slice, of thickness dz, of the circular plate element at a distance z from the middle surface (Fig. 2.5), we see that it is in a two-dimensional state of stress. For this slice we can write

$$\varepsilon_r = \frac{1}{E}(\sigma_r - \nu\sigma_\theta)$$

and

$$\varepsilon_\theta = \frac{1}{E}(\sigma_\theta - \nu\sigma_r)$$

where ε_r and ε_θ, σ_r and σ_θ are functions of z, and ν is Poisson's ratio. By solving these two equations for σ_r and σ_θ, we get

$$\sigma_r = \frac{E}{1 - \nu^2}(\varepsilon_r + \nu\varepsilon_\theta)$$

$$\sigma_\theta = \frac{E}{1 - \nu^2}(\varepsilon_\theta + \nu\varepsilon_r)$$

Substituting from Eqs. (2.4a,b), these become*

$$\sigma_r = \frac{Ez}{1 - \nu^2}(\kappa_r + \nu\kappa_\theta)$$

$$\sigma_\theta = \frac{Ez}{1 - \nu^2}(\kappa_\theta + \nu\kappa_r)$$

Now taking moments about the middle surface, we have per unit length

$$M_r = \int_{-t/2}^{+t/2} \sigma_r \cdot 1 \cdot dz \cdot z$$

$$= \frac{E(\kappa_r + \nu\kappa_\theta)}{1 - \nu^2} \int_{-t/2}^{+t/2} z^2\, dz$$

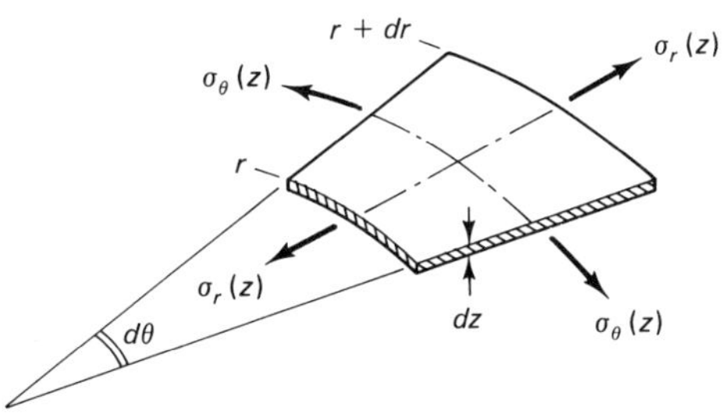

Figure 2.5 Stresses acting on a thin slice (of thickness dz) of the element in Fig. 2.1(b).

*Note that for beams we had a similar expression, $\sigma_x = Ez\kappa_x$ (see Section 1.3), which can be obtained by putting $\nu = 0$ into these equations.

$$= \frac{Et^3}{12(1 - \nu^2)} (\kappa_r + \nu\kappa_\theta)$$

or

$$M_r = K(\kappa_r + \nu\kappa_\theta) \tag{2.5a}$$

where

$$K = \frac{Et^3}{12(1 - \nu^2)}$$

The term K is called the flexural rigidity of the plate. Proceeding in a similar manner, we can also derive

$$M_\theta = K(\kappa_\theta + \nu\kappa_r) \tag{2.5b}$$

We can now summarize our known equations as follows:

Equilibrium equations

$$(Q_r \cdot r)' = -p_z \cdot r \tag{2.1a}$$

$$(M_r \cdot r)' - M_\theta = (Q_r \cdot r) \tag{2.1b}$$

Kinematic relations

$$\kappa_r = -w'' \tag{2.3a}$$

$$\kappa_\theta = -\frac{w'}{r} \tag{2.3b}$$

Hooke's law relations

$$M_r = K(\kappa_r + \nu\kappa_\theta) \tag{2.5a}$$

$$M_\theta = K(\kappa_\theta + \nu\kappa_r) \tag{2.5b}$$

$$K = \frac{Et^3}{12(1 - \nu^2)}$$

We now have six equations for the six unknowns M_r, M_θ, Q_r, κ_r, κ_θ, and w, and hence circular plate problems can be solved using these equations. We first obtain a single governing differential equation by eliminating all unknowns, except the deflection w, from these equations. To do this, we proceed as follows. Eliminating Q_r from Eqs. (2.1a,b), we obtain

$$(M_r \cdot r)'' - M_\theta' = -p_z \cdot r \tag{2.6}$$

Substituting Eqs. (2.3a,b) into Eqs. (2.5a,b), we obtain

$$M_r = -K\left(w'' + \nu \frac{w'}{r}\right) \tag{2.7a}$$

$$M_\theta = -K\left(\frac{w'}{r} + \nu w''\right) \tag{2.7b}$$

Substituting Eqs. (2.7a,b) into Eq. (2.6), we have

$$[K(w''r + \nu w')]'' - \left[K\left(\frac{w'}{r} + \nu w''\right)\right]' = p_z \cdot r$$

Assuming K (i.e., E and t) as constant in r, and expanding the equation above, we finally obtain the following equation governing the deflection w of the plate:

$$w'''' + \frac{2}{r} w''' - \frac{w''}{r^2} + \frac{w'}{r^3} = \frac{p_z}{K} \tag{2.8}$$

This is a fourth-order ordinary differential equation with variable coefficients and is called an equidimensional equation. It can also be written as

$$\left(\frac{d^2}{dr^2} + \frac{1}{r}\frac{d}{dr}\right)\left(\frac{d^2}{dr^2} + \frac{1}{r}\frac{d}{dr}\right) w = \frac{p_z}{K}$$

or

$$\nabla^2\nabla^2 w = \frac{p_z}{K} \tag{2.9}$$

where

$$\nabla^2(\cdot) = \frac{d^2}{dr^2}(\cdot) + \frac{1}{r}\frac{d}{dr}(\cdot)$$

∇^2 is the Laplacian operator of the well-known Laplace equation, $\nabla^2 w = 0$. Equation (2.9) is called the *bi-Laplace* or *biharmonic equation*. Before proceeding to solve this equation, we summarize all the basic equations that are required for the solution of problems of axisymmetric plates with constant thicknesses. These are

$$w'''' + \frac{2}{r} w''' - \frac{w''}{r^2} + \frac{w'}{r^3} = \frac{p_z}{K} \tag{2.8}$$

or

$$\nabla^2\nabla^2 w = \frac{p_z}{K} \tag{2.9}$$

Also,

$$M_r = -K\left(w'' + \nu \frac{w'}{r}\right) \tag{2.7a}$$

$$M_\theta = -K\left(\frac{w'}{r} + \nu\, w''\right) \tag{2.7b}$$

$$Q_r = -K\left(w''' + \frac{w''}{r} - \frac{w'}{r^2}\right) \tag{2.7c}$$

where

$$K = \frac{Et^3}{12(1 - \nu^2)}$$

Equation (2.7c) for Q_r has been obtained by substituting the expressions for M_r and M_θ from Eqs. (2.7a,b) into Eq. (2.1b).

2.3 SOLUTION OF THE DIFFERENTIAL EQUATION

For the nonhomogeneous equation (2.8), the complete solution for w is given by

$$w = w_h + w_p \tag{2.10}$$

In this equation, w_p is a particular solution of the equation for a given loading $p_z(r)$, while w_h is the solution of the homogeneous equation for zero loading, i.e., the solution that contains the integration constants. To obtain w_h for the equidimensional equation, we assume it in the form [3]

$$w_h = Ar^m$$

where A and m are constants to be determined. Substituting this expression for w_h into Eq. (2.8) gives

$$m^2(m - 2)^2Ar^{m-4} = 0$$

The four roots of this equation are $m = 0, 0, 2$, and 2. Thus there are only two independent roots and hence two independent solutions of the equation. Two other linearly independent solutions are obtained by multiplying each of the two solutions by $\ln(r)$, as per standard procedures for solving such equations [3]. The complete solution for w_h is then a linear combination of the four linearly independent solutions. Thus, we have*

$$w_h = C_1 + C_2r^2 + C_3 \ln r + C_4r^2 \ln r \tag{2.11}$$

*A simpler way of obtaining this solution is to consider first the Laplace equation $\nabla^2 w = 0$ or $[d^2w/dr^2 + (1/r)dw/dr] = 0$, which has the general solution $w = C_1 + C_3 \ln r$, obtained by assuming that $w = Ar^m$. Now it is known that if $f(x, y)$ is a solution of the Laplace equation (in Cartesian coordinates), then $f(x, y)$, $y \cdot f(x, y)$, $x \cdot f(x, y)$, and $(x^2 + y^2) \cdot f(x, y)$ are linearly independent solutions of the biharmonic equation. Thus in polar coordinates, we have $w = r^0$ and $w = \ln r$ as two solutions of the Laplace equation, and hence for the biharmonic equation we get two additional solutions as r^2 and $r^2 \ln r$.

This solution, together with w_p, forms the complete solution for w. The constants C_1, C_2, C_3, and C_4 are determined from the boundary conditions. For a plate with a concentric hole (annular plate), for example, we shall have two boundary conditions each at both the inner and outer edges $r = b$ and $r = a$, respectively (see Problem 2.1).

Example 2.1

As a first example, we consider a plate simply supported at $r = a$ and carrying a UDL (uniformly distributed load) of p_o per unit area (Fig. 2.6). Equation (2.9) now becomes

$$\nabla^2\nabla^2 w = \frac{p_o}{K}$$

Here we have a constant term on the right-hand side. Hence we can assume the particular solution in the form $w_p = Ar^4$, where A is a constant to be determined. Substituting into Eq. (2.8), we get

$$w_p = \frac{p_o r^4}{64K}$$

The complete solution is obtained when we add this solution to the homogeneous solution of Eq. (2.11). The boundary conditions at $r = a$ are $w = 0$ and $M_r = 0$. Owing to the fact that there is no concentric hole in the plate, we have no other edge (boundary) on which to prescribe boundary conditions. Two other equations are, however, obtained from conditions of finiteness of w, M_r, M_θ, Q_r, etc., at $r = 0$. We note from Eq. (2.11) that the term $C_3 \ln r$ tends to $-\infty$ as $r \to 0$. Since we expect $w(r = 0)$ to be a finite quantity for our problem, we take $C_3 = 0$. We also note that the term $C_4 r^2 \ln r$ becomes $(0) \cdot (-\infty)$, i.e., indeterminate, when we substitute $r = 0$. By using L'Hôpital's rule for taking limits, we obtain

$$\lim_{r\to 0} (r^2 \ln r) = \lim_{r\to 0} \left(\frac{\ln r}{1/r^2}\right) = \lim_{r\to 0} \left(\frac{1/r}{-2/r^3}\right) = \lim_{r\to 0} \left(-\frac{r^2}{2}\right) = 0$$

Hence this term gives w as finite at $r = 0$. By substituting this term into Eqs. (2.7a–c), however, the reader should verify that it gives infinite values for M_r, M_θ, and Q_r at $r = 0$. Since we also expect these quantities to be finite at $r = 0$, we take C_4 to be zero. Hence the solution for w becomes

$$w = C_1 + C_2 r^2 + \frac{p_o r^4}{64K}$$

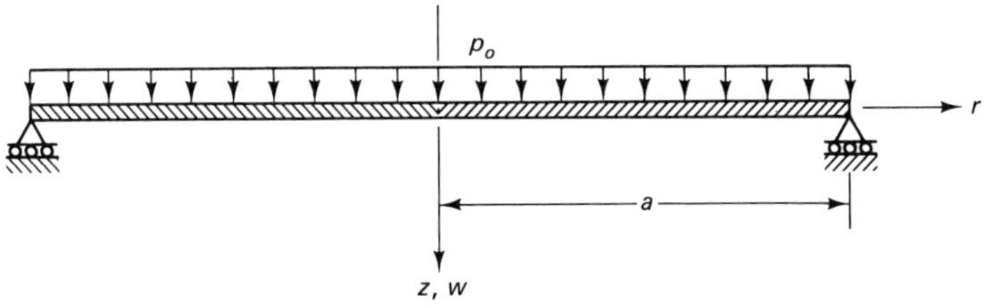

Figure 2.6 Simply supported circular plate subjected to a UDL.

Substituting the boundary conditions at $r = a$ into the expression above, we can now solve for C_1 and C_2. From $w(a) = 0$ and $M_r(a) = 0$, we thus obtain the following two equations:

$$C_1 + C_2 a^2 + \frac{p_o a^4}{64K} = 0$$

and

$$2C_2(1 + \nu) + \frac{4p_o a^2}{64K}(3 + \nu) = 0$$

which give

$$C_1 = \frac{p_o a^4}{64K}\left(\frac{5 + \nu}{1 + \nu}\right)$$

and

$$C_2 = -\frac{2p_o a^2}{64K}\left(\frac{3 + \nu}{1 + \nu}\right)$$

Then the final solution becomes

$$w = \frac{p_o a^4}{64K}\left[\frac{5 + \nu}{1 + \nu} - \frac{2(3 + \nu)}{1 + \nu}\frac{r^2}{a^2} + \frac{r^4}{a^4}\right]$$

From Eqs. (2.7a,b), we obtain the following for M_r and M_θ:

$$M_r = \frac{p_o a^2}{16}(3 + \nu)\left(1 - \frac{r^2}{a^2}\right)$$

and

$$M_\theta = \frac{p_o a^2}{16}\left[(3 + \nu) - (1 + 3\nu)\frac{r^2}{a^2}\right]$$

Maximum values of w, M_r, and M_θ occur at $r = 0$, and they are

$$w_{\text{max}} = \left(\frac{5 + \nu}{1 + \nu}\right)\frac{p_o a^4}{64K}$$

$$M_{r_{\text{max}}} = M_{\theta_{\text{max}}} = \frac{p_o a^2}{16}(3 + \nu)$$

If, instead of being hinged, this plate were fixed at $r = a$, results could readily be obtained in a similar manner by taking $w = 0$ and $w' = 0$ as the boundary conditions at $r = $ a. For an annular plate with a UDL, we would have $M_r = 0$ and $Q_r = 0$ at $r = b$ as the two additional boundary conditions (if the inner edge were free) instead of the finiteness conditions. In this case the solution becomes more complicated, since we have to solve four equations for constants C_1, C_2, C_3, and C_4.

Example 2.2

Problems such as the one shown in Fig. 2.7 become further complicated. Here $p_z(r)$ is not a continuous function of r from $r = 0$ to $r = a$. Hence the plate has to be considered in two parts. We use w_1 to denote the deflection of the plate in the portion $b \leqslant r \leqslant a$, and w_2 to denote the deflection in $0 \leqslant r \leqslant b$. The outer annular portion of the plate, $b \leqslant r \leqslant a$, carries no load, and hence w_1 is governed by the equation

$$\nabla^2\nabla^2 w_1 = 0$$

the solution of which is

$$w_1 = C_1 + C_2 r^2 + C_3 \ln r + C_4 r^2 \ln r$$

For the inner portion, $0 \leqslant r \leqslant b$, we have the differential equation

$$\nabla^2\nabla^2 w_2 = \frac{p_o}{K}$$

whose solution, assuming finiteness conditions at $r = 0$, was just obtained as

$$w_2 = C_5 + C_6 r^2 + \frac{p_o r^4}{64K}$$

Constants $C_1, \ldots, C_6$ should be determined from the two boundary conditions at $r = a$ and the four continuity conditions at $r = b$. These can be written as

$$\text{at } r = a\text{:} \quad w_1 = 0 \quad \text{and} \quad M_{r_1} = 0$$

$$\text{at } r = b\text{:} \quad w_1 = w_2 \quad \text{and} \quad w_1' = w_2'$$

$$M_{r_1} = M_{r_2} \quad \text{and} \quad Q_{r_1} = Q_{r_2}$$

Substituting for w_1, w_2, w_1', w_2', M_{r_1}, M_{r_2}, Q_{r_1}, and Q_{r_2}, we can solve the six equations so obtained for the constants $C_1, \ldots, C_6$, and thus obtain final expressions for w_1 and w_2. For $\nu = 0$ it can easily be verified that these solutions are

$$w_1 = \frac{p_o a^2 b^2}{64K}\left[\frac{4b^2}{a^2}\ln\frac{r}{a} - \frac{2b^2}{a^2} + 12 + \frac{r^2}{a^2}\left(8\ln\frac{r}{a} + \frac{2b^2}{a^2} - 12\right)\right]$$

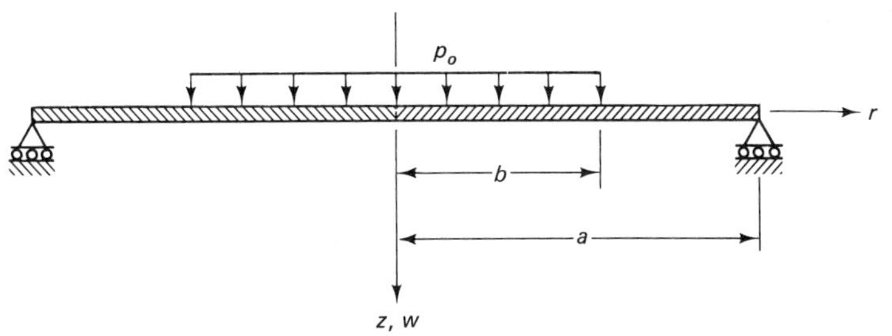

Figure 2.7 Simply supported circular plate carrying a UDL over the region $0 \leqslant r \leqslant b$.

$$w_2 = \frac{p_o a^2 b^2}{64K}\left[\frac{4b^2}{a^2}\ln\frac{b}{a} - \frac{7b^2}{a^2} + 12 + \frac{r^2}{a^2}\left(8\ln\frac{b}{a} + \frac{2b^2}{a^2} - 8\right) + \frac{r^4}{a^2b^2}\right]$$

From these expressions we can also obtain the solutions for M_r, M_θ, and Q_r. The variations of these quantities can then be plotted versus r for a given value of b/a.

Circular plate problems can be solved for a variety of other boundary conditions and loadings. Ready tables and formulas for some cases can be found in references such as [1] and [4].

2.4 ROTATIONAL STIFFNESS

Consider the plate shown in Fig. 2.8, to which a moment M_o is applied per unit length all around its periphery. Since there is no other loading on the plate, we have $w_p = 0$, and then the solution for w becomes

$$w = C_1 + C_2r^2$$

The boundary conditions are at $r = a$, $w = 0$* and $M_r = M_o$. After solving for C_1 and C_2, we get

$$w = \frac{M_o a^2}{2K(1 + \nu)}\left(1 - \frac{r^2}{a^2}\right)$$

and

$$w' = -\frac{M_o r}{K(1 + \nu)}$$

Hence, from Eqs. (2.7a–c), we obtain

$$M_r = M_o \qquad M_\theta = M_o \qquad Q_r = 0$$

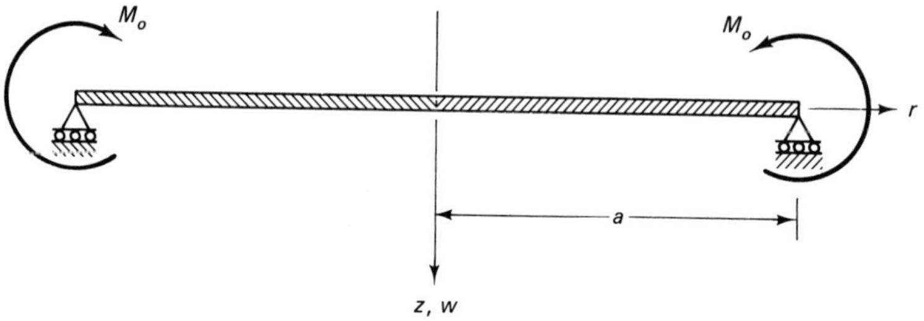

Figure 2.8 Simply supported circular plate loaded with an axisymmetric edge moment M_o.

*Since moments M_o are self-equilibrating, the plate does not need a support at $r = a$. Hence this boundary condition need not be prescribed, in which case C_1 will be indeterminate. Constant C_1 corresponds to a rigid-body displacement, whose value does not affect M_r, M_θ, or Q_r in the plate; i.e., a solution $w = C_1$ gives M_r, M_θ, and Q_r as zero, as seen from Eqs. (2.7a–c).

From these expressions, it is seen that there is a "pure bending" of the plate, with moments M_r and M_θ as constants and equal to the applied edge moment M_o throughout. Rotation of the plate at the edge, in the direction of M_o, will be equal to $-w'$ (since a negative edge rotation $-w'$ has the same direction as that of positive M_o). Hence we can write the rotational stiffness of the plate, S_p, as

$$S_p = \frac{M_o}{-w'} = \frac{K(1 + \nu)}{a}$$

This rotational stiffness can be used for analyzing the distribution of *fixed-end moments* at the junction between a cylindrical shell wall and a circular plate slab, the same as is done in typical moment distribution of portal frames. We shall make use of this stiffness later when solving such problems.

2.5 SINGULARITY PROBLEMS

Let us assume the solution of a circular plate problem as

$$w = C_4 r^2 \ln r$$

which is obtained by assuming zero distributed loading on the plate and constants C_1, C_2, and C_3 [Eq. (2.11)] as zero. Substituting this solution into Eq. (2.7c), we have

$$Q_r = -\frac{4K}{r} C_4$$

Now consider a freebody, of radius r, of an interior portion of the plate (Fig. 2.9). The resultant vertical force V, due to Q_r on this section, is

$$V = 2\pi r Q_r = -8\pi K C_4$$

We note that V is nonzero and is also independent of r. Since there is no distributed load on the plate, this nonzero and constant value of V must come from a concen-

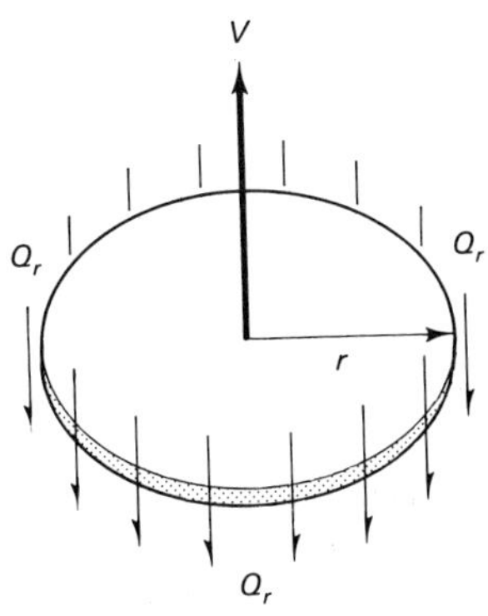

Figure 2.9 Interior freebody diagram of a circular plate subjected to a concentrated central load V.

trated load equal to $-(8\pi KC_4)$ applied at the center of the plate. Alternatively, for a known vertically upward load V, applied at the center of the plate, we can directly determine the constant C_4 in Eq. (2.11) as*

$$C_4 = \frac{-V}{8\pi K}$$

where V is positive when acting upward.

Example 2.3

The result just obtained can be used to solve problems such as the one shown in Fig. 2.10. We have a "mushroom" slab, supported on a single column at its center, and carrying a uniformly distributed load q. The solution for w can be taken as

$$w = C_1 + C_2r^2 + C_3 \ln r + C_4r^2 \ln r + \frac{qr^4}{64K}$$

where the last term is the particular solution of the differential equation, corresponding to the distributed loading q. For finiteness of w at $r = 0$, we must take $C_3 = 0$. Also, by the boundary condition $w = 0$ at $r = 0$, we get $C_1 = 0$. From statics, we know that the support reaction is

$$V = \pi a^2 q$$

Hence C_4 can be determined directly as

$$C_4 = -\frac{\pi a^2 q}{8\pi K} = -\frac{qa^2}{8K}$$

The remaining unknown constant, C_2, can be determined by applying the boundary condition $M_r = 0$ at $r = a$ in the following manner. We have

$$w = C_2r^2 + C_4r^2 \ln r + \frac{qr^4}{64K}$$

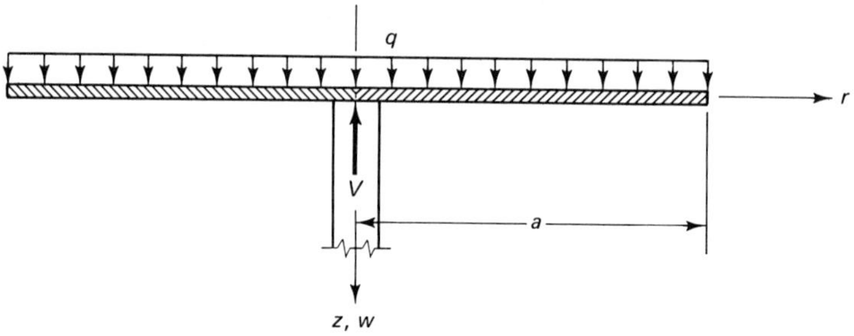

Figure 2.10 Circular slab supported by a central column and carrying a uniform load q.

*The term $C_3 \ln r$ in Eq. (2.11) gives Q_r as zero and $M_r = [(C_3K/r^2)(1 - \nu)]$, which increases to ∞ as $r \to 0$. However, moment M_r around any plate section $r = r$, as in Fig. 2.9, has a zero total resultant. Hence this term in the solution does not correspond to a singular vertical load at $r = 0$.

Hence

$$w' = 2C_2 r + C_4(2r \ln r + r) + \frac{4qr^3}{64K}$$

and

$$w'' = 2C_2 + C_4(2 \ln r + 3) + \frac{12qr^2}{64K}$$

Thus since

$$M_r = -K\left(w'' + \nu\frac{w'}{r}\right)$$

we obtain

$$M_r = -K\left[2C_2(1 + \nu) + 2C_4 \ln r\,(1 + \nu) + C_4(3 + \nu) + \frac{4qr^2}{64K}(3 + \nu)\right]$$

Then, substituting $M_r = 0$ at $r = a$ and the expression for C_4 above into this equation, we find

$$C_2 = \frac{qa^2}{8K} \ln a + \frac{qa^2}{16K}\left(\frac{3 + \nu}{1 + \nu}\right) - \frac{qa^2}{32K}\left(\frac{3 + \nu}{1 + \nu}\right)$$

Hence, the final solution for w can be written as

$$w = \frac{qa^4}{16K}\left[\frac{3 + \nu}{2(1 + \nu)}\frac{r^2}{a^2} - 2\left(\frac{r^2}{a^2}\right) \ln\frac{r}{a} + \frac{r^4}{4a^4}\right]$$

Instead of proceeding as above, we could have determined constant C_4 by directly using the boundary condition $Q_r = 0$ at $r = a$. One may think that there is one more condition, i.e., $w' = 0$ at $r = 0$; however, this condition is inherent when axisymmetric deformation is assumed, and hence will not be useful for obtaining C_1, C_2, C_3, or C_4. Using the expression for w given above, expressions for M_r and M_θ can now be obtained as follows:

$$M_r = -\frac{qa^2}{16}\left[(3 + \nu)\left(\frac{r^2}{a^2} - 1\right) - 4(1 + \nu) \ln\frac{r}{a}\right]$$

$$M_\theta = \frac{-qa^2}{16}\left[(1 - 5\nu) + \frac{r^2}{a^2}(1 + 3\nu) - 4(1 + \nu) \ln\frac{r}{a}\right]$$

We notice from the expressions above that as $r/a \to 0$, M_r and M_θ blow up to negative infinity. Evaluating Q_r for $r/a \to 0$ also produces a solution that blows up to infinity. This is because we have idealized the column as a point support. However, in a practical problem the column will not be a point support but will have a finite radius. Thus M_r, M_θ, and Q_r will be finite at the outer periphery of the column.

Roof slabs of circular tanks sometimes have to be supported by a central column in addition to the peripheral wall support if the tank diameter is large. Such a problem can readily be solved by retaining the C_4-term in the general solution and using the appropriate boundary conditions, as in Example 2.3. The Portland

Cement Association (PCA) [5] gives tables for M_r and M_θ in such slabs. Such tables can be conveniently used for solving practical problems.

2.6 DESIGN CONSIDERATIONS

Reinforced concrete circular slabs without openings can be designed for the values of M_r and M_θ obtained from the solution of the differential equation. Reinforcement in simply supported slabs can be designed for the maximum moments at the center, and can be reduced toward the support. For such slabs it is a common practice to provide reinforcement in a square-mesh pattern instead of providing radial bars and circular rings. For $\nu = 0.1$, the maximum moments obtained in Example 2.1 for a simply supported circular plate with a uniform load p_o become

$$M_{r_{\max}} = M_{\theta_{\max}} = \frac{3.1 p_o a^2}{16} = \frac{p_o D_s^2}{20.65}$$

where D_s is the diameter of the slab. It is a common practice to design the slab for a moment of $p_o D_s^2/20$ (Ref. [6]).

For slabs that are fixed at their edges or develop negative moments due to continuity (with tank walls, say), radial steel at the top should be provided near the edges. Reference [6] gives design moments to be used for a few typical cases.

Figure 2.11 shows two schemes for reinforcing circular concrete slabs. The square-mesh detail in Fig. 2.11(b), however, is often preferred due to the simplicity

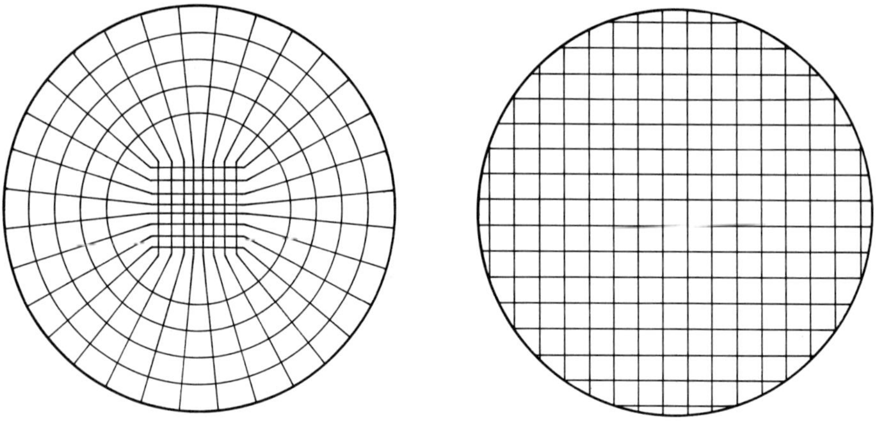

(a) (b)

Figure 2.11 Reinforcing layouts for circular slabs. (a) Radial and circular reinforcement. (b) Square-mesh pattern.

of its construction. Alternate bars could be curtailed near the edges for cases where the design moment becomes small there.

REFERENCES

[1] Timoshenko, S. P., and Woinowski-Krieger, S., *Theory of Plates and Shells,* 2nd ed., McGraw-Hill Book Company, New York, 1959.

[2] Timoshenko, S. P., and Gere, J. M., *Theory of Elastic Stability,* 2nd ed., McGraw-Hill Book Company, New York, 1961.

[3] Coddington, E. A., *An Introduction to Ordinary Differential Equations,* Prentice-Hall, Inc., Englewood Cliffs, N.J., 1961.

[4] Roark, R. J., and Young, W. C., *Formulas for Stress and Strain,* 5th ed., McGraw-Hill Book Company, New York, 1975.

[5] "Circular Concrete Tanks Without Prestressing," Portland Cement Association Publication No. IS072.01D, Portland Cement Association, Skokie, Ill.

[6] Reynolds, C. E., and Steedman, J. C., *Reinforced Concrete Designers Handbook,* 8th ed., Cement and Concrete Association, Slough, England, 1974.

PROBLEMS

2.1. Figure P2.1 shows a circular plate of radius a with a concentric hole of radius b which is subjected to a uniform load p_o. The plate is simply supported at its outer edge and free at its inner edge. Assuming that $b/a = 0.7$ and $\nu = 0$, derive expressions for w, M_r, M_θ, and Q_r for this plate. Plot normalized values of these quantities versus r/a.

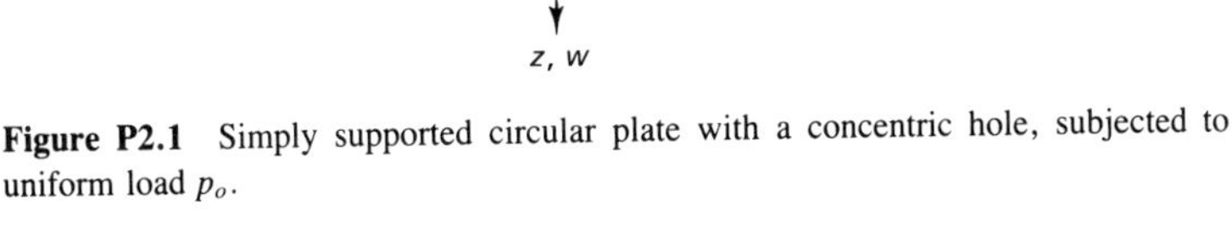

Figure P2.1 Simply supported circular plate with a concentric hole, subjected to a uniform load p_o.

2.2. A circular plate of radius a carries a concentric uniformly distributed load p_o over a radius b (Fig. P2.2). The plate is simply supported at its outer edge. Assuming that $b/a = 0.7$ and $\nu = 0$, derive expressions for w, M_r, M_θ, and Q_r for this plate. Plot normalized values of these quantities versus r/a. (See Example 2.2.)

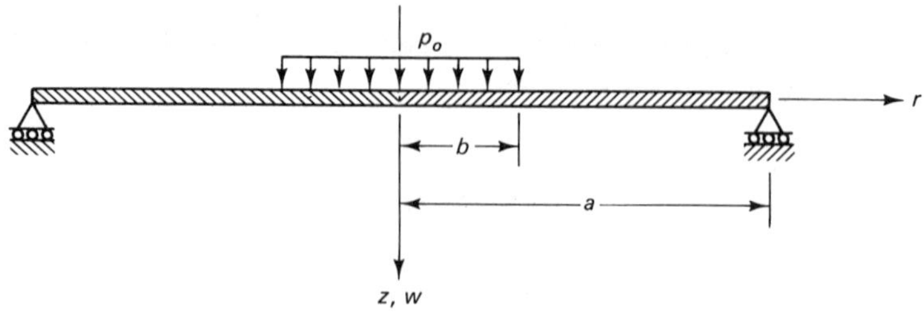

Figure P2.2 Simply supported circular plate subjected to a concentric uniform load p_o over a radius b.

2.3. Figure P2.3 shows a simply supported circular plate of radius a subjected to a linearly varying load of

$$p_z(r) = p_o\left(\frac{r}{a}\right)$$

For this plate, obtain an expression for its deflection w, and plot this versus r/a.

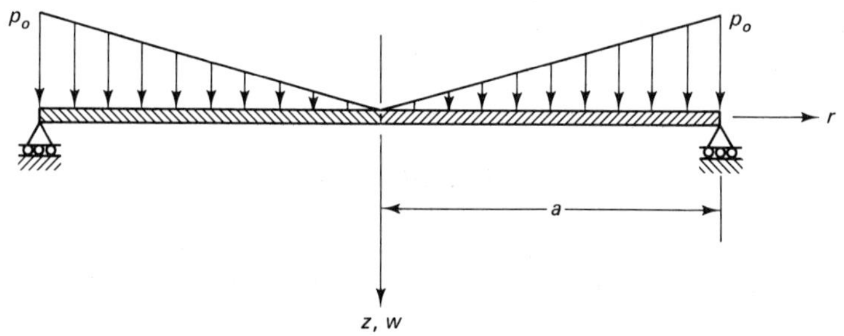

Figure P2.3 Simply supported circular plate subjected to a linearly increasing load.

2.4. A circular slab is fixed at its outer periphery $r = a$, and has a central support (Fig. P2.4). The slab is loaded by a uniformly distributed load p_o. Obtain values of M_r and M_θ for this plate, for r/a equal to 0.1, 0.2, 0.5, 0.7, and 1.0, assuming that $\nu = 0.25$. Compare these values with M_r and M_θ given in the PCA table [5] reproduced here as Table P2.4. Explain any discrepancies between your solution and that in Table P2.4.

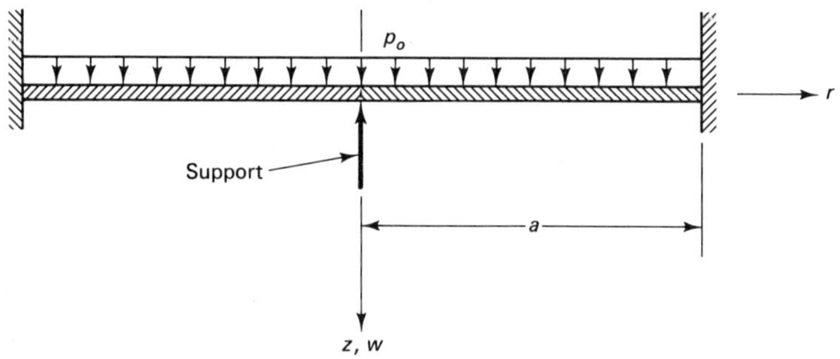

Figure P2.4 Fixed-ended plate with a central column, supporting a uniform load p_o.

TABLE P2.4 Radial and Tangential Moments in Circular Slab with a Center Support and Fixed Edges

Moments in circular slab with center support
Uniform load
Fixed edge

Mom. = coef. × pa^2 ft–lb/ft
Positive sign indicates compression in surface loaded

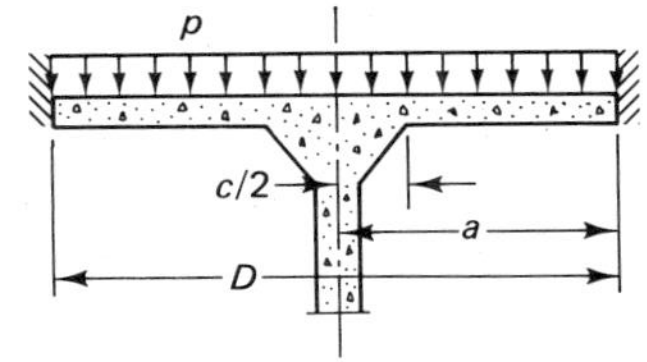

c/D	0.05a	0.10a	0.15a	0.20a	0.25a	0.30a	0.40a	0.50a	0.60a	0.70a	0.80a	0.90a	1.00a
	Coefficients at point												
	Radial moments, M_r												
0.05	−0.2100	−0.0729	−0.0275	−0.0026	+0.0133	+0.0238	+0.0342	+0.0347	+0.0277	+0.0142	−0.0049	−0.0294	−0.0589
0.10		−0.1433	−0.0624	−0.0239	−0.0011	+0.0136	+0.0290	+0.0326	+0.0276	+0.0158	−0.0021	−0.0255	−0.0541
0.15			−0.1089	−0.0521	−0.0200	+0.0002	+0.0220	+0.0293	+0.0269	+0.0169	+0.0006	−0.0216	−0.0490
0.20				−0.0862	−0.0429	−0.0161	+0.0133	+0.0249	+0.0254	+0.0176	+0.0029	−0.0178	−0.0441
0.25					−0.0698	−0.0351	+0.0029	+0.0194	+0.0231	+0.0177	+0.0049	−0.0143	−0.0393
	Tangential moments, M_θ												
0.05	−0.0417	−0.0700	−0.0541	−0.0381	−0.0251	−0.0145	+0.0002	+0.0085	+0.0118	+0.0109	+0.0065	−0.0003	−0.0118
0.10		−0.0287	−0.0421	−0.0354	−0.0258	−0.0168	−0.0027	+0.0059	+0.0099	+0.0098	+0.0061	−0.0009	−0.0108
0.15			−0.0218	−0.0284	−0.0243	−0.0177	−0.0051	+0.0031	+0.0080	+0.0086	+0.0057	−0.0006	−0.0098
0.20				−0.0172	−0.0203	−0.0171	−0.0070	+0.0013	+0.0063	+0.0075	+0.0052	−0.0003	−0.0088
0.25					−0.0140	−0.0150	−0.0083	−0.0005	+0.0046	+0.0064	+0.0048	0.0000	−0.0078

Source: Ref. [5].

2.5. Figure P2.5 shows a circular plate that is simply supported at its outer periphery ($r = a$). The plate also has additional linear vertical spring supports, of stiffness α, at radius $r = b$. Assume that the plate carries a uniform load p_o. Write down the boundary conditions and compatibility equations necessary for solving the differential equation of deflection for this plate. Outline how to obtain expressions for w, M_r, M_θ, and Q_r.

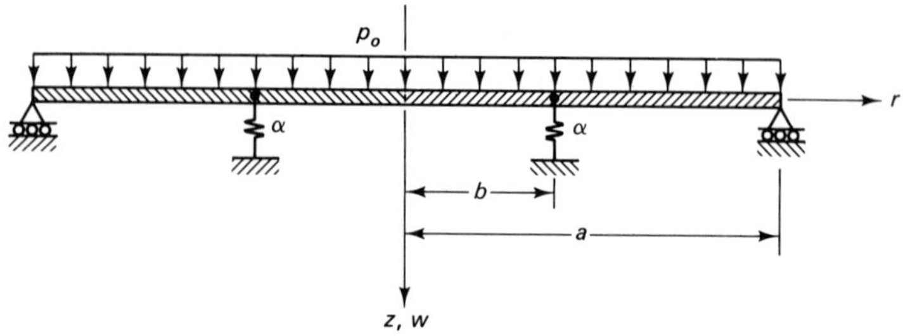

Figure P2.5 Circular plate carrying a uniform load p_o. The plate is simply supported at its outer edge and has linear spring supports of stiffness α at radius b.

2.6. For a circular plate of radius a, which has a central column support, obtain the rotational stiffness for a moment applied at the outer periphery.

2.7. Assume that the mushroom slab of Example 2.3 carries a fascia panel at its outer periphery, which weighs P pounds per foot of circumference. Obtain the deflection w and the moment M_r in the slab under this loading.

2.8. The annular plate shown in Fig. P2.8 is subjected to a uniformly distributed load of magnitude q. Obtain an expression for the slope dw/dr at $r = b$, assuming that $\nu = 0$. Also, show that the flexibility coefficient δ_{22} (the rotation at $r = b$ due to a unit moment applied around the circumference at $r = b$), is given by

$$\delta_{22} = \frac{(a^2 + b^2)b}{K(a^2 - b^2)}$$

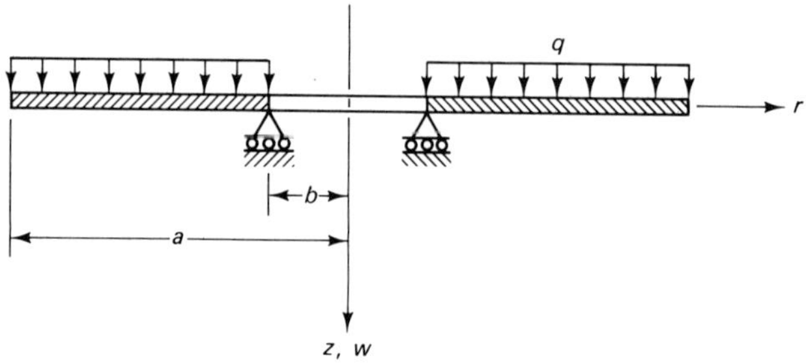

Figure P2.8 Annular plate, simply supported at its inner edge and subjected to a uniformly distributed loading q.

3

Theory of Surfaces, Shell Classification, and Membrane versus Bending Theory

3.1 THEORY OF SURFACES OR DIFFERENTIAL GEOMETRY

The geometry of a shell is completely defined if we know the shape of its middle surface and its thickness at all points. Hence it is desirable at this stage that we discuss briefly a few basic concepts of the *theory of surfaces* or *differential geometry*.

A surface in space can be represented by an equation

$$z = f(x,y)$$

where x and y are two independent variables and z is the dependent variable (Fig. 3.1). The coordinates x, y, and z define a point on the surface. Alternatively, we can represent the surface by its *parametric equations*,

$$x = f_1(\alpha_1, \alpha_2)$$

$$y = f_2(\alpha_1, \alpha_2)$$

$$z = f_3(\alpha_1, \alpha_2)$$

where α_1 and α_2 are the two independent variables, which are also referred to as the Gaussian or curvilinear coordinates of a point on the surface. If α_1 or α_2 is kept constant, we get a curve on the surface. A curve in space is thus defined by one independent variable; say α_1 or α_2. The intersection of two surfaces; $z = g_1(x, y)$ and $z = g_2(x, y)$, also gives a curve in space.

For a point on a surface, one could conceive an infinite number of curves that pass through this point and also lie entirely on the surface. These curves could be obtained, as well, by intersection of the surface with other surfaces, and would

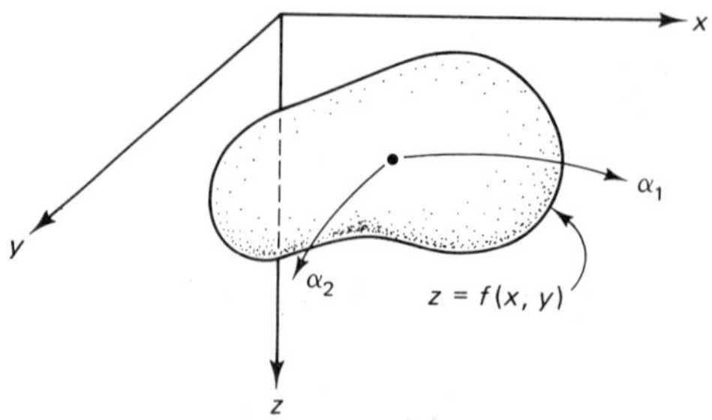

Figure 3.1 A surface in space characterized by either Cartesian coordinates x and y, or curvilinear coordinates α_1 and α_2.

be space curves in general. However, if the intersecting surface is a plane, the intersection curve would be a *plane curve*. Through a given point, A', on a surface (Fig. 3.2), several such plane curves (lying on the surface) may be obtained by intersections of different planes with the surface.

We can also imagine a plane tangent to the surface at a point A (Fig. 3.2). A normal to this plane through A would be a normal to the surface at A. Out of the infinite plane curves passing through A, some curves could be formed by intersection of the surface with planes containing the normal to the surface at A. Such plane curves are called *normal sections* of the surface at that point.

There can be several normal sections through A, each of these having a different curvature κ_s at A. Out of these, there will be one curve having a maximum value of κ_s at A and another having a minimum value of κ_s. These two curves (normal sections) are called the *principal sections*, and their curvatures, denoted by κ_1 and κ_2 (or vice versa), are called the *principal curvatures* of the surface at A. It can be shown that the two principal sections are orthogonal [1]. The product $\kappa = \kappa_1 \cdot \kappa_2$ is called the *Gaussian curvature* of the surface at A. Depending on

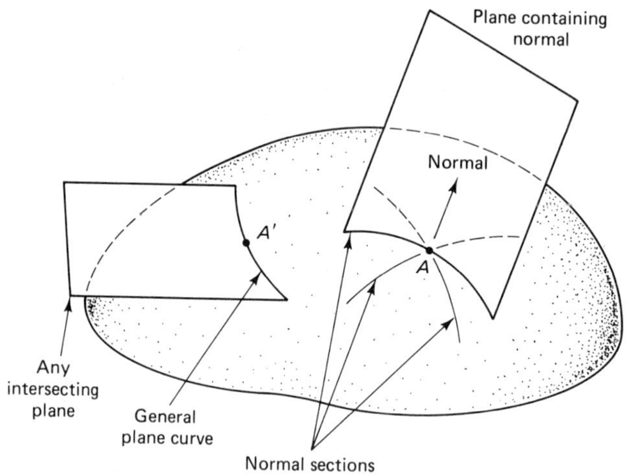

Figure 3.2 Intersection of surface with a plane to form a plane curve. If the intersecting plane contains the normal to the surface at a point, the curve formed is called a normal section.

whether κ is positive, negative, or zero at a point, the surface is called, respectively, *synclastic, anticlastic,* or *developable* at that point.

Knowing the parametric equations of a surface, formulas for κ_1 and κ_2 can be derived using the theory of differential geometry.

3.2 TYPES OF SURFACES

We shall define here some of the surfaces that are commonly used for shell structures.

Surfaces of Revolution

Surfaces of revolution are generated by revolving a plane curve, called the *meridian,* about an axis not necessarily intersecting the meridian. Circular cylinders, cones, spherical or elliptical domes, hyperboloids of revolution, and toroids (Fig. 3.3) are some examples of surfaces of revolution.

It can be proven that for surfaces of revolution, the meridian is one of the principal sections and its curvature at any point is one of the principal curvatures, κ_1. The corresponding radius of curvature of the meridian may be denoted as r_1, so that $r_1 = 1/\kappa_1$. The other orthogonal principal section is obtained by intersection of the surface with a plane that is at right angles to the plane of the meridian and that also contains the normal (Fig. 3.4). This second principal section has curvature κ_2 and radius of curvature $r_2 = 1/\kappa_2$.

For better visualization of this second principal section, one can take a normal section by cutting a rubber ball or a lemon with a knife, and observe the surface curve thus generated. It may be noted that although the tangent to the second principal section and the *parallel circle** are the same, the parallel circle is not the second principal section because it is not a normal section; i.e., its plane does not contain, generally, the normal to the surface.

It can also be shown that for surfaces of revolution, if r_1 is the radius of curvature of the meridian, then r_2 (the radius of curvature of the second principal section) is the distance from the point on the surface to the point of intersection of the line containing r_1 with the axis of revolution. The radii of curvature r_1 and r_2 for typical surfaces of revolution are shown in Fig. 3.3. It can be seen that for the circular cylinder and the cone, the meridian is a straight line, and hence, $\kappa_1 = 0$, which gives $\kappa = 0$. These are shells of zero Gaussian curvatures, and hence are developable surfaces. For ellipsoids of revolution, the sphere, or paraboloids of revolution, both the principal curvatures are in the same direction, and thus these surfaces have positive Gaussian curvatures (i.e., they are synclastic). On the other hand, for the hyperboloid of revolution [Fig. 3.3.(d)], the curvatures of the meridian and the second principal section are in opposite directions (i.e., r_1 and r_2 lie on

*The parallel circle is the circular curve formed by the intersection of a horizontal plane with the surface of revolution.

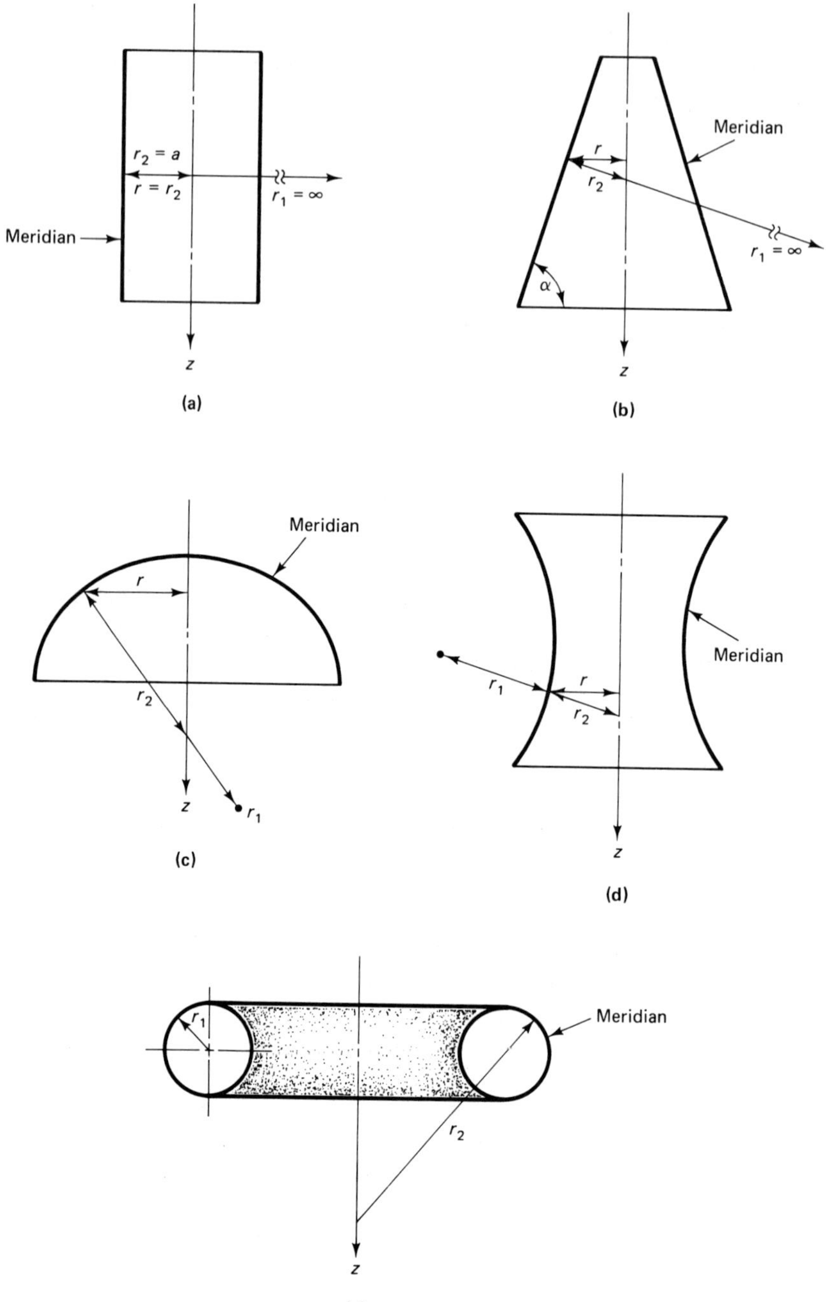

Figure 3.3 Some examples of surfaces of revolution. (a) Circular cylinder. (b) Cone. (c) Elliptical dome. (d) Hyperboloid of revolution. (e) Toroid.

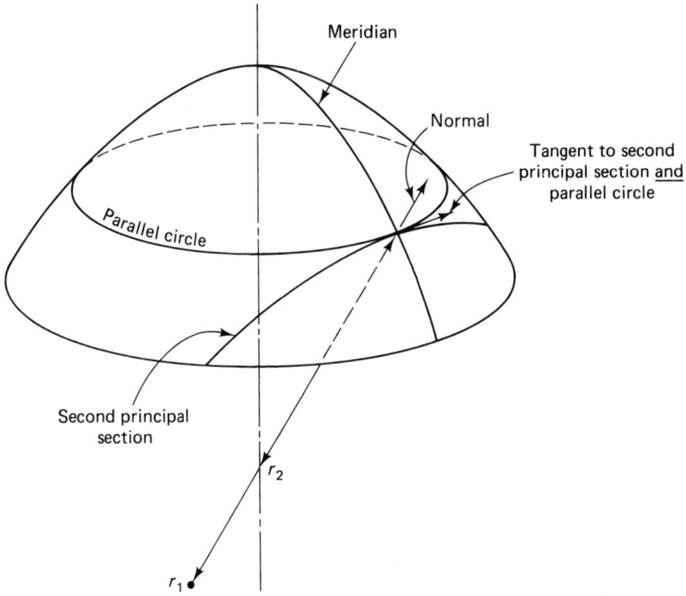

Figure 3.4 Surface of revolution, showing the meridian, parallel circle, and the second principal section.

opposite sides of the surface) at all points on the surface. Therefore, this shell has a negative Gaussian curvature and is thus anticlastic. For the toroid [Fig. 3.3(e)] the Gaussian curvature changes from positive to negative as we move along the surface.

If the equation of the meridian, taking rectangular coordinate axes as r and z, is written as $r = f(z)$, then for a surface of revolution formed with this meridian, it can be proven that*

$$\kappa_1 = \frac{-r''}{[1 + (r')^2]^{3/2}} \qquad \text{and} \qquad \kappa_2 = \frac{1}{r\,[1 + (r')^2]^{1/2}}$$

where $(\cdot)'$ denotes $d(\cdot)/dz$. Thus for the circular cylinder [Fig. 3.3(a)], we can write the equation of the meridian as $r = a$, which gives r' and r'' as zero. Hence substituting into the formulas above, we get, as expected,

$$\kappa_1 = 0 \qquad \text{and} \qquad \kappa_2 = \frac{1}{a}$$

or

*This expression for κ_1 is the well-known "exact" expression for curvature of a plane curve. Indeed, for the curves formed by deflections of beams/plates, we use only the numerator in this term [see Eqs. (1.10) and (2.3a)], since r' in the denominator is usually small compared to unity (due to the shallow curves formed by small deflections).

$$r_1 = \infty \qquad \text{and} \qquad r_2 = a$$

For the conical surface [Fig. 3.3(b)], the equation of the meridian can be written as $r = z \cot \alpha$, which gives $r' = \cot \alpha$ and $r'' = 0$. Substituting, we get

$$\kappa_1 = 0 \qquad \text{and} \qquad \kappa_2 = \frac{1}{r \csc \alpha}$$

or

$$r_1 = \infty \qquad \text{and} \qquad r_2 = r \csc \alpha$$

For a sphere, the equation of the meridian is

$$r^2 + z^2 = a^2 \qquad \text{or} \qquad r = \sqrt{a^2 - z^2}$$

Using this, the reader can check that

$$\kappa_1 = \kappa_2 = \frac{1}{a}$$

Similarly, we can obtain expressions for κ_1 and κ_2, in terms of the parameters defining the meridian, for other surfaces of revolution, such as ellipsoids, paraboloids, and hyperboloids of revolution.

Surfaces of Translation and Ruled Surfaces

A surface of translation is defined as the surface generated by keeping a plane curve parallel to its initial plane as we move it along another plane curve. The two planes containing the two curves are at right angles to each other. Thus an *elliptic paraboloid* is obtained by translation of a parabola on another parabola at right angles, both parabolas having their curvatures in the same direction [Fig. 3.5(a)]. Therefore, the Gaussian curvature κ is positive for such a surface. For this surface, sections $x = constant$ or $y = constant$ are parabolas. A section $z = constant$ would give an ellipse, hence the name "elliptic paraboloid." If the two parabolas are similar, the ellipse becomes a circle and the surface also becomes a surface of revolution, called a *paraboloid of revolution.*

The surface shown in Fig. 3.5(b) is generated by the translation of one parabola along another parabola having an opposite curvature. Hence the surface will have a negative Gaussian curvature. For this surface, sections $z = constant$ will give conjugate hyperbolas, and thus such surfaces are called *hyperbolic paraboloids,* or simply *hypars*. These surfaces can also be generated by translation of straight lines as shown in Fig. 3.5(c).

Ruled surfaces are obtained by the translation of straight lines over two end curves (Fig. 3.6). The straight lines are not necessarily at right angles to the planes containing the end curves. The frustum of a cone can thus be considered as a ruled surface, since it can be generated by translation of a straight line (the generator) over two circles at its ends. It is also, of course, a shell of revolution.

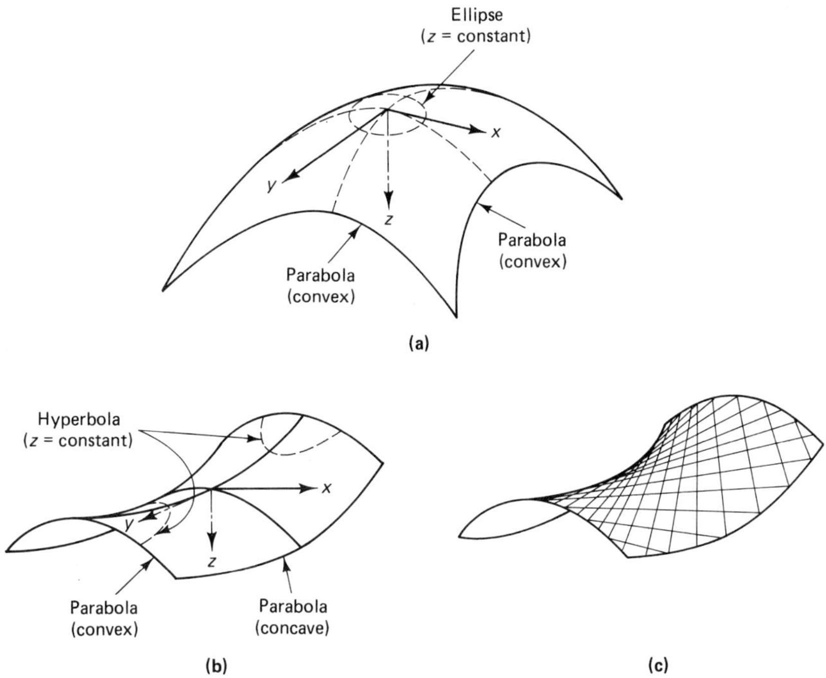

Figure 3.5 Surfaces of translation. (a) Elliptic paraboloid. (b) Hyperbolic paraboloid. (c) Hypar and its straight-line generators.

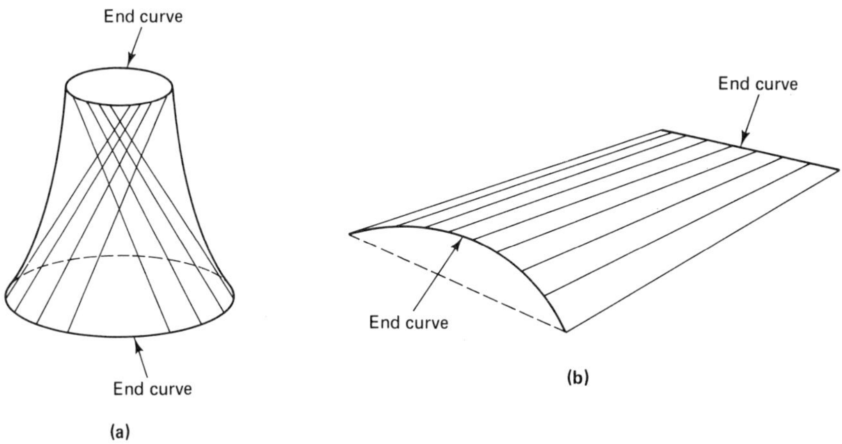

Figure 3.6 Ruled surfaces formed by translation of straight lines over two end curves. (a) Hyperboloid of revolution of one sheet. (b) Conoid.

The *hyperboloid of revolution of one sheet,* shown in Fig. 3.6(a), can also be generated by the translation of a straight line over two circles at its ends. Indeed, sitting stools of such shapes, generated by fixing cane or bamboo sticks between top and bottom rims, are commonly seen.

Figure 3.6(b) shows a surface generated by translation of a straight line on a circular curve at one end, and on a straight line at the other end. Such surfaces are commonly known as *conoids*. Both surfaces shown in Fig. 3.6 have negative Gaussian curvatures.

By knowing the equations of a surface of translation or of a ruled surface, we can obtain expressions for the principal radii of curvature and principal curvatures by using the theory of differential geometry. However, these procedures are more complex than for shells of revolution, and are not detailed here.

3.3 SUMMARY AND CLASSIFICATION OF SHELL SURFACES

Based on our discussion so far, we can now summarize and broadly classify the shell surfaces which are commonly used in engineering practice [2,3]. It may be noted that the same type of shell may appear in more than one category.

1. *Singly curved or shells of zero Gaussian curvature* (developable shells)
   - *Shells of revolution*: circular cylinders, cones
   - *Shells of translation or ruled surfaces*: circular or noncircular cylinders, cones
2. *Doubly curved shells having positive* κ (synclastic shells)
   - *Shells of revolution*: circular domes, ellipsoids of revolution, paraboloids of revolution
   - *Shells of translation or ruled surfaces*: elliptic paraboloids, paraboloids of revolution
3. *Doubly curved shells having negative* κ (anticlastic shells)
   - *Shells of revolution*: hyperboloids of revolution of one sheet
   - *Shells of translation or ruled surfaces*: hyperbolic paraboloids, conoids, hyperboloids of revolution of one sheet

In the following chapters we study the stress analysis of some of the foregoing categories of shell structures commonly used in practice.

3.4 MEMBRANE VERSUS BENDING THEORY FOR SHELLS

Before we undertake the study of analysis of the different types of shells, we consider the complexities involved in such analyses and what simplifications need be made to obtain a solution. We had seen in Chapter 1 (Fig. 1.15) that a shell element will have, in general, 10 unknown internal stress resultants acting on it: N_x, N_y, N_{xy}, N_{yx}, M_x, M_y, M_{xy}, M_{yx}, Q_x, and Q_y. To obtain the variation of these 10 stress resultants for a given loading and given boundary conditions would be a complex task. However, by making suitable assumptions, we try to obtain a simpler solution that will give satisfactory results for practical purposes.

We first assume that the shell is thin. Such shells are then capable of resisting in-plane stresses quite well, but are very flexible for resisting bending moments and shearing forces. Considerable deformations have to take place when loads causing large moments and shears are applied on the shell. A piece of paper rolled in the form of a cylindrical barrel can support its own weight without much deformation, since such loads can be supported by development of in-plane stresses alone. However, it cannot carry any bending loads without considerable deformation. Similarly, although a balloon can resist large internal pressures, a prick of a pin that causes bending moments and large deformations locally can easily burst it.

We would always prefer the shell to resist any loading by development of in-plane forces only rather than by development of bending and shear. For a rectangular cross section, the in-plane stresses are uniformly distributed over the entire thickness, and hence the section is fully utilized. On the other hand, a bending moment gives stresses that vary linearly over the thickness so that the portions near the middle surface remain understressed, and hence under-utilized.

We therefore first try to explore the possibility of the shell resisting a given load by development of in-plane stresses only; i.e., we assume that the moments M_x, M_y, M_{xy}, and M_{yx} are zero in the shell. Then, by taking moments about the x- and y-axes (Fig. 1.15), we conclude that Q_y and Q_x must also be zero, and by taking moments about the z-axis, we get $N_{xy} = N_{yx}$. Thus there remain only three unknown internal stress resultants, N_x, N_y, and N_{xy}, to support a given loading. Also, there still remain three equations of equilibrium to be utilized,

$$\sum F_x = 0 \qquad \sum F_y = 0 \qquad \sum F_z = 0$$

These three equations should be adequate to determine the three unknowns. The problem thus becomes statically determinate, and hence much simplified.

Such simplified theory for solving shell problems is called the *membrane theory,* as opposed to the more general and complex *bending theory,* which does not neglect the moments and shears. Although the membrane theory gives good results for many practical problems and is much easier to use than the bending theory, we are unable to satisfy the compatibility conditions at boundaries with the membrane theory solution alone, as we shall see later. To satisfy the boundary conditions, we have to apply certain corrective edge loads at the boundaries. Since these edge loads themselves will generally consist of bending moments and shears on the shell, we must necessarily use the bending theory for obtaining the edge-load solutions. The edge-load solutions (bending solutions), together with the membrane solution, will yield the complete solution of a problem.

In mathematical terms, we can consider the general equation for the deflection w of a shell as

$$\mathscr{L}\,[w(x, y)] = f[p(x, y)]$$

where $\mathscr{L}$ is a differential operator and $f[p]$ is some function of the given loading p. The general solution of this equation will be

$$w = w_h + w_p$$

where w_p is the particular solution satisfying the differential equation for a given load $p(x, y)$, but not necessarily satisfying the boundary conditions. Thus w_p corresponds to a solution that satisfies equilibrium and compatibility at all internal points in the shell for a given loading but by itself does not necessarily satisfy the boundary conditions. The term w_h is the solution of the homogeneous equation with $p = 0$; physically, it corresponds to an edge-load solution (since with $p = 0$ only edge loads can be present). Together, w_h and w_p satisfy both the differential equation and the boundary conditions. In other words, the complete solution of a shell problem can be obtained by superimposing on a particular solution w_p an edge-load solution w_h.

It is not always easy to obtain w_p of the exact* differential equation. Instead, for solving practical problems, we can use w_p of a much simplified differential equation obtained by assuming moments and shears to be zero. This is the "membrane solution." It is found that this solution is still very close to the solution w_p of the exact differential equation. It is still an approximation that is similar to obtaining (fairly correctly) the forces in truss members by assuming moments and shears in the members as zero (i.e., assuming the joints as perfect pins), although we know that in reality there will be some rigidity at the joints which can give rise to small moments and shears in all members [4].

For obtaining w_h, however, we must necessarily use the exact differential equation since the simplified equation will not permit us the freedom of prescribing general boundary conditions. The w_h solutions are therefore called the "bending theory" solutions and are naturally more difficult to obtain. Fortunately, for most types of shells, they die out quickly as we move away from the boundaries. In the remaining major portion of the shell, the simple membrane solutions can be considered to be valid.

Thus, for solving shell problems, our general procedure will be to obtain *membrane forces* under a given loading, and then superimpose on these the bending theory solutions for *edge loads*. The edge loads are determined so that the membrane and edge-load solutions together satisfy the boundary conditions; i.e., the edge loads are obtained by solving equations of compatibility at the boundaries, as we do for obtaining redundant forces in statically indeterminate structures.

*By "exact" differential equation we mean an equation derived without assuming the moments and shears to be zero in the shell.

REFERENCES

[1] O'Neill, Barrett, *Elementary Differential Geometry*, Academic Press, Inc., New York, 1966.

[2] "Recommendations for Reinforced Concrete Shells and Folded Plates," International Association for Shell and Spatial Structures, Madrid, 1979.

[3] "Criteria for the Design of Reinforced Concrete Shell Structures and Folded Plates," Indian Standard 2210-1962, Indian Standards Institution, New Delhi, 1963.

[4] Billington, D. P., *Thin Shell Concrete Structures*, 2nd ed., McGraw-Hill Book Company, New York, 1982.

PROBLEMS

3.1. The equation of the meridian of an ellipsoid of revolution is

$$\frac{r^2}{a^2} + \frac{z^2}{b^2} = 1$$

where a and b are lengths of the major and minor axes, respectively. Show that the principal radii of curvature are given by

$$r_1 = \frac{1}{\kappa_1} = \frac{(b^4 + e^2z^2)^{3/2}}{ab^4}$$

$$r_2 = \frac{1}{\kappa_2} = \frac{a(b^4 + e^2z^2)^{1/2}}{b^2}$$

where $e^2 = a^2 - b^2$.

3.2. Draw sketches of the following meridians, and evaluate κ_1, κ_2, and κ in each case.

(a) $$z = Ar^2$$

(b) $$(r^2/a^2) - (z^2/b^2) = 1$$

3.3. Figure P3.3 shows a semicircular arch subjected to a uniform radial load p_r per unit length. Find N_θ, M_θ, and Q_θ as functions of θ. Now if the arch is fixed at the supports,

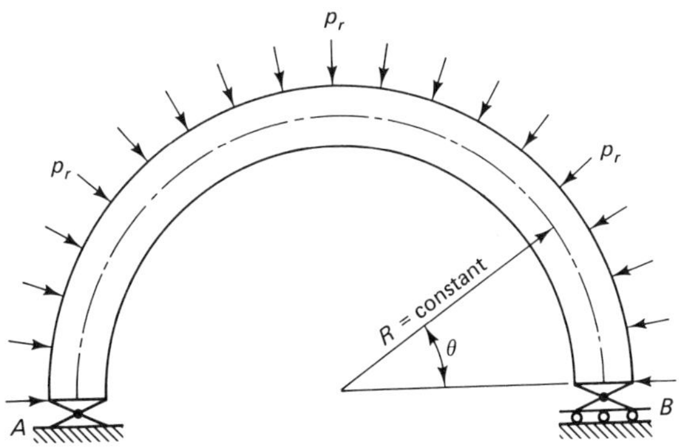

Figure P3.3 Arch problem. Simply supported arch subjected to uniform load intensity.

explain how you can obtain the stress resultants by superimposing on the earlier solution an edge-load solution. The reader may note (from the known solution of such a fixed arch in the theory of structures) that the stress resultants due to the edge loads do not die out as we move away from the edge.

3.4. Figure P3.4 shows a semi-infinite beam on an elastic foundation which gives an upward reaction of αw per unit length. The beam carries an edge load P, as shown. Formulate the differential equation for deflection w of the beam, and solve this equation for the given loading. Plot the variations of w and M_x versus x, and note how these quantities die out away from the edge $x = 0$. Discover which parameters affect the rapidity with which the values die out.

(Comparing the variations of stress resultants as we move away from the edge, in Problems 3.3. and 3.4, we note that when the springs resist the deformations as in Problem 3.4, the effects of edge loads die out quickly as we move away from the edge. We shall see in Chapter 5 that in cylindrical tanks, elemental circular rings resist the radial deformations of the shell by acting as "springs," and hence edge-load effects die out away from the edge. The same analogy can be drawn concerning the behavior of a spherical shell subjected to edge loads.)

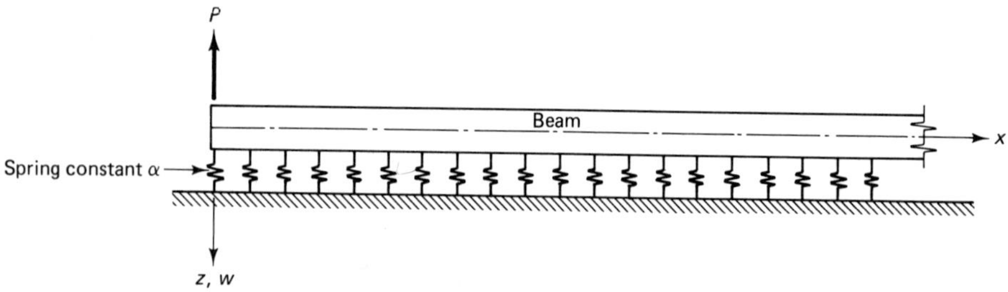

Figure P3.4 Semi-infinite beam on an elastic foundation.

4

Membrane Analysis: Shells of Revolution of Zero Gaussian Curvature

4.1 INTRODUCTION

We shall now consider the membrane theory for the simplest types of shells, the circular cylinder and the cone (singly curved shells). These shells have zero Gaussian curvatures and are shells of revolution. We shall consider only axisymmetric loading, and hence the problems become further simplified.

4.2 CIRCULAR CYLINDRICAL SHELLS: AXISYMMETRIC LOADING

Figure 4.1(a) shows a circular cylindrical shell of radius a, with x and θ as, respectively, the axial and circumferential coordinates. We shall consider equilibrium of an element bounded by sections x, $x + dx$ and θ, $\theta + d\theta$. An enlarged view of this element is shown in Fig. 4.1(b). The loads p_x and p_r are defined per unit area of the middle surface of the shell. Since we are considering only the membrane analysis, the moments and transverse shears on the element are assumed to be zero. Because of axisymmetry, loading p_θ in the circumferential direction and the shear $N_{x\theta}$ will also be zero. We therefore have only two unknown internal stress resultants: N_x and N_θ, which are defined as forces per unit length along the middle surface. We can readily write the following two equations of equilibrium in the x- and radial directions:

$$\frac{d}{dx}(N_x a d\theta)dx + p_x(a d\theta)dx = 0$$

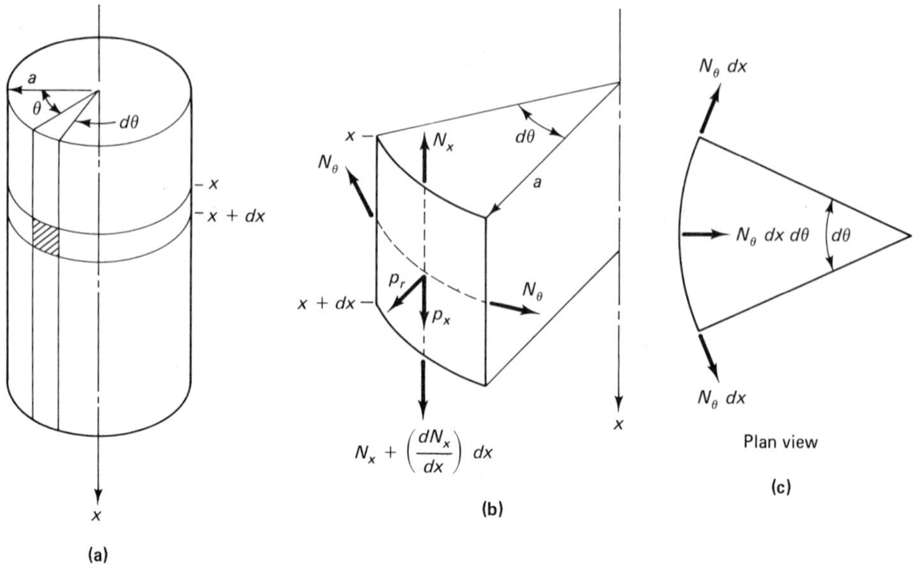

Figure 4.1 Circular cylinder under axisymmetric loading. (a) Coordinate system. (b) Shell element and membrane stress resultants. (c) Radial resultant of hoop stress in plan view.

and

$$-(N_\theta dx)\, d\theta + p_r(a d\theta) dx = 0$$

From the first equation we get

$$\frac{dN_x}{dx} = -p_x$$

or

$$N_x = \int -p_x\, dx + C_1 \tag{4.1}$$

where C_1 is a constant. Also, from the second equation, we get

$$N_\theta = p_r \cdot a \qquad \text{or} \qquad \frac{N_\theta}{a} = p_r \tag{4.2}$$

Equations (4.1) and (4.2) give the membrane stress resultants in the shell under a given loading p_r and p_x. Equation (4.1) has only one constant, which is to be determined from a boundary condition. Thus although the shell has two boundaries x = *constant,* we have the freedom of prescribing N_x on only one boundary, and we must accept and provide for whatever N_x that Eq. (4.1) then gives at the other boundary. Equation (4.2) is similar to that for circular rings [Eq. (1.16)]. Indeed, as we shall see later, the normal loads at a point on a shell are resisted by the "hoop" forces in the two orthogonal rings of principal curves at that point. For the cylinder, one of the principal curves—the meridian—is a straight line with $r_1 = \infty$, and hence the contribution to Eq. (4.2) of the "hoop" force N_x in this ring [i.e., $N_x/(r_1 = \infty)$], is zero.

Membrane Displacements: Cylindrical Shell

We shall denote the axial and radial displacements by u and w, respectively. Then, under membrane conditions, we have $\varepsilon_x = du/dx$. We also know from the theory of elasticity the stress-strain relation

$$\varepsilon_x = \frac{\sigma_x}{E} - \nu\frac{\sigma_\theta}{E}$$

So, in terms of stress resultants per unit width, we can write

$$\frac{du}{dx} = \frac{1}{Et}(N_x - \nu N_\theta)$$

Integrating, we have

$$u = \frac{1}{Et}\int (N_x - \nu N_\theta)dx + C_2 \tag{4.3}$$

Also, using the same expression for circumferential strain as for rings, we can write

$$w = a \cdot \varepsilon_\theta = \frac{a}{Et}(N_\theta - \nu N_x) \tag{4.4}$$

If we choose not to specify N_x at any of the two boundaries, then for a given loading, Eq. (4.1) will give N_x in terms of the constant C_1. On substituting this as well as N_θ from Eq. (4.2) into Eq. (4.3), we shall obtain an expression for u containing two constants C_1 and C_2 which can be determined from the prescribed values of u at the two boundaries. This means that when N_x is not prescribed at any boundary, we have the freedom of prescribing u at both the boundaries. Equation (4.4) will also contain C_1 when N_x is substituted from Eq. (4.1). Thus there are a total of six quantities—N_x, u, and w, at each of the two boundaries—which can be prescribed as boundary conditions. However, due to the availability of only two constants C_1 and C_2, to be determined, we have a freedom of prescribing only two suitable quantities out of these six. These two quantities fix the values of C_1 and C_2, and then we must accept whatever is given by the equations for the remaining quantities. This was also the case for bars (see Chapter 1); when we use the equilibrium equation (1.1) alone, we can only solve statically determinate problems involving the specification of N_x at one edge. For more general boundary conditions, we must use the general equation (1.5) for bars. The same can also be said for beams.

Example 4.1

The cylindrical shell shown in Fig. 4.2 is suspended from its top edge. Under self-weight, if γ is the density of the shell material, we shall have $p_x = \gamma t$ and $p_r = 0$. Hence from Eqs. (4.2) and (4.1) we get

$$N_\theta = 0$$

and

$$N_x = -\int \gamma t \, dx + C_1 = -\gamma t x + C_1$$

Substituting the boundary condition $N_x = 0$ at $x = L$, we obtain

$$N_x = \gamma t(L - x)$$

Also, substituting this result into Eq. (4.3), we have

$$u = \frac{\gamma}{E}\left(Lx - \frac{x^2}{2}\right) + C_2$$

Since at $x = 0$ we have $u = 0$, we get $C_2 = 0$, and hence

$$u = \frac{\gamma L^2}{E}\left(\frac{x}{L} - \frac{x^2}{2L^2}\right)$$

Also, from Eq. (4.4), we find

$$w = -\nu \frac{\gamma a L}{E}\left(1 - \frac{x}{L}\right)$$

Thus, in this example, we used the two boundary conditions $N_x = 0$ at $x = L$ and $u = 0$ at $x = 0$ to solve for the two constants C_1 and C_2.

Example 4.2: Water Tank

For the water tank shown in Fig. 4.3, neglecting self-weight, we have $p_x = 0$ and $p_r = \gamma_w x$, where γ_w is the density of water. This gives us simply $N_x = 0$ and $N_\theta = \gamma_w a x$, and from Eq. (4.4) we can calculate w as

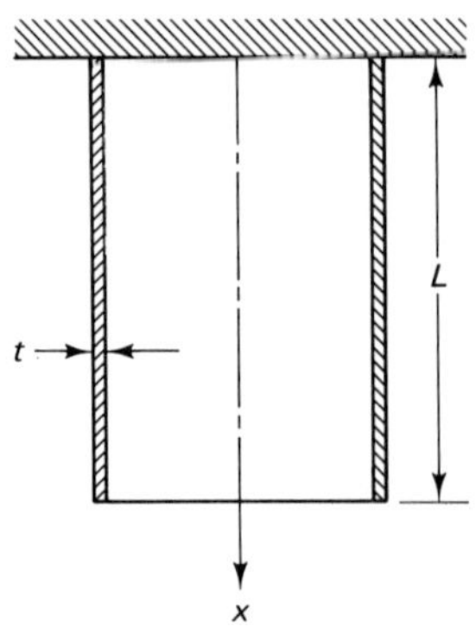

Figure 4.2 Circular cylinder under self-weight, suspended from its top edge.

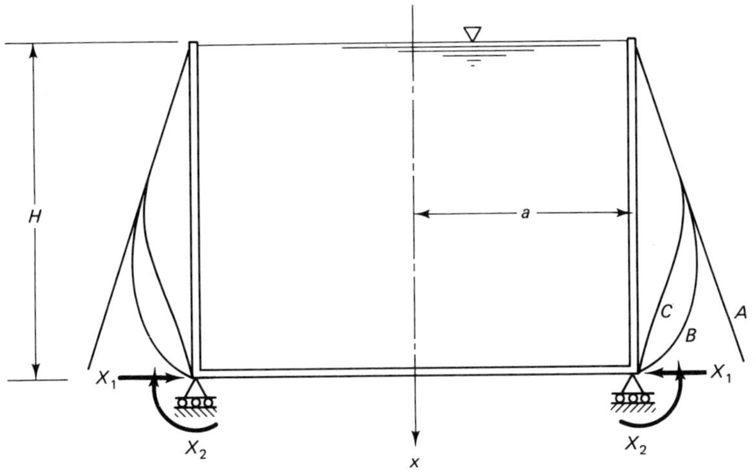

A = Membrane displacement curve
B = Displacement after X_1 is applied
(At $x = H$: $w = 0$, $w' \neq 0$)
C = Displacement after X_1 and X_2 are applied
(At $x = H$: $w = 0$, $w' = 0$)

Figure 4.3 Cylindrical tank under water load.

$$w = \frac{\gamma_w a^2 x}{Et} \tag{4.5a}$$

and hence

$$\frac{dw}{dx} = \frac{\gamma_w a^2}{Et} \tag{4.5b}$$

4.3 DISPLACEMENT INCOMPATIBILITY

It is seen in Example 4.2 that w increases linearly to a maximum at the base, where it should have the necessary freedom to deform. Although in Fig. 4.3 we have shown rollers at the base, in practice, a tank wall will be connected to a base slab. The base slab can be considered as almost infinitely rigid for radial deformations. If the base slab is very thick, it will also restrain the rotation of the wall. Thus the shell will not be free to deform at its base as assumed by the membrane theory. This incompatibility of displacements at the base cannot, therefore, be solved by membrane theory alone. We must apply corrective edge loads X_1 and X_2 (Fig. 4.3), which will make w and dw/dx equal to zero at the base (assuming a very stiff base slab). The final solution will be given by the membrane solution plus the solution for edge loads X_1 and X_2—the values of X_1 and X_2 being obtained by writing compatibility equations at the base. These equations can be derived as follows.

Let δ_{10} and δ_{20} be the displacement and rotation, respectively, at the base of the *primary structure* (i.e., the shell without any constraints at its base). In other words, δ_{10} and δ_{20} are the displacements obtained by the membrane solution. In Chapter 5 we consider how to solve the shell problem under edge loads such as X_1 and X_2. For the present, let us assume that δ_{11} and δ_{21} are the displacements and rotations at the base caused by the application of $X_1 = 1$, while δ_{12} and δ_{22} are those caused by the application of $X_2 = 1$. All δ_{ij}'s are assumed to be positive in the respective assumed positive directions of X_1 and X_2. Then the compatibility equations can be written as

$$\delta_{11}X_1 + \delta_{12}X_2 + \delta_{10} = 0 \tag{4.6a}$$

$$\delta_{21}X_1 + \delta_{22}X_2 + \delta_{20} = 0 \tag{4.6b}$$

These equations are the same as *flexibility equations* written in the flexibility method of structural analysis, with which the reader may be familiar [1]. Equations (4.6) can be solved for X_1 and X_2, and a final solution consisting of the membrane solution plus the effects of X_1 and X_2 can be obtained.

Similar incompatibility problems will arise in Example 4.1 if the shell is assumed to be fixed against horizontal deformations at $x = 0$, where the membrane solution gives w as nonzero. This will also then require the application of edge loads X_1 and X_2 and determining the solution of Eqs. (4.6a,b) for compatibility of displacements at the boundary.

4.4 CONICAL SHELLS: AXISYMMETRIC LOADING

We next consider a conical shell (Fig. 4.4), which has a base angle α. The meridian has zero curvature, while the radius of curvature of the second principal curve is r_2. The radius of the parallel circle is denoted by r. For coordinates, we choose s, measured from the apex along the meridian, and θ, the angle in the horizontal plane measured on the parallel circle. An enlarged view of an element formed by sections s, $s + ds$, and θ, $\theta + d\theta$ is shown in Fig. 4.4(b). It may be noted that the planes $\theta =$ *constant* (meridional plane) and $s =$ *constant* (horizontal plane containing the parallel circle) are not at right angles here. Instead of using θ measured along the parallel circle, we could have used as the second coordinate a variable ψ measured along the second principal curve, which is at right angles to the meridian. However, such a coordinate, although orthogonal to s, would be difficult to visualize. By adopting the coordinate θ, we can also use the axisymmetry of the shell and loading in the θ-direction, and eliminate $N_{s\theta}$ from the equilibrium equations. Hence we prefer to use θ as the second coordinate. Thus, as shown in Fig. 4.4(b), there are only two unknown internal stress resultants, N_s and N_θ, under axisymmetric loads p_s and p_r. Resultants N_s and N_θ are, as usual, defined per unit length along the middle surface, whereas p_s and p_r are loads per unit area of the surface.

The resultant of the forces ($N_\theta ds$) acting on the two sides θ and

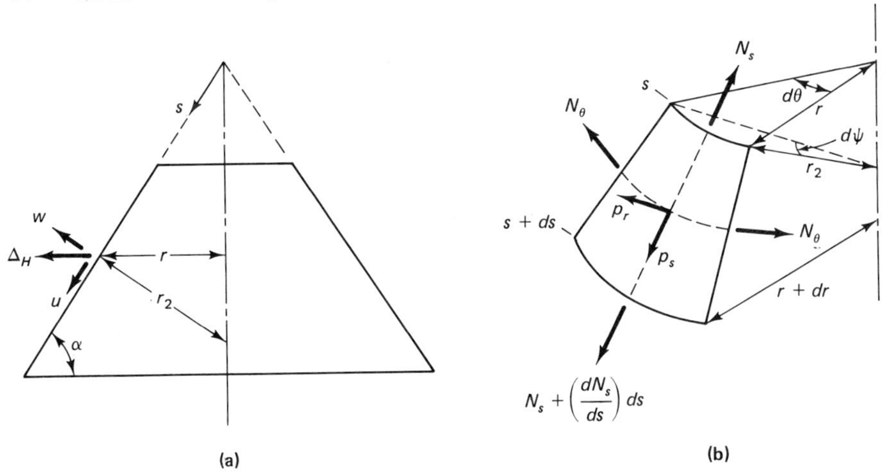

Figure 4.4 Conical shell under axisymmetric loading. (a) Coordinate system. (b) Shell element and membrane stress resultants.

$\theta + d\theta$ = *constant* will be a radially inward force $(N_\theta \, ds)d\theta$ in the horizontal plane [Fig. 4.5(a)]. This resultant can be resolved into two components along the meridional and normal directions, as shown in Fig. 4.5(b). We shall now derive the equations of equilibrium of forces in these two directions. As seen, $(N_s \, r \, d\theta)$ is the force in the s-direction acting on the edge s = *constant*, and hence the difference of the forces in the s-direction on the edges s and $s + ds$ = *constant* can be written directly as

$$\frac{d}{ds}(N_s \cdot r \, d\theta)ds$$

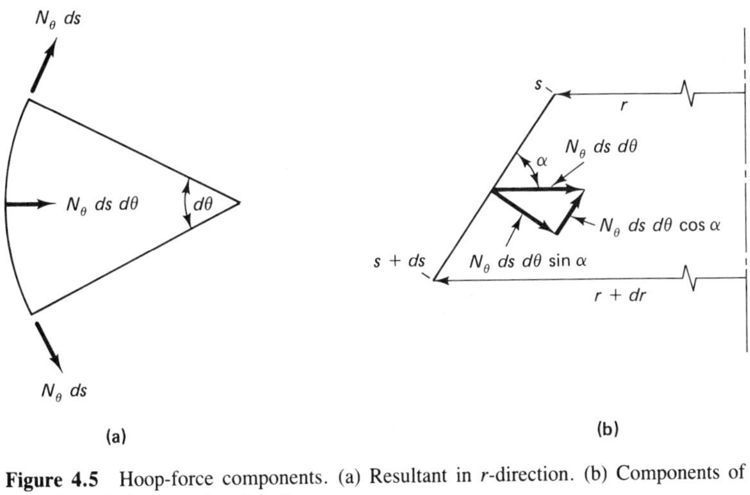

Figure 4.5 Hoop-force components. (a) Resultant in r-direction. (b) Components of hoop force in the s- and r_2-directions.

Here both N_s and r have to be kept inside the derivative, since both quantities vary with s. The equilibrium equation for forces in the s-direction can now be written as

$$\frac{d}{ds}(N_s \cdot r\, d\theta)ds - N_\theta \cdot ds\, d\theta \cos\alpha + p_s\, ds \cdot r\, d\theta = 0$$

or

$$(N_s \cdot r)' - N_\theta \cos\alpha = -p_s \cdot r$$

where $(\cdot)'$ denotes $d(\cdot)/ds$. Now equilibrium of forces in the normal direction gives

$$-N_\theta \cdot ds\, d\theta \sin\alpha + p_r \cdot ds\, r\, d\theta = 0$$

or

$$N_\theta = \frac{p_r \cdot r}{\sin\alpha} = p_r \cdot r_2 = p_r \cdot s \cot\alpha$$

We can substitute this expression for N_θ into the first equilibrium equation, and also substitute $r = s\cos\alpha$, to get

$$(N_s \cdot s)' = N_\theta - p_s \cdot s$$

or

$$(N_s \cdot s)' = (p_r \cdot s \cot\alpha) - (p_s \cdot s)$$

Integrating, we have

$$N_s = \frac{1}{s}\int (p_r \cot\alpha - p_s)s\, ds + \frac{C}{s}$$

where C is a constant of integration. We can now summarize the expressions for the membrane stress resultants below:

$$N_s = \frac{1}{s}\int (p_r \cot\alpha - p_s)s\, ds + \frac{C}{s} \tag{4.7a}$$

$$N_\theta = p_r \cdot r_2 \qquad \text{or} \qquad \frac{N_\theta}{r_2} = p_r \tag{4.7b}$$

As for cylindrical shells, Eq. (4.7b) shows that the normal pressure is balanced by development of hoop forces in a "ring" in the second principal plane with radius of curvature r_2. The contribution of N_s (i.e., the hoop force in the "ring" in the first principal plane defined by the meridian) to this equation is zero, because $r_1 = \infty$. Again, we also see that Eq. 4.7(a) gives us the freedom of prescribing N_s on only one boundary, and we have to accept whatever we get at the other boundary.

4.5 MEMBRANE DISPLACEMENTS: CONICAL SHELL

Let us denote the displacements along the meridional and normal directions by u and w, respectively [Fig. 4.4(a)]. Usually, we would be more interested in obtaining the horizontal displacement and the rotation of the meridian at a point, since these are required for writing equations for compatibility of displacements at the supports. Hence we define Δ_H as the outward horizontal displacement and χ as the clockwise rotation of the meridian at any point.* We can then write

$$\Delta_H = \varepsilon_\theta \cdot r = \frac{r}{Et}(N_\theta - \nu N_s) \tag{4.8a}$$

and

$$\chi = \frac{dw}{ds}$$

To obtain χ in terms of N_s and N_θ, we first write

$$\Delta_H = w \sin \alpha + u \cos \alpha$$

and hence

$$\varepsilon_\theta = \frac{w \sin \alpha}{r} + \frac{u \cos \alpha}{r} = \frac{w}{s} \tan \alpha + \frac{u}{s}$$

or

$$w \tan \alpha = s\, \varepsilon_\theta - u$$

Now differentiating this with respect to s, we obtain

$$\chi = \frac{dw}{ds} = \frac{1}{\tan \alpha}\left(\varepsilon_\theta + s\frac{d\varepsilon_\theta}{ds} - \frac{du}{ds}\right)$$

Substituting $\varepsilon_s = du/ds$, we get

$$\chi = \frac{1}{\tan \alpha}\left(\varepsilon_\theta - \varepsilon_s + s\frac{d\varepsilon_\theta}{ds}\right)$$

Finally, substituting for ε_s and ε_θ in terms of N_s and N_θ, we have

$$\chi = \frac{1}{\tan \alpha}\left[\frac{1+\nu}{Et}(N_\theta - N_s) + \frac{s}{Et}\frac{d}{ds}(N_\theta - \nu N_s)\right]$$

or

*Strictly speaking, by positive χ we mean an opening out of the shell. This corresponds to a clockwise rotation of the meridian on the left-hand side of the axis of revolution and an anticlockwise rotation of the meridian on the right-hand side.

$$\chi = \frac{1}{Et \tan \alpha} \left[(1 + \nu)(N_\theta - N_s) + s \frac{d}{ds} (N_\theta - \nu N_s) \right] \tag{4.8b}$$

Values of N_s and N_θ may be substituted from Eqs. (4.7a,b) into Eqs. (4.8a,b) to obtain Δ_H and χ for a given loading p_r and p_s. A conical shell frustum has two boundaries s = *constant*, at each of which N_s, Δ_H, and χ can be prescribed. Since the expressions for N_s and Δ_H contain only one constant C, we have the freedom of prescribing only one quantity at one of the boundaries, and we must accept the others as given by the resulting equations.* We can now solve some examples of practical interest.

Example 4.3

Let us obtain the internal stress resultants in a conical shell roof, shown in Fig. 4.6, under its own weight. If γ is the density of the material of the shell, the vertical load per unit area of the shell surface will be $p = \gamma t$. Hence we can write

$$p_r = -p \cos \alpha \qquad \text{and} \qquad p_s = p \sin \alpha \tag{4.9a,b}$$

Substituting into Eqs. (4.7a,b), we get

$$N_\theta = -pr_2 \cos \alpha = -p \cos \alpha \cdot s \cot \alpha \tag{4.10a}$$

and

$$N_s = -\frac{p}{s} \int (\cot \alpha \cos \alpha + \sin \alpha) s \, ds + \frac{C}{s}$$

or

$$N_s = -\frac{ps}{2 \sin \alpha} + \frac{C}{s}$$

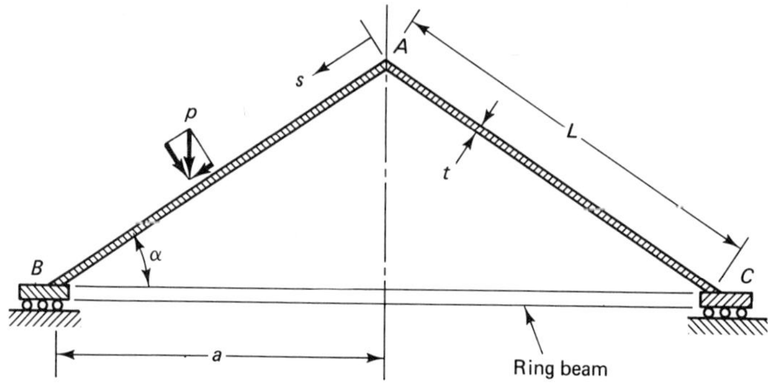

Figure 4.6 Conical shell roof under its own weight $p = \gamma t$.

*Writing similar expressions as Eqs. (4.3) and (4.4) for u and w, we shall obtain one more constant of integration in the expressions for u and w. Also, Eq. (4.8b) for χ contains derivatives of N_θ and N_s. When integrated, this equation will also yield one more undetermined constant.

The second term above gives an infinite value of N_s at $s = 0$. This is not expected for a conical shell roof under its own weight, and hence we take $C = 0$, which gives us

$$N_s = \frac{-ps}{2 \sin \alpha} \tag{4.10b}$$

By substituting for N_s and N_θ in Eqs. (4.8a,b), we can readily obtain

$$\Delta_H = -\frac{psr}{2Et \sin \alpha}(2 \cos^2 \alpha - \nu) \tag{4.11a}$$

and

$$\chi = \frac{-ps}{2Et \tan \alpha \sin \alpha}[4 \cos^2 \alpha - 1 - 2\nu(1 - \cos^2 \alpha)] \tag{4.11b}$$

For most practical civil engineering roof structures, α will be less than 45° and ν less than 0.3, and hence the expressions in brackets in the two equations above will always be positive. Thus Δ_H and χ will be negative; i.e., the shell will move inward with an anticlockwise rotation of the meridian AB (Fig. 4.6). This is as expected, because Δ_H depends mainly on N_θ, which is compressive [Eq. (4.10a)]. Also, since N_θ increases with s, the inward deflection will also increase with s; i.e., the meridian will rotate counterclockwise.

At the base of the shell for $s = L$, we get from Eq. (4.10b),

$$N_s = -\frac{pL}{2 \sin \alpha}$$

Let us denote the numerical value of this force as F [Fig. 4.7(a)]. The shell will exert on the support a force F in the opposite direction, as shown. The vertical component of F, namely $(F \cdot \sin \alpha)$, can be well resisted by the support (which would be a cylindrical wall or a ring beam supported on columns). For resisting the horizontal component $(F \cdot \cos \alpha)$, generally a ring beam is provided at the base as shown. The hoop tension T in the ring beam caused by this force will be

$$T = (F \cos \alpha)a = \frac{pa^2}{2 \sin \alpha} \tag{4.12}$$

The ring beam will tend to deform outward because of this tension. However, we have seen by Eq. (4.11a) that the shell deflects inward under membrane forces. There is thus an incompatibility of horizontal displacements at the base, between the shell and the ring beam. Depending on the stiffness of the ring beam to resist rotational deformation, there will be an incompatibility of rotations also between the shell and the ring beam. Since the shell and the ring beam would be monolithically connected, such incompatibilities are not permissible. There should be no relative displacement or rotation between the shell and the ring beam. Interaction forces, such as X_1 and X_2 shown in Fig. (4.7b), will therefore develop between the shell and the ring beam. Values of these redundant forces can be obtained by solving compatibility equations (4.6a,b), where δ_{ij} will now represent the relative displacement* between the shell and the ring beam. We can write

$$\delta_{ij} = \delta_{ij}^C + \delta_{ij}^R$$

*Displacement here means both horizontal displacement or a rotation.

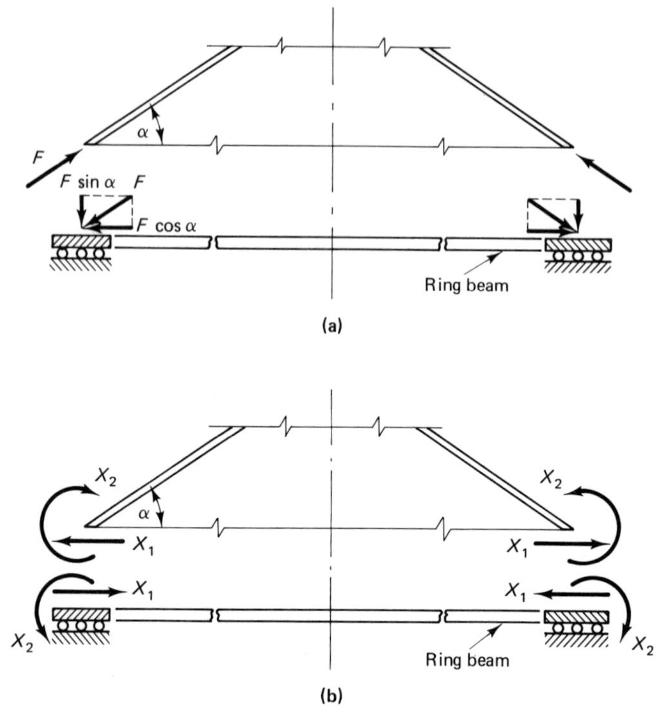

Figure 4.7 Shell-ring interaction. (a) Membrane forces exerted on shell and ring. (b) Edge loads required for displacement compatibility of differential shell and ring membrane deformations.

where δ_{ij}^C and δ_{ij}^R are the displacements of the cone and the ring beam, respectively. Each displacement is positive in the chosen positive directions of X_1 and X_2.* The final solution will be obtained by superimposing the influence of edge loads X_1 and X_2 on the results of the membrane solution.

Example 4.4

We shall next calculate the membrane stresses in the conical umbrella roof (Fig. 4.8), under its own weight. Proceeding as in Example 4.3, we shall get exactly the same expression for N_θ, as in Eq. (4.10a). For N_s we shall also obtain the following expression, as before:

$$N_s = -\frac{ps}{2 \sin \alpha} + \frac{C}{s}$$

Substituting the boundary condition at $s = L$, $N_s = 0$, we get

$$C = \frac{pL^2}{2 \sin \alpha}$$

*Thus for the assumed directions of X_1 and X_2 shown in Fig. 4.7(b), displacement of the primary structure (membrane displacement) δ_{10} will be positive outward for the shell and positive inward for the ring beam.

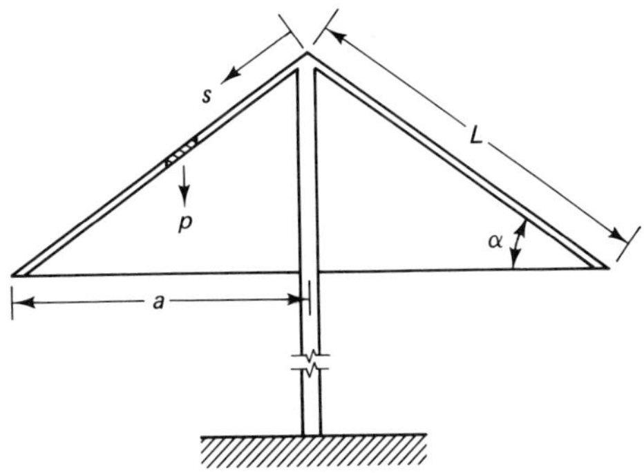

Figure 4.8 Conical umbrella roof under self-weight.

and hence

$$N_s = \frac{p}{2 \sin \alpha} \frac{L^2 - s^2}{s} \tag{4.13}$$

Expressions for the membrane displacements can also be obtained now from Eqs. (4.8a,b).

4.6 METHOD OF SECTIONS

We have seen that under membrane conditions, the shell problem becomes statically determinate, and we could obtain expressions for the membrane forces by integrating only the equilibrium equations for a shell element. Alternatively, we can consider equilibrium of a large section (freebody) of the shell, and derive the same expression for the meridional stress resultant.

Thus for the conical shell in Example 4.3, we can take a section $s = constant$ and consider equilibrium of the entire shell above this section (Fig. 4.9). The exposed internal force is simply N_s, which is shown in its positive sense. The total weight of the shell above the section is found by multiplying p by the surface area (πrs) of the conical cap and is equal to

$$W = \pi rsp$$

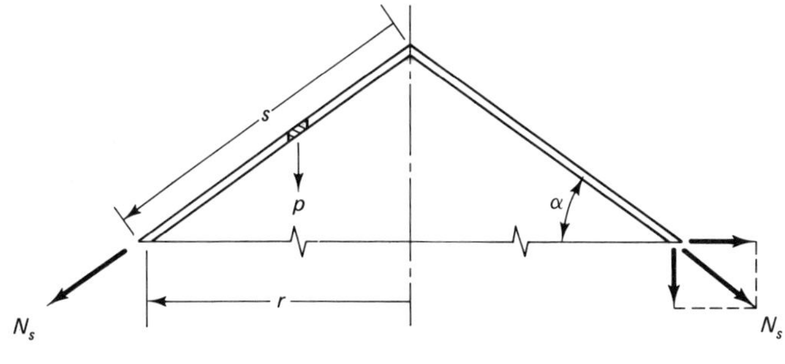

Figure 4.9 Entire section of a conical shell above the coordinate $s = constant$.

Writing equilibrium of vertical forces, we have

$$N_s \cdot 2\pi\, r \sin\alpha + \pi r s p = 0$$

Hence

$$N_s = \frac{-ps}{2 \sin\alpha}$$

It can be seen that this expression is the same as Eq. (4.10b) obtained earlier. Equation (4.13) in Example 4.4 can also be obtained in a similar manner.

The method of sections does not give N_θ directly. However, as we shall see in Chapter 6, the normal pressure p_r on the shell is always supported by hoop forces in the two rings in the directions of the two principal curves. This gives a simple relation between the two membrane forces and p_r. Thus, knowing the meridional force by the method of sections, the expression for N_θ can readily be obtained. In the present chapter we are concerned only with shells of zero κ. Hence, as discussed earlier, p_r is supported only by N_θ, as seen from Eqs. (4.2) and (4.7b). Thus N_θ can be obtained directly from p_r.

The method of sections is useful for better visualization of shell behavior. It also gives an improved physical insight into the derivation of the expression for meridional stress, as compared to the earlier more mathematical approach.

4.7 SINGULARITY PROBLEMS

For conical shells, in the expression for N_s [Eq. (4.7a)], we had a term C/s which tends to infinity as $s \to 0$. We can easily find out what this singularity means in a physical sense. Let us consider a conical shell with no loading on it. Hence, substituting $p_r = p_s = 0$ into Eqs. (4.7a,b), we get $N_\theta = 0$ and $N_s = C/s$. Let us now consider the equilibrium of a cap of this shell, obtained by taking a section $s = \textit{constant}$. The freebody is shown in Fig. 4.10.

The horizontal components of N_s acting on this section around the parallel circle are self-equilibrating. However, the vertical components give a resultant downward load, given by

$$V = 2\pi r\left(\frac{C}{s}\sin\alpha\right) = \pi C \sin 2\alpha$$

This force is a constant, independent of s. Since there is no distributed loading on the shell, for equilibrium we must have an opposite force V concentrated at the apex of the cone. In other words, if there is a concentrated load V (positive upward) at $s = 0$, the constant C in Eq. (4.7a) can be directly determined as

$$C = \frac{V}{\pi \sin 2\alpha}$$

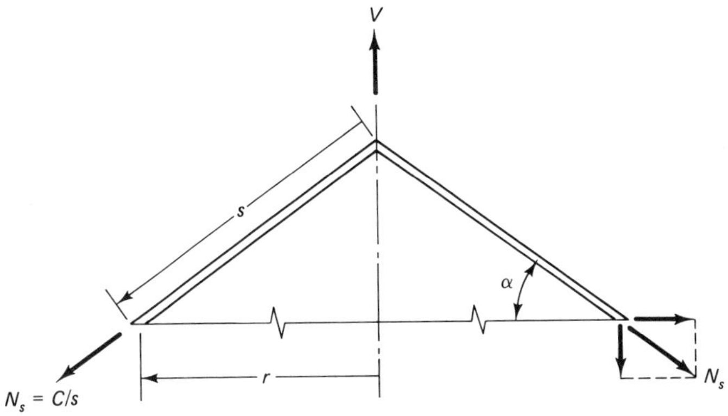

Figure 4.10 Freebody diagram of a conical shell cap above s = *constant*, illustrating the meaning of the singular term in Eq. (4.7a).

Using this result, the value of C in Example 4.4 can be obtained in an alternative manner. The total weight of the shell in this case is

$$W = \pi a L p$$

where L is the inclined length of the shell along s and a is the radius of the parallel circle at $s = L$. This weight must be equal to the upward support reaction, and hence the value of V at the apex is $V = \pi a L p$. We then get directly

$$C = \frac{\pi a L p}{\pi \sin 2\alpha} = \frac{pL^2}{2 \sin \alpha}$$

as obtained before.

If, instead of a concentrated load at the apex, we have a line load q acting around a circumference r = *constant*, with a vertical upward resultant force V, then using similar arguments we can conclude that for finding N_s below the line load, we must take

$$C = \frac{V}{\pi \sin 2\alpha} = \frac{2qr}{\sin 2\alpha}$$

C will be zero for membrane stresses in the portion above the line load, assuming that the shell is supported at its base.

4.8 DESIGN EXAMPLE: WATER TANK

Although we have seen that the membrane theory does not give the complete solution, it gives reasonable values for stresses away from points of discontinuity, such as supports, junctions with other shells, etc. As such, it could be used for preparing a preliminary design. Let us consider a conical water tank, constructed of reinforced concrete, as shown in Fig. 4.11(a).

The tank is assumed to be filled with water, as shown. We shall find the membrane forces under self-weight, water load, and a line load of $q = 1000$ lb/ft, which may be considered as the load from a dome roof above.

The line load can be resolved into the two components

$$H = q \text{ lb/ft}$$

and

$$|N_{s_o}| = q\sqrt{2} \text{ lb/ft}$$

N_{s_o} gives a compressive meridional stress in the shell, which the shell can resist easily. However, the shell cannot resist the load H (which causes shear and bending) without considerable deformations and stresses. Hence a ring beam needs to be provided at this location to resist H.* Again, as discussed in Example 4.3, there will be incompatibility of displacements at the junction of the cone and the ring beam. We shall presently not consider the effects of this incompatibility.

Thus, under membrane conditions, we can write

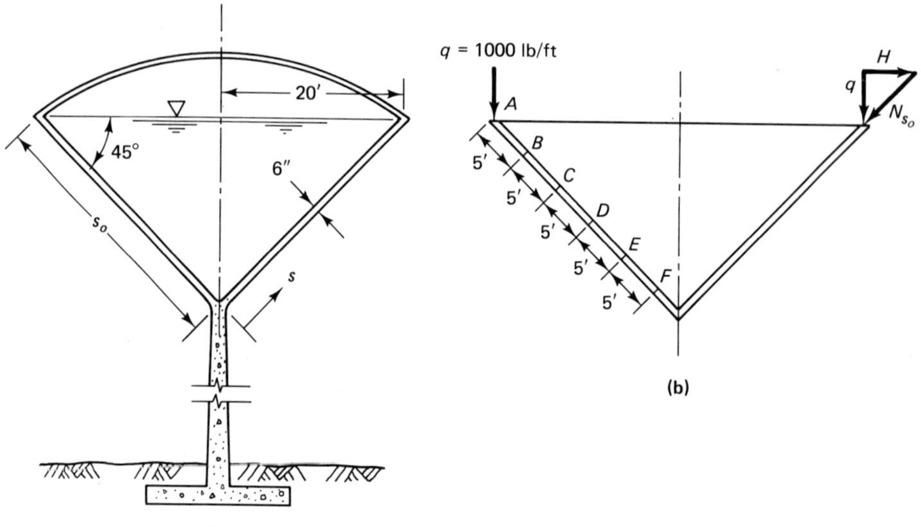

Figure 4.11 "Wine glass" conical water tank. (a) Loaded by self-weight, water, and the roof dome weight. (b) Various rings considered for designing hoop reinforcement.

*Such a ring beam can even be seen at the top of an ordinary paper or Styrofoam cup (which is a conical shell) used for holding drinks. Without such a ring beam, the cup can resist the water load easily without much deformation. However, when the cup is lifted, it is held by fingertips which act as horizontal point loads that cause significant shears and bending moments. In this case, the cup will undergo considerable deformation without a ring beam. Therefore, a stiffening ring is usually provided at the top of the cup.

$$p_s = -\gamma_c t \sin\alpha$$

and

$$p_r = \gamma_w(s_o - s)\sin\alpha + \gamma_c t \cos\alpha$$

where γ_w (62.4 lb/ft^3) and γ_c (150 lb/ft^3) are the densities of water and concrete, respectively. The thickness of the shell, t, we shall assume as 6 in. Substituting for p_s and p_r, with $\alpha = 45°$ and $t = 0.5$ ft, into Eq. (4.7a), we get

$$N_s = \frac{1}{s}\left[\frac{\gamma_w}{\sqrt{2}}\left(s_o\frac{s^2}{2} - \frac{s^3}{3}\right) + \frac{\gamma_c\, s^2}{2\sqrt{2}}\right] + \frac{C}{s}$$

Now we can use the boundary condition at $s = s_o$, namely,

$$N_s\,(s = s_o) = N_{s_o} = -\sqrt{2}\,q$$

which gives us

$$C = \sqrt{2}\,s_o\left(-q - \frac{\gamma_w s_o^2}{12} - \frac{\gamma_c s_o}{4}\right)$$

and hence

$$N_s = \frac{\gamma_w s^2}{6\sqrt{2}}\left[-\left(\frac{s_o}{s}\right)^3 + 3\left(\frac{s_o}{s}\right) - 2\right] + \frac{\gamma_c s}{2\sqrt{2}}\left[1 - \left(\frac{s_o}{s}\right)^2\right] - \sqrt{2}\,q\left(\frac{s_o}{s}\right)$$

Also, from Eq. (4.7b), we have

$$N_\theta = p_r \cdot r_2 = p_r \cdot s\cot\alpha$$

or

$$N_\theta = \frac{\gamma_w s^2}{\sqrt{2}}\left(\frac{s_o}{s} - 1\right) + \frac{\gamma_c s}{2\sqrt{2}}$$

From the equations above, we can see that N_s is compressive throughout. This is expected, as we can physically visualize by taking a section $s = constant$ and considering equilibrium of the shell above this section. It can also be seen that N_θ is tensile throughout. Since p_r is radially outward for both self-weight and water pressure, N_θ, which depends purely on p_r, is expected to be tensile. Numerical values of N_s and N_θ can now be calculated for various values of s, by substituting appropriate values of γ_w, γ_c, q, and s_o.

For designing the reinforcement, we note that the meridional stresses due to N_s are generally small,* and only temperature and shrinkage steel needs to be provided in this direction. Some codes [2] specify that the minimum amount of this steel be 0.6% of the area of the cross section, with a minimum of 0.2% in any

*From the expression for N_s, we see that $N_s \to \infty$ as $s \to 0$. In reality, however, the support will have some finite diameter and will not be a point support as assumed here. Thus stresses due to N_s will take on finite values near the support.

one direction. We may assume an average of 0.3% as the minimum steel for temperature and shrinkage in each direction. Thus for a 6-in. wall thickness, the minimum required steel area would be 0.216 in.^{2}/ft, i.e., 0.108 in.^{2}/ft on each face. We can therefore provide No. 3 bars, at 12-in. center-to-center spacing, in the meridional direction. To maintain the same spacing, alternate bars can be curtailed from the top, say at $s = s_o/2$, $s = s_o/4$, etc. Some codes [3,4] specify that a different percentage of temperature and shrinkage reinforcement should be used for liquid-retaining structures. In such a case, the local code should govern the design.

For calculating reinforcement in the circumferential direction, we divide the shell into five rings: *AB*, *BC*, . . . , *EF* [Fig. 4.11(b)]. The average value of N_θ is calculated for each ring, from which the hoop tensile force in each ring can be obtained, as shown in a tabular form in Table 4.1.

Assuming deformed bars with a yield stress of 40,000 psi, the allowable stress in tension in the bars is taken as 20,000 psi. (To control cracking in water-retaining structures, some codes [3,4] specify much lower allowable stresses for tension in steel reinforcement.) Circumferential steel areas A_s and corresponding bars for the various rings are given in Table 4.1.

The given areas of steel are subject to a minimum required value for temperature and shrinkage, which is the same as that obtained in the meridional direction. The hoop reinforcement is placed as rings on both faces. A typical reinforcement arrangement is shown in Fig. 4.12. The results, of course, will have to be corrected near the ends to account for edge-load effects.

From Table 4.1 it is seen that the maximum value of N_θ is about 9500 lb/ft. For a 6-in. wall thickness, the maximum tensile stress in concrete (neglecting the effect of steel) would thus be about 132 lb/in.^{2}. The tensile stress of concrete in water-retaining structures is normally kept low, to keep cracking at a minimum. For concrete with a 3000-psi cylinder strength, it is desirable to keep the tensile stress below 300 psi. Hence, in the present example, the concrete thickness of 6 in. should be satisfactory.

TABLE 4.1 Circumferential Reinforcement for the Water Tank of Fig. 4.11(a)

Point	s (ft)	N_θ (lb/ft)	Length (ft)	$(N_\theta)_{avg}$ (lb/ft)	Hoop Force (lb)	A_s (in.^{2})	Rebar Design
A	28.28	1500					
			5	3936	19,680	0.984	6-No. 4[a]
B	23.28	6372					
			5	7705	38,525	1.926	10-No. 4
C	18.28	9037					
			5	9267	46,335	2.317	12-No. 4
D	13.28	9497					
			5	8624	43,120	2.156	12-No. 4
E	8.28	7750					
			5	5774	28,870	1.444	8-No. 4
F	3.28	3797					

[a]Minimum steel governs.

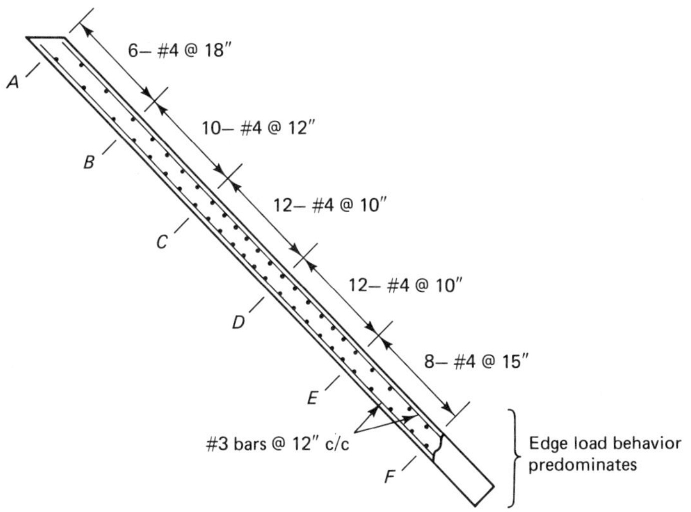

Figure 4.12 Reinforcement arrangement for "wine glass" tank.

4.9 CODE DETAILING REQUIREMENTS: BRIEF REVIEW

To help us in solving practical problems of shell design considered in the following chapters, we review very briefly a few important detailing recommendations found in codes of practice for concrete shells.

The International Association for Shells and Spatial Structures (IASS) has published detailed recommendations for the design and construction of reinforced concrete shells and folded plates [2]. The ACI Code [5] also devotes one chapter to shell design and construction. In addition, reference may also be made to the two Indian Codes [6,7].

We consider below the use of concrete shells, mainly in the construction of roof structures and liquid-retaining tanks.

Roof Structures

Thickness. The stresses in major portions of roof shells are generally small, and hence the thickness would not be governed by strength considerations. Rather, requirements for ease of construction and suitable placement of reinforcing with proper cover usually dictate the design thickness. Generally, such shells would be constructed with a thickness ranging from 2 to 4 in. In regions near junctions with edge members/supports, the thickness is gradually increased to cater to the higher bending stresses that generally develop there. The edge beams that have higher concentration of stresses may, of course, have larger dimensions—similar to those of typical beams.

Reinforcement. Reinforcement in roof shells is provided mainly as membrane reinforcement and bending reinforcement. Membrane reinforcement is distributed throughout the shell to fully resist tensile stresses. It also serves as reinforcement for temperature and shrinkage. Such reinforcement is placed near the middle surface of the shell. IASS [2] recommends a minimum membrane reinforcement of 0.6% of the cross-sectional area of the shell, to be provided in two orthogonal directions, with not less than 0.2% in any one direction. This percentage appears to be much larger than that used for slab structures. However, considering the thinness of the shell, the actual quantity of steel is small and it helps to reduce crack widths. Assuming 0.3% reinforcement in each direction, for a 3-in.-thick shell, the minimum reinforcement works out to be No. 3 bars at 12-in. center-to-center in each direction. The ACI code requires the minimum reinforcement to be the same as that for temperature and shrinkage steel for slabs. This would generally work out to be 0.2% in the two orthogonal directions.

The bending reinforcement is provided to resist the moments that develop near the boundaries and points of discontinuity. It would be placed near the tension surface of the shell, which would depend on the direction of the bending moment. Due to the fluctuating sign of bending moments along the shell, ACI code recommends that equal steel be placed on both faces of the shell, even when the analysis does not indicate reversal of bending moments.

It would be ideal to place reinforcement along the stress trajectories in the two principal directions. But this may not always be convenient in practice, and in such cases, reinforcement is placed in two convenient orthogonal directions, such that the resulting arrangement is simple for construction.

The size of reinforcing bars will generally vary from No. 2 to No. 5. Smaller-diameter deformed bars are helpful in minimizing crack widths, but they are also very flexible and tend to be permanently bent during construction. Hence adequate supports should be provided for such reinforcement in the field. Welded wire mesh, with a minimum bar diameter of $\frac{1}{4}$ in. (6 mm) may also be used as reinforcement for shells with shallow curvatures. For large curvature shells, the mesh has a tendency to spring back, and hence the use of wire mesh should be avoided in such cases. Larger-diameter bars may be used for the bending reinforcement near the edges.

IASS recommends that the use of high-strength bars with a yield stress greater than 40,000 psi be avoided, due to the increased tensile strains experienced in the surrounding concrete and the resultant cracking. However, it is felt that the stresses in the membrane reinforcement are generally so low that the use of high-strength bars as membrane reinforcement does not cause major problems. When used as bending reinforcement, the allowable stresses may be restricted in such bars, in order to minimize cracking.

IASS specifies that the spacing of the membrane reinforcing bars should be less than five times the shell thickness but not more than 12 in. (ACI code permits a spacing of up to 18 in.). Welded-wire-mesh bars should not be spaced at more than four times the shell thickness, and never more than 8 in. A clear cover of $\frac{1}{2}$-in. is generally sufficient for most reinforcement and could be increased suitably where larger-diameter bars are used. Due to the small thickness of most shells, hooks should generally be avoided or provided only in the plane of the shell surface.

Liquid-Retaining Structures

Cylindrical shells, conical shells, and domes are commonly adopted for economical design and construction of water tanks and other similar structures. These shells are generally much thicker than roof shells—their thicknesses usually being greater than 5 in. Membrane reinforcement would therefore be placed on each face of the shell in two orthogonal directions. Thus, totally, there will be four layers of reinforcement. In addition, bending reinforcement near the edges will also be provided, usually combined with the membrane reinforcement there. Considering a minimum of 0.3% steel in each direction, for a 6-in.-thick shell, the minimum reinforcement works out to be No. 3 bars at 12 in. center-to-center on each face.

In designing reinforcement for direct tension or bending, some codes [3] restrict the allowable stresses in reinforcement to low values which depend on whether the reinforcement is on the liquid face or away from it. The British Code [4] requires design of this reinforcement such that the crack width is restricted between 0.1 and 0.3 mm, the allowable crack width being determined by the class of the structure, the liquid stored, the location of the crack, etc. ACI Code [5] does not recommend any reduction in stresses for designing reinforcement in liquid-retaining structures, and hence in Table 4.1 we have used an allowable value of 20,000 psi in designing bars for resisting hoop forces.

Due to the rather large thicknesses of tank walls, larger-diameter bars will be used in these structures, and the clear cover will also be not less than 1 in. (although the British Code [4] recommends a minimum cover of $1\frac{1}{2}$ in.).

The foregoing summary of practical considerations for the design of shell reinforcement will be referred to again and again when we solve numerical design examples in the following chapters. In solving the examples in this text, we generally assume concrete with a 28-day cylinder strength of 3000 psi and deformed reinforcing bars with a yield strength of 40,000 psi. Since bending moments are usually small, we shall calculate reinforcement for resisting bending moments from simple "elastic theory," i.e., dividing the moment by the allowable steel stress and the lever arm [8]. Usually, the moment is also coupled with an axial force. In such cases the axial force could be transferred to the tensile steel with corresponding modification in the bending moment, or it may be neglected altogether, since usually the axial forces and bending moments are very small.

For shells constructed of materials other than concrete, practical design requirements from the appropriate codes, such as the ASME Boiler and Pressure Vessel Code [9], have to be considered.

REFERENCES

[1] Weaver, W., Jr., and Gere, J. M., *Matrix Analysis of Framed Structures,* 2nd ed., Van Nostrand Reinhold Company, Inc., New York, 1980.

[2] "Recommendations for Reinforced Concrete Shells and Folded Plates," International Association for Shell and Spatial Structures, Madrid, 1979.

[3] "Code of Practice for Concrete Structures for the Storage of Liquids," Indian Standard 3370-1965 (Parts I to IV), Indian Standards Institution, New Delhi, 1966.

[4] "The Structural Use of Concrete for Retaining Aqueous Liquids," British Standard BS 5337, 1976, British Standards Institution.

[5] "Building Code Requirements for Reinforced Concrete—ACI 318-84," ACI Committee 318, American Concrete Institute, Detroit, 1984.

[6] "Criteria for the Design of Reinforced Concrete Shell Structures and Folded Plates," Indian Standard 2210-1962, Indian Standards Institution, New Delhi, 1963.

[7] "Code of Practice for Construction of Reinforced Concrete Shell Roofs," Indian Standard 2204-1962, Indian Standards Institution, New Delhi, 1962.

[8] Wang, C., and Salmon, C. G., *Reinforced Concrete Design,* 3rd ed., Harper & Row, Publishers Inc., New York, 1979.

[9] *ASME Boiler and Pressure Vessel Code,* Section VIII. Pressure Vessels—Divisions 1 and 2, American Society of Mechanical Engineers, New York.

PROBLEMS

4.1. A vertical cylindrical shell, under its own weight, is fixed against axial deformation at both its ends, $x = 0$ and $x = L$. Find the force N_x in the shell.

4.2. A circular cylindrical grain silo is suspended from its top and has a conical bottom as shown in Figure P4.2. The thickness of both shells is *t,* and the unit weight of the shell

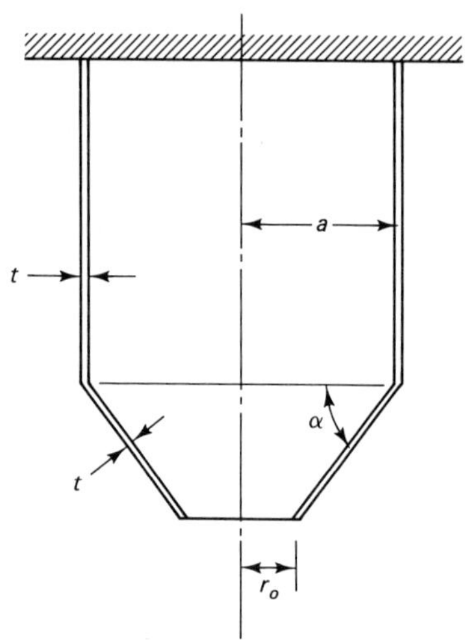

Figure P4.2 Grain silo consisting of cylindrical and conical portions.

material is γ. The cone is open at its bottom. Find membrane stresses in the shells under self-weight only. Explain why a ring is required at the junction of the cylinder and cone. Find the hoop force in the ring. Find the horizontal displacements (under membrane conditions) of the cone, ring, and cylinder at their junction. Assume A to be the cross-sectional area of the ring.

4.3. Shown in Figure P4.3 is a proposed design for an underwater observatory used to view marine life at the new amusement park, "Ocean World." The observatory is a conical shell with a dome roof. The roof is constructed of a very lightweight fabric and is subjected to an internal pressure which is 10 lb/ft^2 greater than the ambient air pressure. A circular ring is provided at the junction of the cone and the roof, and also serves as a walkway for "water-shy" guests.

The entire structure is very light and can be assumed weightless. In order to anchor the structure in its position, tension piles are connected to a circular ring that distributes the tension in a uniform manner to the cone. Find the membrane stresses in the conical shell. Specify appropriate materials to resist these stresses. Comment subjectively on the structural aspects of this design.

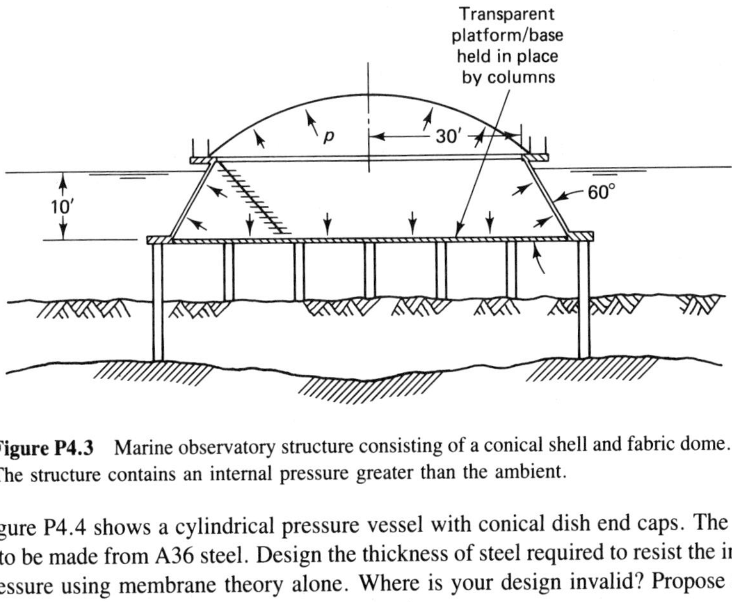

Figure P4.3 Marine observatory structure consisting of a conical shell and fabric dome. The structure contains an internal pressure greater than the ambient.

4.4. Figure P4.4 shows a cylindrical pressure vessel with conical dish end caps. The vessel is to be made from A36 steel. Design the thickness of steel required to resist the internal pressure using membrane theory alone. Where is your design invalid? Propose a conservative design to resist stresses in these regions.

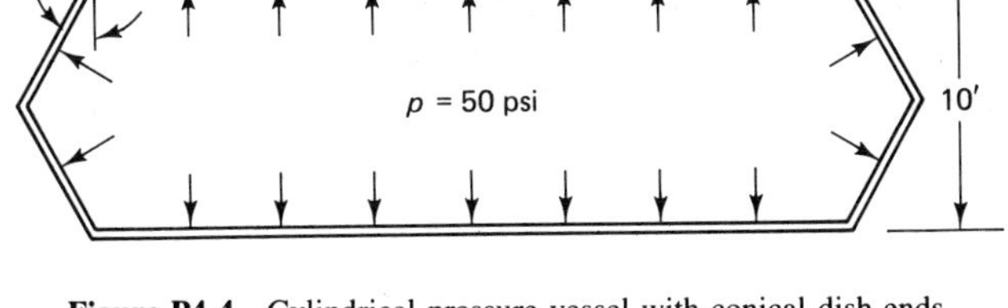

Figure P4.4 Cylindrical pressure vessel with conical dish ends.

4.5. For Examples 4.1 to 4.3 solved in this chapter, assume that the support or ring in each case is infinitely rigid. Write down, in matrix form, the set of simultaneous equations required to solve for redundant edge loads X_1 and X_2. Use generic notation for displacements due to edge loads.

4.6. For the shell in Example 4.3 (Fig. 4.6), obtain expressions for N_s and N_θ in the cone under a roof live load or snow load of p_o. (*Note:* This load is to be taken as uniformly distributed over the plan area.)

4.7. For the shells considered in Examples 4.3 and 4.4 (Figs. 4.6 and 4.8, respectively), if $\alpha = 30°$ and the radius of the parallel circle at the base is 15 ft, design the reinforcement in the cone under self-weight plus a finish load of 10 lb/ft² over the surface area. Assume a suitable thickness for the shell.

Also, for the shell of Example 4.3, obtain horizontal displacements of the cone at its base and of the ring beam under membrane conditions. Assume that $\nu = 0.1$, $E = 2.5 \times 10^6$ psi, and that the cross section of the ring beam is 12 in. by 6 in.

4.8. A clarifier tank of a wastewater treatment plant is shown in Figure P4.8. It consists of a cylindrical wall, annular ring beam, and a shallow conical bottom. When the tank is empty, it will be held down against uplift by its own weight plus the counterweight of the earth over the ring beam. Hence point *A* can be considered to have zero vertical deflection under this condition, or if necessary, a vertical support can be assumed here. The loads acting on the empty tank (including self-weight) under this uplift condition are as shown.

Find the membrane stresses in the conical shell and the cylindrical wall. Also determine the bending moments and in-plane force in the ring beam. Design the reinforcement in this tank. (Assume loads to be applied on the midsurfaces of the shells.)

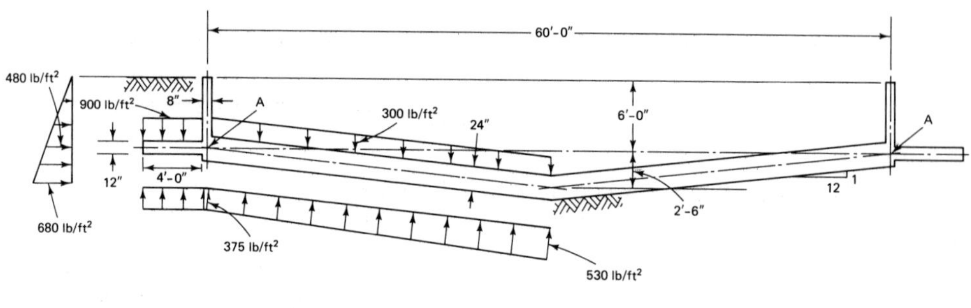

Figure P4.8 Clarifier tank used in wastewater treatment.

4.9. A conical water tank is filled with water and is supported at its midheight by a circumferencial ring support, as shown in Fig. P4.9. The tank has a circular slab/plate at its base. Plot the variations of membrane stress resultants in the tank under **(a)** self-weight and **(b)** water load. Discuss design problems that arise at the ring support and at the base.

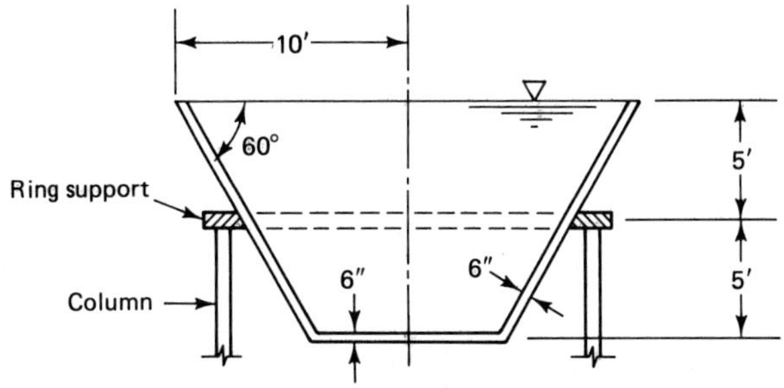

Figure P4.9 A conical water tank supported at its midheight by a ring beam.

5

Bending Theory of Circular Cylindrical Shells: Axisymmetric Problems

5.1 DERIVATION OF GOVERNING DIFFERENTIAL EQUATION

In Chapter 4 we had seen that the membrane solution, by itself, could not satisfy compatibility conditions at the boundaries. Thus we have to superimpose the effect of edge loads on the membrane solution, such loads being determined from equations defining compatibility of displacements at the shell edges. Since the edge loads resolve into shears and bending moments on the shell, we shall now develop the theory describing bending behavior of closed circular cylindrical shells, which will enable us to solve such edge-load problems. We shall here consider axisymmetric edge loads only, and assume distributed loads p_r and p_x to be zero. However, when distributed loads do act on the shell, results of the membrane solution are used to closely approximate the shell behavior away from the edges.

Shown in Figure 5.1 is an element of a circular cylindrical shell bounded by sections x, $x + dx$, and θ, $\theta + d\theta$. Distributed loads p_x and p_r are taken as zero, since we are considering only the edge-load problem. Under axisymmetric loading, only five internal stress resultants need to be determined; these are N_x, N_θ, M_x, M_θ, and Q_x.

We make assumptions similar to those noted earlier for plate and beam theories. These are as follows:

1. The material of the shell is homogeneous, isotropic, and linearly elastic, following Hooke's law.
2. The loads are applied on the middle surface of the shell.

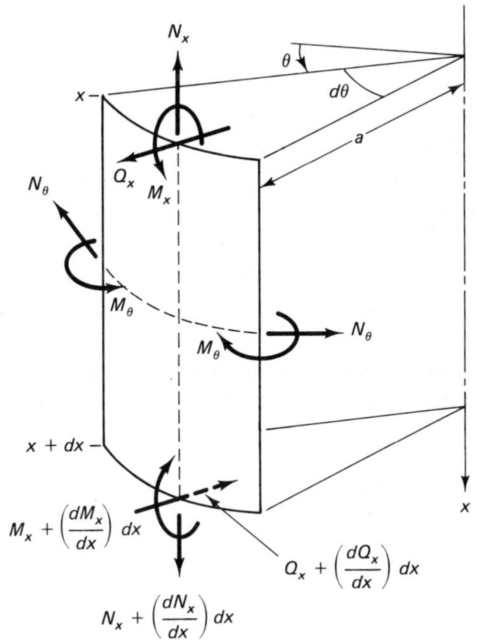

Figure 5.1 Element of a circular cylindrical shell.

3. The shell is thin, its thickness t being small compared to its radii of curvature r_1 and r_2. Therefore, quantities such as t/r_1 and t/r_2 can be neglected compared to unity.
4. Deflection w is small, so that the slope dw/dx is small compared to unity. Thus an approximate expression for curvature κ_x can be used.
5. Plane sections remain plane after bending; i.e., shear deformations are neglected.

We can now proceed to derive equations of equilibrium, kinematics, and Hooke's law, based on these simplifying assumptions.

Equilibrium Equations

Referring to Fig. 5.1, and considering equilibrium of forces in the x-direction, we can write

$$\frac{d}{dx}(N_x \cdot a\, d\theta)\, dx = 0$$

or

$$\frac{dN_x}{dx} = 0$$

On integration, this gives us $N_x = constant$. This constant is to be determined from the value of N_x applied at either of the two boundaries of the cylinder. For our present solution, we assume that $N_x = 0$ on the boundaries, so that we simply obtain $N_x = 0$ throughout the shell. For any other axial load on the boundaries, we can always superimpose the solution for N_x obtained above to the solution that we are about to derive. This superposition is possible since the equation for N_x is degenerated from the other equations, so that N_x will not appear in any other equilibrium equations of the shell.

We can now derive the equations of equilibrium for forces in the radial direction and for moments about a section $x + dx = constant$. From Fig. 5.1, we write

$$-\left(\frac{dQ_x}{dx}\right) dx \cdot a\,d\theta - N_\theta\,dx\,d\theta = 0$$

or

$$\frac{dQ_x}{dx} = -\frac{N_\theta}{a}$$

Also, from moment equilibrium, we have

$$\left(\frac{dM_x}{dx}\right) dx \cdot a\,d\theta - Q_x \cdot a\,d\theta\,dx = 0$$

or

$$\frac{dM_x}{dx} = Q_x$$

We desire to express these equations in terms of the nondimensional variable x/a. We write

$$\frac{d(\cdot)}{dx} = \frac{1}{a}\frac{d(\cdot)}{d(x/a)} = \frac{1}{a}(\cdot)'$$

where $(\cdot)'$ denotes differentiation with respect to x/a. The two equilibrium equations above can now be expressed in the following form:

$$Q_x' + N_\theta = 0 \tag{5.1a}$$

$$M_x' = Q_x \cdot a \tag{5.1b}$$

Thus we have three unknowns (N_θ, Q_x, M_x) and only two equations [Eqs. (5.1a,b)]. Hence we need to consider deformations of the shell, in terms of kinematics and Hooke's law, in order to obtain a solution.

Kinematics

We denote w as the outward radial deflection of the shell midsurface, and χ as the angle of rotation of the tangent (assumed positive clockwise).* Then, from Fig. 5.2(a), we can write

$$\varepsilon_\theta = \frac{(a + w)\,d\theta - a\,d\theta}{a\,d\theta} = \frac{w}{a} \tag{5.2a}$$

Also, from Fig. 5.2(b), we obtain

$$\chi = \frac{dw}{dx}$$

and

$$\kappa_x = \frac{\chi + (d\chi/dx)\,dx - \chi}{dx} = \frac{d\chi}{dx}$$

Hence we have

$$\kappa_x = \frac{d^2w}{dx^2} = \frac{w''}{a^2} \tag{5.2b}$$

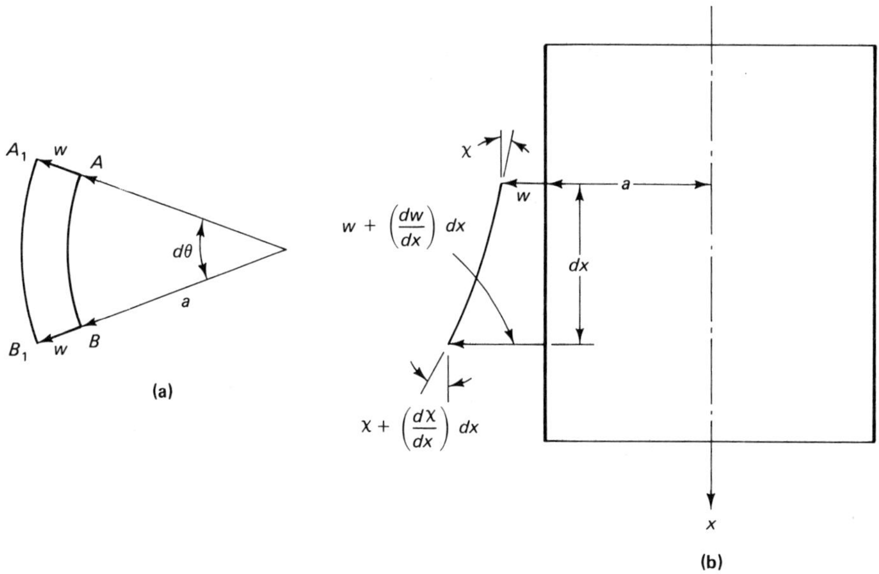

Figure 5.2 Deformations of a circular cylindrical element. (a) Radial deformation. (b) Bending deformation.

*Strictly speaking, χ is positive corresponding to an opening up of the shell. This corresponds to a clockwise rotation of the meridian on the left-hand side of the axis of revolution.

where κ_x is the change of curvature caused by M_x, and is taken as positive when caused by positive M_x. From Fig. 5.2(a), we can also write an expression for κ_θ, the change of curvature in the circumferential direction. Thus we can derive

$$\kappa_\theta = \frac{d\theta}{(a + w)\,d\theta} - \frac{d\theta}{a\,d\theta}$$

$$= \frac{1}{a + w} - \frac{1}{a}$$

$$= \frac{-w}{a^2 + wa}$$

or

$$\kappa_\theta \approx -\frac{w}{a^2} \tag{5.2c}$$

It may be noted from Fig. 5.2(a) that there is no change of the angle between the tangents at points A and B after deformation. However, a small change of curvature κ_θ is still obtained, due to the increased arc length caused by deflection w.

Hooke's Law Relations

We can now easily write the following expressions from elasticity theory:

$$\varepsilon_x = \frac{1}{Et}(N_x - \nu N_\theta)$$

$$\varepsilon_\theta = \frac{1}{Et}(N_\theta - \nu N_x)$$

And again taking $N_x = 0$, we have*

$$N_\theta = Et\varepsilon_\theta \tag{5.3a}$$

Here ε_θ is the *membrane* or in-plane strain, which is assumed to be constant over the thickness of the shell. For *bending* strains, which vary linearly over the shell thickness, we define the terms $\bar{\varepsilon}_x$ and $\bar{\varepsilon}_\theta$, which are caused by bending moments. Thus $\bar{\varepsilon}_x$ and $\bar{\varepsilon}_\theta$ will be functions of z, which we measure from the middle surface (positive inward).

From Fig. 5.2(b), if χ and $\chi + (d\chi/dx)\,dx$ are the rotations of the sections $x = constant$ and $x + dx = constant$, the relative rotation (Fig. 5.3) between these

*It can easily be verified that if $N_x = N_{x_0}$ is nonzero at one edge of the shell, we obtain $w = -(\nu N_{x_0} a/Et)$ for $N_\theta = 0$. This solution can always be linearly superimposed on the edge-load solution.

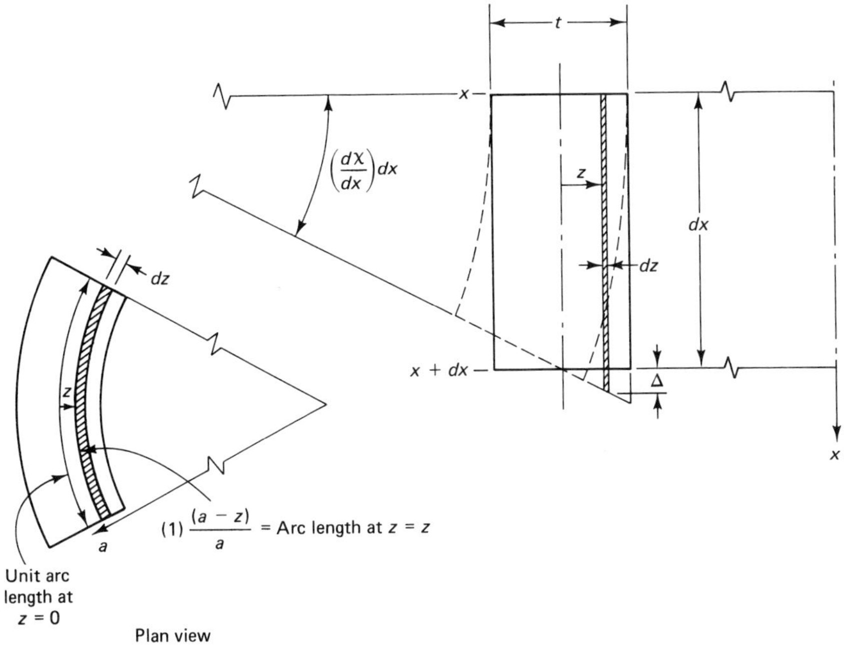

Figure 5.3 Axial strain caused by shell bending and circumferential element showing variation of width with z.

two sections is $(d\chi/dx)\ dx$. This rotation causes an extension, Δ, of a fiber of thickness dz at a distance z from the middle surface. This extension is given by

$$\Delta = z\left(\frac{d\chi}{dx}\right) dx = z\kappa_x\ dx$$

So we have

$$\bar{\varepsilon}_x(z) = \frac{\Delta}{dx} = z\kappa_x$$

In the circumferential direction, we have seen that there is no relative rotation between the tangents at A and B, after an axisymmetric deformation w [see Fig. 5.2(a)]. However, a fiber of length $(a - z)\ d\theta$, at a distance z from the middle surface, will still undergo an elongation which varies with z, due to the initial curvature of the shell in the θ-direction. For a fiber at a distance z from the middle surface, we can write the sum of the membrane strain (constant with z) and the bending strain, due to deformation w, as

$$\varepsilon_\theta + \bar{\varepsilon}_\theta(z) = \frac{(a - z + w)\ d\theta - (a - z)\ d\theta}{(a - z)\ d\theta}$$

$$= \frac{w}{a - z}$$

or

$$\varepsilon_\theta + \bar{\varepsilon}_\theta(z) = \frac{w}{a(1 - z/a)} = \frac{w}{a}\left(\frac{1 + z/a}{1 - (z/a)^2}\right) \approx \frac{w}{a}\left(1 + \frac{z}{a}\right)$$

The first term (w/a) on the right-hand side above is constant in z and corresponds to the membrane strain ε_θ, while the second term varies linearly in z and corresponds to the bending strain $\bar{\varepsilon}_\theta$. Using Eq. (5.2c), we can also see that $\bar{\varepsilon}_\theta = -(z\kappa_\theta)$. For thin shells, z/a is small compared to unity, and hence we can neglect the second term above and take $\bar{\varepsilon}_\theta \approx 0$. We already neglect such terms when we assume the trapezoidal shape of the elemental cross section to be rectangular (see Section 1.6), and this is consistent with our assumption 3 (Section 5.1). This simplification corresponds to taking the *change* in curvature, κ_θ, to be zero due to the deformation w [Eq. (5.2c)].*

For a slice of thickness dz of the element, we can now readily write the following bending stress-strain relationship:

$$\bar{\sigma}_x(z) = \frac{E}{1 - \nu^2}\left[\bar{\varepsilon}_x(z) + \nu\bar{\varepsilon}_\theta(z)\right]$$

or

$$\bar{\sigma}_x(z) = \frac{E}{1 - \nu^2} z\kappa_x$$

And hence for the bending moment, we obtain†

$$M_x = \int_{-t/2}^{+t/2} \bar{\sigma}_x(z) \cdot 1 \cdot z\, dz$$

$$= \frac{E\kappa_x}{1 - \nu^2}\int_{-t/2}^{+t/2} z^2\, dz$$

or

$$M_x = K\kappa_x \tag{5.3b}$$

where

$$K = \frac{Et^3}{12(1 - \nu^2)}$$

K is called the flexural rigidity of the shell.

*We had $\kappa_\theta = -(w/a^2)$, while $\kappa_x = w''/a^2 = d^2w/dx^2$. We neglect κ_θ, but not κ_x, since w/a^2 is a small quantity but d^2w/dx^2 is not necessarily small.

†Actually, for the equation below, the width of a circumferential strip, at a distance z from the middle surface is not unity, but rather is equal to $1 \cdot (a - z)/a = 1 - z/a$. However, as discussed in Section 1.6, we neglect z/a compared to unity.

Now, looking at stresses in the circumferential direction, we can write

$$\bar{\sigma}_\theta(z) = \frac{E}{1 - \nu^2}[\bar{\varepsilon}_\theta(z) + \nu\bar{\varepsilon}_x(z)]$$

or

$$\bar{\sigma}_\theta(z) = \frac{\nu E}{1 - \nu^2}(z\kappa_x)$$

And hence for the bending moment, we obtain

$$M_\theta = \int_{-t/2}^{+t/2} \bar{\sigma}_\theta(z) z\, dz$$

$$= \frac{\nu E \kappa_x}{1 - \nu^2}\int_{-t/2}^{+t/2} z^2\, dz$$

or

$$M_\theta = \nu K \kappa_x \tag{5.3c}$$

Before obtaining the final differential equation for w, we shall now summarize the various equations thus far derived.

Equilibrium

$$Q_x' + N_\theta = 0 \tag{5.1a}$$

$$M_x' = Q_x \cdot a \tag{5.1b}$$

Kinematics

$$\varepsilon_\theta = \frac{w}{a} \tag{5.2a}$$

$$\kappa_x = \frac{w''}{a^2} \tag{5.2b}$$

$$\kappa_\theta = -\frac{w}{a^2} \approx 0 \tag{5.2c}$$

Hooke's Law

$$N_\theta = Et\varepsilon_\theta \tag{5.3a}$$

$$M_x = K\kappa_x \tag{5.3b}$$

$$M_\theta = \nu K \kappa_x \tag{5.3c}$$

where $K = Et^3/12(1 - \nu^2)$ and $(\cdot)'$ denotes derivative with respect to the dimensionless variable x/a.

Now we proceed to derive the differential equation for the deflection w. First, eliminating Q_x from Eqs. (5.1a,b), we obtain

$$(M_x)'' + N_\theta \cdot a = 0 \tag{5.4}$$

Substituting Eqs. (5.2a–c) into Eqs. (5.3a–c), we can write

$$N_\theta = \frac{Etw}{a} \tag{5.5a}$$

$$M_x = \frac{Kw''}{a^2} \tag{5.5b}$$

and

$$M_\theta = \frac{\nu K w''}{a^2} \tag{5.5c}$$

Then, upon substituting Eqs. (5.5a–c) into Eq. (5.4), and assuming K as constant in x, we obtain the following differential equation for w:

$$(w)'''' + \left(\frac{Eta^2}{K}\right) w = 0 \tag{5.6}$$

or

$$(w)'''' + 4\lambda^4 w = 0 \tag{5.7a}$$

where

$$4\lambda^4 = \frac{Eta^2}{K} = \frac{12(1 - \nu^2)a^2}{t^2}$$

or

$$\lambda^4 = \frac{3(1 - \nu^2)\, a^2}{t^2} \tag{5.7b}$$

Another interesting approach to the derivation of this differential equation for w comes from the theory for a beam on an elastic foundation [1]. We can consider as our beam an axial strip of unit width taken from a cylinder (Fig. 5.4). Any deformation w of this beam (which must be the same for any such strip, due to axisymmetry) causes a hoop tension $N_\theta = Etw/a$ in the circumferential rings. These rings then give a radially inward resisting force of $N_\theta/a = Etw/a^2$, as seen in Fig. 5.4(b). Since this force is linearly varying with w, we can consider the axial strip as a beam supported on a bed of linear springs of stiffness Et/a^2. Such a foundation is known as a "Winkler"-type foundation [1]. The equation for the deflection of a beam supported on such a foundation, which has a spring reaction αw per unit length, is obtained by replacing the loading term in the beam equation (see Chapter 1) by $q = p_z - \alpha w$. Thus the differential equation for a beam on an elastic foundation becomes

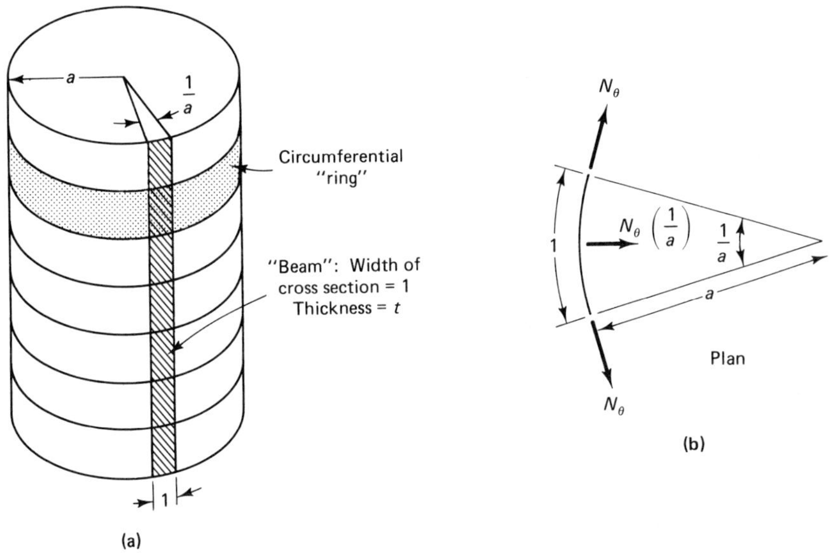

Figure 5.4 Circular cylinder, showing an axial strip that behaves as a beam on an elastic foundation.

$$K\frac{d^4w}{dx^4} = p_z - \alpha w$$

Substituting, for our problem, $p_z = 0$ and $\alpha = Et/a^2$, we get

$$K\frac{d^4w}{dx^4} + \frac{Etw}{a^2} = 0$$

where, in this case, $K = Et^3/12$. This term is independent of Poisson's ratio, because only unidirectional stresses are considered in beam theory. Recognizing this fact, and performing the variable transformation from x to x/a, the correspondence of the equation above to Eq. (5.6) for the cylindrical shell is easily seen.

5.2 SOLUTION OF THE DIFFERENTIAL EQUATION

Equation (5.7a) is a fourth-order ordinary differential equation with constant coefficients. The solution to such an equation is well documented in texts on ordinary differential equations. First, we assume a solution in the form

$$w = Ae^{mx/a}$$

where A and m are constants to be determined. Then, substituting this expression into Eq. (5.7a), gives

$$(m^4 + 4\lambda^4)Ae^{mx/a} = 0$$

Hence, for a nontrivial solution, we must have

$$m^4 = -4\lambda^4$$

or

$$m = \sqrt{2}\,\lambda(-1)^{1/4} \tag{5.8}$$

The four roots of (-1) can be obtained by using the well-known *deMoivre's* theorem from the *theory of complex variables* [2]. By writing (-1) in its complex form, we obtain

$$\begin{aligned}(-1)^{1/4} &= (\cos\pi + i\sin\pi)^{1/4} \\ &= \cos\frac{\pi + 2\pi k}{4} + i\sin\frac{\pi + 2\pi k}{4} \qquad k = 0, 1, 2, 3\end{aligned}$$

Hence

$$(-1)^{1/4} = \pm\frac{1}{\sqrt{2}} \pm \frac{i}{\sqrt{2}}$$

Thus, from Eq. (5.8), we get

$$m = \pm\lambda \pm i\lambda$$

The solution for w can now be written as a linear combination of the four independent solutions obtained from the four roots of m above. Thus we obtain

$$w = \overline{C}_1 e^{m_1 x/a} + \overline{C}_2 e^{m_2 x/a} + \overline{C}_3 e^{m_3 x/a} + \overline{C}_4 e^{m_4 x/a}$$

where $\overline{C}_1$, $\overline{C}_2$, $\overline{C}_3$, and $\overline{C}_4$ are constants. Substituting the appropriate values for m_1, m_2, m_3, and m_4 and rearranging, we can write

$$w = e^{-\lambda x/a}\left(C_1\cos\frac{\lambda x}{a} + C_2\sin\frac{\lambda x}{a}\right) + e^{\lambda x/a}\left(C_3\cos\frac{\lambda x}{a} + C_4\sin\frac{\lambda x}{a}\right) \tag{5.9}$$

where C_1, C_2, C_3, and C_4 are some new constants. They are to be determined from the boundary conditions of the shell at its two edges, say $x = 0$ and $x = L$. At each of these boundaries, we can specify two of the following quantities: w, w', M_x, or Q_x.

We can also simplify the solution further. We see from Eq. (5.7b) that since for thin shells a/t is a large quantity, then $\lambda \gg 1$. Thus the first two terms in Eq. (5.9), which are multiplied by $e^{-\lambda x/a}$, will die out quickly as we go away from the edge $x = 0$. Hence if we are dealing with a long thin cylinder, these first two terms have a negligible effect at the edge $x = L$. Therefore, the constants C_3 and C_4 can be determined by edge-load boundary conditions at $x = L$ alone.

On the other hand, we can see that if the last two terms involving $e^{\lambda x/a}$ have any significant values at the edge $x = 0$, they must have very large values at $x = L$. This is not expected if we are applying edge loads at $x = 0$ on a long cylinder.

Hence the last two terms must have negligible values at $x = 0$. This can also be seen by substituting $\xi = L - x$ into the last two terms of Eq. (5.9), which will transform them to the form

$$e^{-\lambda\xi/a}\left(C_3' \cos\frac{\lambda\xi}{a} + C_4' \sin\frac{\lambda\xi}{a}\right)$$

where ξ is now measured from the edge $x = L$, and C_3' and C_4' are new constants. We can see that these terms will die out as we approach $\xi = L$ (i.e., $x = 0$). We therefore conclude that the constants C_1 and C_2 in Eq. (5.9) will be determined only from the boundary conditions at the edge $x = 0$. The solutions for M_x, Q_x, and w' so obtained from the edge loads at $x = 0$ will also die out as we go away from the edge $x = 0$, due to their also possessing the term $e^{-\lambda x/a}$.

We conclude that for solving problems with edge loads at $x = 0$, we need to consider only the first two terms in the solution [Eq. (5.9)]. These can be put into a more convenient form as

$$w = e^{-\lambda x/a} \cdot A \sin\left(\frac{\lambda x}{a} + \psi\right) \tag{5.10a}$$

where A and ψ are two new constants. Using this expression for w, we can derive expressions for other important quantities, as follows:

$$w' = -\lambda e^{-(\lambda x/a)} A\left[\sin\left(\frac{\lambda x}{a} + \psi\right) - \cos\left(\frac{\lambda x}{a} + \psi\right)\right]$$

We multiply the first and second terms inside the brackets by cos $(\pi/4)$ and sin $(\pi/4)$, respectively, and multiply by $\sqrt{2}$ outside the brackets. Then, combining the terms inside the brackets, we can write

$$w' = -\lambda\sqrt{2}\, e^{-\lambda x/a} \cdot A \sin\left(\frac{\lambda x}{a} + \psi - \frac{\pi}{4}\right) \tag{5.10b}$$

We can similarly derive

$$M_x = \frac{Kw''}{a^2} = -\frac{2K\lambda^2}{a^2}\, e^{-\lambda x/a} \cdot A \cos\left(\frac{\lambda x}{a} + \psi\right) \tag{5.10c}$$

$$Q_x = \frac{M_x'}{a} = \frac{2\sqrt{2}K\lambda^3}{a^3}\, e^{-\lambda x/a} \cdot A \sin\left(\frac{\lambda x}{a} + \psi + \frac{\pi}{4}\right) \tag{5.10d}$$

Also, we have from Eq. (5.5c),

$$M_\theta = \nu M_x \tag{5.10e}$$

and from Eq. (5.5a),

$$N_\theta = \frac{Etw}{a} \tag{5.10f}$$

Using these expressions, we now proceed to solve two basic and important problems, the solutions of which we shall be using again and again for solving more difficult future problems.

5.3 TWO BASIC PROBLEMS

We consider a long cylinder with edge loads,* shears H_o and moments M_o, applied separately at the edge $x = 0$ (Fig. 5.5). First, for solving the problem involving edge shear loads, shown in Fig. 5.5(a), we consider the following boundary conditions at $x = 0$:

$$M_x(0) = 0 \qquad \text{and} \qquad Q_x(0) = H_o$$

Substituting the first condition into Eq. (5.10c) gives $\psi = \pi/2$. Also, using the second boundary condition in Eq. (5.10d), we obtain

$$A = \frac{H_o a^3}{2K\lambda^3}$$

Incorporating these values for ψ and A into Eqs. (5.10a–f), and simplifying, we can easily obtain the following solution for this shear edge-load problem:

$$M_x = \frac{H_o a}{\lambda} e^{-\lambda x/a} \sin \frac{\lambda x}{a} \tag{5.11a}$$

$$Q_x = \sqrt{2}\, H_o e^{-\lambda x/a} \cos\left(\frac{\lambda x}{a} + \frac{\pi}{4}\right) \tag{5.11b}$$

$$N_\theta = 2H_o\lambda\, e^{-\lambda x/a} \cos \frac{\lambda x}{a} \tag{5.11c}$$

$$M_\theta = \nu M_x \tag{5.11d}$$

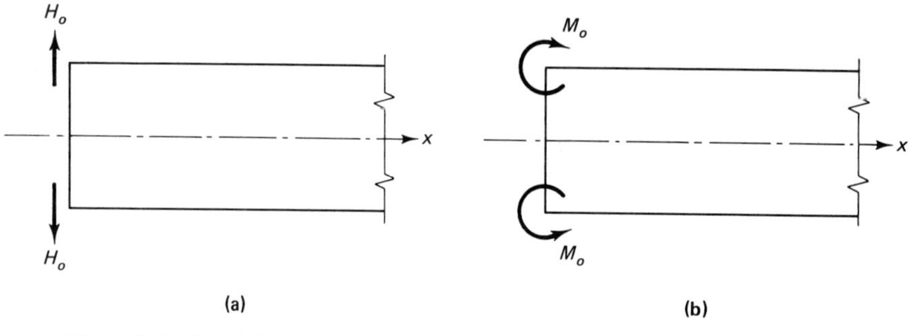

Figure 5.5 Two basic edge-load problems. (a) Circular cylinder with edge shears H_o. (b) Circular cylinder with edge moments M_o.

*It may be noted that these loads are self-equilibrating.

$$w = \frac{H_o a^3}{2K\lambda^3} e^{-\lambda x/a} \cos \frac{\lambda x}{a} \tag{5.11e}$$

$$\frac{dw}{dx} = \frac{-H_o a^2}{\sqrt{2}\, K\lambda^2} e^{-\lambda x/a} \sin\left(\frac{\lambda x}{a} + \frac{\pi}{4}\right) \tag{5.11f}$$

Evaluating Eqs. (5.11e,f) at $x = 0$, we have

$$w(0) = \frac{H_o a^3}{2K\lambda^3} \tag{5.11g}$$

and

$$\frac{dw}{dx}(0) = \frac{-H_o a^2}{2K\lambda^2} \tag{5.11h}$$

We now consider the second basic problem involving the application of an edge moment at $x = 0$ [see Fig. 5.5(b)]. We have the following boundary conditions at $x = 0$:

$$Q_x(0) = 0 \qquad \text{and} \qquad M_x(0) = M_o$$

Substituting the first condition into Eq. (5.10d) gives $\psi = -\pi/4$. Also, using the moment boundary condition in Eq. (5.10c), we obtain

$$A = \frac{-M_o a^2}{\sqrt{2}\, K\lambda^2}$$

Substituting these values for ψ and A into Eqs. (5.10a–f), we can write the following solution for this edge moment problem:

$$M_x = \sqrt{2}\, M_o e^{-\lambda x/a} \cos\left(\frac{\lambda x}{a} - \frac{\pi}{4}\right) \tag{5.12a}$$

$$Q_x = \frac{-2M_o\lambda}{a} e^{-\lambda x/a} \sin \frac{\lambda x}{a} \tag{5.12b}$$

$$N_\theta = \frac{-2\sqrt{2}\, M_o\lambda^2}{a} e^{-\lambda x/a} \sin\left(\frac{\lambda x}{a} - \frac{\pi}{4}\right) \tag{5.12c}$$

$$M_\theta = \nu M_x \tag{5.12d}$$

$$w = \frac{-M_o a^2}{\sqrt{2}\, K\lambda^2} e^{-\lambda x/a} \sin\left(\frac{\lambda x}{a} - \frac{\pi}{4}\right) \tag{5.12e}$$

$$\frac{dw}{dx} = \frac{M_o a}{K\lambda} e^{-\lambda x/a} \sin\left(\frac{\lambda x}{a} - \frac{\pi}{2}\right) \tag{5.12f}$$

Evaluating the equations for w and dw/dx at $x = 0$, we find

$$w(0) = \frac{M_o a^2}{2K\lambda^2} \tag{5.12g}$$

and

$$\frac{dw}{dx}(0) = \frac{-M_o a}{K\lambda} \tag{5.12h}$$

As a check on the results, from Eqs. (5.11h) and (5.12g), it is seen that the rotation in the direction of M_o (i.e., $-dw/dx$) due to $H_o = 1$ is equal to w due to $M_o = 1$. This is a verification of *Maxwell's reciprocal theorem*, which is applicable to such linear structures.

It is seen from Eqs. (5.11a–h) and Eqs. (5.12a–h) that all quantities decay exponentially as we go away from the loaded edge, and thus the effects of the edge loads H_o and M_o are felt only very locally. This observation is expected from *St.-Venant's principle* in the theory of elasticity [3], which states that the effect of redistribution of any loading into another statically equivalent loading is felt only near the loaded region. For our present example, both H_o and M_o are self-equilibrating, and thus could be replaced (in a statically equivalent sense) by a zero edge load. Such a zero edge load will not cause any deformations or stresses in the shell. Hence, applying St.-Venant's principle in a converse manner, we see that redistribution of a zero edge load into self-equilibrating loads H_o or M_o will have an effect only near the edge.

How rapidly the effects of the edge loads decay with distance will be dependent on the value of λ. As an example, for applied edge moments M_o, let us consider the variation of M_x with x/a [Eq. (5.12a)]. At $x/a = 0$, we have $M_x = M_o$, while at $x/a = \pi/\lambda$, we find

$$M_x = -M_o e^{-\pi} = -0.043M_o$$

Thus, within the distance $x/a = \pi/\lambda$, the moment M_x dies down to a mere -4.3% of its maximum value at the edge $x/a = 0$. It is clearly seen that the larger the value of λ, the shorter the distance $x/a = \pi/\lambda$ becomes. This distance is sometimes called the *characteristic length*. The value of λ is given by Eq. (5.7b), and it is a function of the ratio a/t. For shells of aircraft structures and pressure vessels, ratios of a/t are on the order of 100 or so. For civil engineering structures, such as concrete water tanks, dome roofs, etc., a/t ratios are smaller, and hence the characteristic lengths are somewhat longer.

We shall next consider the solutions of more complicated problems of practical interest, using the results obtained from the two previous basic problems.

5.4 WATER TANK FIXED AT ITS BASE

Consider the cylindrical water tank of uniform thickness t in Fig. 5.6. When the wall of such a tank is monolithically connected to a *thick* raft slab foundation, it can be considered as practically fixed against both translation and rotation at the base. Under water pressure, the horizontal membrane displacement increases linearly from the top of the structure [see Figure 5.6(b)]. Due to fixed conditions at the base, redundant forces X_1 and X_2 will develop, such that the net displacement and rotation at the foundation will be equal to zero. The results for the structure are thus obtained by superimposing on the results of a "primary" structure, shown in Fig. 5.6(b), the effects of the self-equilibrating edge loads [Fig. 5.6(c)]. The primary structure is obtained by releasing the original structure such that membrane deformations are freely allowed, but at the same time a stable structure is maintained.

We let δ_{11} and δ_{21} be, respectively, the horizontal displacement and the rotation caused by $X_1 = 1$. Similarly, δ_{12} and δ_{22} are the corresponding quantities caused by $X_2 = 1$. We also denote the membrane base displacement and rotation, for the primary structure, as δ_{10} and δ_{20}. All δ_{ij}'s are considered positive in accordance with the assumed positive conventions for X_1 and X_2. We can now write

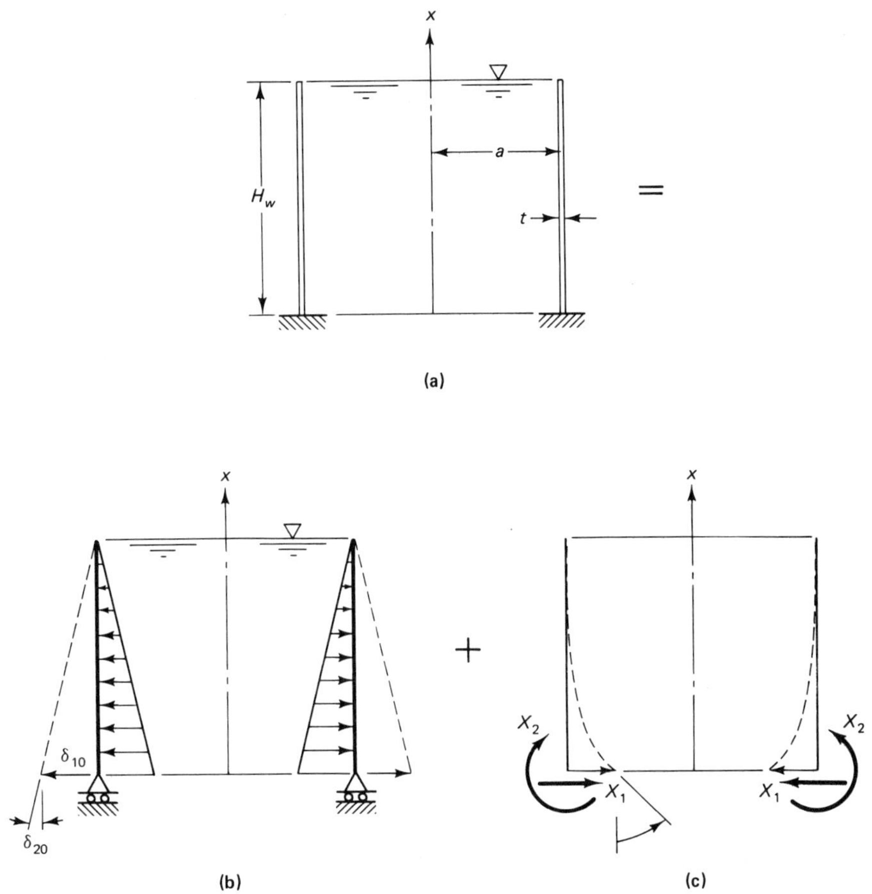

Figure 5.6 Water tank fixed at its base, showing membrane deformations and corrective edge-load deformations.

down the compatibility (or flexibility) equations for the redundants X_1 and X_2 (see also Section 4.3). These are

$$X_1\delta_{11} + X_2\delta_{12} + \delta_{10} = 0 \tag{5.13a}$$

$$X_1\delta_{21} + X_2\delta_{22} + \delta_{20} = 0 \tag{5.13b}$$

We must first calculate the appropriate δ_{ij}'s for substitution into these equations. Considering the membrane response of the structure to water load, we had obtained earlier, in Eqs. (5.2a) and (5.3a), the following (also refer to Example 4.2):

$$w = a\varepsilon_\theta = \frac{aN_\theta}{Et} = \frac{\gamma a^2(H_w - x)}{Et}$$

where γ is the density of water. Hence we also obtain

$$\frac{dw}{dx} = \frac{-\gamma a^2}{Et}$$

Comparing these quantities with the assumed positive directions of X_1 and X_2, we find for $x = 0$,

$$\delta_{10} = -w(0) = \frac{-\gamma a^2 H_w}{Et} \tag{5.14a}$$

and

$$\delta_{20} = -\frac{dw}{dx}(0) = \frac{\gamma a^2}{Et} \tag{5.14b}$$

Also, from Eqs. (5.11g,h) and (5.12g,h), we can write

$$\delta_{11} = \frac{a^3}{2K\lambda^3} \quad \text{and} \quad \delta_{21} = \frac{-a^2}{2K\lambda^2} \tag{5.15a,b}$$

$$\delta_{12} = \frac{-a^2}{2K\lambda^2} \quad \text{and} \quad \delta_{22} = \frac{a}{K\lambda} \tag{5.15c,d}$$

Now, substituting all values for δ_{ij} into Eqs. (5.13a,b), we have the following equations for X_1 and X_2:

$$\frac{a^3}{2K\lambda^3}X_1 - \frac{a^2}{2K\lambda^2}X_2 = \frac{\gamma a^2 H_w}{Et} \tag{5.16a}$$

and

$$\frac{-a^2}{2K\lambda^2}X_1 + \frac{a}{K\lambda}X_2 = \frac{-\gamma a^2}{Et} \tag{5.16b}$$

Thus we have a set of two linear algebraic equations for X_1 and X_2. The solution of these equations is easily found to be

$$X_1 = \frac{2K\lambda^3\gamma}{Eta}\left(2H_w - \frac{a}{\lambda}\right) \tag{5.17a}$$

and

$$X_2 = \frac{2K\lambda^2\gamma}{Et}\left(H_w - \frac{a}{\lambda}\right) \tag{5.17b}$$

Finally, the complete solution for M_x, Q_x, N_θ, M_θ, w, and dw/dx can be obtained by substituting $H_o = -X_1$ and $M_o = X_2$ into Eqs. (5.11a–f) and (5.12a–f), respectively, and superimposing the results on the corresponding membrane solutions. Of course, the membrane solution will be nonzero for only N_θ, w, and dw/dx.

In engineering practice, it is not feasible to solve each individual problem separately in the foregoing manner. Hence ready tables depicting a wide variety of problems are available in the literature, such as in References [4–7]. These tables give coefficients for values of hoop tension N_θ and moment M_x in the tank wall for various values of x/H_w and for various loadings. The values are tabulated for different ratios of $H_w^2/2at$ and different support conditions. Table 5.1 lists four such tables reproduced from Ref. [4] for tanks retaining fluid loads. (Note the difference in notation between that of the tables and that used in this text.)

We shall now compare some numerical values of M_x, obtained by the preceding theoretical solution, with those obtained from the tables.

Example 5.1

Consider a water tank, with an inner diameter of 24 ft, a height of 15 ft, and a wall thickness of 8 in., as shown in Figure 5.7. We assume that the tank is filled with water and is fixed at its base due to the presence of a thick raft slab foundation. Assuming that $\nu = 0.1$ for concrete, we obtain

$$\begin{aligned}\lambda^4 &= 3(1 - \nu^2)\left(\frac{a}{t}\right)^2 \\ &= \frac{3(0.99)(12.33)^2}{(0.67)^2} \\ &= 1005.85\end{aligned}$$

or

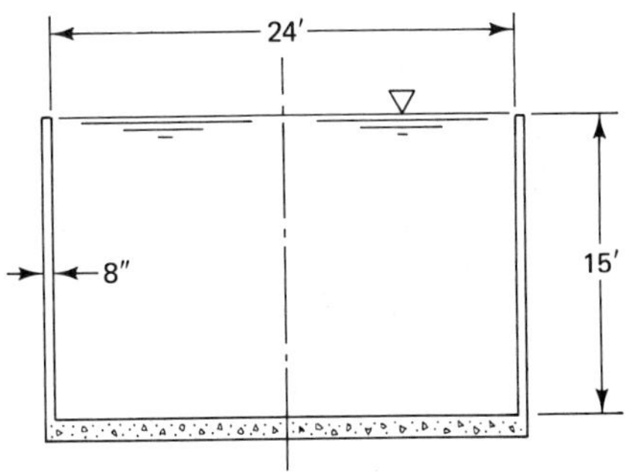

Figure 5.7 Cylindrical concrete tank fixed at its base and filled with water.

TABLE 5.1 Moments M_x and Hoop Forces N_θ in Cylindrical Shell Walls for Various Loadings and Boundary Conditions

Moments in cylindrical wall, M_x
Triangular load
Fixed base, free top

$M_x = -$ coef. $\times wH^3$ ft. lb. per ft.
Negative sign indicates tension in the outside

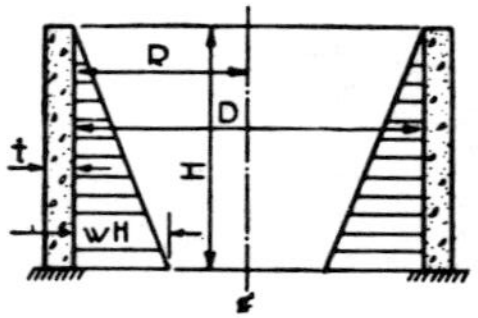

$\frac{H^2}{Dt}$	Coefficients at point									Bottom
	0.1H	0.2H	0.3H	0.4H	0.5H	0.6H	0.7H	0.8H	0.9H	1.0H
0.4	+.0005	+.0014	+.0021	+.0007	−.0042	−.0150	−.0302	−.0529	−.0816	−.1205
0.8	+.0011	+.0037	+.0063	+.0080	+.0070	+.0023	−.0068	−.0224	−.0465	−.0795
1.2	+.0012	+.0042	+.0077	+.0103	+.0112	+.0090	+.0022	−.0108	−.0311	−.0602
1.6	+.0011	+.0041	+.0075	+.0107	+.0121	+.0111	+.0058	−.0051	−.0232	−.0505
2.0	+.0010	+.0035	+.0068	+.0099	+.0120	+.0115	+.0075	−.0021	−.0185	−.0436
3.0	+.0006	+.0024	+.0047	+.0071	+.0090	+.0097	+.0077	+.0012	−.0119	−.0333
4.0	+.0003	+.0015	+.0028	+.0047	+.0066	+.0077	+.0069	+.0023	−.0080	−.0268
5.0	+.0002	+.0008	+.0016	+.0029	+.0046	+.0059	+.0059	+.0028	−.0058	−.0222
6.0	+.0001	+.0003	+.0008	+.0019	+.0032	+.0046	+.0051	+.0029	−.0041	−.0187
8.0	.0000	+.0001	+.0002	+.0008	+.0016	+.0028	+.0038	+.0029	−.0022	−.0146
10.0	.0000	.0000	+.0001	+.0004	+.0007	+.0019	+.0029	+.0028	−.0012	−.0122
12.0	.0000	−.0001	+.0001	+.0002	+.0003	+.0013	+.0023	+.0026	−.0005	−.0104
14.0	.0000	.0000	.0000	.0000	+.0001	+.0008	+.0019	+.0023	−.0001	−.0090
16.0	.0000	.0000	−.0001	−.0002	−.0001	+.0004	+.0013	+.0019	+.0001	−.0079

(a)

Moments in cylindrical wall, M_x
Trapezoidal load
Hinged base, free top

$M_x = -$ coef. $\times (wH^3 + pH^2)$ ft.lb. per ft.
Negative sign indicates tension in the outside

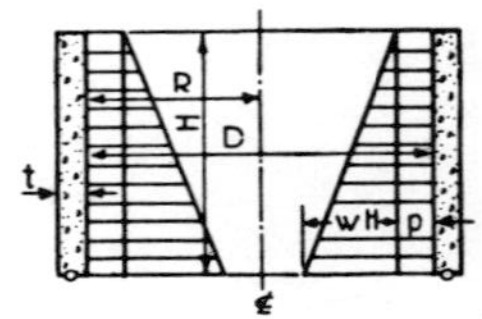

$\frac{H^2}{Dt}$	Coefficients at point									Bottom
	0.1H	0.2H	0.3H	0.4H	0.5H	0.6H	0.7H	0.8H	0.9H	1.0H
0.4	+.0020	+.0072	+.0151	+.0230	+.0301	+.0348	+.0357	+.0312	+.0197	0
0.8	+.0019	+.0064	+.0133	+.0207	+.0271	+.0319	+.0329	+.0292	+.0187	0
1.2	+.0016	+.0058	+.0111	+.0177	+.0237	+.0280	+.0296	+.0263	+.0171	0
1.6	+.0012	+.0044	+.0091	+.0145	+.0195	+.0236	+.0255	+.0232	+.0155	0
2.0	+.0009	+.0033	+.0073	+.0114	+.0158	+.0199	+.0219	+.0205	+.0145	0
3.0	+.0004	+.0018	+.0040	+.0063	+.0092	+.0127	+.0152	+.0153	+.0111	0
4.0	+.0001	+.0007	+.0016	+.0033	+.0057	+.0083	+.0109	+.0118	+.0092	0
5.0	.0000	+.0001	+.0006	+.0016	+.0034	+.0057	+.0080	+.0094	+.0078	0
6.0	.0000	.0000	+.0002	+.0008	+.0019	+.0039	+.0062	+.0078	+.0068	0
8.0	.0000	.0000	−.0002	.0000	+.0007	+.0020	+.0038	+.0057	+.0054	0
10.0	.0000	.0000	−.0002	−.0001	+.0002	+.0011	+.0025	+.0043	+.0045	0
12.0	.0000	.0000	−.0001	−.0002	.0000	+.0005	+.0017	+.0032	+.0039	0
14.0	.0000	.0000	−.0001	−.0001	−.0001	.0000	+.0012	+.0026	+.0033	0
16.0	.0000	.0000	.0000	−.0001	−.0002	−.0004	+.0008	+.0022	+.0029	0

(b)

TABLE 5.1 Moments M_x and Hoop Forces N_θ in Cylindrical Shell Walls for Various Loadings and Boundary Conditions *(continued)*

Tension in circular rings, N_θ
Triangular load
Fixed base, free top
N_θ = coef. × wHR lb. per ft.
Positive sign indicates tension

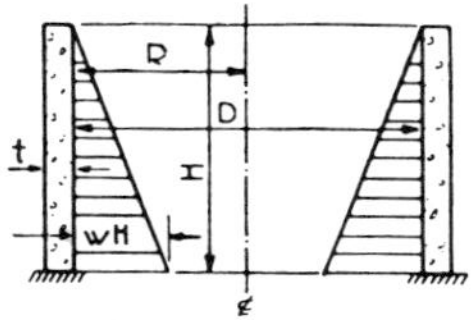

$\frac{H^2}{Dt}$	Top	Coefficients at point								
	0.0H	0.1H	0.2H	0.3H	0.4H	0.5H	0.6H	0.7H	0.8H	0.9H
0.4	+0.149	+0.134	+0.120	+0.101	+0.082	+0.066	+0.049	+0.029	+0.014	+0.004
0.8	+0.263	+0.239	+0.215	+0.190	+0.160	+0.130	+0.096	+0.063	+0.034	+0.010
1.2	+0.283	+0.271	+0.254	+0.234	+0.209	+0.180	+0.142	+0.099	+0.054	+0.016
1.6	+0.265	+0.268	+0.268	+0.266	+0.250	+0.226	+0.185	+0.134	+0.075	+0.023
2.0	+0.234	+0.251	+0.273	+0.285	+0.285	+0.274	+0.232	+0.172	+0.104	+0.031
3.0	+0.134	+0.203	+0.267	+0.322	+0.357	+0.362	+0.330	+0.262	+0.157	+0.052
4.0	+0.067	+0.164	+0.256	+0.339	+0.403	+0.429	+0.409	+0.334	+0.210	+0.073
5.0	+0.025	+0.137	+0.245	+0.346	+0.428	+0.477	+0.469	+0.398	+0.259	+0.092
6.0	+0.018	+0.119	+0.234	+0.344	+0.441	+0.504	+0.514	+0.447	+0.301	+0.112
8.0	−0.011	+0.104	+0.218	+0.335	+0.443	+0.534	+0.575	+0.530	+0.381	+0.151
10.0	−0.011	+0.098	+0.208	+0.323	+0.437	+0.542	+0.608	+0.589	+0.440	+0.179
12.0	−0.005	+0.097	+0.202	+0.312	+0.429	+0.543	+0.628	+0.633	+0.494	+0.211
14.0	−0.002	+0.098	+0.200	+0.306	+0.420	+0.539	+0.639	+0.666	+0.541	+0.241
16.0	0.000	+0.099	+0.199	+0.304	+0.412	+0.531	+0.641	+0.687	+0.582	+0.265

(c)

Tension in circular rings, N_θ
Triangular load
Hinged base, free top
N_θ = coef. × wHR lb. per ft.
Positive sign indicates tension

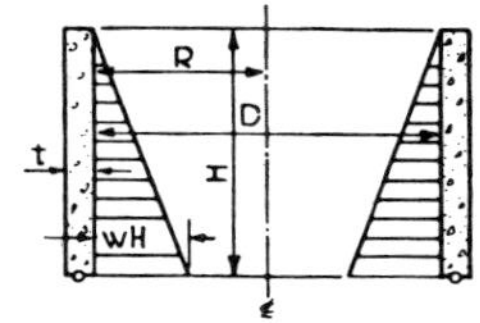

$\frac{H^2}{Dt}$	Top	Coefficients at point								
	0.0H	0.1H	0.2H	0.3H	0.4H	0.5H	0.6H	0.7H	0.8H	0.9H
0.4	+0.474	+0.440	+0.395	+0.352	+0.308	+0.264	+0.215	+0.165	+0.111	+0.057
0.8	+0.423	+0.402	+0.381	+0.358	+0.330	+0.297	+0.249	+0.202	+0.145	+0.076
1.2	+0.350	+0.355	+0.361	+0.362	+0.358	+0.343	+0.309	+0.256	+0.186	+0.098
1.6	+0.271	+0.303	+0.341	+0.369	+0.385	+0.385	+0.362	+0.314	+0.233	+0.124
2.0	+0.205	+0.260	+0.321	+0.373	+0.411	+0.434	+0.419	+0.369	+0.280	+0.151
3.0	+0.074	+0.179	+0.281	+0.375	+0.449	+0.506	+0.519	+0.479	+0.375	+0.210
4.0	+0.017	+0.137	+0.253	+0.367	+0.469	+0.545	+0.579	+0.553	+0.447	+0.256
5.0	−0.008	+0.114	+0.235	+0.356	+0.469	+0.562	+0.617	+0.606	+0.503	+0.294
6.0	−0.011	+0.103	+0.223	+0.343	+0.463	+0.566	+0.639	+0.643	+0.547	+0.327
8.0	−0.015	+0.096	+0.208	+0.324	+0.443	+0.564	+0.661	+0.697	+0.621	+0.386
10.0	−0.008	+0.095	+0.200	+0.311	+0.428	+0.552	+0.666	+0.730	+0.678	+0.433
12.0	−0.002	+0.097	+0.197	+0.302	+0.417	+0.541	+0.664	+0.750	+0.720	+0.477
14.0	0.000	+0.098	+0.197	+0.299	+0.408	+0.531	+0.659	+0.761	+0.752	+0.513
16.0	+0.002	+0.100	+0.198	+0.299	+0.403	+0.521	+0.650	+0.764	+0.776	+0.536

(d)

Source: Ref. [4].

$$\lambda = 5.63$$

Hence, from Eq. (5.17b), substituting $K = Et^3/12(1 - \nu^2)$, we obtain the following value for M_x at the base:

$$M_x = X_2 = \frac{(62.4)(2)(0.67)^2(5.63)^2}{12(0.99)}\left(15 - \frac{12.33}{5.63}\right)$$

Thus

$$M_x = 1914.75 \text{ ft-lb/ft}$$

Now, referring to Table 5.1(a), we have

$$\frac{H_w^2}{Dt} = \frac{225.0}{24(0.67)} = 14.0$$

where D is the internal diameter of the tank. Thus we find from the table,

$$\begin{aligned} M_x &= 0.009\gamma_w H^3 \\ &= 0.009(62.4)(15)^3 \\ &= 1895.4 \text{ ft-lb/ft} \end{aligned}$$

The difference between the value of M_x obtained by us and that from Table 5.1 is only about 1%. Reference [4] does not stipulate the value of ν used in preparing its tables, and undoubtedly this is the source of the minor discrepancy. From the tables one can also easily see how the moment M_x dies out with distance from the bottom edge. For the present example we obtain a characteristic length of

$$x = \frac{\pi a}{\lambda} = 6.88 \text{ ft}$$

This length is approximately $0.46H_w$ from the base of the tank or $0.54H_w$ from the top. From the table it is seen that M_x is almost equal to zero at a value $x = 0.5H_w$, which agrees well with the value of the characteristic length derived here.

Example 5.2

As a second example, consider a cylindrical tank of internal diameter 60 ft and height 20 ft fixed at its base and filled with water. Assuming that $\nu = 0.1$ and a wall thickness of 12 in., we have

$$\lambda^4 = \frac{3(0.99)(30.5)^2}{(1.0)^2} = 2762.84$$

or

$$\lambda = 7.25$$

Therefore, from Eq. (5.17b), we obtain the following value for M_x at the base:

$$M_x = X_2 = \frac{(62.4)(2)(1.0)^2(7.25)^2}{12(0.99)}\left(20.0 - \frac{30.5}{7.25}\right)$$

or

$$M_x = 8720.5 \text{ ft-lb/ft}$$

Alternatively, for this tank, $H_w^2/Dt = 6.67$. Thus, using linear interpolation in Table 5.1(a), we find that M_x at the base of the tank is given by

$$M_x = 0.0173(62.4)(20)^3 = 8636.2 \text{ ft-lb/ft}$$

This value is also very close to that obtained by shell theory. Again, for this example, we find the characteristic length to be

$$x = \frac{\pi a}{\lambda} = 13.2 \text{ ft}$$

This length is about $0.66H_w$ from the bottom of the tank, which agrees well with that observed in Table 5.1(a).

5.5 WATER TANK HINGED AT ITS BASE

Many engineers prefer to design the walls of a tank for a hinged condition at the base, since the effects of any soil settlements are felt lesser in such tanks than in fixed-base tanks. For a hinged base, the moment X_2 will be zero at the base, and we have to satisfy only compatibility of displacement w. For this case the appropriate compatibility equation can be written as

$$X_1\delta_{11} + \delta_{10} = 0$$

The values of δ_{10} and δ_{11} are, respectively, the same as those obtained in Eqs. (5.14a) and (5.15a) for the fixed-base example. Substituting these into the equation above, we have

$$X_1 \frac{a^3}{2K\lambda^3} - \frac{\gamma a^2 H_w}{Et} = 0$$

or

$$X_1 = \frac{2\gamma H_w K\lambda^3}{Eta}$$

If we substitute $K = Et^3/12(1 - \nu^2)$ into this expression, we can rewrite it as

$$X_1 = \frac{\gamma H_w \lambda^3 t^2}{6(1 - \nu^2)a}$$

Then, from Eq. (5.11a), we find the following expression for M_x:

$$M_x = -X_1 \frac{a}{\lambda} e^{-\lambda x/a} \sin \frac{\lambda x}{a}$$

For design, we are interested in the maximum value of M_x. To obtain this value, we take $dM_x/dx = 0$, i.e.,

$$-X_1 \frac{a}{\lambda} e^{-\lambda x/a} \left(-\frac{\lambda}{a} \sin \frac{\lambda x}{a} + \frac{\lambda}{a} \cos \frac{\lambda x}{a} \right) = 0$$

which gives

$$\frac{\lambda x}{a} = \frac{\pi}{4}$$

Substituting this value into the expression for M_x above, the maximum value is found to be

$$(M_x)_{\max} = -\frac{X_1}{\sqrt{2}} \frac{a}{\lambda} e^{-\pi/4}$$

We now consider again the water tank of Example 5.2, this time for the case of a hinged base. Substituting the appropriate numerical values for this tank into the expressions above, we obtain

$$X_1 = 2625.08 \text{ lb/ft}$$

and

$$(M_x)_{\max} = -3560.4 \text{ ft-lb/ft}$$

Again, these values compare well, respectively, with the values of 2632.5 lb/ft and −3544.3 ft-lb/ft obtained readily by interpolation in the tables of Reference [4]. The second value can be readily checked from Table 5.1(b) reproduced from this reference.

5.6 GENERAL FLEXIBILITY SOLUTION: SHELL WALL AND SLAB INTERACTION

When the wall of a cylindrical tank is constructed monolithically with a foundation/raft slab or a roof slab, the connection, in reality, will be neither perfectly fixed nor perfectly hinged. Rather, the resistance given by the slab to the deformations of the wall will depend on thc relative stiffnesses of both the wall and the slab for deflection and rotation.

For solving such statically indeterminate problems, we separate the wall and slab into two "primary" structures by taking a cut at the point of discontinuity of displacements. These primary structures are such that they, by themselves, are in static equilibrium under the action of distributed loads, and they are also free to deform without restraint from one another. For example, we consider the problem of a filled water tank that rests on a foundation raft slab of finite stiffness.

The primary structures, under membrane action, are shown in Fig. 5.8(b). The wall is under the self-equilibrating water pressure and is free to deform at its base, while the raft slab is in equilibrium under water load and an equal and opposite

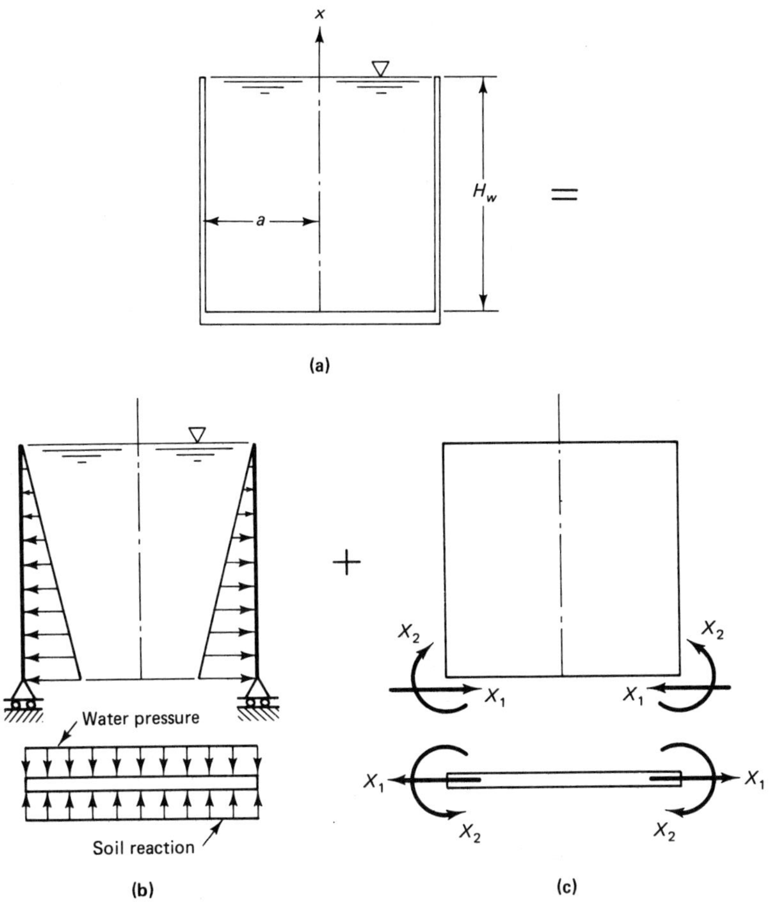

Figure 5.8 Cylindrical shell and raft slab interaction under water load. (a) Tank configuration and coordinate system. (b) Loadings on the primary structures. (c) Edge loads due to shell/raft interaction.

soil reaction. Since the deformations of the primary structures do not match each other at the junction, we shall have to apply unknown forces X_1 and X_2, as shown in Fig. 5.8(c). The values of X_1 and X_2 will be such that the final relative rotation and horizontal deflection between the wall and the plate equal zero at the junction. Hence X_1 and X_2 will be given by the compatibility Eqs. (5.13a,b), where now the δ_{ij}'s (i, j = 1, 2) will be relative displacements between the wall and the plate. These relative displacements can be obtained by the addition of corresponding displacements of the wall and the plate, which are taken as positive in the respective assumed positive directions of X_1 and X_2. As an example, for the direction of X_1 assumed in Fig. 5.8(c), δ_{11} will be positive inward for the wall and positive outward for the plate; and so on for other displacements. Similarly, δ_{10} and δ_{20} in Eqs. (5.13a,b) will be the relative displacements of the primary structures. These are also obtained by summation of the corresponding displacements of the two primary structures, which are taken as positive in the respective assumed positive directions of X_1 and X_2. Thus we can write

$$\delta_{ij} = \delta_{ij}^W + \delta_{ij}^P \tag{5.18a}$$

and

$$\delta_{i0} = \delta_{i0}^W + \delta_{i0}^P \tag{5.18b}$$

where the superscripts W and P refer to the wall and plate, respectively.

In matrix form we can write Eqs. (5.13a,b) and (5.18a,b) as

$$\boldsymbol{\delta}\mathbf{X} + \boldsymbol{\delta}_0 = \mathbf{0} \tag{5.13a,b}$$

where

$$\boldsymbol{\delta}_0 = \begin{bmatrix} \delta_{10} \\ \delta_{20} \end{bmatrix} \qquad \boldsymbol{\delta} = \begin{bmatrix} \delta_{11} & \delta_{12} \\ \delta_{21} & \delta_{22} \end{bmatrix}$$

$$\mathbf{X} = \begin{bmatrix} X_1 \\ X_2 \end{bmatrix}$$

and

$$\boldsymbol{\delta} = \boldsymbol{\delta}^W + \boldsymbol{\delta}^P \tag{5.18a}$$

$$\boldsymbol{\delta}_0 = \boldsymbol{\delta}_0^W + \boldsymbol{\delta}_0^P \tag{5.18b}$$

Flexibility Coefficients for the Cylindrical Shell Wall

We have previously derived the flexibility coefficients for a cylindrical shell wall in Eqs. (5.15a–d). Assuming positive directions for X_1 and X_2, as in Fig. 5.8(c), we rewrite these coefficients here in matrix form:

$$\boldsymbol{\delta}^W = \begin{bmatrix} \dfrac{a^3}{2K\lambda^3} & \dfrac{-a^2}{2K\lambda^2} \\ \dfrac{-a^2}{2K\lambda^2} & \dfrac{a}{K\lambda} \end{bmatrix} \tag{5.19a–d}$$

We multiply and divide the expression for δ_{11}^W by λ, thus obtaining

$$\delta_{11}^W = \frac{a^3\lambda}{2K\lambda^4}$$

Then, substituting $4\lambda^4 = Eta^2/K$, we find

$$\delta_{11}^W = \frac{2a\lambda}{Et}$$

Similar expressions can be obtained for the other coefficients δ_{12}^W, δ_{21}^W, and δ_{22}^W, by eliminating K in Eqs. 5.19(b–d). Thus we can summarize the matrix $\boldsymbol{\delta}^W$ in the following alternative form:

$$\boldsymbol{\delta}^W = \begin{bmatrix} \dfrac{2a\lambda}{Et} & \dfrac{-2\lambda^2}{Et} \\ \dfrac{-2\lambda^2}{Et} & \dfrac{4\lambda^3}{Eta} \end{bmatrix} \tag{5.20a–d}$$

This form will be more convenient to use when solving subsequent problems of displacement compatibility at the junction of a dome roof and a cylindrical wall, as will be discussed in Chapter 7. Obviously, in Eqs. (5.19a–d) or (5.20a–d), δ_{11}^W and δ_{22}^W will always be positive since they represent displacements caused by *corresponding* unit actions. However, the signs for δ_{12}^W and δ_{21}^W will have to be modified for each specific problem if the assumed directions of X_1 and X_2 are different from those in Fig. 5.8(c).

Flexibility Coefficients for the Plate

We now consider flexibilities for the plate or slab attached to a cylindrical wall. We expect that such a plate will have a very large stiffness for in-plane deformations. Hence we can assume approximately that $\delta_{11}^P = 0$.* Considering, as in Chapter 2, the plate deflections to be small, we can neglect axial-flexural interaction for the plate (i.e., in-plane loads do not cause a plate rotation, and similarly, a moment will not cause an in-plane plate deformation). Hence δ_{12}^P and δ_{21}^P are also approximately equal to zero.

In Section 2.4 we had obtained the solution for a circular plate with a moment M_o applied around its circumference. There we found the rotational stiffness or the moment required to produce a unit edge rotation of the plate. This stiffness was given by

$$S^P = \frac{K(1 + \nu)}{a}$$

Hence the flexibility δ_{22}^P is found as the inverse of S^P, or

$$\delta_{22}^P = \frac{a}{K(1 + \nu)}$$

Thus we can summarize the plate flexibility matrix as follows:

$$\boldsymbol{\delta}^P = \begin{bmatrix} 0 & 0 \\ 0 & \dfrac{a}{K(1 + \nu)} \end{bmatrix} \tag{5.21a–d}$$

*If greater accuracy is required, one can obtain the actual flexibility δ_{11}^P by solving the plane stress equations in the theory of elasticity [3].

Example 5.3: Cylindrical Wall and Raft Slab

Let us return to the solution of the problem of the cylindrical water tank on a raft slab, as shown in Fig. 5.8(a). The primary structure and edge loads X_1 and X_2 are assumed as in Fig. 5.8(b) and (c). The displacements of the primary wall structure are given by Eqs. (5.14a,b). Hence we summarize these as

$$\boldsymbol{\delta}_0^W = \begin{bmatrix} -\dfrac{\gamma a^2 H_w}{E t_w} \\ \dfrac{\gamma a^2}{E t_w} \end{bmatrix}$$

For the plate primary structure, we see that since the water load on the plate will be balanced by an equal and oppositely distributed soil reaction, there will be no horizontal displacement or rotation. Hence we obtain

$$\boldsymbol{\delta}_0^P = \mathbf{0}$$

Thus we have

$$\boldsymbol{\delta}_0 = \boldsymbol{\delta}_0^W$$

Also, by adding Eqs. (5.19a–d) and (5.21a–d), we find

$$\boldsymbol{\delta} = \boldsymbol{\delta}^W + \boldsymbol{\delta}^P = \begin{bmatrix} \dfrac{a^3}{2K_w\lambda^3} & \dfrac{-a^2}{2K_w\lambda^2} \\ \dfrac{-a^2}{2K_w\lambda^2} & \dfrac{a}{K_w\lambda} + \dfrac{a}{K_p(1+\nu)} \end{bmatrix} \tag{5.22}$$

where the subscripts w and p are used to distinguish between constants for the wall and plate, respectively.

Substituting the above into the compatibility equations (5.13a,b), we obtain explicitly the following two equations for X_1 and X_2:

$$\left(\frac{a^3}{2K_w\lambda^3}\right)X_1 - \left(\frac{a^2}{2K_w\lambda^2}\right)X_2 = \frac{\gamma a^2 H_w}{E t_w} \tag{5.23a}$$

$$-\left(\frac{a^2}{2K_w\lambda^2}\right)X_1 + \left[\frac{a}{K_w\lambda} + \frac{a}{K_p(1+\nu)}\right]X_2 = -\left(\frac{\gamma a^2}{E t_w}\right) \tag{5.23b}$$

These equations are similar to Eqs. (5.16a,b) for the wall with a fixed base, with the exception of the coefficient of X_2 in the second equation. Upon solving the above, we can write the solution for X_2 in the form

$$X_2 = \frac{X_{2f}}{1 + 2(K_w/K_p)\,[\lambda/(1+\nu)]}$$

where X_{2f} is the value of X_2 obtained for the case of a fixed base, given by Eq. (5.17b). The second term in the denominator of the equation above gives the effect of the relative stiffnesses of the shell and plate. Note that as $K_p \to \infty$, X_2 approaches X_{2f}, and as $K_p \to 0$,

X_2 approaches zero, as expected. Solving for X_1 from Eqs. (5.23a,b), we can obtain the complete solution for the wall by superimposing the effects of X_1 and X_2 on the solution for the primary structure.

For the plate, as noted previously, in-plane stresses (caused by X_1) may be calculated from the plane stress solution in the theory of elasticity [3]. Due to X_2, we have a case of pure bending in the plate, with $M_r = M_\theta = -X_2$ everywhere, as derived in Section 2.4. These results have to be superimposed on the solution for the primary structure of the plate, which in this case has zero moments throughout.

Example 5.4: Self-Weight Loading

The previous cylindrical tank problem can be solved in a similar manner for self-weight loading [Fig. 5.9(a)]. The self-weight of the raft slab will simply create an equal and opposite reaction from the soil, which results in zero moments throughout the plate. However, the self-weight of the wall does create stresses in both the wall and the plate.

If γ_c is the density of concrete, the total weight of the wall is

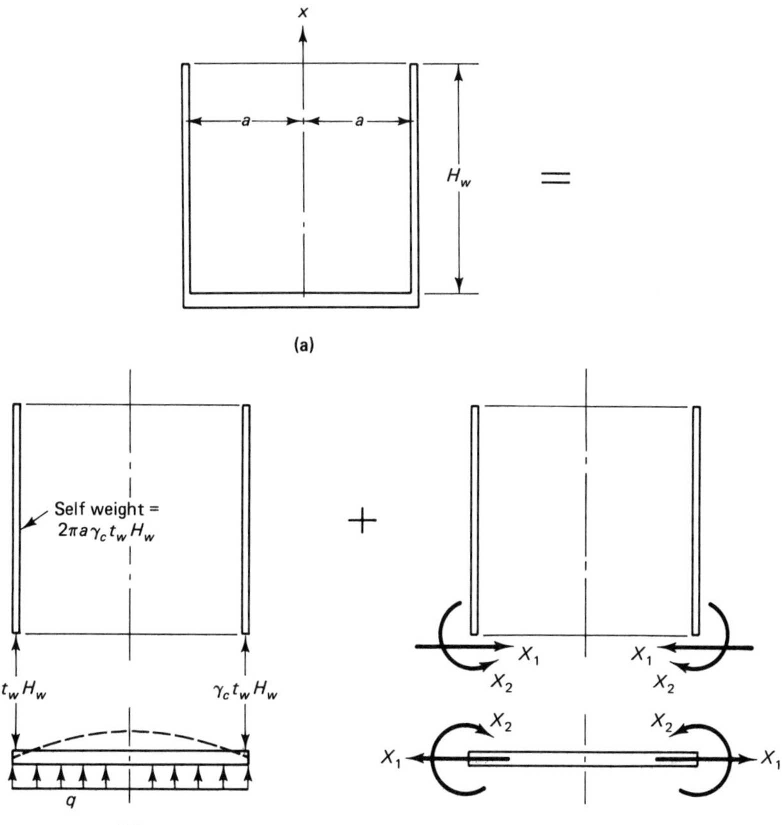

Figure 5.9 Cylindrical tank and raft slab interaction under self-weight loading. (a) Tank configuration and coordinate system. (b) Loadings on the primary structures. (c) Edge loads due to shell/raft interaction.

$$W = 2\pi a\gamma_c t_w H_w$$

This weight will be transferred to the soil through the raft slab. As per normal foundation design practice, the raft slab is assumed to be sufficiently rigid that the upward soil reaction for this load can be taken as uniformly distributed [Fig. 5.9(b)]. Hence the soil reaction is approximated by a uniformly distributed load of

$$q = \frac{2\pi a\gamma_c t_w H_w}{\pi a^2}$$

Under self-weight, the cylindrical shell has the following stresses (see Example 4.1):

$$N_\theta = 0 \qquad \text{and} \qquad N_x = -\gamma_c t_w (H_w - x)$$

Hence we have

$$w = a\varepsilon_\theta = \frac{a}{Et_w}(N_\theta - \nu N_x)$$

or

$$w = \frac{\nu\gamma_c a}{E}(H_w - x) \tag{5.24a}$$

and thus

$$\frac{dw}{dx} = -\frac{\nu\gamma_c a}{E} \tag{5.24b}$$

Therefore, at the base of the wall, we find

$$\delta_{10}^W = -w(0) = -\frac{\nu\gamma_c a H_w}{E}$$

$$\delta_{20}^W = \frac{dw}{dx}(0) = -\frac{\nu\gamma_c a}{E}$$

For the plate we can again assume that $\delta_{10}^P \approx 0$. Flexibility δ_{20}^P can be obtained from the solution of the plate problem derived in Example 2.1, wherein we substitute $p_o = -q$, to get

$$w = -\frac{qa^4}{64K_p}\left[\frac{5+\nu}{1+\nu} - \frac{2(3+\nu)}{1+\nu}\left(\frac{r}{a}\right)^2 + \left(\frac{r}{a}\right)^4\right]$$

and

$$\frac{dw}{dr} = -\frac{qa^3}{16K_p}\left[-\frac{3+\nu}{1+\nu}\left(\frac{r}{a}\right) + \left(\frac{r}{a}\right)^3\right]$$

Hence we obtain δ_{20}^P as

$$\delta_{20}^P = \frac{dw}{dr}(a) = \frac{qa^3}{8(1+\nu)K_p}$$

Since the structure is the same as for Example 5.3, the flexibility matrix remains the same as in Eqs. (5.22), except that now the signs of δ_{12} and δ_{21} will be positive since the assumed positive direction for X_2 in Fig. 5.9(c) is opposite to that shown in Fig. 5.8(c) for water load.

Substituting all the values above into the compatibility equations (5.13a,b), we can derive equations similar to Eqs. (5.23a,b), where the coefficients of X_1 and X_2 are the same as written previously. Solving for X_1 and X_2, and superimposing the edge-load solutions on the membrane solutions, we can obtain the complete solution of this problem. Finally, we can superimpose the solution for water load to the solution just obtained. Alternatively, we could, as well, solve the water load and self-weight cases in one stage, instead of solving the two problems separately.

A similar flexibility solution can be obtained for the interaction between the cylindrical tank wall and a roof slab that carries self-weight plus some superimposed load. However, we shall solve this problem by an alternative approach that uses the relative *stiffnesses* of the wall and the slab, as discussed in the next section.

5.7 SHELL AND SLAB INTERACTION: STIFFNESS APPROACH

We now introduce the well-known *stiffness method* [8] of structural analysis for obtaining the moments and shears that act at the junction of a circular slab and a long circular cylinder. Referring to Fig. 5.10, let us denote u_1 and u_2 as the horizontal displacement and the rotation, respectively, of the joint* between the cylindrical shell and the circular plate. We initially assume the joint to be fixed against translation and rotation. Under this fixed condition, we define $\text{FER}_1^{W,P}$ and $\text{FER}_2^{W,P}$ as, respectively, the unbalanced force in the u_1-direction and unbalanced moment in the u_2-direction, acting on the shell or plate.† We can express the unbalanced force or moment *acting on the joint* as $-\text{FER}_i$, where

$$-\text{FER}_i = -(\text{FER}_i^W + \text{FER}_i^P) \qquad i = 1, 2$$

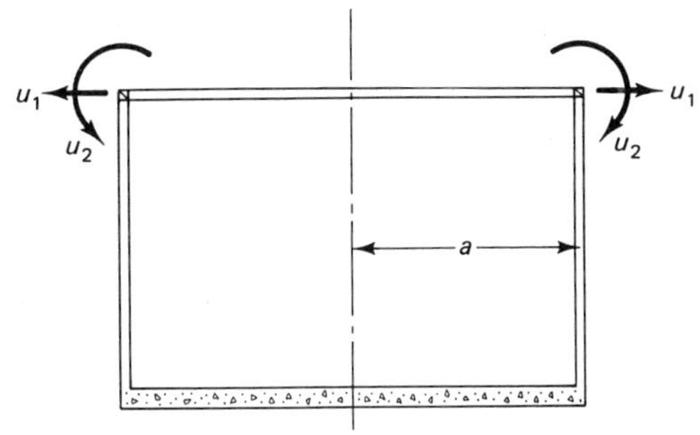

Figure 5.10 Circular cylindrical tank and roof plate interaction, indicating joint displacements.

*In this case the term "joint" refers to the cylindrical ring where the shell and slab meet.

†The notation FER is here used as shorthand for "fixed-end reaction," taken positive in the positive directions of u_1 and u_2.

The superscripts W and P distinguish between reactions attributed to the cylindrical wall and circular plate, respectively.

We also wish to define *stiffness coefficients* S_{11}^W and S_{11}^P as the forces in the u_1-direction acting on the shell and the plate, respectively, and caused by a unit joint displacement $u_1 = 1$. Thus the net force acting on the joint due to $u_1 = 1$ will be $-S_{11}$, where

$$S_{11} = S_{11}^W + S_{11}^P$$

In like manner, we can obtain the net moment acting on the joint, due to a unit displacement $u_1 = 1$, as $-S_{21}$, where

$$S_{21} = S_{21}^W + S_{21}^P$$

Similarly, $-S_{12}$ and $-S_{22}$ can be obtained as the net force and moment, respectively, acting on the joint due to $u_2 = 1$. From Maxwell's reciprocity theorem, of course, S_{21} must equal S_{12}.

If we suppose that we release the joint from its formerly fixed condition, for joint equilibrium to be maintained, we must satisfy the following equilibrium equations for the final displacements u_1 and u_2.

$$-S_{11}u_1 - S_{12}u_2 - \text{FER}_1 = 0$$

or

$$S_{11}u_1 + S_{12}u_2 + \text{FER}_1 = 0 \tag{5.25a}$$

and similarly,

$$S_{21}u_1 + S_{22}u_2 + \text{FER}_2 = 0 \tag{5.25b}$$

where

$$S_{ij} = S_{ij}^W + S_{ij}^P \qquad i, j = 1, 2$$

In matrix form, we can condense the equations above to

$$\mathbf{SU} + \mathbf{FER} = \mathbf{0} \tag{5.25a,b}$$

$$\mathbf{S} = \mathbf{S}^W + \mathbf{S}^P$$

where $\mathbf{S}$ is the 2×2 stiffness matrix, and **FER** is the load vector of fixed-end reactions. Upon solving these equations, we can obtain displacements u_1 and u_2. The joint force and moment in the plate are then given by, respectively,

$$X_1^P = S_{11}^P u_1 + S_{12}^P u_2 + \text{FER}_1^P$$

and

$$X_2^P = S_{21}^P u_1 + S_{22}^P u_2 + \text{FER}_2^P$$

The joint shear force and bending moment in the shell are given by similar expressions, where the superscript P is replaced by W. In matrix form, we write these relations as

$$\mathbf{X}^P = \mathbf{S}^P\mathbf{U} + \mathbf{FER}^P \qquad (5.26a,b)$$

and

$$\mathbf{X}^W = \mathbf{S}^W\mathbf{U} + \mathbf{FER}^W \qquad (5.26c,d)$$

Having thus obtained the joint forces, complete solutions for the plate and the shell can now be obtained using the previously derived edge-load formulas. Before actually solving any particular problem, we shall first obtain explicit expressions for the stiffness coefficients of the shell and plate.

Stiffness Coefficients of the Shell Wall

Figure 5.11 illustrates the two cases of deformation when unit positive values of u_1 or u_2 are applied on the shell, one at a time, while the other remains constrained. The positive directions of u_1 and u_2 are taken as shown in Fig. 5.10.

Case I. For the case shown in Fig. 5.11(a), we allow the shell to have a unit displacement $w = 1$ (corresponding to $u_1 = 1$) at the edge $x = 0$, while we keep the joint rotation u_2 restrained. Returning to Eqs. (5.10a–d) and substituting the boundary conditions at $x = 0$, $dw/dx = 0$ and $w = 1$, we find that

$$\psi = \frac{\pi}{4} \qquad \text{and} \qquad A = \sqrt{2}$$

Hence to maintain these boundary conditions, we have (at $x = 0$)

$$Q_x(0) = \frac{4K_w\lambda^3}{a^3}$$

and also

$$M_x(0) = -\frac{2K_w\lambda^2}{a^2}$$

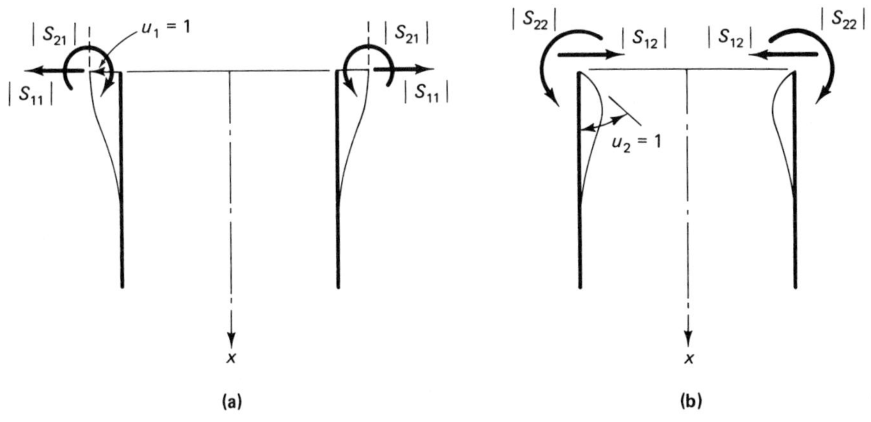

Figure 5.11 Stiffness coefficients for a circular cylinder. (a) Edge stiffnesses due to displacement u_1. (b) Edge stiffnesses due to rotation u_2.

From these expressions, and using the definitions for stiffnesses, we can write

$$S_{11}^{W} = \frac{4K_w\lambda^3}{a^3} \tag{5.27a}$$

$$S_{21}^{W} = -\frac{2K_w\lambda^2}{a^2} \tag{5.27b}$$

Case II. For the case shown in Fig. 5.11(b), we allow the shell to have a unit edge rotation $dw/dx = -1$, (corresponding to $u_2 = +1$), while keeping the horizontal deflection u_1 restrained. Hence substituting into Eqs. (5.10a–d), the boundary conditions at $x = 0$, $w = 0$ and $dw/dx = w'/a = -1$, we obtain the following:

$$\psi = 0 \qquad \text{and} \qquad A = -\frac{a}{\lambda}$$

Thus for $x = 0$, we obtain

$$Q_x(0) = -\frac{2K_w\lambda^2}{a^2}$$

and also

$$M_x(0) = \frac{2K_w\lambda}{a}$$

From these expressions, and using the definitions for stiffnesses, we can write

$$S_{12}^{W} = -\frac{2K_w\lambda^2}{a^2} \tag{5.27c}$$

$$S_{22}^{W} = \frac{2K_w\lambda}{a} \tag{5.27d}$$

As expected, we note that $S_{21}^{W} = S_{12}^{W}$.

The shell stiffness coefficients may now be summarized in matrix form, as follows:

$$\mathbf{S}^{W} = \begin{bmatrix} \dfrac{4K_w\lambda^3}{a^3} & \dfrac{-2K_w\lambda^2}{a^2} \\ \dfrac{-2K_w\lambda^2}{a^2} & \dfrac{2K_w\lambda}{a} \end{bmatrix} \tag{5.27a–d}$$

It is of interest to note that we could have also obtained this stiffness matrix by inversion of the flexibility matrix in Eq. (5.19). Indeed, the reader should independently verify, by matrix multiplication, that

$$\boldsymbol{\delta}^W \mathbf{S}^W = \mathbf{I}$$

where $\mathbf{I}$ is the identity matrix. In performing this multiplication, however, we must note that the direction of X_1 formerly assumed for deriving $\boldsymbol{\delta}^W$ [see Fig. 5.8(c)] is opposite to that of u_1 assumed for deriving $\mathbf{S}^W$. Hence the algebraic signs of the off-diagonal members in either $\boldsymbol{\delta}^W$ or $\mathbf{S}^W$ should be changed for notational correspondence.

If we substitute $4\lambda^4 = Et_w a^2/K$ into Eqs. (5.27a–d), we can express $\mathbf{S}^W$ in an alternative form, similar to that in Eqs. (5.20a–d) for $\boldsymbol{\delta}^W$. Thus we obtain

$$\mathbf{S}^W = \begin{bmatrix} \dfrac{Et_w}{a\lambda} & \dfrac{-Et_w}{2\lambda^2} \\ \\ \dfrac{-Et_w}{2\lambda^2} & \dfrac{Et_w a}{2\lambda^3} \end{bmatrix} \qquad (5.28\text{a–d})$$

Plate Stiffness Coefficients

We now consider two cases of deformation of the circular plate, i.e., for $u_1 = 1$ and for $u_2 = 1$, in order to obtain its stiffness coefficients (see Fig. 5.12). For practical problems, we can assume the plate to be infinitely rigid for any in-plane deformations.* Also, a displacement u_1 will not cause any moment corresponding to u_2, and vice versa. Hence we can immediately write

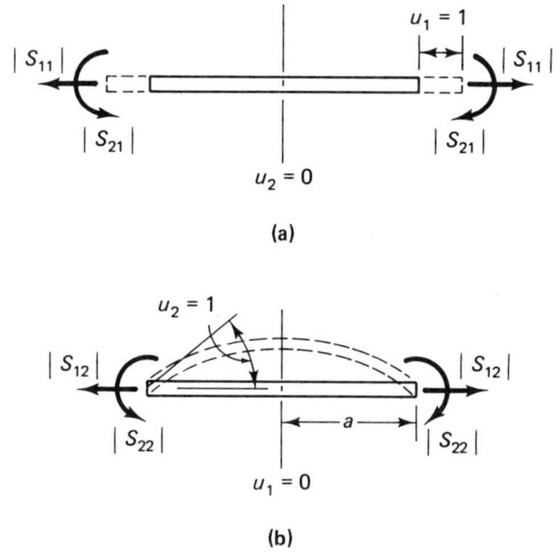

Figure 5.12 Stiffness coefficients for a circular plate. (a) Edge stiffnesses due to displacement u_1. (b) Edge stiffnesses due to displacement u_2.

*Otherwise, in-plane deformations can be obtained by the plane stress solution in polar coordinates, given in textbooks dealing with the theory of elasticity [3].

$$S_{11}^P = \infty \tag{5.29a}$$

$$S_{12}^P = S_{21}^P = 0 \tag{5.29b,c}$$

In Section 2.4 we have already obtained the rotational stiffness for a plate as

$$S_{22}^P = \frac{K_p\,(1 + \nu)}{a} \tag{5.29d}$$

Thus we can summarize stiffness coefficients of the plate into matrix form as follows:

$$\mathbf{S}^P = \begin{bmatrix} \infty & 0 \\ 0 & \dfrac{K_p\,(1 + \nu)}{a} \end{bmatrix} \tag{5.29a–d}$$

Having available now the joint stiffness matrices for the shell and the plate, we proceed to solve a specific example problem.

Stiffness Analysis of Slab-Wall Junction

Let us assume that the tank of Example 5.1 now has a 9-in.-thick roof slab which carries a total distributed load of $q = 200$ 1b/ft^2. We also assume that the tank is empty, and thus carries no water load (i.e., it carries only the wall self-weight and the load from the roof slab).

Slab fixed-end reactions. By using plate theory (Chapter 2), the solution for the fixed-end slab with a uniform dead load of q can readily be obtained. For the axisymmetric moment $M_r(r)$, we find

$$M_r(r) = \frac{qa^2}{16}\left[(1 + \nu) - (3 + \nu)\left(\frac{r}{a}\right)^2\right]$$

Thus at $r = a$, we have

$$M_o = M_r(a) = -\frac{qa^2}{8}$$

Also, at $r = 0$, we obtain

$$M_c = M_r(0) = \frac{qa^2}{16}(1 + \nu)$$

M_c and M_o are, respectively, positive and negative (i.e., sagging and hogging), as per our former sign convention used in plate theory. Now FER_2^P is the fixed-edge plate moment, positive in the positive direction of u_2. Hence

$$\mathrm{FER}_2^P = -M_o = \frac{qa^2}{8}$$

We can also write that under load q,

$$\text{FER}_1^P = 0$$

Shell fixed-end reactions. Due to the loading q on the plate, the shell will experience a membrane compressive stress given by

$$N_x = -\frac{qa}{2}$$

Due to the absence of a radial load p_r, N_θ will be zero, and hence membrane displacements w and dw/dx are given as

$$w = a\varepsilon_\theta = \frac{a}{Et_w}(-\nu N_x) = \frac{\nu}{Et_w}\frac{qa^2}{2}$$

and

$$\frac{dw}{dx} = 0$$

Under self-weight of the shell, membrane displacements given by Eqs. (5.24a,b) are incurred. Hence at the top of the shell, we have the total displacements as

$$w = \frac{\nu}{Et_w}\frac{qa^2}{2}$$

and

$$\frac{dw}{dx} = \frac{-\nu\gamma_c a}{E}$$

Using these values, and solving the compatibility equations (5.13a,b), we can assess the fixed-end shear and moment for the shell. However, we note that the values of w and dw/dx above contain a factor ν, and hence will have negligible effects in comparison to that caused by dw/dr for the plate. Thus, with reasonable accuracy, we assume for simplicity that

$$\text{FER}_1^W = \text{FER}_2^W = 0$$

Solution for joint displacement and moment. The total values of FER and S_{ij} for this problem become

$$\text{FER}_1 = 0 \quad \text{and} \quad \text{FER}_2 = \frac{qa^2}{8}$$

$$S_{11} = S_{11}^W + S_{11}^P = \infty$$

$$S_{21} = S_{12} = \frac{-2K_w\lambda^2}{a^2}$$

$$S_{22} = \frac{2K_w\lambda}{a} + \frac{K_p\,(1 + \nu)}{a}$$

Substituting these values into Eqs. (5.25a,b) and solving, we obtain $u_1 = 0$, as expected, since we had assumed the plate to be infinitely rigid for in-plane deformations. We also obtain the solution for u_2 as

$$u_2 = \frac{-qa^2/8}{\dfrac{2K_w\lambda}{a} + \dfrac{K_p\,(1 + \nu)}{a}} \tag{5.30}$$

The plate and shell joint moments, due to u_2, will be simply

$$M^P = S_{22}^P u_2 = -\frac{qa^2}{8}\left(\frac{K_p\,(1 + \nu)/a}{\dfrac{2K_w\lambda}{a} + \dfrac{K_p\,(1 + \nu)}{a}}\right)$$

and

$$M^W = S_{22}^W u_2 = -\frac{qa^2}{8}\left(\frac{2K_w\lambda/a}{\dfrac{2K_w\lambda}{a} + \dfrac{K_p\,(1 + \nu)}{a}}\right)$$

The final moments in the plate are obtained by superimposing on the fixed-edge condition moments due to the relieving edge action $M_r(a) = -M^P$.*

It may be noted from the above that the unbalanced joint moment is simply distributed to the plate or shell in direct proportion to the respective rotational stiffnesses. This is similar to what is done in a one-step *moment distribution* for analysis of joint equilibrium in continuous beams or frames in ordinary structural analysis. This is possible here only because, due to the infinite in-plane plate stiffness, no horizontal joint displacement occurs. Thus rotational and displacement stiffnesses are uncoupled. This makes Eq. (5.25a) trivial, and the term $S_{21}u_1$ in Eq. (5.25b) equal to zero. This decoupling will be true for all such plate and shell problems, and the joint moment can be obtained by one-step distribution of the unbalanced joint moment. We shall see later that such a simplification cannot be introduced when considering similar problems for dome-cylinder junctions.

Numerical results. Substituting the appropriate numerical values for the tank under consideration (Example 5.1), we obtain for the plate

*As per our plate theory derivations of Chapter 2, M_r is positive when the plate is sagging. M^P and FER_2^P are positive in the positive direction of u_2 (i.e., for a hogging curvature).

$$M_o = -\frac{0.2(12.33)^2}{8} = -3.8 \text{ k-ft/ft}$$

and

$$M_c = \frac{(0.2)(12.33)^2}{16}(1.1) = 2.09 \text{ k-ft/ft}$$

Thus since $\text{FER}_2 = -M_o$, we have

$$\text{FER} = 3.8 \text{ k-ft/ft}$$

A plot of M_r versus r for the plate, under a fixed-end condition, would be as shown in Fig. 5.13. Now we have

$$S_{22}^P = \frac{(0.75)^3E}{12(0.99)}\left(\frac{1.1}{12.33}\right) = 0.00317E$$

$$S_{22}^W = \frac{2(0.67)^3E}{12(0.99)}\left(\frac{5.63}{12.33}\right) = 0.02312E$$

Hence, due to the joint rotation u_2, the unbalanced joint moments, which are distributed in proportion to rotational stiffnesses, are

$$M^P = -3.8\left(\frac{0.00317}{0.00317 + 0.02312}\right) = -0.46 \text{ k-ft/ft}$$

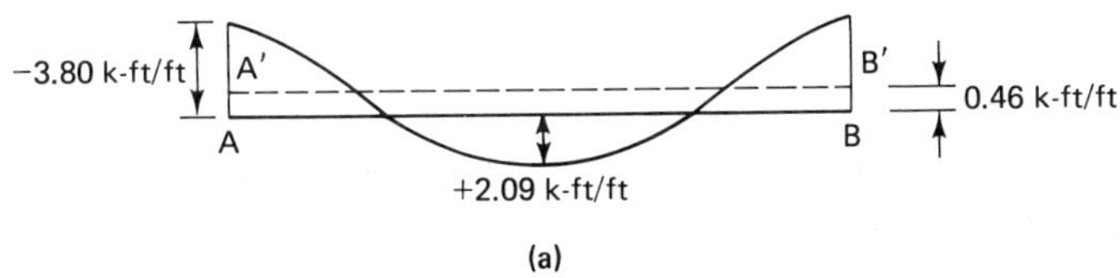

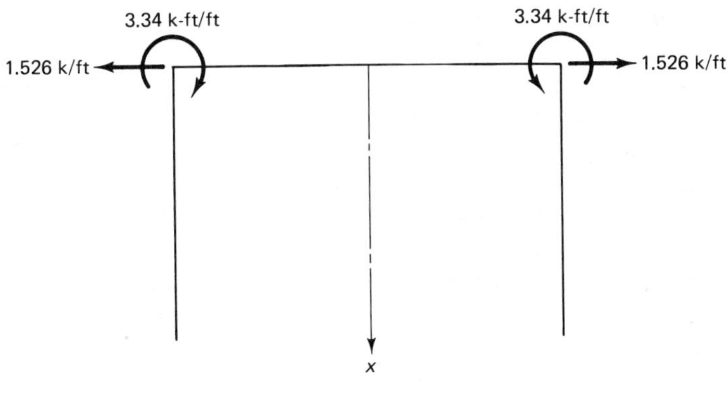

Figure 5.13 Plot of M_r for the roof plate, showing the effect of the edge moment obtained from the stiffness analysis.

$$M^W = -3.8\left(\frac{0.02312}{0.00317 + 0.02312}\right) = -3.34 \text{ k-ft/ft}$$

The direction of these moments are opposite to that of u_2 in Fig. 5.10, i.e., sagging for the plate and tension on the outer fibers of the shell. The final moment $M_r(r)$ in the plate is obtained by superimposing results of the application of $(-M^P) = 0.46$ k-ft/ft on the boundary $r = a$ to the fixed-edge results plotted in Fig. 5.13. We learned from Chapter 2 that the application of a moment on the boundary of a plate gives only a constant moment throughout; i.e., $M_r = M_\theta = 0.46$ k-ft/ft. Hence the final moment diagram for the plate is obtained by simple addition of a positive (sagging) moment of 0.46 k-ft/ft to the fixed-edge condition. Thus, shifting the datum line in Fig. 5.13(a) up from AB to $A'B'$ gives the correct final moments acting within the plate.

The final edge moment and shear in the shell will be given by Eqs. (5.26c,d). We first calculate, from Eq. (5.30),

$$u_2 = \frac{-3.8}{(0.00317 + 0.02312)E} = -\frac{144.54}{E}$$

We can also calculate

$$S_{12}^W = -\frac{2K_w\lambda^2}{a^2} = -0.01056E$$

Hence

$$X_1^W = S_{12}^W u_2 = 1.526 \text{ kips/ft}$$

The values of edge moment and shear on the wall are shown in Fig. 5.13(b). Forces and displacements at other locations within the shell are obtained by superimposing edge load solutions due to the final shear and moment acting at $x = 0$.

For the practical case when a tank has a very large diameter and the roof slab needs to be supported by a central column, the plate stiffness coefficients derived above will have to be modified. These new coefficients, as well as the fixed-edge moment, can easily be worked out using the theory developed in Chapter 2. Reference [4] also gives convenient tables for such centrally supported plates, which significantly simplify the solutions to practical problems.

Similarly, for annular roof slabs, there will be *carry-over factors* to and from the inner circular boundary that must be assessed. For this case, new stiffness coefficients for the plate can be derived, and the problem can then be solved using the conventional moment distribution procedures in ordinary structural analysis.

So far, we have been assuming that the cylindrical shell is "long," i.e., the effects of edge loads at one edge die out before they reach the other edge. Hence we used only the first two terms in the general solution [Eq. (5.9)] of the differential equation. For the case when the cylindrical shell is short, a solution having all four constants of integration in Eq. (5.9) must be used. The solution becomes further complicated if axial deformations of the shell are also considered. For a computer solution, a ring element of the shell, of length L, can be considered. This element will possess three displacement degrees of freedom at each edge. Hence a general 6×6 stiffness matrix for the shell element can be obtained. Once this matrix is formulated, a variety of cylindrical shell problems can be solved on the computer, by suitably dividing the shell into such ring elements [9,10]. A practical example of such an application arises when a shell of variable thickness carries any arbitrary axisymmetric load distribution. In such a case, small elements of constant thickness can be used to approximate the actual variation in shell thickness.

5.8 SOME DESIGN CONSIDERATIONS

Once an analysis for the internal shears and bending moments is performed, the wall of a cylindrical tank (such as the one in Example 5.1) can then be appropriately designed to resist the maximum stress resultants. For the preceding example, the vertical compressive stresses caused by self-weight plus roof loads are, in general, very small and do not govern the design in most cases. The wall should be designed mainly to resist the hoop tension N_θ and the meridional moment M_x, the circumferential moment M_θ usually being small. Values of N_θ and M_x, at varying levels in a circular cylindrical tank under water load, can be obtained from standard tables in References [4–7] once the condition of the base is known (i.e., hinged or fixed). In cases where the interaction of the shell and the raft or roof slab is to be considered, a more detailed analysis, as illustrated previously, should be undertaken.

Design for the thickness of the wall is generally decided so as to maintain allowable tensile stresses in concrete. Thus, for ordinary concrete with a compressive strength of 3000 psi, direct tensile stresses due to N_θ should be restricted to about 300 psi. For tension due to bending moment M_x, however, stresses can be designed for a maximum allowable value of around 400 psi.*

We also design reinforcement to resist the entire tensile forces due to N_θ and M_x. Both the outer and inner faces of the tank wall are provided with both vertical and horizontal steel reinforcing. Such reinforcing must at least meet the minimum code requirement for temperature and shrinkage, which is usually about 0.3% of the pertinent cross-sectional area. For water tanks, some codes recommend reduced allowable stresses for reinforcement in direct and bending tension, as compared to those for normal beams and slabs, in order to control cracking. American practice, however, is to use full allowable stresses. Hence, using the appropriate allowable stresses in steel reinforcement, design for resistance of N_θ may be tabulated, as illustrated in Table 4.1 of Chapter 4 for a conical tank. Vertical reinforcement, for resisting the resultant M_x, can also be calculated and designed similarly. This is illustrated in Table 5.2 for the water tank of Example 5.1, assuming a hinged base and no roof slab.

*Allowable stresses are code dependent and are normally functions of the concrete strength, type of loading, etc. The values used in this text are reasonable and representative of most codes. However, the designer should, of course, become familiar with and obey the code in his or her local jurisdiction.

TABLE 5.2 Reinforcement Design for Water Tank of Example 5.1 with Hinged Base

x From Base (ft)	x/H_w	N_θ[a] (lb/ft)	$(N_\theta)_{avg}$ (lb/ft)	Hoop Tension $T = (N_\theta)_{avg} \times 1.5'$ (lbs)	Horizontal A_S (Both Faces) per Element $= T/20{,}000$	M_x[a] (ft-lb/ft)	Vertical A_S Outer Face
0.0	0.0	0.0				−694.98	↑
			550.35	825.5	4 No. 3[b]		
1.5	0.1	1,100.7				−547.56	
			1,656.70	2,485.1	4 No. 3[b]		
3.0	0.2	2,212.7				−252.72	
			2,785.55	4,178.3	4 No. 3[b]		
4.5	0.3	3,358.4				0.	
			3,970.55	5,955.8	4 No. 3[b]		
6.0	0.4	4,582.7				+21.1	No. 3 at 9 in. c/c for all[c]
			5,273.45	7,910.2	4 No. 3[b]		
7.5	0.5	5,964.2				+21.1	
			6,683.05	10,024.6	5 No. 3		
9.0	0.6	7,401.9				+21.1	
			7,974.70	11,962.1	6 No. 3		
10.5	0.7	8,547.5				0.	
			8,497.00	12,745.5	6 No. 3		
12.0	0.8	8,446.5				0.	
			7,104.25	10,656.4	5 No. 3		
13.5	0.9	5,762.0				0.	↓

[a]Values for N_θ and M_x based on values in PCA tables [4].
[b]Minimum steel governs.
[c]Minimum vertical steel of No. 3 at 9 in. center-to-center is provided on inner face also.

5.9 OTHER APPLICATIONS: TUBES AND PIPES

The bending theory for axisymmetric edge loading, developed earlier in this chapter, can also be applied to other miscellaneous problems of circular cylinders; some applications involve analysis of tubes, pressure vessels, etc. For example, consider a long tube with a radial line load P per unit length applied around the circumference at its center, as shown in Fig. 5.14(a).

We look at a freebody of one-half of the tube, which will have the boundary conditions; at $x = 0$, $Q_x = -P/2$ and $dw/dx = 0$. We can substitute these conditions into Eqs. (5.10a–f) to obtain the final solution directly from the calculated constants A and ψ. Alternatively, we can use the results in Eqs. (5.11a–h) and (5.12a–h), and use superposition of the two basic edge-load problems to obtain the same solution. For the latter approach, we use the boundary condition that at $x = 0$, the slope dw/dx caused by the edge load $Q_x = -P/2$ plus an unknown internal moment M_o must be equal to zero. We thus obtain the following equation for M_o:

$$\left(\frac{dw}{dx}\right)_{x=0} = \frac{(P/2)a^2}{2K\lambda^2} - \frac{M_o a}{K\lambda} = 0$$

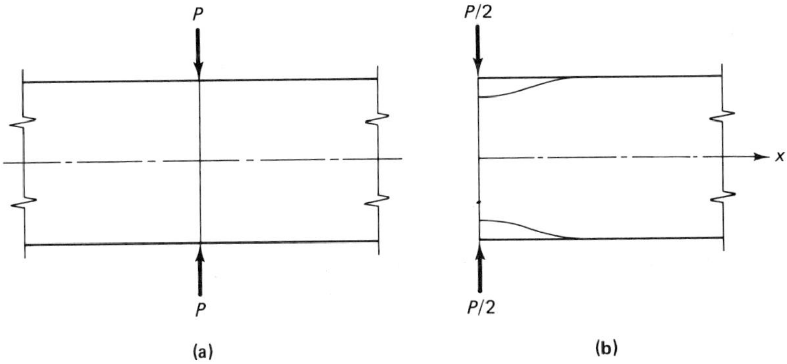

Figure 5.14 Circular tube problem. (a) Infinite cylinder with a radial line load. (b) Symmetrical right-half of tube with a radial line load of P/2.

which gives

$$M_o = \frac{Pa}{4\lambda}$$

The final stress resultants, displacements, etc., are then found by addition of the individual edge-load solutions for $H_o = -(P/2)$ and for $M_o = Pa/4\lambda$ from Eqs. (5.11a–h) and (5.12a–h).

If the tube happened to be of a short length, say $L = 2\ell$, which was smaller than the characteristic length $(\pi a/\lambda)$, the solution above would yield nonzero values for M_x and Q_x at the ends $x = \pm\ell$ which might not be negligible. For the case of the ends $x = \pm\ell$ being free, in order to make these stress resultants equal zero, we must apply the appropriate opposite values of M_x and Q_x at the ends. Since the tube is short, however, we must use all four terms which contain the four constants in Eq. (5.9), in order to solve this second portion of the problem. However, this solution scheme is rather complicated to perform, and a better approach would be to rewrite Eq. (5.9) in the following form:

$$w = A_1 \cosh\frac{\lambda x}{a}\cos\frac{\lambda x}{a} + A_2 \sinh\frac{\lambda x}{a}\sin\frac{\lambda x}{a} + A_3 \cosh\frac{\lambda x}{a}\sin\frac{\lambda x}{a} + A_4 \sinh\frac{\lambda x}{a}\cos\frac{\lambda x}{a} \tag{5.31}$$

where A_1, A_2, A_3, and A_4 are a set of new constants.

The first two terms above are symmetric in x, whereas the last two terms are antisymmetric in x. Since the problem of the tube under consideration is obviously symmetric in x, for the solution of the second part, we can directly take $A_3 = A_4 = 0$. Thus only the constants A_1 and A_2 need be determined, and the solution for the short tube becomes somewhat simpler.

In some cases, for economy and weight considerations, instead of using a thick tube, a thinner tube with stiffening rings is employed to resist internal or

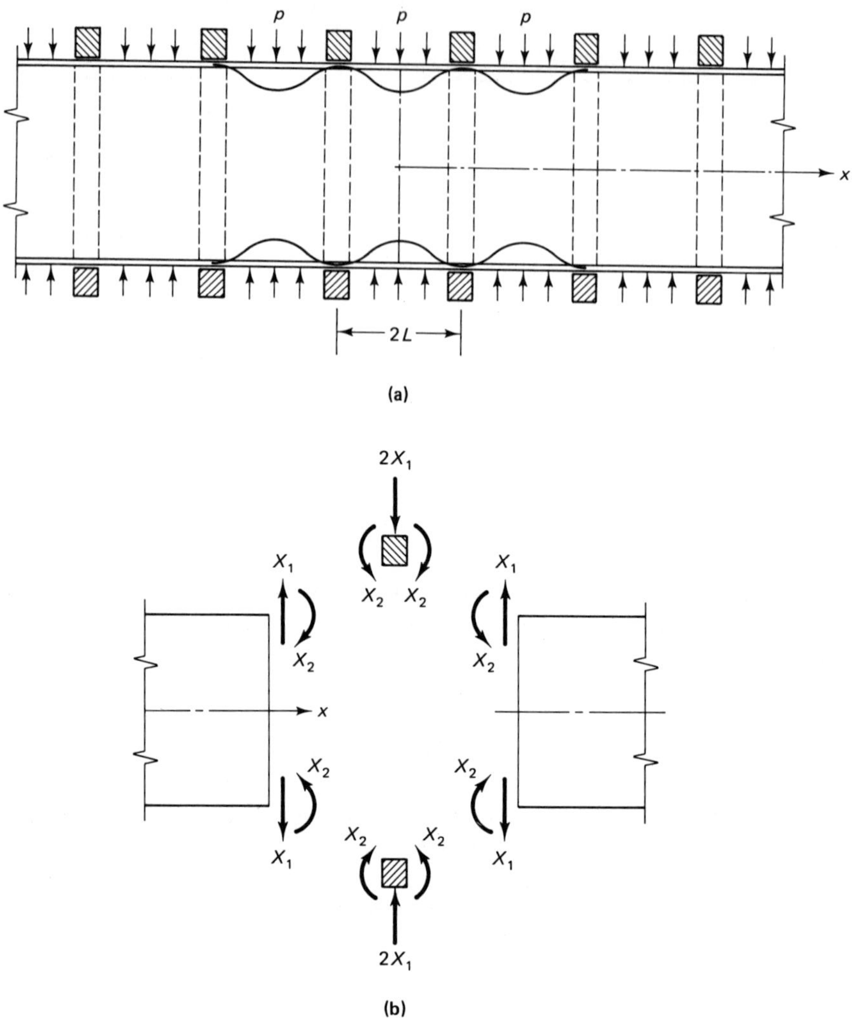

Figure 5.15 Stiffened pipe problem. (a) Deformations under uniform pressure. (b) Corrective edge loads for cylinder and stiffener.

external pressure loads (Fig. 5.15). Such problems arise in practice for pipelines subjected to internal pressure, which would require external rings. Similarly, in submarine hulls, which carry external pressure, internal rings are needed. Also, airplane fuselages, which carry internal pressure, usually are stiffened with internal rings. Let us now consider the general approach taken in solving such a problem in light of our available theory.

Example 5.5: Stiffened Pipe

Consider a long pipe with rings spaced equidistant at $2L$ centers and subjected to a uniform external pressure p. The deformed shape of the shell will be as shown in Fig. 5.15(a), and can be considered as symmetric about $x = 0$. Hence only the first two terms of Eq. (5.31) need to be involved in the analysis. The situation at $x = \pm L$ (i.e., at the junction of the pipe and ring) is indicated by the freebody diagram in Fig. 5.15(b). Unknown edge loads X_1 and X_2 will develop between the two adjacent shell portions and the ring. Solution of the compatibility equations (5.13a,b) will give X_1 and X_2, where

$$\delta_{ij} = \delta^w_{ij} + \delta^r_{ij} \qquad i, j = 1, 2$$

The superscripts w and r denote, respectively, identifiers for the cylindrical shell wall and the ring beam.

For the primary structures, under membrane external pressure loading, we can write*

$$\delta^w_{10} = \frac{-pa^2}{Et} \qquad \text{and} \qquad \delta^r_{10} = 0$$

Also,

$$\delta^w_{20} = 0 \qquad \text{and} \qquad \delta^r_{20} = 0$$

Hence we have

$$\delta_{10} = \delta^w_{10} + \delta^r_{10} = -\frac{pa^2}{Et}$$

and

$$\delta_{20} = \delta^w_{20} + \delta^r_{20} = 0$$

Now, under an edge load $X_1 = 1$, for the ring, we have from Section 1.4

$$\delta^r_{11} = \frac{2a^2}{AE}$$

where A is the cross-sectional area of the ring. Also, since loads X_2 on either side of the ring are equal and opposing, we shall find

$$\delta^r_{22} = \delta^r_{12} = \delta^r_{21} = 0$$

To the above we add coefficients δ^w_{ij} $(i, j = 1, 2)$, which can be obtained by using the first two terms of the solution to Eq. (5.31) and proceeding as done for the long cylinder solution considered earlier in this chapter. Finally, complete compatibility conditions can be written and solved for X_1 and X_2.

For detailed solutions of problems similar to those noted in this chapter, reference may be made to Flügge [11]. Also, various handbooks, such as that of Roark [12], give readily accessible formulas for different problems involving tubes.

*If the shell is closed at its ends, due to external pressure p, we find $N_x = -(\pi a^2 p/2\pi a) = -(pa/2)$. In this case we shall have $\delta^w_{10} = w = a\varepsilon_\theta = (a/Et)(N_\theta - \nu N_x) = -(pa^2/Et)(1 - \nu/2)$.

REFERENCES

[1] Timoshenko, S. P., *Strength of Materials,* 3rd ed. Pt. II, D. Van Nostrand Company, Princeton, N.J., 1955.

[2] Churchill, R. V., and Brown, J. W., *Complex Variables and Applications,* 4th ed., McGraw-Hill Book Company, New York, 1984.

[3] Timoshenko, S. P., and Goodier, J. N., *Theory of Elasticity,* 3rd ed., McGraw-Hill Book Company, New York, 1970.

[4] "Circular Concrete Tanks without Prestressing," Portland Cement Association Publication No. IS072.01D, Portland Cement Association, Skokie, Ill.

[5] "Code of Practice for Concrete Structures for the Storage of Liquids," Indian Standard 3370-1965, (Part IV), Indian Standards Institution, New Delhi, 1966.

[6] Arya, A. S., "Data for Analysis and Design of Circular Shell Structures," 1, 2, 3, 4, and 5, *Indian Concr. J.*, Aug. 1969, Jan. 1970, Feb. 1970, June 1970, and Dec. 1971.

[7] Olander, H. C., "Design of Cylindrical Water Tanks," *ASCE J. Struct. Div.*, May 1970, pp. 947–973.

[8] Weaver, W., Jr., and Gere, J. M., *Matrix Analysis of Framed Structures,* 2nd ed., Van Nostrand Reinhold Company, Inc., New York, 1980.

[9] Zienkiewicz, O. C., *The Finite Element Method,* 3rd ed., McGraw-Hill Book Company, New York, 1977.

[10] Popov, E. P., Penzien, J., and Lu, Z. A., "Finite Element Solution for Axisymmetric Shells," *Proc. Am. Soc. Civ. Eng., EM Div.*, 1964, pp. 119–145.

[11] Flügge, W., *Stresses in Shells,* 2nd ed., Springer-Verlag, New York, 1973.

[12] Roark, R. J., and Young, W. C., *Formulas for Stress and Strain,* 5th ed., McGraw-Hill Book Company, New York, 1975.

PROBLEMS

5.1. Assume that the circular cylindrical tank of Example 5.1 (Fig. 5.7) has a monolithic roof slab of 9 in. thickness. The structure is loaded by self-weight only. We have solved this problem in Example 5.4 by the stiffness method. Now solve for the redundant moment which acts at the shell/roof junction, using the flexibility method.

5.2. For the annular plate shown in Fig. P5.2, obtain the rotational stiffnesses (for fixed-edge boundary conditions) corresponding to rotational displacements u_2 and u_4 at the outer and inner edges. Hence, obtain a 2×2 stiffness matrix giving the stiffnesses and carry-over factors for these displacements. Also show that the rotational stiffness for the inner edge when the outer edge is free (and $\nu = 0.0$) is given by

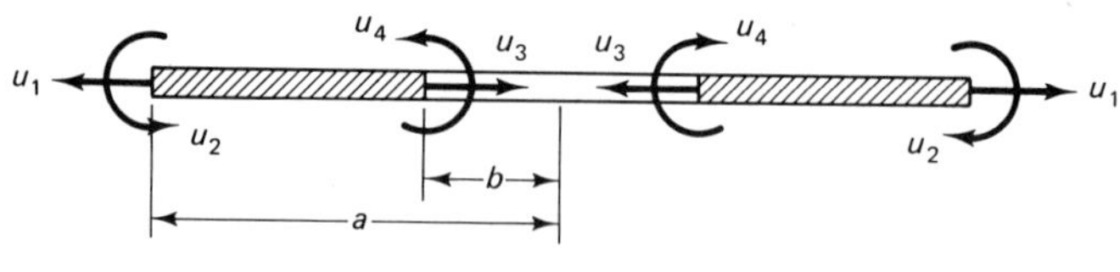

Figure P5.2 Rotational stiffness of an annular plate.

$K(a^2 - b^2)/[b(a^2 + b^2)]$. We shall use this result as the rotational stiffness of a gallery ring beam when solving Intze tank problems in Chapter 7.

5.3. Solve Problem 5.1 using the stiffness method if now the roof is an annular plate with inner radius $b = 8$ ft (see Problem 5.2).

5.4. Figure P5.4 shows a circular concrete tank filled with water and subjected to a uniform roof load of 200 lb/ft^2, including self-weight. The roof is supported by a central column as well as by the tank walls.

Evaluate the redundants acting at the shell/roof-slab junction separately under dead load and water load, using the flexibility approach. From these results and the results of Example 5.2, design the required reinforcing for the wall. Assume that $\nu = 0.1$, and neglect axial deformations of the shell wall.

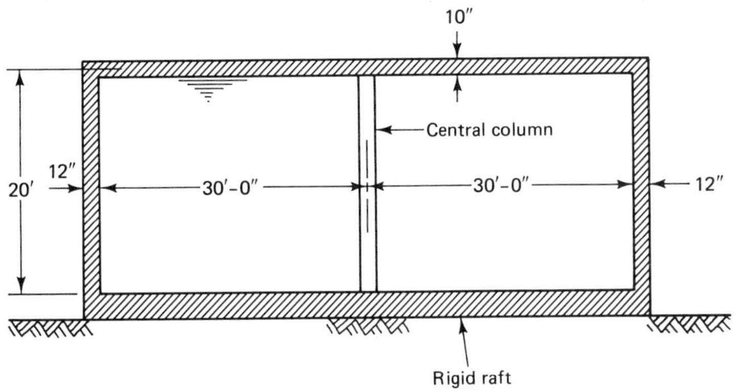

Figure P5.4 Cylindrical concrete tank with a central support, under water and roof loadings.

5.5. Figure P5.5 shows a structure that has a cylindrical wall, dome roof, and an annular plate meeting at a single junction. Explain how this problem can be solved using the stiffness method. Write down all appropriate formulas in generic form. Also, consider how this problem can be solved by the flexibility method, and check that when more than two members meet at a joint, the stiffness method becomes much simpler.

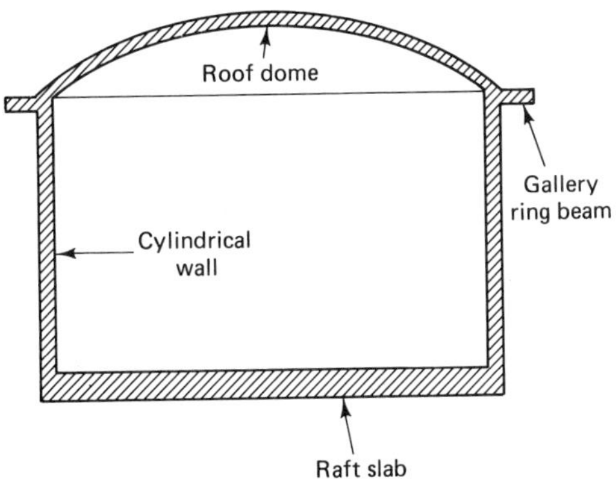

Figure P5.5 Wall/dome/plate junction to be analyzed by the stiffness method.

5.6. Explain step by step how you would derive the 6×6 stiffness matrix of a circular cylindrical shell axisymmetric ring element, assuming that only axisymmetric loads act on the structure. Derive expressions for some of the stiffness coefficients. State which of them will be zero (see also Ref. [10]). Write a computer flowchart to analyze the response of a cylindrical shell structure with varying K (i.e., thickness and modulus of elasticity).

5.7. For the tank shown in Figure P5.4, obtain the moments at the shell/roof-slab junction separately, under dead load and water load, using the stiffness method. Check that the solution matches with the corresponding values found in Problem 5.4. (*Hint:* For the wall, assume FER_1 and FER_2 as zero under dead load. Under water load, find FER_1 and FER_2 for the wall by obtaining the slope at its top using membrane theory and then solving the flexibility problem to make this slope equal to zero.)

6

Membrane Theory for Shells of Revolution of Nonzero Gaussian Curvature: Axisymmetric Loading

6.1 INTRODUCTION

Having derived equations for the membrane behavior of shells of revolution of zero Gaussian curvature in Chapter 4, let us now consider the membrane analysis of shells of revolution of nonzero Gaussian curvature under axisymmetric loading. Such an analysis would be useful for the design of roof domes, water tanks, pressure vessel heads, cooling towers, and other structures that are commonly constructed in practice.

Since these shells have nonzero curvatures in both principal directions, they are also referred to as *double-curvature* shells. Because of their double curvature, the construction of such shells—fixing the formwork to generate the correct geometry of the shell surface—becomes more difficult than for singly curved shells. However, these shells resist normal loads by generation of hoop forces in the two "rings" in their two principal directions, and hence they are generally more efficient in resisting loads than are the singly curved shells. Such is especially true if these shells are synclastic, i.e., if they have positive Gaussian curvatures. Hence such doubly curved synclastic shells can span large distances with very little concrete thickness and/or steel reinforcement and are therefore commonly used for covering large stadiums, planetariums, water tanks, etc.

6.2 DERIVATION OF EQUATIONS

Let us consider a shell of revolution of double curvature as shown in Fig. 6.1. The radius of curvature of the meridian is denoted by r_1, and r_2 is the radius of curvature of the second principal curve, whose plane is at right angles to the meridional

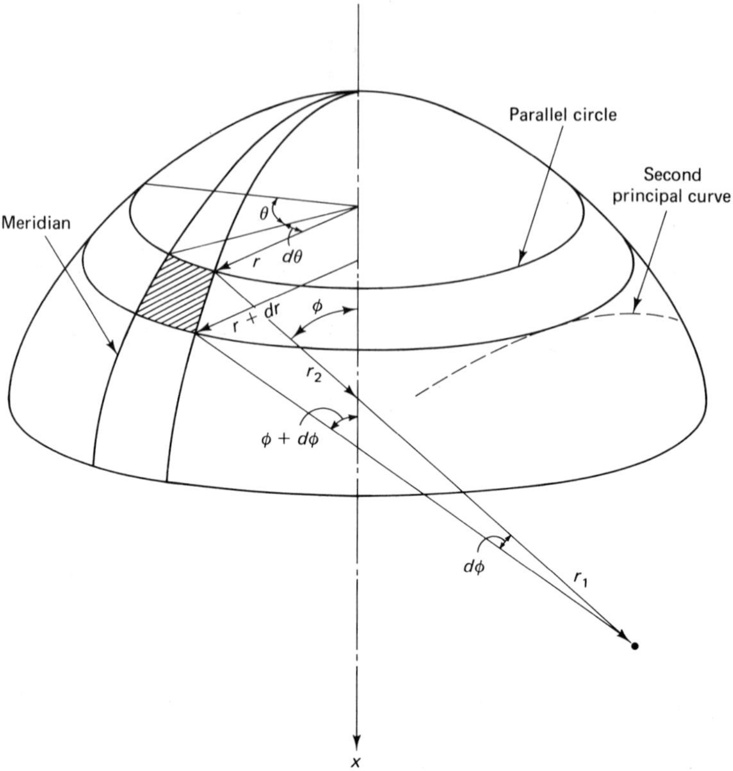

Figure 6.1 Shell of revolution of double curvature, showing the parallel circle, principal curves, and shell element.

plane.* For defining a point on the shell surface, we use as one coordinate the angle ϕ which the radius vector r_1 makes with the axis of revolution. This angle increases as we move along the meridian from the top of the shell. We would normally choose the other coordinate such that it varies as we move along the second principal curve, which is in a plane orthogonal to the meridional plane. However, it is more convenient to use as the second coordinate the angle θ, which defines the location of a point as we move along the parallel circle of radius r. This also permits us to simplify our solution by use of axisymmetry in θ. It should be noted that the plane containing the parallel circle and the plane containing the meridian are, in general, not at right angles to each other.

We can now consider the equilibrium of an element bounded by the sections ϕ, $\phi + d\phi$ and θ, $\theta + d\theta$, shown hatched in Fig. 6.1. An enlarged view of this

*By "meridional plane" and "plane of the second principal curve" we mean, respectively, the plane containing the meridian and the plane containing the second principal curve at a point (see Section 3.1).

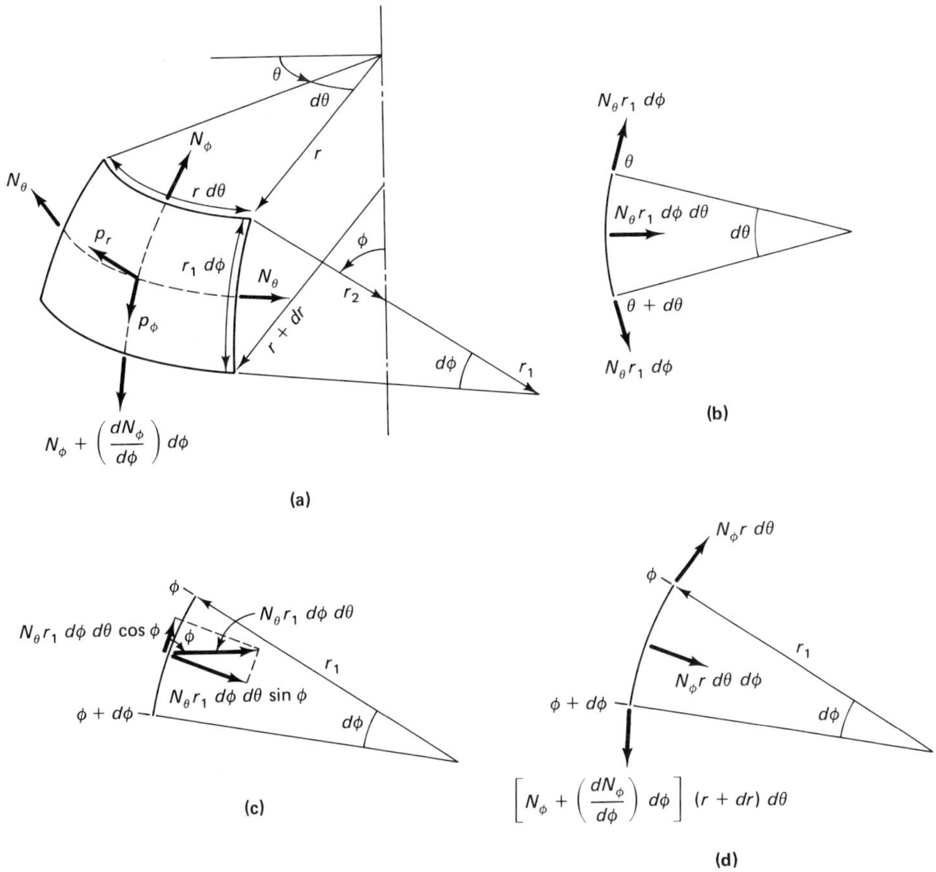

Figure 6.2 Axisymmetric shell element. (a) Geometry and membrane stress resultants acting on element. (b) Hoop forces and their radial resultant. (c) Components of hoop force radial resultant. (d) Normal component due to meridional forces.

element is shown in Fig. 6.2. Because of axisymmetry, N_ϕ and N_θ are the only nonzero membrane stress resultants; the shear $N_{\phi\theta}$ is clearly zero. Also, the distributed loading p_θ in the θ-direction has to be zero, and p_ϕ and p_r in the meridional and the normal directions, respectively, (p_ϕ and p_r being defined per unit area of the middle surface), are the only loads acting on the shell. These loadings, as well as N_ϕ and N_θ, will be constant in θ.

We see from Fig. 6.2(b) that the hoop forces ($N_\theta\, r_1 d\phi$) acting on the faces $\theta = constant$ and $\theta + d\theta = constant$ will have a resultant ($N_\theta\, r_1 d\phi\; d\theta$) acting radially inward in the horizontal plane. This resultant force can be resolved into two components in directions tangential and normal to the meridian, as shown in Fig. 6.2(c). Also, the meridional force ($N_\phi\, r d\theta$) acting on the edge $\phi = constant$ will have a component in the normal direction given by

$$(N_\phi \, r d\theta) \sin \frac{d\phi}{2} \approx (N_\phi \, r d\theta) \frac{d\phi}{2}$$

Similarly, the force $(N_\phi + (dN_\phi/d\phi)\, d\phi)(r + dr)\, d\theta$ on the edge $\phi + d\phi =$ *constant* will have a normal component

$$\left(N_\phi + \frac{dN_\phi}{d\phi} d\phi\right)(r + dr)\, d\theta \frac{d\phi}{2} \approx (N_\phi \, r d\theta) \frac{d\phi}{2}$$

Here we neglect all terms containing a product of more than two differentials ($d\phi$, $d\theta$, or dr), which are small compared to the product ($d\phi \, d\theta$). Hence the total normal component due to the two meridional forces will become, by summation, $N_\phi r d\theta \, d\phi$, as shown in Fig. 6.2(d). We can now write down the equation of equilibrium of forces in the meridional direction. The increment of the N_ϕ forces in the meridional direction from edge $\phi =$ *constant* to edge $\phi + d\phi =$ *constant* is

$$\frac{d}{d\phi}(N_\phi \cdot r d\theta)\, d\phi$$

The term r is kept inside the differentiation sign since it also varies with ϕ. We can thus write

$$\frac{d}{d\phi}(N_\phi \cdot r \, d\theta)\, d\phi - N_\theta \, r_1 d\phi \, d\theta \cos \phi + p_\phi \, r d\theta \, r_1 \, d\phi = 0$$

Dropping $d\phi$ and $d\theta$, we obtain

$$\frac{d}{d\phi}(N_\phi r) - N_\theta \cdot r_1 \cos \phi = -p_\phi \cdot r r_1 \tag{6.1}$$

Considering equilibrium of forces in the direction of the normal, we can write

$$-N_\phi \, r d\theta \, d\phi - N_\theta \cdot r_1 \, d\phi \, d\theta \sin \phi + p_r \cdot r \, d\theta \, r_1 \, d\phi = 0$$

From Fig. 6.2(a) it can easily be seen that $r = r_2 \sin \phi$. Substituting this relation, and dividing by $(r_1 r_2 \sin \phi)\, d\theta d\phi$ throughout in the equation above, we obtain

$$\frac{N_\phi}{r_1} + \frac{N_\theta}{r_2} = p_r \tag{6.2}$$

This equation represents an important relationship between the membrane stress resultants and the normal load. Let us try to interpret it in a physical sense.

We had seen in Chapter 1 that for circular rings of radius a, the radial pressure p_r is resisted by development of hoop forces N such that $N/a = p_r$. In the doubly curved shell, we have two such rings in orthogonal directions: one along the meridian with a hoop force N_ϕ, and the second along the second principal curve with a hoop force which we shall call N_ψ [Fig. 6.3(a)]. The normal load is resisted by the

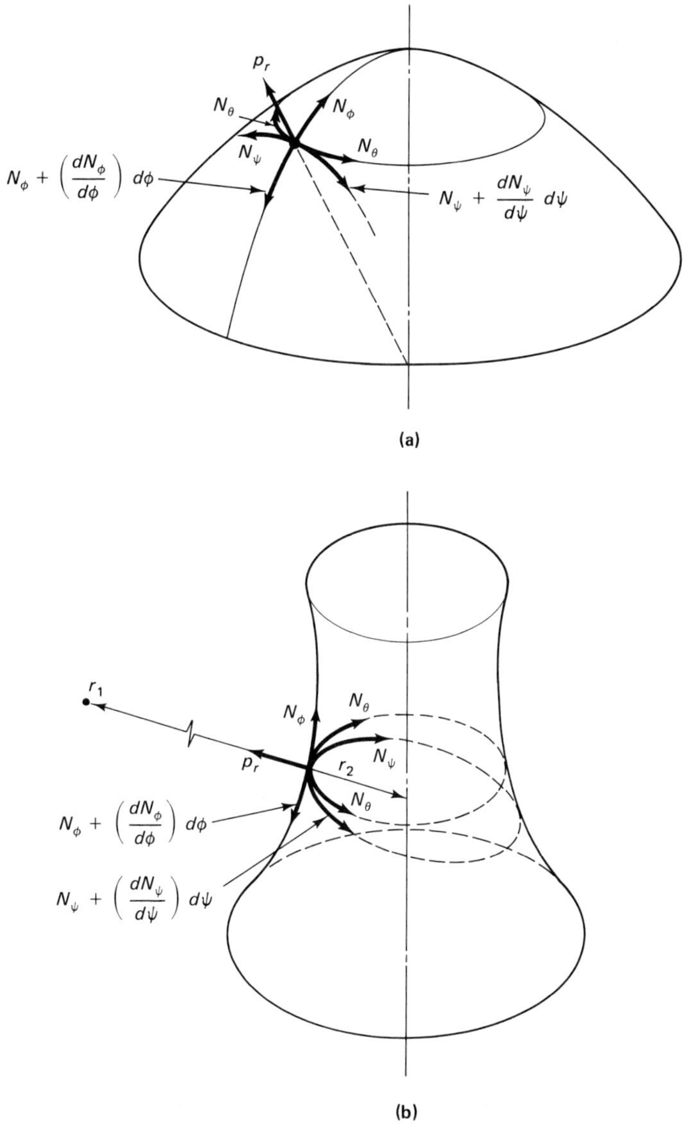

Figure 6.3 Orthogonal "rings" that resist shell loading in doubly curved shells. (a) Shell with positive Gaussian curvature. (b) Shell with negative Gaussian curvature.

contribution from both of these rings, which have radii of curvature r_1 and r_2, respectively. Hence, as for simple rings, we can write

$$\frac{N_\phi}{r_1} + \frac{N_\psi}{r_2} = p_r$$

At their junction, the parallel circle and the second principal curve have a common tangent, and hence* $N_\psi = N_\theta$, which reduces the equation above to Eq. (6.2). It may be noted that for shells of negative Gaussian curvature, the two rings at a point are in opposing directions [Fig. 6.3(b)]. In this case, radii of curvature r_1 and r_2 lay on opposite sides of the meridian. Hence r_1 will have to be taken as negative when using Eq. (6.2). It is also clearly seen from Fig. 6.3(b) that when N_ϕ and N_θ are both positive (tensile), the normal component of N_ϕ will be in the same direction as p_r. Hence the normal component of N_θ (i.e., N_ψ) will have to balance this component of N_ϕ plus p_r, as also seen in Eq. (6.2) when r_1 takes on a negative value.

Now we can obtain from Eq. (6.2)

$$r_1 N_\theta = (p_r \cdot r_1 r_2) - (N_\phi \cdot r_2)$$

Substituting this expression for N_θ into Eq. (6.1) and letting $r = (r_2 \sin \phi)$, we have

$$\frac{d}{d\phi}(N_\phi r_2 \sin \phi) + N_\phi r_2 \cos \phi = (p_r r_1 r_2 \cos \phi) - (p_\phi r_1 r_2 \sin \phi)$$

Multiplying by $\sin \phi$ throughout, we reduce the left-hand side to one derivative. Hence we can write

$$\frac{d}{d\phi}(N_\phi r_2 \sin^2 \phi) = (p_r \cdot r_1 r_2 \sin \phi \cos \phi) - (p_\phi \cdot r_1 r_2 \sin^2 \phi) \tag{6.3}$$

Upon integrating, we obtain†

$$N_\phi r_2 \sin^2 \phi = \int (p_r \cos \phi - p_\phi \sin \phi) r_1 r_2 \sin \phi \, d\phi + C$$

or

$$N_\phi = \frac{1}{r_2 \sin^2 \phi} \int (p_r \cos \phi - p_\phi \sin \phi) r_1 r_2 \sin \phi \, d\phi + \frac{C}{r_2 \sin^2 \phi} \tag{6.4}$$

The equation above can be used to calculate stress resultant N_ϕ in a shell, according to the following procedure. Knowing the equation of the meridian $r = f(\phi)$, we can calculate r_1 and r_2 as functions of ϕ, as explained in Chapter 3. Then, for a given loading, the integral on the right-hand side of Eq. (6.4) can be evaluated. The constant C is determined from the value of N_ϕ prescribed at the boundary. Here again we have the freedom of prescribing N_ϕ at only one edge, say $\phi = \phi_o$, and we must accept whatever N_ϕ we get at the other boundary, say $\phi = \phi_1$.

*If ψ is an angular coordinate along the second principal curve, then at a given point $r_2\, d\psi \approx r\, d\theta$. Considering a unit length along the meridian, the normal component of N_ψ will be $(N_\psi \cdot 1 \cdot d\psi)$. Equating this to the normal component of N_θ given by $(N_\theta \cdot d\theta) \sin \phi$, and substituting $r_2\, d\psi \approx r\, d\theta$ and $r = r_2 \sin \phi$, we also get $N_\psi \approx N_\theta$. Coordinate θ is preferred over ψ since it is easier to visualize, and because we can take advantage of axisymmetry in θ to eliminate the shear terms in our equations.

†If we here substitute $r_1\, d\phi = ds$, $\phi = \alpha = constant$, and $s = r_2 \tan \alpha$, we can obtain the equation for N_s in a conical shell, as derived in Chapter 4.

Having calculated N_ϕ, we can obtain N_θ from Eq. (6.2). Membrane stress resultants in a given shell can thus be determined easily. Let us summarize the two formulas for N_ϕ and N_θ:

$$N_\phi = \frac{1}{r_2 \sin^2 \phi} \int (p_r \cos \phi - p_\phi \sin \phi) r_1 r_2 \sin \phi \, d\phi + \frac{C}{r_2 \sin^2 \phi} \tag{6.4}$$

$$\frac{N_\phi}{r_1} + \frac{N_\theta}{r_2} = p_r \tag{6.2}$$

6.3 SPHERICAL DOMES: SELF-WEIGHT

As an example of the application of the foregoing derivations, let us obtain expressions for the membrane stresses in a spherical dome under its own weight, which we define as p per unit area of the surface (Fig. 6.4). We know that for the dome $r_1 = r_2 = a$ (this can be verified as explained in Chapter 3, by taking the equation of the meridian as $r^2 + x^2 = a^2$). Resolving the load p into components tangential and normal to the shell, we have

$$p_r = -p \cos \phi \qquad \text{and} \qquad p_\phi = p \sin \phi \tag{6.5a,b}$$

Substituting these into Eq. (6.4) and simplifying, we obtain

$$N_\phi = -\frac{pa}{\sin^2 \phi} \int (\cos^2 \phi + \sin^2 \phi) \sin \phi \, d\phi + \frac{C}{a \sin^2 \phi}$$

or, after integrating,

$$N_\phi = \frac{pa^2 \cos \phi + C}{a \sin^2 \phi}$$

To determine C, we note that the shell does not contain an opening at its top,

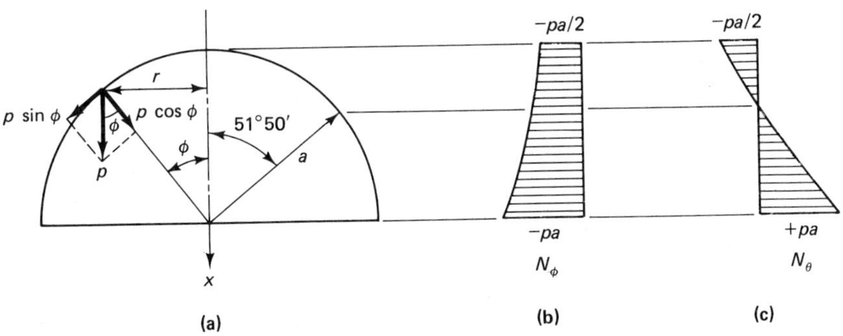

Figure 6.4 Spherical dome under self-weight. (a) Configuration and coordinate system. (b) Plot of resultant N_ϕ versus x. (c) Plot of resultant N_θ versus x.

and at its bottom edge we presume that supports will be provided. We also note, however, that the expression for N_ϕ above becomes infinite at $\phi = 0$. Since we do not expect the shell to have infinite stresses under its own weight, stresses for this case must be finite at $\phi = 0$; i.e., we have a "finiteness" condition at $\phi = 0$ instead of a boundary condition. For any finite value of C, N_ϕ will be infinite at $\phi = 0$ except for when we have $C = (-pa^2)$, in which case we have a chance of finiteness since N_ϕ then becomes (0/0), i.e., indeterminate at $\phi = 0$. Hence we assume that $C = -pa^2$, which gives us

$$N_\phi = -pa\left(\frac{1 - \cos\phi}{\sin^2\phi}\right)$$

Now taking the limit by L'Hôpital's rule, we can check that

$$\begin{aligned}\lim_{\phi\to 0} N_\phi &= \lim_{\phi\to 0}\left[-\frac{pa(1 - \cos\phi)}{\sin^2\phi}\right] \\ &= \lim_{\phi\to 0}\frac{-pa\sin\phi}{2\sin\phi\cos\phi} \\ &= -\frac{pa}{2} \\ &= \text{finite}\end{aligned}$$

Hence our assumption of $C = -pa^2$ is correct. Now taking $\sin^2\phi = (1 - \cos\phi)(1 + \cos\phi)$, we can rewrite the solution for N_ϕ as

$$N_\phi = \frac{-pa}{1 + \cos\phi} \tag{6.6a}$$

Substituting this expression into Eq. (6.2), with $r_1 = r_2 = a$, we have

$$N_\theta = -pa\left(\cos\phi - \frac{1}{1 + \cos\phi}\right) \tag{6.6b}$$

Variations of N_ϕ and N_θ as ϕ changes from 0 to 90° are shown in Fig. 6.4(b) and (c). It may be noted that N_ϕ is compressive throughout, whereas N_θ becomes tensile in the lower portion of the shell. By letting the expression in the parentheses of Eq. (6.6b) equal zero, we find that N_θ is zero at $\phi = 51.83°$. For concrete dome roofs, we would like to have no tensile stresses in the structure, since concrete is weak in tension. Hence such domes are usually constructed as shallow domes, with the bottom edge angle ϕ_1 being less than 45°. On the other hand, for steel pressure vessels, the tensile stresses can easily be accommodated; however, compressive stresses may give rise to buckling problems. Hence domes of pressure vessels are not necessarily shallow. Of course, the self-weight stresses in pressure vessel domes are not significant compared to those due to design internal pressure, the effect of which we shall see later.

6.4 DEFINITE INTEGRAL APPROACH

For general distributed loads and for shells with openings at the top, determination of the constant C in Eq. (6.4), as illustrated in the preceding section, can become rather involved. As an alternative, we can integrate both sides of Eq. (6.3), from ϕ_o to ϕ, and eliminate the constant C altogether, $\phi = \phi_o$ being a boundary of the shell. We would then obtain

$$[N_\phi r_2 \sin^2 \phi]_{\phi_o}^{\phi} = \int_{\phi_o}^{\phi} (p_r \cos \phi - p_\phi \sin \phi) r_1 r_2 \sin \phi \, d\phi \qquad (6.7)$$

On both sides of the equation above, it is understood that ϕ is a dummy variable, which we could have indicated by any other variable name.

We can now use Eq. (6.7) to obtain N_ϕ directly for the problem of a spherical dome under self-weight considered in the preceding section. Since the shell is closed at its top, we take the lower integration limit to be $\phi_o = 0$. Substituting the limits of integration on the left-hand side of Eq. (6.7), we note that at the lower limit $\phi = 0$, the expression vanishes since $\sin \phi$ is zero. Also, substituting values of r_1, r_2 and p_r, p_ϕ, as before, we find

$$N_\phi \cdot a \sin^2 \phi - 0 = -pa^2 \int_0^{\phi} \sin \phi \, d\phi$$

or

$$N_\phi = \frac{-pa}{\sin^2 \phi} [-\cos \phi]_0^{\phi} = \frac{-pa}{1 + \cos \phi}$$

which is the same result as that obtained previously.

We can thus either use Eq. (6.4) or (6.7) for obtaining N_ϕ in a given problem. We shall generally use Eq. (6.7) in the problems that follow, although Eq. (6.4) can be used with equal ease.

6.5 SPHERICAL DOME: LIVE LOAD OR SNOW LOAD

Live loads and snow loads, by their very nature of application, are uniformly distributed over the plan area of the shell surface, as opposed to self-weight loading, which is uniformly distributed over the surface area (assuming constant thickness). If q is the loading defined on the plan area, then referring to Fig. 6.5(a), we can obtain an equivalent load p defined on the surface area from the relation

$$p \cdot a \, d\phi = (q \cdot a \, d\phi) \cos \phi$$

Hence

$$p = q \cos \phi$$

Thus, from Eqs. (6.5a,b), we have

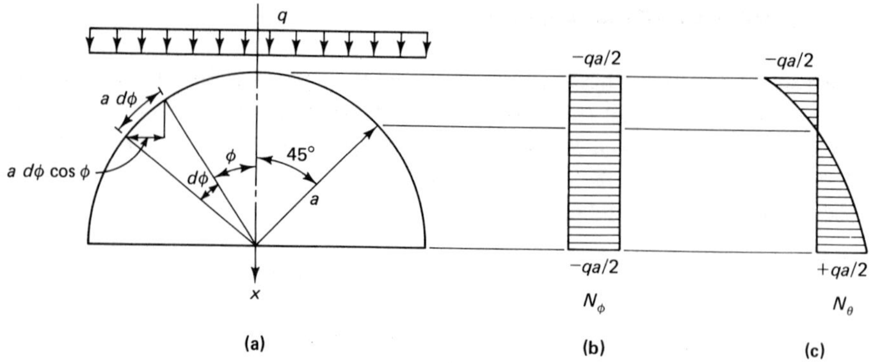

Figure 6.5 Spherical dome under live load uniformly distributed over the plan area. (a) Coordinate system. (b) Plot of N_ϕ versus x. (c) Plot of N_θ versus x.

$$p_r = -q\cos^2\phi \quad \text{and} \quad p_\phi = q\sin\phi\cos\phi \tag{6.8a,b}$$

Substituting these values into Eq. (6.7), we obtain

$$N_\phi a\sin^2\phi - 0 = -qa^2\int_0^\phi (\cos^3\phi + \cos\phi\sin^2\phi)\sin\phi\, d\phi$$

Hence

$$N_\phi = \frac{-qa}{\sin^2\phi}\int_0^\phi \frac{\sin 2\phi}{2}\, d\phi$$

or

$$N_\phi = \frac{-qa}{2} \tag{6.8c}$$

Substituting this into Eq. (6.2) and simplifying, we obtain

$$N_\theta = \frac{-qa}{2}\cos 2\phi \tag{6.8d}$$

Curves for N_ϕ and N_θ are shown in Fig. 6.5(b) and (c). Again, N_ϕ is compressive throughout, from $\phi = 0°$ to $\phi = 90°$, while N_θ is tensile beyond $\phi = 45°$. As discussed previously, if a concrete dome is constructed with a base angle less than 45°, then under membrane conditions we will have compressive stresses throughout, as desired.

6.6 METHOD OF SECTIONS

Let us consider a frustum of a shell carrying distributed loads p_r and p_ϕ. The frustum, shown in Fig. 6.6, is defined by the upper boundary $\phi = \phi_o$ and a section $\phi = constant$ where we desire to evaluate N_ϕ. We can see that the net vertical

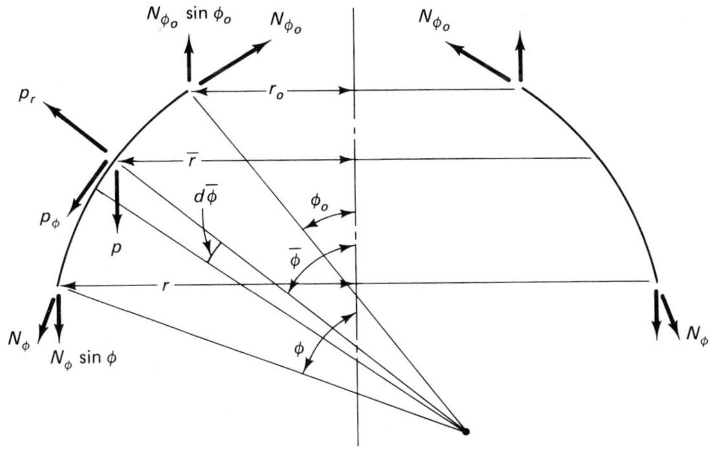

Figure 6.6 Frustum section of a dome, showing the geometry and internal and applied loads.

load intensity at a point is given by $p = (p_\phi \sin \phi - p_r \cos \phi)$. So the total vertical load on the shell, between ϕ_o and ϕ, is given by

$$\int_{\phi_o}^{\phi} (p_{\bar{\phi}} \sin \bar{\phi} - p_{\bar{r}} \cos \bar{\phi}) \cdot (2\pi \bar{r} \bar{r}_1 \, d\bar{\phi})$$

where all bar superscripts indicate corresponding quantities at $\phi = \bar{\phi}$, $\bar{\phi}$ being a dummy variable. The vertically downward force due to edge forces N_ϕ and N_{ϕ_o} can be written as

$$(N_\phi \sin \phi) 2\pi r - (N_{\phi_o} \sin \phi_o) 2\pi r_o$$

This force, together with the downward pressure load, must satisfy equilibrium, thus giving us

$$(N_\phi \sin \phi)\, 2\pi r - (N_{\phi_o} \sin \phi_o) 2\pi r_o + \int_{\phi_o}^{\phi} (p_{\bar{\phi}} \sin \bar{\phi} - p_{\bar{r}} \cos \bar{\phi}) 2\pi \bar{r} \bar{r}_1 \, d\bar{\phi} = 0$$

Substituting $r = r_2 \sin \phi$, etc., canceling the term 2π, and dropping bars for the dummy variables, we can rewrite the above as

$$[N_\phi r_2 \sin^2 \phi]_{\phi_o}^{\phi} = \int_{\phi_o}^{\phi} (p_r \cos \phi - p_\phi \sin \phi) r_1 r_2 \sin \phi \, d\phi$$

This equation is the same as Eq. (6.7). Thus we see that N_ϕ can be calculated simply by taking a section $\phi = \textit{constant}$ and considering the equilibrium of vertical forces acting on the freebody above (or below) the section.

Using this approach, calculation of N_ϕ for the case of a snow or live load acting on a dome of any shape, not necessarily spherical, becomes very easy. As an example, consider a dome as shown in Fig. 6.7, with a load q uniformly distributed over its plan area. Then considering the equilibrium of forces on the freebody above the section ϕ = *constant*, we have

$$N_\phi 2\pi r \sin\phi + q\pi r^2 = 0$$

Hence

$$N_\phi = \frac{-qr}{2\sin\phi} = \frac{-qr_2}{2} \tag{6.9}$$

Substituting $r_2 = a$ in the above, we obtain Eq. (6.8c) of Section 6.5 for the spherical dome. An expression for N_θ for the *general* dome can be written by substituting Eq. (6.9) into Eq. (6.2). This direct approach of obtaining N_ϕ gives a more enhanced physical insight into the shell behavior than do the mathematical formulas of Eq. (6.4) or (6.7) alone. Thus we shall be using this approach in several instances in the following sections.

6.7 SINGULARITY PROBLEMS

In Eq. (6.4) for N_ϕ, both terms on the right-hand side become infinite when $\phi = 0$. Indeed, however, for the case of self-weight loading on a spherical dome, we evaluated C to yield a finite N_ϕ at $\phi = 0$ (Section 6.3). Let us now consider what happens when we have no distributed loads on the shell. Substituting $p_r = p_\phi = 0$ into Eq. (6.4), we have

$$N_\phi = \frac{C}{r_2 \sin^2\phi}$$

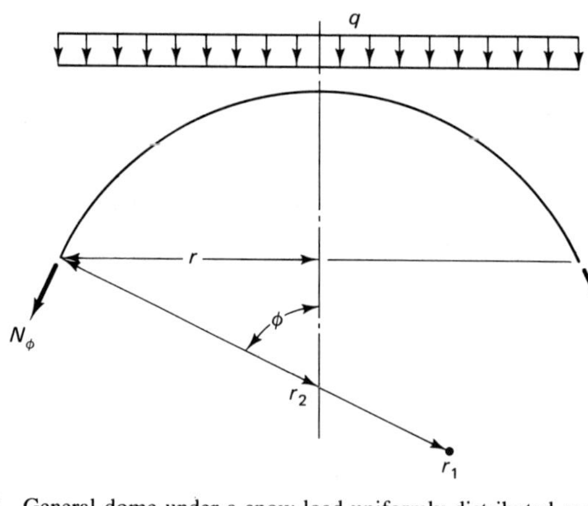

Figure 6.7 General dome under a snow load uniformly distributed over the plan area.

On any section $\phi = constant$ [Fig. 6.8(a)], the summation of the horizontal components of N_ϕ around the circumference will balance to zero. However, the vertical components will have a resultant force V, given by

$$V = (N_\phi \sin \phi)2\pi r = \frac{C}{r_2 \sin^2 \phi} 2\pi r_2 \sin^2 \phi = 2\pi C$$

This force V is a constant for any value of ϕ, and for equilibrium we must have an equal and opposite concentrated load at the apex, as shown. Hence we can conclude that whenever we have a concentrated load V at the apex of the shell (positive acting upward), then the corresponding value of N_ϕ is given by

$$N_\phi = \frac{V/2\pi}{r_2 \sin^2 \phi} \tag{6.10}$$

Of course, the corresponding value for N_θ is given by Eq. (6.2).

In the case of self-weight on a spherical dome, as in Section 6.3, we had used the constant C to make stresses finite at $\phi = 0$. If this dome also carries an upward concentrated load V at the apex, we will have to superimpose on the solution of Eq. (6.6a) another solution due to V, given by Eq. (6.10). If we are using the definite integral of Eq. (6.7), then on the right-hand side, we must add a term $V/2\pi$ whenever a concentrated load V exists.

The arguments above also apply if we have a line (ring) load at $\phi = \alpha$, with a total upward resultant V, as shown in Fig. 6.8(b). In this case it is obvious, by considering equilibrium of vertical forces, that in the expression for N_ϕ with $\phi \geqslant \alpha$ we would add a term given by Eq. (6.10). However, for a section $0 \leqslant \phi < \alpha$, we shall not add this term since there is no concentrated load on sections above $\phi = \alpha$.

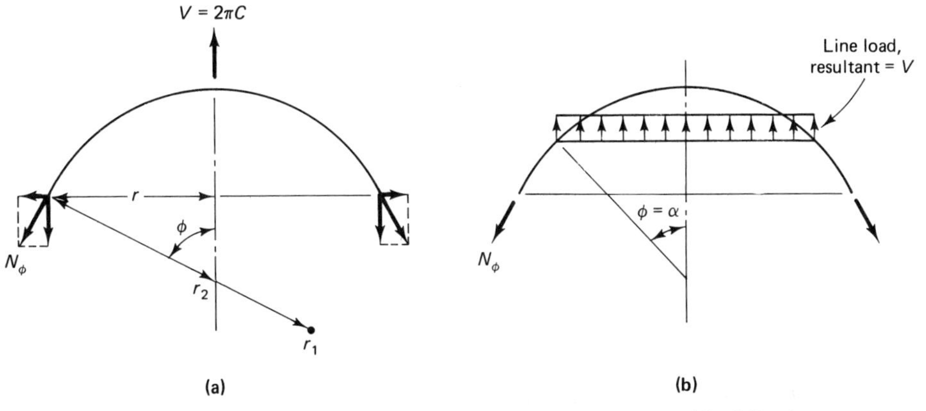

Figure 6.8 Singularity problems for domes. (a) Dome with concentrated load V at its apex. (b) Dome with a line load that has a resultant equal to V.

6.8 OTHER PROBLEMS OF SPHERICAL DOME ROOFS

Dome with a Skylight

Let us find the stress resultants in a spherical dome under self-weight, which has a skylight at its top [Fig. 6.9(a)]. We shall assume that p is the weight per unit area of the surface of the shell. In addition, there will be a line load of P per unit length at the top of the dome, due to the weight of the skylight, hood, etc. To find N_ϕ, we substitute $r_1 = r_2 = a$, $\phi_o = \alpha$, $p_\phi = p \sin \phi$, and $p_r = -p \cos \phi$ into Eq. (6.7), to obtain

$$[N_\phi a \sin^2 \phi]_\alpha^\phi = -pa^2 \int_\alpha^\phi \sin \phi \, d\phi = -pa^2(\cos \alpha - \cos \phi) \quad (6.11)$$

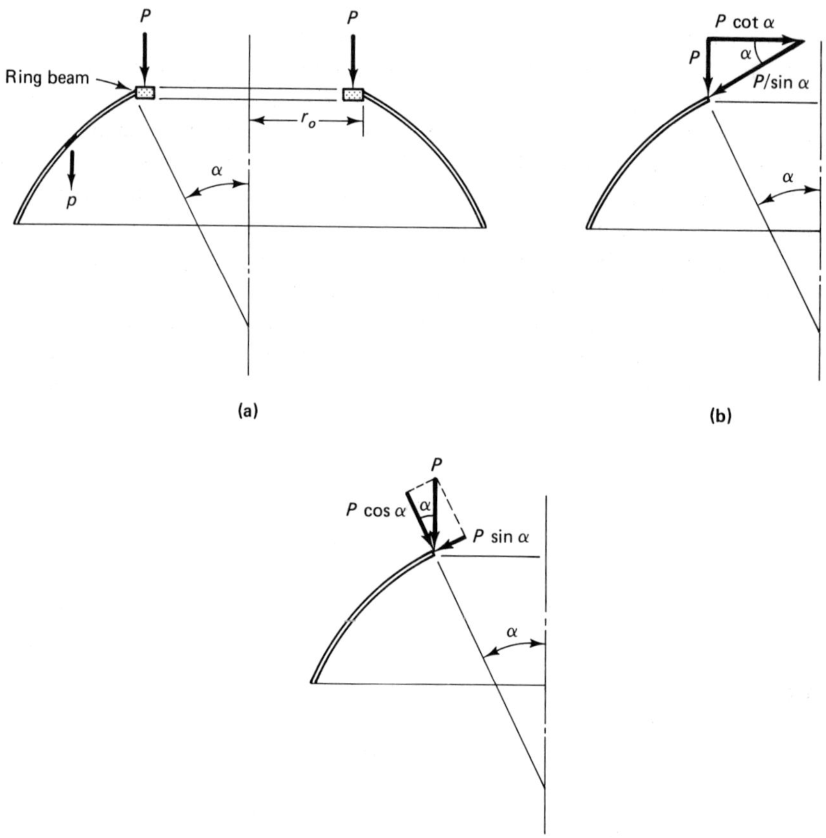

Figure 6.9 Spherical dome with a skylight. (a) Dome under self-weight with a skylight load of P around the top circumference. (b) Horizontal and meridional components of the load P. (c) Normal and meridional components of P.

If the line load of the skylight is first assumed to be zero, N_ϕ will also be zero at the free edge $\phi = \alpha$. Hence the product of terms for the lower limit on the left-hand side of the equation above will be zero. So we shall obtain

$$N_\phi = -pa \frac{\cos\alpha - \cos\phi}{\sin^2\phi}$$

To this resultant we must add the value of N_ϕ due to the line load from the skylight, which can be obtained by substituting $V = -2\pi r_o P$ into Eq. (6.10). Thus the complete solution for N_ϕ becomes

$$N_\phi = -pa \frac{\cos\alpha - \cos\phi}{\sin^2\phi} - \frac{Pr_o}{a\sin^2\phi}$$

or, since $r_o = a \sin\alpha$, we have

$$N_\phi = -pa \frac{\cos\alpha - \cos\phi}{\sin^2\phi} - \frac{P\sin\alpha}{\sin^2\phi}$$

From Eq. (6.2), we then obtain N_θ as

$$N_\theta = -pa\left(\cos\phi - \frac{\cos\alpha - \cos\phi}{\sin^2\phi}\right) + \frac{P\sin\alpha}{\sin^2\phi}$$

As an alternative approach, we can also obtain N_ϕ by resolving P suitably into two components, one equal to $P/\sin\alpha$ along the meridian at $\phi = \alpha$, and the other equal to $P\cot\alpha$ in the horizontal direction, as shown in Fig. 6.9(b). Hence in Eq. (6.11), on the left-hand side, we can substitute the boundary condition $N_\phi = -(P/\sin\alpha)$ at the lower limit $\phi = \alpha$, and thus obtain

$$N_\phi a \sin^2\phi + \frac{Pa\sin^2\alpha}{\sin\alpha} = -pa^2(\cos\alpha - \cos\phi)$$

or

$$N_\phi = -pa \frac{\cos\alpha - \cos\phi}{\sin^2\phi} - \frac{P\sin\alpha}{\sin^2\phi}$$

which is the same expression as that obtained earlier. The horizontal force $P\cot\alpha$ is resisted by hoop compression in a ring beam which must be provided at the top. This hoop compression in the ring beam would be equal to $Pr_o\cot\alpha$.

It may be explained here that we could have also resolved P into meridional and normal components $P\sin\alpha$ and $P\cos\alpha$, respectively, as shown in Fig. 6.9(c). The normal component $P\cos\alpha$ is a shear edge load on the shell which will give rise to shears and bending moments within the shell. The shell cannot resist these resultants by membrane action or by bending action without considerable deformations. Hence to resist the line load P efficiently, with minimal deformation, we resolve this load into components as shown in Fig. 6.9(b), and provide a ring beam at the top to take the horizontal component. Of course, there will arise some problems due to incompatibility of displacements at the junction of the shell and the ring beam, which we shall discuss later.

Dome with Varying Thickness

In practice, because of bending moments that develop near the supports, we have to increase the dome thickness locally, for a short distance near the lower edge. The membrane stress resultants can be readily calculated in such cases, where now the self-weight loading will vary with ϕ (see Ref. [1]).

Let us consider a spherical dome (Fig. 6.10), which has a constant thickness $t = t_1$ for $0 \leq \phi \leq \phi_o$. For the region $\phi_o < \phi \leq \phi_1$ (ϕ_1 is at the supported boundary), the thickness increases linearly with ϕ, from t_1 at ϕ_o to t_2 at ϕ_1. For the self-weight due to thickness t_1, from $\phi = 0$ to $\phi = \phi_1$, the stresses can be calculated by substituting $p = \gamma t_1$ into Eqs. (6.6a,b) derived previously. Because of the varying thickness, additional stresses will be incurred only over the region $\phi_o < \phi \leq \phi_1$. Since the thickness increases linearly in this portion, the additional load will also increase linearly with ϕ. We denote this rate of increase of load per radian as q, and it is given by

$$q = \frac{\gamma(t_2 - t_1)}{\phi_1 - \phi_o}$$

where ϕ_o and ϕ_1 are in radians. Considering the freebody above a section defined by the coordinate ϕ (where $\phi > \phi_o$), the total additional load due to the thickness increase is given by

$$\begin{aligned} W &= \int_{\phi_o}^{\phi} q(\phi - \phi_o)\, 2\pi(a \sin\phi) \cdot a\, d\phi \\ &= 2\pi a^2 q \int_{\phi_o}^{\phi} (\phi \sin\phi - \phi_o \sin\phi)\, d\phi \end{aligned}$$

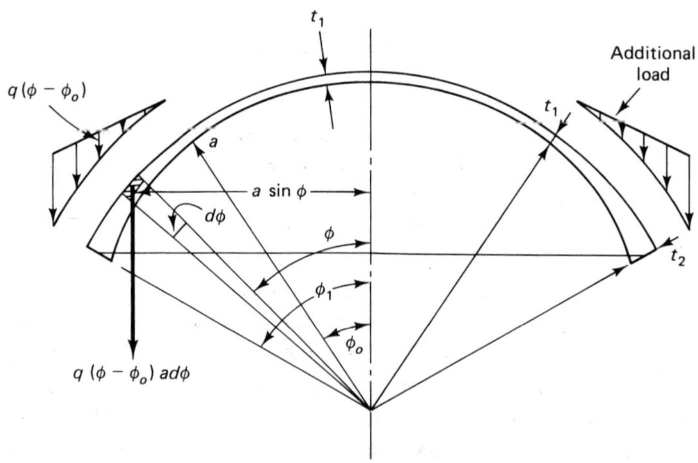

Figure 6.10 Dome with varying thickness under self-weight.

The first term in the integral on the right-hand side above can be evaluated using *integration by parts;* i.e., $\int u\,dv = uv - \int v\,du$, wherein we take $u = \phi$ and $dv = \sin\phi$. Hence

$$W = 2\pi a^2 q[\sin\phi - \sin\phi_o - (\phi - \phi_o)\cos\phi]$$

Thus, knowing W, the increase in the meridional stress, ΔN_ϕ, due to the increased thickness over $\phi_o < \phi \leqslant \phi_1$, is given by

$$\Delta N_\phi \sin\phi\, 2\pi(a\sin\phi) = W$$

or

$$\Delta N_\phi = \frac{qa}{\sin^2\phi}[\sin\phi - \sin\phi_o - (\phi - \phi_o)\cos\phi]$$

Resultant N_θ is then obtained from Eq. 6.2.

6.9 MEMBRANE DISPLACEMENTS

We shall denote v and w to be the deflections along the tangent and the normal to the meridian, respectively, as shown in Fig. 6.11(a). The third displacement, u, along the θ-direction, is absent due to axisymmetric loading. Referring to Fig. 6.11(b), we see that because of deflections v and $v + (dv/d\phi)\,d\phi$, of the points ϕ and $\phi + d\phi$ on the shell, the increase in the length $r_1 d\phi$ of an infinitesimal element is equal to $(dv/d\phi)\,d\phi$. Also, due to deflection w, an additional increase in the length of this element will be equal to $wd\phi$, as seen from Fig. 6.11(c). Hence the meridional strain ε_ϕ is given by

$$\varepsilon_\phi = \frac{(dv/d\phi)\,d\phi + w\,d\phi}{r_1\,d\phi}$$

or

$$\varepsilon_\phi = \frac{(v)^\cdot + w}{r_1} \tag{6.12a}$$

where $(\cdot)^\cdot$ denotes $d(\cdot)/d\phi$. Now, from Fig. 6.11(a), we see that the total radially outward deflection in the plane of the parallel circle is given by the sum of the components of v and w, as $(v\cos\phi + w\sin\phi)$. Hence, as for rings (Section 1.4), the circumferential strain ε_θ will be obtained by dividing this deflection by the radius of the parallel circle, r. Therefore,

$$\varepsilon_\theta = \frac{v\cos\phi + w\sin\phi}{r}$$

or substituting $r = r_2\sin\phi$, we obtain

$$\varepsilon_\theta = \frac{v\cot\phi + w}{r_2} \tag{6.12b}$$

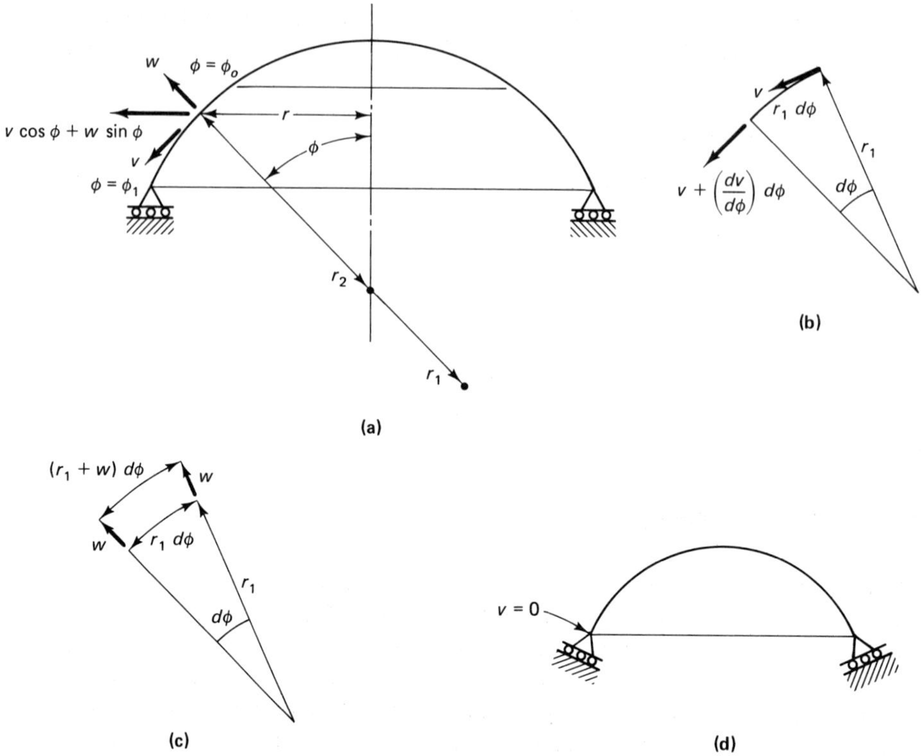

Figure 6.11 Membrane displacements of a dome. (a) Normal and tangential displacements w and v. (b) Meridional strain due to v. (c) Meridional strain due to w. (d) Dome supported on rollers such that displacement v is zero at the edge.

Strains ε_ϕ and ε_θ are related to the stress resultants N_ϕ and N_θ by the following simple relations:

$$\varepsilon_\phi = \frac{1}{Et}(N_\phi - \nu N_\theta) \tag{6.13a}$$

and

$$\varepsilon_\theta = \frac{1}{Et}(N_\theta - \nu N_\phi) \tag{6.13b}$$

Knowing N_ϕ and N_θ for a given problem, we can obtain ε_ϕ and ε_θ as functions of ϕ from these relationships, and then from Eqs. (6.12a,b) we can evaluate v and w. For this purpose we multiply Eq. (6.12a) by r_1/r_2 and subtract Eq. (6.12b), to obtain

$$\frac{(v)^{\bullet} - v \cot \phi}{r_2} = \left(\frac{r_1}{r_2}\right)\varepsilon_\phi - \varepsilon_\theta$$

Multiplying both sides by $r_2 \csc \phi$, we have

$$(v)^{\cdot} \csc \phi - v \cot \phi \csc \phi = \frac{r_2}{\sin \phi}\left(\frac{r_1}{r_2}\varepsilon_\phi - \varepsilon_\theta\right)$$

or

$$(v \csc \phi)^{\cdot} = f_1(\phi)$$

where

$$f_1(\phi) = \frac{r_2}{\sin \phi}\left(\frac{r_1}{r_2}\varepsilon_\phi - \varepsilon_\theta\right)$$

The function $f_1(\phi)$ can be found when N_ϕ and N_θ are known. Integrating, we get the following expression for v:

$$v = \sin \phi \int f_1(\phi)\, d\phi + C_2 \sin \phi \tag{6.14a}$$

where C_2 is a constant. Then substituting into Eq. (6.12b), we obtain the following expression for w:

$$w = r_2\varepsilon_\theta - \cos \phi \int f_1(\phi)\, d\phi - C_2 \cos \phi \tag{6.14b}$$

with

$$f_1(\phi) = \frac{r_2}{\sin \phi}\left(\frac{r_1}{r_2}\varepsilon_\phi - \varepsilon_\theta\right) \tag{6.14c}$$

The constant C_2 is determined from the boundary condition at either $\phi = \phi_o$ or $\phi = \phi_1$. Thus we have the freedom of prescribing v or w at only one of the boundaries, and we must permit any other displacements yielded by our analysis; otherwise, incompatibility problems will arise. We may also choose to provide a combination of v and w as a boundary condition. For example, if we have a vertical rigid support at $\phi = \phi_1$, we can prescribe that the vertical deflection given by $(v \sin \phi - w \cos \phi)$ equals zero at $\phi = \phi_1$, and thus obtain C_2. However, since this does not mean that either v or w will be zero, we must provide rollers at the bottom edge, as shown in Fig. 6.11(a), thus permitting any horizontal movement (i.e., nonzero v and w). Also, the top edge must then be kept completely free, to allow for whatever values of v and w we obtain there. Alternatively, if we prescribe $v = 0$ at the lower edge, we must provide rollers along inclined planes, as shown in Fig. 6.11(d), to allow the displacement w. In all such cases the values of N_ϕ and N_θ are as determined by Eqs. (6.4) and (6.2).

If, however, we choose not to prescribe N_ϕ at any of the edges $\phi = \phi_o$ or $\phi = \phi_1$, the constant C in Eq. (6.4) will not be determined and will appear in Eqs. (6.14a–c). We shall then have two constants in Eqs. (6.14a,b), and hence the freedom of prescribing a total of two displacement conditions at the boundaries. In this case, we can prescribe $v = 0$ at both $\phi = \phi_o$ and $\phi = \phi_1$ (by providing rollers along inclined planes) to determine C and C_2, and thus obtain v and w for a given p_ϕ and p_r. Then, knowing C, we can determine N_ϕ and N_θ from Eqs. (6.4) and (6.2). Similar situations were encountered for bars (Chapter 1) and shells of zero Gaussian curvature (Chapter 4), where we could prescribe displacements or a combination of displacement and force boundary conditions at the two edges.

In practice, we are generally interested in the horizontal deflection Δ_H and rotation of the tangent to the meridian, χ, at the boundaries, as these quantities are required for satisfying compatibility relations for ring beams and supports. Let χ be defined positive clockwise (when considering a meridian on the left-hand side of the axis of revolution). We have, as for rings,

$$\Delta_H = r\varepsilon_\theta = \frac{r}{Et}(N_\theta - \nu N_\phi) \tag{6.15a}$$

In Fig. 6.12(a) we see that if w and $w + (dw/d\phi)\, d\phi$ are the normal deflections of the neighboring points A and B on the shell, the clockwise rotation of the tangent at A is given by $(dw/d\phi)(1/r_1)$. Also, we see in Fig. 6.12(b) that due to the deflection v of point A, the tangent rotates an amount v/r_1. Since this rotation is anticlockwise for positive v, we can write

$$\chi = \frac{(w)^{\cdot} - v}{r_1} \tag{6.15b}$$

By differentiating Eq. (6.14b), we have

$$(w)^{\cdot} = (r_2)^{\cdot}\, \varepsilon_\theta + (\varepsilon_\theta)^{\cdot}\, r_2 - \cos\phi\, f_1(\phi) + \sin\phi \int f_1(\phi)\, d\phi + C_2 \sin\phi$$

The last two terms above add up to the displacement v given by Eq. (6.14a). Hence

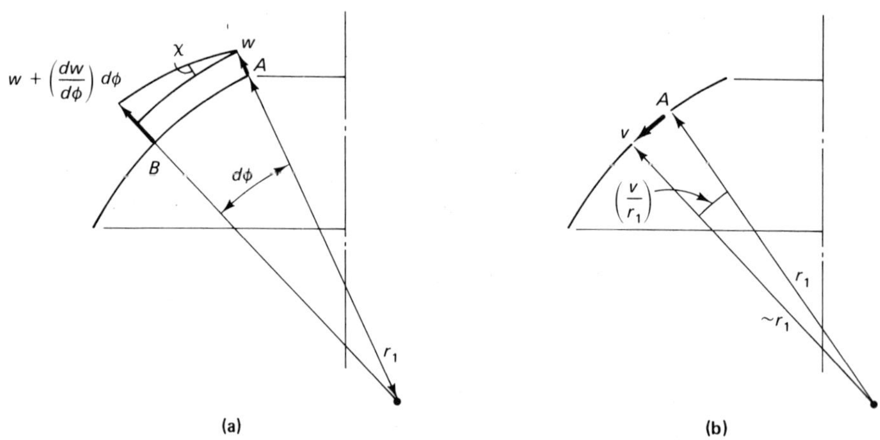

Figure 6.12 Displacement and rotation of shell meridian. (a) Rotation caused by variation in w. (b) Rotation caused by displacement v.

$$(w)^{\cdot} - v = (r_2)^{\cdot}\, \varepsilon_\theta + (\varepsilon_\theta)^{\cdot}\, r_2 - \cos\phi\, f_1(\phi)$$

or from Eq. (6.15b), we have

$$\chi = \frac{(r_2)^{\cdot}}{r_1}\, \varepsilon_\theta + (\varepsilon_\theta)^{\cdot}\, \frac{r_2}{r_1} - \cot\phi\left(\varepsilon_\phi - \frac{r_2}{r_1}\, \varepsilon_\theta\right) \tag{6.15c}$$

where we have substituted for $f_1(\phi)$ from Eq. 6.14(c). Inserting expressions in terms of N_ϕ and N_θ from Eqs. (6.13a,b) into Eq. (6.15c), we can find a general expression for χ. For the present we write this expression only for the special case of spherical domes. Thus we substitute $r_1 = r_2 = a$ into Eq. 6.15(c), to obtain

$$\chi = (\varepsilon_\theta)^{\cdot} + \cot\phi\, (\varepsilon_\theta - \varepsilon_\phi) \tag{6.16a}$$

Substituting from Eqs. (6.13a,b), we have for the sphere,

$$Et\, \chi = N_\theta^{\cdot} - \nu N_\phi^{\cdot} + \cot\phi(1 + \nu)(N_\theta - N_\phi) \tag{6.16b}$$

Also, Eq. (6.15a) becomes for a sphere,

$$\Delta_H = \frac{a \sin\phi}{Et}\, (N_\theta - \nu N_\phi) \tag{6.16c}$$

Equations (6.16b,c) can be used to obtain horizontal deflections and the rotation of the tangent of a spherical dome when N_ϕ and N_θ are known. Deflection Δ_H is defined as positive outward, while χ is clockwise positive when considering the left-hand portion of the meridian. We shall be using these expressions later for deriving equations of bending theory and for solving compatibility problems.

Example 6.1

Let us assess the membrane stresses and displacements for the roof dome shown in Fig. 6.13(a). The radius of the dome is $a = 60$ ft, and we shall assume that its thickness is 3 in. throughout. At its lower edge, the dome is connected to a ring beam. We shall assume that $\nu = 0$ for concrete. The loading on the shell will be

3 in. concrete thickness	= 37.5 lb/ft^2
Finishes/insulation, etc.	= 7.5
Total	= 45.0 lb/ft^2

Although live loads or snow loads are distributed uniformly on the *plan* area, for simplicity of calculation we can assume (without significant errors for practical purposes) this load to be 10.0 lb/ft^2 distributed uniformly over the *surface* area of the shell. Hence, adding this load to the dead load above, we obtain the total design load as $p = 55.0$ lb/ft^2. Substituting this value into Eqs. (6.6a,b), we obtain

$$N_\phi = -\frac{55.0(60.0)}{1 + \cos\phi} = -\frac{3300.0}{1 + \cos\phi} \text{ lb/ft} \tag{6.17a}$$

$$N_\theta = -3300.0\left(\cos\phi - \frac{1}{1 + \cos\phi}\right) \text{ lb/ft} \tag{6.17b}$$

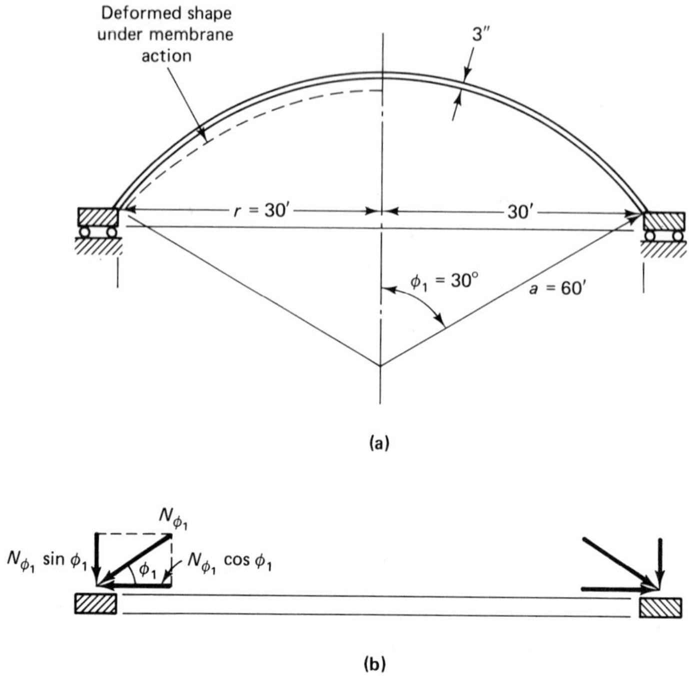

Figure 6.13 Simply supported spherical dome and ring beam. (a) Configuration of dome and its deformed shape under loading. (b) Resultant forces on the ring beam due to membrane stresses.

The values of N_{ϕ_1} and N_{θ_1} at the base are found by substituting $\phi = 30°$ into the expressions above. Thus we have

$$N_{\phi_1} = -1768.46 \text{ lb/ft}$$

and

$$N_{\theta_1} = -1089.42 \text{ lb/ft}$$

The maximum compressive meridional stress resultant N_ϕ occurs at the base, as obtained above. The maximum compressive stress resultant N_θ occurs at the apex $\phi = 0$ and is equal to $-(1650.0)$ lb/ft. Substituting $\nu = 0$ into Eqs. (6.16b,c), we obtain the displacements as

$$Et\chi = N_\theta^{\bullet} + \cot\phi\,(N_\theta - N_\phi)$$

Hence

$$Et\chi = -3300.0\left\{\left[-\sin\phi - \frac{\sin\phi}{(1+\cos\phi)^2}\right] + \cot\phi\left(\cos\phi - \frac{2}{1+\cos\phi}\right)\right\}$$

$$= 3300.0\left\{\sin\phi\left[1 + \frac{1}{(1+\cos\phi)^2}\right] - \cot\phi\left(\cos\phi - \frac{2}{1+\cos\phi}\right)\right\}$$

Also,

$$
\begin{aligned}
Et\,\Delta_H &= a \sin\phi\, N_\theta \\
&= -60.0 \sin\phi \left[3300.0\left(\cos\phi - \frac{1}{1+\cos\phi}\right)\right] \\
&= -198{,}000.0 \sin\phi\left(\cos\phi - \frac{1}{1+\cos\phi}\right)
\end{aligned}
$$

We are generally interested in finding the displacements at the base. Substituting $t = 3$ in. and $\phi = 30°$ into the equations above and simplifying, we obtain

$$E\chi_1 = +13{,}200.0 \text{ lb/ft}^2 \tag{6.18a}$$

and

$$E\Delta_{H_1} = -130{,}730.3 \text{ lb/ft} \tag{6.18b}$$

The value of E for concrete can be substituted to obtain the absolute values of the displacements. As expected, since N_θ is compressive at the base, Δ_{H_1} is negative (i.e., inward). Also, $|N_\theta|$ increases as we go toward the apex; i.e., the inward deflection ($a\varepsilon_\theta$) along the normal will also increase as we move toward the apex. This means that the rotation of the tangent is clockwise;* i.e., χ is positive, as obtained above.

The shell will exert on the supports a force equal and opposite to N_{ϕ_1}, as shown in Fig. 6.13(b). The vertical component of this force can be taken by the support, which may be a cylindrical shell wall. The horizontal component can be better resisted by a ring beam in hoop tension. This hoop tension is given by†

$$
\begin{aligned}
T &= N_{\phi_1}\, r \cos\phi_1 \\
&= 1768.46\ (30.0) \cos 30° \\
&= 45{,}945.94 \text{ lb}
\end{aligned}
$$

The ring beam can easily be designed to carry this hoop tension. Assuming the size of the ring beam to be 1.5 ft × 1.0 ft, we can calculate the radially outward deflection Δ_H^R of the ring beam as its hoop strain multiplied by r. Hence

$$\Delta_H^R = \frac{Tr}{AE} = \frac{45{,}945.94}{1.5(1.0)}\,\frac{1}{E}\,30.0$$

or

$$E\Delta_H^R = 918{,}918.80 \text{ lb/ft} \tag{6.18c}$$

It may be noted from the above that the ring beam deflects outward, whereas the shell deflects inward at its base. Since the shell and ring beam will be cast monolithically, there cannot be any relative displacement between them at their junction. Hence this discontinuity will give rise to additional stresses near the junction. We shall discuss these discontinuity problems in the next section.

*This rotation is clockwise if we are looking at the meridian on the left-hand side of the axis of revolution.

†Here we assume that the centroidal radius of the ring beam is nearly the same as r.

6.10 DISCONTINUITY PROBLEMS

In Example 6.1, we saw that under membrane conditions, the shell deflects inward at its base while the ring beam deflects outward. This incompatibility of deflections will give rise to unknown forces X_1 (per unit length of circumference) at the junction of the ring beam and the shell, as shown in Fig. 6.14(a). We have also seen that the membrane rotation χ of the shell, at the base, is clockwise. If we assume that the ring is so supported that it cannot rotate, then due to membrane stresses there will be a change of angle between the shell and the ring beam from the original angle. Assuming that the original angle has to be maintained (due to monolithicity), redundant moments X_2 as shown will also develop at the junction.

We determine X_1 and X_2 such that the net relative deflection and rotation between the shell and ring beam are zero, and hence these forces are given by the compatibility equations

$$\delta_{11}X_1 + \delta_{12}X_2 + \delta_{10} = 0 \tag{6.19a}$$

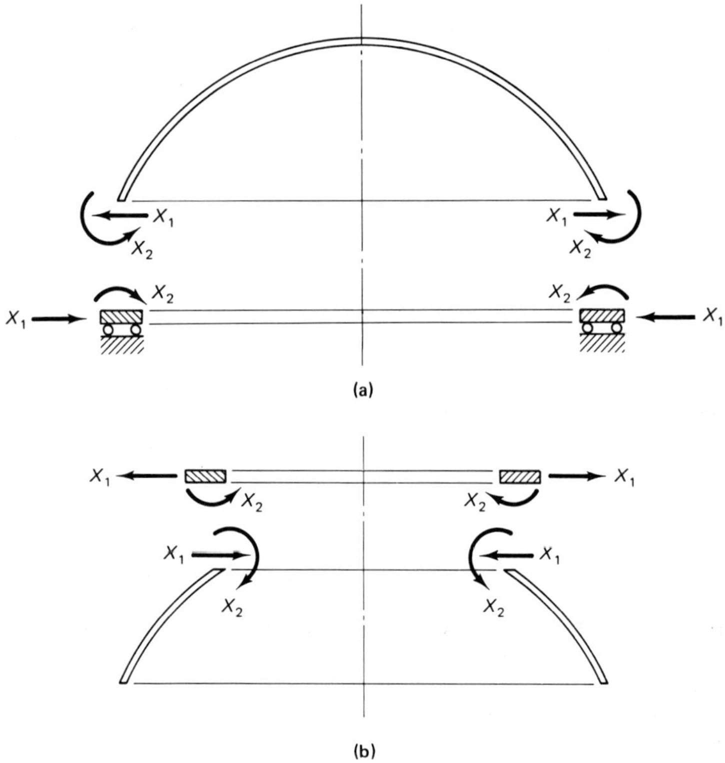

Figure 6.14 Edge loads required to satisfy displacement compatibility. (a) Edge loads at a dome/ring-beam junction. (b) Edge loads in a dome/ring junction of a roof with skylight.

$$\delta_{21}X_1 + \delta_{22}X_2 + \delta_{20} = 0 \tag{6.19b}$$

where δ_{ij} $(i, j = 1, 2)$ are the relative displacements between the shell and ring beam under unit values of X_1 and X_2, and δ_{i0} $(i = 1, 2)$ are the relative displacements of the primary structures under membrane conditions. We can therefore write

$$\delta_{ij} = \delta_{ij}^D + \delta_{ij}^R \qquad i = 1, 2 \qquad j = 0, 1, 2$$

where superscripts D and R represent displacements of the dome and ring beam, respectively, and are taken positive when in the positive directions assumed for X_1 and X_2.

In the case of the dome with a skylight (Fig. 6.9), we see that at $\phi = \alpha$,

$$N_{\phi_o} = -\frac{P}{\sin \alpha}$$

Hence from the relation $N_\phi + N_\theta = p_r a$, after substituting for N_{ϕ_o} and $p_r = -p \cos \alpha$, we obtain

$$N_{\theta_o} = -pa \cos \alpha + \frac{P}{\sin \alpha}$$

The membrane deflection of the shell at the top will be given by

$$\Delta_{H_o} = r_o \varepsilon_{\theta_o} = \frac{r_o}{Et}(N_{\theta_o} - \nu N_{\phi_o})$$

or

$$\Delta_{H_o} = \frac{r_o}{Et}\left[-pa \cos \alpha + (1 + \nu)\frac{P}{\sin \alpha}\right]$$

Whether this deflection will be outward or inward will depend on the relative values of the loads p and P. Now the ring beam is acted on by a radially inward load equal to $P \cot \alpha$. Hence the inward deflection of the ring beam will be

$$\Delta_H^R = \frac{P \cot \alpha \, r_o^2}{EA_r}$$

where A_r is the cross-sectional area of the ring. It is unlikely that Δ_{H_o} and Δ_H^R will be equal, and thus again, unknown forces X_1 and X_2 [Fig 6.14(b)] will develop at the junction, such that the resultant relative displacements are zero.

Since edge loads X_1 and X_2 cause shears and moments in the shell, we must use the bending theory for shells under axisymmetric edge loads in order to solve such problems. We shall develop this theory in Chapter 7. It will be seen that, fortunately, the effects of edge loads die out rapidly as we go away from the edge. Hence the simple membrane solutions remain valid in the remaining portion of the shell.

6.11 LIQUID STORAGE TANKS

In industrial refineries, metal tanks of various types may be used for the storage of liquids. Let us consider one such tank, as shown in Fig. 6.15. The tank has a hemispherical bottom portion of radius a, a cylindrical portion AB of height H and a hemispherical top roof. The top roof could also have been a conical shell or a flat plate. The tank would usually be supported on columns with a ring beam; however, for simplicity we shall assume here that the tank is supported on a cylindrical wall which is located at an angle $\phi = \alpha$. We shall also assume that the tank is filled with a liquid of density γ. We want to determine membrane stress resultants in the hemispherical bottom under the load of the liquid.

The liquid will exert only a normal pressure p_r. In the bottom hemispherical portion, we have

$$p_\phi = 0 \qquad p_r = \gamma a\left(\frac{H}{a} + 1 - \cos\phi\right) \qquad \phi \geqslant \frac{\pi}{2}$$

Substituting these values in Eq. (6.7), and integrating from $\phi = \pi/2$ to $\phi = \phi$, we obtain

$$[N_\phi a \sin^2\phi]^{\phi}_{\pi/2} = \gamma a^3 \int_{\pi/2}^{\phi}\left(\frac{H}{a} + 1 - \cos\phi\right)\cos\phi\sin\phi\, d\phi \qquad \text{(a1)}$$

After integration, this gives

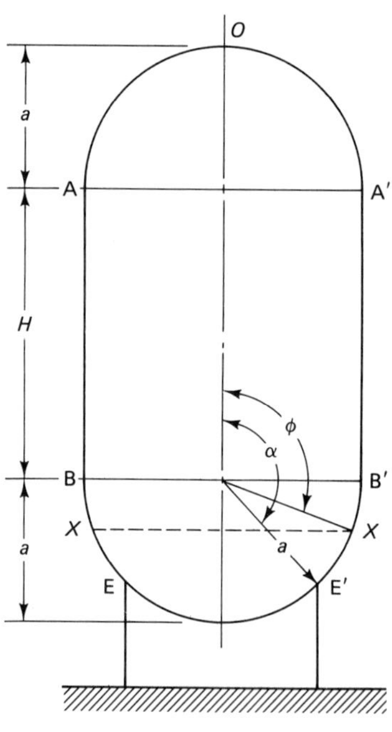

Figure 6.15 Liquid retaining tank.

$$N_\phi = \frac{\gamma a^2}{\sin^2\phi}\left(\frac{\cos^3\phi}{3} - \frac{\cos^2\phi}{2} - \frac{H}{a}\frac{\cos^2\phi}{2}\right) + \frac{N_{\phi_o}}{\sin^2\phi} \tag{a2}$$

where N_{ϕ_o} is N_ϕ evaluated at $\phi = \pi/2$, which will be the same as N_ϕ at A for the upper dome and can be found by integrating an equation similar to Eq. (a1) with $H = 0$, from $\phi = 0$ to $\phi = \phi$. Alternatively, we can determine N_{ϕ_o} by the method of sections. The total reaction of the support, V, is equal to the weight of the liquid in the tank. Hence

$$V = \left(\frac{2}{3}\pi a^3 + \pi a^2 H + \frac{2}{3}\pi a^3\right)\gamma = \left(\frac{4}{3}\pi a^3 + \pi a^2 H\right)\gamma$$

When we take any section X–X [Fig. 6.16(a)] and consider equilibrium of the portion of the tank below, we see that the vertical component of N_ϕ balances the support reaction minus the weight of the liquid acting on the freebody of the tank portion X-E-E'-X. We must assess this weight correctly. For the section X-X, the weight must not be just the weight of the liquid up to the top dome DOD', but rather, that up to the level COC'. This is because at any point T_1, the pressure p_r is due to a head of the liquid from a level at point O, i.e., from T_3 and not from T_2. Hence to consider the weight of the liquid on an element at T_1, we must consider the weight of the liquid column from T_1 to T_3. Now if we take a section at B-B',

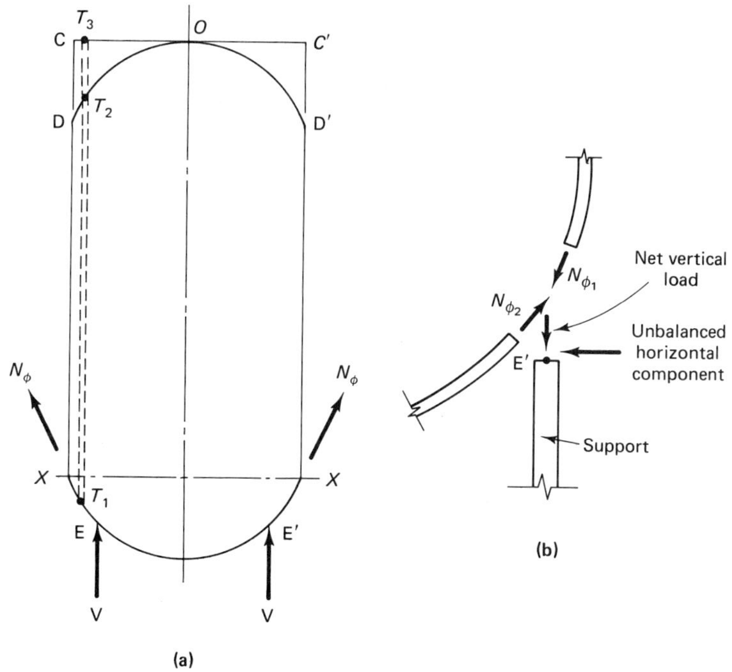

Figure 6.16 (a) Freebody of liquid storage tank. (b) Equilibrium of forces at support.

i.e., for $\phi = 90°$, and write the equation of equilibrium of vertical forces, we shall have

$$N_{\phi_o}\, 2\pi a = \frac{2}{3}\pi a^3\gamma + \pi a^2 H\gamma + (\pi a^2)(a)\,\gamma - V$$

Substituting for V from above and simplifying, we obtain

$$N_{\phi_o} = \frac{\gamma a^2}{6}$$

Hence, from Eq. (a2), we obtain

$$N_\phi = \frac{\gamma a^2}{\sin^2\phi}\left(\frac{\cos^3\phi}{3} - \frac{\cos^2\phi}{2} + \frac{1}{6} - \frac{H}{a}\frac{\cos^2\phi}{2}\right) \tag{a3}$$

which can also be written, after simplification, as

$$N_\phi = \frac{\gamma a^2}{6}\frac{(1 - \cos\phi)(1 + 2\cos\phi)}{1 + \cos\phi} - \frac{\gamma a^2}{2}\frac{H}{a}\frac{\cos^2\phi}{\sin^2\phi} \qquad \frac{\pi}{2} \leqslant \phi \leqslant \alpha \tag{a4}$$

Also, from Eq. (6.2), we have

$$N_\theta = \gamma a^2\left(\frac{H}{a} + 1 - \cos\phi\right) - N_\phi$$

Substituting for N_ϕ from Eq. (a3) and simplifying, we can write

$$N_\theta = \frac{\gamma a^2}{\sin^2\phi}\left(\frac{5 - 6\cos\phi - 3\cos^2\phi + 4\cos^3\phi}{6}\right) + \frac{\gamma a^2}{\sin^2\phi}\left(\frac{2 - \cos^2\phi}{2}\right)\frac{H}{a} \tag{b1}$$

or

$$N_\theta = \frac{\gamma a^2}{6}\frac{(1 - \cos\phi)(5 + 4\cos\phi)}{1 + \cos\phi} + \frac{\gamma a^2}{2}\frac{2 - \cos^2\phi}{\sin^2\phi}\frac{H}{a} \qquad \frac{\pi}{2} \leqslant \phi \leqslant \alpha \tag{b2}$$

These equations give values of N_ϕ and N_θ for $\pi/2 \leqslant \phi \leqslant \alpha$. For $\phi > \alpha$, we must incorporate the effect of the line load V, the support reaction. Hence, using Eq. (6.10), we add a term to the expression for N_ϕ of Eq. (a3) above, to find (for $\phi > \alpha$)

$$N_\phi = \frac{\gamma a^2}{\sin^2\phi}\left(\frac{\cos^3\phi}{3} - \frac{\cos^2\phi}{2} + \frac{1}{6} - \frac{H}{a}\frac{\cos^2\phi}{2}\right) + \frac{4}{3}\frac{\pi a^3\gamma}{2\pi a\sin^2\phi} + \frac{\pi a^2\gamma H}{2\pi a\sin^2\phi}$$

$$= \frac{\gamma a^2}{\sin^2 \phi}\left(\frac{\cos^3 \phi}{3} - \frac{\cos^2 \phi}{2} + \frac{5}{6}\right) + \frac{\gamma a^2}{2}\frac{H}{a}\left(\frac{1 - \cos^2 \phi}{\sin^2 \phi}\right)$$

or

$$N_\phi = \frac{\gamma a^2}{6}\left(\frac{2 \cos^2 \phi - 5 \cos \phi + 5}{1 - \cos \phi}\right) + \frac{\gamma a^2}{2}\frac{H}{a} \qquad \phi > \alpha \tag{a5}$$

We could have also obtained this expression directly from Eq. (6.7), by integrating from $\phi = \pi$ to $\phi = \phi$. In this case, the lower limit term on the left-hand side of Eq. (6.7) vanishes, giving us

$$N_\phi a \sin^2 \phi = \gamma a^3 \int_\pi^\phi \left(\frac{H}{a} + 1 - \cos \phi\right) \cos \phi \sin \phi \, d\phi$$

which yields, upon integration, the same expression for N_ϕ as above. N_θ is now obtained as

$$N_\theta = \frac{\gamma a^2}{6}\left(\frac{4 \cos^2 \phi - 7 \cos \phi + 1}{1 - \cos \phi}\right) + \frac{\gamma a^2}{2}\frac{H}{a} \qquad \phi > \alpha \tag{b3}$$

From Eqs. (a5) and (b3) we see that N_ϕ and N_θ are always positive in the region $\phi > \alpha$. This is to be expected, since the shell below the support will carry the load in tensile stresses. For $\phi < \alpha$, N_ϕ and N_θ are given by Eqs. (a4) and (b2). We shall consider the following three cases: $H/a = 0$, $H/a = 1$ and $H/a \gg 1$. When $H/a = 0$, we have a spherical tank. In this case N_θ is always positive (i.e., tensile), whereas N_ϕ is negative if $\phi > 120°$, as seen from Eq. (a4). For metal tanks we would prefer to have tensile stresses to avoid buckling problems. Hence it would be desirable to locate supports at an angle $\alpha < 120°$ so that N_ϕ and N_θ are tensile throughout.

From Eq. (a3), we see that for $H/a = 1$, N_ϕ is compressive if $\phi > 112.5°$ and for $H/a \gg 1$, N_ϕ is compressive throughout (i.e., for $\phi > \pi/2$). Thus, depending on the ratio H/a, the angle α determining the location of the support should be varied from 90° for $H/a \gg 1$ to 120° for $H/a = 0$. Fixing the support location in this manner will ensure a tensile N_ϕ throughout the bottom shell. It may be easily seen from Eq. (b2) that the corresponding N_θ stresses will also be tensile.

At the junction of the tank with the support, the value of N_ϕ for $\phi = \alpha^-$ (i.e., just above the support) will not be the same as N_ϕ calculated for $\phi = \alpha^+$ (i.e., just below the support) [Fig. 6.16(b)]. Due to the line load of the support reaction, there will be a jump in N_ϕ at this location. Hence the horizontal components of the two N_ϕ values will not cancel each other, and a ring beam will have to be provided at the support to resist hoop forces due to this unbalanced horizontal load.

6.12 CONOIDAL AND ELLIPTICAL DOMES

As the names suggest, conoidal or ogival domes have a pointed apex and are formed when the meridian does not meet the axis of revolution at a right angle. Figure 6.17(a) shows such a dome, whose meridian consists of an arc of a circle with radius a. As seen from the figure, the second principal radius r_2 now varies with ϕ. We let b denote the horizontal distance between the axis of revolution and the center of curvature (point O) of the meridian. We can then write the following geometric relations:

$$r = a \sin \phi - b$$

$$b = a \sin \phi_o$$

and

$$r_2 = \frac{r}{\sin \phi} = a - \frac{b}{\sin \phi} = a\left(1 - \frac{\sin \phi_o}{\sin \phi}\right)$$

Under a self-weight load of p per unit area of the shell surface, we can substitute $p_r = -(p \cos \phi)$ and $p_\phi = (p \sin \phi)$, along with the relations above, into Eq. (6.7). The lower limit on the left-hand side of this equation will vanish since r_2 is zero at $\phi = \phi_o$. Hence we have

$$N_\phi r_2 \sin^2 \phi = -pa \int_{\phi_o}^{\phi} r_2 \sin \phi \, d\phi$$

Therefore,

$$N_\phi = -\frac{pa^2}{r_2 \sin^2 \phi} \int_{\phi_o}^{\phi} \left(1 - \frac{\sin \phi_o}{\sin \phi}\right) \sin \phi \, d\phi$$

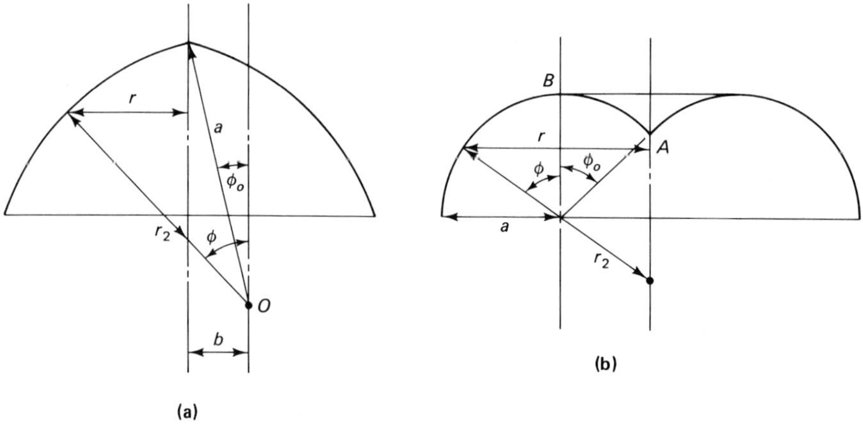

Figure 6.17 Other doubly curved shells of revolution. (a) Conoidal dome with pointed apex. (b) Shell generated by a meridian with negative ϕ_o.

or

$$N_\phi = -\frac{pa}{(\sin\phi - \sin\phi_o)\sin\phi}[(\cos\phi_o - \cos\phi) - (\phi - \phi_o)\sin\phi_o]$$

The reader may now easily obtain the expression for N_θ from Eq. (6.2). Reference may also be made to the PCA pamphlet [1] for formulas derived by the method of sections, for self-weight, as well as for other loads acting on these shells.

It is interesting to note that if ϕ_o is negative, we shall have a shell as shown in Fig. 6.17(b). In this case, we could take the limits of integration from $\phi = -\phi_o$ to ϕ, and assume that $N_\phi = 0$ at $\phi = -\phi_o$ on the left side of the shell. The resulting expression for N_ϕ will be similar to that above, except that we shall now have $(\phi + \phi_o)$ inside the brackets instead of $(\phi - \phi_o)$. Alternatively, we can start with Eq. (6.7). For the portion *AB*, we should substitute $p_\phi = -p \sin\phi$ (since p_ϕ is positive in the increasing ϕ-direction), $p_r = p\cos\phi$ (because p_r is positive when directed away from the axis of revolution), and $r_1 = -a$. We then integrate from $\phi = -\phi_o$ to ϕ.*

The expression for N_ϕ gives $N_\phi = 0$ at point *A* (where $\phi = -\phi_o$), as expected since we had used this as the boundary condition. However, it gives N_ϕ equal to infinity at $\phi = 0$, and hence we must provide a support there. On the other hand, starting from Eq. (6.4) or otherwise, and taking the boundary condition as $N_\phi = 0$ at $\phi = 0$, we shall obtain an expression for N_ϕ that will give an infinite value at $\phi = -\phi_o$ and zero value at $\phi = 0$. For this case, therefore, we must provide a support at point *A*. From our indefinite integral solution [Eq. (6.4)], we see that we have only one constant of integration. Hence we have the freedom of prescribing N_ϕ (as finite or zero) at only *A* or *B*, and we must accept whatever we get at the other location, by providing a support. We can also easily visualize that since N_ϕ at *B* will be horizontal, the weight of the shell portion between $\phi = -\phi_o$ to $\phi = 0$ has to be taken up by a support at *A* or *B*.

Elliptical domes or ellipsoids of revolution are not commonly used as roof structures, because they present analytical and construction difficulties which do not give them any additional advantages over corresponding spherical roof domes. For analysis, expressions for principal curvatures, and hence the principal radii of curvature of these shells can be determined as explained in Chapter 3 once the equation of the meridian is known [1,2]. This equation will have the form $r^2/a^2 + x^2/b^2 = 1$, where r is the radius of the parallel circle, and a, b are the half-lengths of the major and minor axes, respectively. Expressions for N_ϕ and N_θ can then be obtained, for a given loading, by integration of Eq. (6.7) or by the method of sections. The PCA pamphlet of Reference [1] gives such formulas and ready tables for self-weight, snow, and ring loadings.

Semi-ellipsoidal heads may, however, be commonly used for dish ends of pressure vessels, as well as for bottoms of vertical "columns" (vessels). They meet

*When in doubt about the proper signs to use, the reader should start again with the equilibrium of the shell element and rederive Eqs. (6.1), (6.2), and Eq. (6.4) or (6.7).

tangentially to the cylindrical vessel (which reduces discontinuity stresses), and at the same time give a reduced head height as compared to similar hemispherical heads. For shallow heads, with the ratio $a/b > \sqrt{2}$, both N_ϕ and N_θ are tensile throughout under internal pressure load, as desired [3,4]. The analysis for the case of internal pressure loading becomes relatively simple, as we shall see in a later section dealing with pressure vessels.

6.13 SHELLS OF REVOLUTION WITH NEGATIVE κ

So far we have considered membrane stresses in shells of revolution with only positive or zero Gaussian curvature. Shells with negative κ, such as the hyperboloid of revolution, are also of practical interest, since they are commonly used for the construction of natural-draft cooling towers and are also occasionally used as water tower structures. Let us consider the stresses in a hyperboloid of revolution of one sheet,* shown in Fig. 6.18(a), obtained by revolving the hyperbola ABC about the z-axis. The angle ϕ is measured from the top toward the bottom; from $\phi = \phi_o$ at point A to $\phi = 90°$ at the equator B, and then to $\phi = \phi_1$ at the lower edge C.† It is also seen that the meridian is now curved away from the axis of revolution; r_1 thus being directed outward from the meridian and not toward the axis of revolution, as is r_2. Hence the value of r_1 will have to be taken as negative in Eqs. (6.1), (6.2), and (6.7). Equation (6.2) will now take the form

$$\frac{N_\theta}{r_2} - \frac{N_\phi}{|r_1|} = p_r \tag{6.20}$$

The physical interpretation of this is seen from Fig. 6.18(b), which shows that the two rings, one along the meridian and the other along the parallel circle, now have opposing curvatures. Therefore, although the normal component of the resultant N_θ resists the pressure p_r, the normal component of N_ϕ is in the same direction as p_r.

The equation of the surface can be written as

$$\frac{x^2 + y^2}{a^2} - \frac{z^2}{b^2} = 1$$

where x and y are the coordinate axes in the horizontal plane. Intersection of planes $x = constant$ or $y = constant$ with this surface gives different hyperbolas, except when $x = \pm a$ or $y = \pm a$, in which case we obtain straight lines, $z = \pm(b/a)y$

*A hyperboloid of revolution of *two* sheets is the surface generated by revolving the conjugate hyperbolas about the z-axis. Naturally, such a surface will appear as two "cups" inverted with respect to each other.

†Note that it does not matter how we measure ϕ for calculating the weight W in finding N_ϕ by the method of sections. If we want to find N_ϕ by using the integrals of Eq. (6.4) or (6.7), we must return to the original shell element and use proper signs.

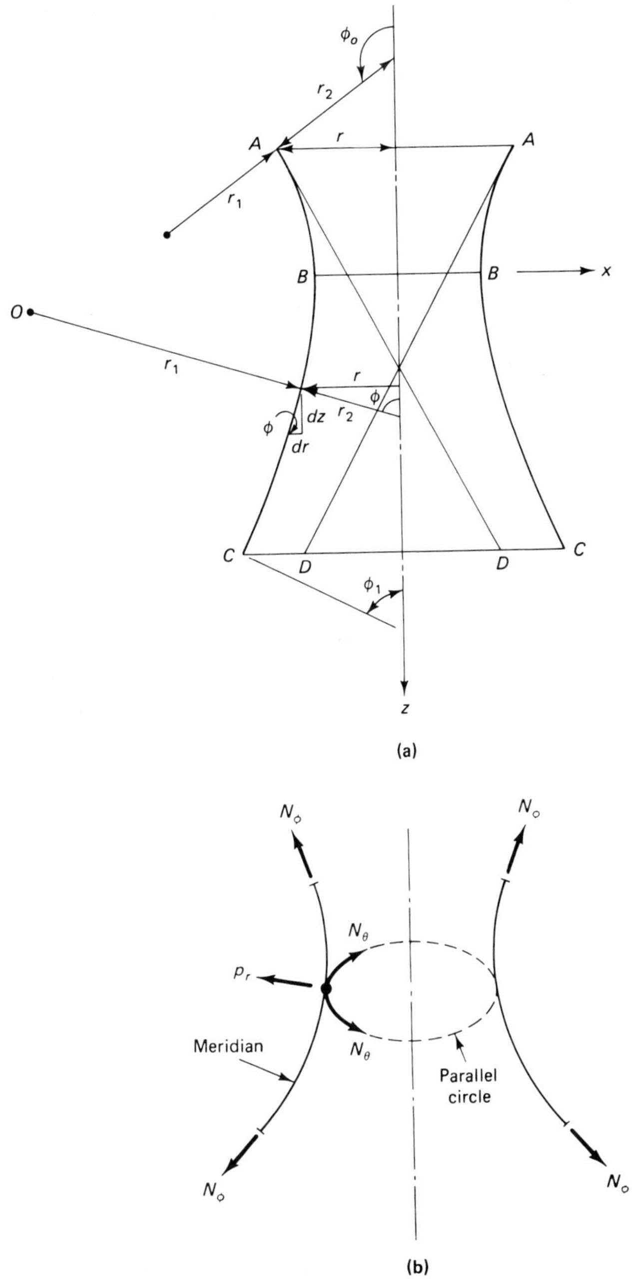

Figure 6.18 Shell of revolution with negative Gaussian curvature. (a) Hyperboloid of revolution of one sheet. (b) Opposing "rings" in the hyperboloid of revolution.

or $z = \pm(b/a)x$, respectively.* In polar form, we can also write the equation of the shell surface as

$$\frac{r^2}{a^2} - \frac{z^2}{b^2} = 1$$

which gives us r as a function of z:

$$r = \frac{a}{b}\sqrt{b^2 + z^2} \tag{c}$$

Upon determining dr/dz and d^2r/dz^2, and substituting into the equations for principal curvatures of a surface of revolution, as given in Chapter 3, we can readily obtain expressions for the principal radii of curvature, as

$$r_1 = \frac{1}{\kappa_1} = -a^2b^2\left(\frac{r^2}{a^4} + \frac{z^2}{b^4}\right)^{3/2}$$

and

$$r_2 = \frac{1}{\kappa_2} = a\left[1 + \frac{z^2}{a^2}\left(\frac{a^2}{b^2} + \frac{a^4}{b^4}\right)\right]^{1/2}$$

It is seen that r_1 is negative, as expected. We would prefer to have r_1 and r_2 as functions of ϕ, and hence we must eliminate z and r from the above. To do this, we consider a triangle formed by the sides ds (along the meridian), dr, and dz. We can write

$$\tan\phi = \frac{dz}{dr}$$

From Eq. (c) we can write

$$z = \pm\frac{b}{a}\sqrt{r^2 - a^2} \tag{d}$$

Hence

$$\frac{dz}{dr} = \pm\frac{b}{a}\left(\frac{r^2}{r^2 - a^2}\right)^{1/2} = \tan\phi$$

or

$$\tan^2\phi = \frac{b^2}{a^2}\frac{r^2}{r^2 - a^2}$$

*This indicates that this doubly curved surface can also be generated by translation of straight lines, such as AD [Fig. 6.18(a)], on parallel circles at A and D. Indeed, such a curved surface is generated for the commonly used sitting stools, which are made of a number of bamboo or cane sticks crossing each other. Such a surface is also generated when a pencil is rotated in a sharpener, against the slanting edge of the blade, which acts as the straight-line generator.

Solving for r, we get

$$r = \frac{a^2 \sin \phi}{(a^2 \sin^2 \phi - b^2 \cos^2 \phi)^{1/2}} \tag{e1}$$

Substituting this expression into Eq. (d) above and simplifying, we have

$$z = \pm \frac{b^2 \cos \phi}{(a^2 \sin^2 \phi - b^2 \cos^2 \phi)^{1/2}} \tag{e2}$$

It may be noted that z is zero at the equator ($\phi = 90°$) and negative above this point, since ϕ is greater than 90° above the equator. Substituting for r and z from Eqs. (e1) and (e2) into the expression for r_1 obtained earlier, we have

$$r_1 = \frac{-a^2b^2}{(a^2 \sin^2 \phi - b^2 \cos^2 \phi)^{3/2}} \tag{e3}$$

Also,

$$r_2 = \frac{r}{\sin \phi} = \frac{a^2}{(a^2 \sin^2 \phi - b^2 \cos^2 \phi)^{1/2}} \tag{e4}$$

It may be noted from Eq. (e3) and (e4) that we shall always find r_1 and r_2 to be negative and positive, respectively. This is because for given values of a and b, the range of variation of ϕ for the shell will be such that the expression inside the parentheses in the denominators will always be positive. For example, if $a = b$, the denominators will always be positive for $\phi > 45°$. Now, when $a = b$, the generator hyperbolas ABC have asymptotes* that make angles of 45° with the horizontal. Hence ϕ for the shell can vary only between 135 and 45°; i.e., it is always greater than 45°.

We can now readily obtain the resultant N_ϕ for the shell under self-weight, by the method of sections. If p is the load per unit area of the shell surface, then the total weight W of the shell, above the section ϕ, is

$$W = p \int_{\phi_o}^{\phi} (2\pi r|r_1|)\, d\phi$$

$$= 2\pi p a^4 b^2 \int_{\phi_o}^{\phi} \frac{\sin \phi}{(a^2 \sin^2 \phi - b^2 \cos^2 \phi)^2}\, d\phi$$

This integral can be evaluated by making the transformation

$$\xi = \frac{\sqrt{a^2 + b^2}}{a} \cos \phi$$

so

*For definitions of asymptotes and derivation of their slopes, refer to the discussion on hyperbolic paraboloid shells in Chapter 9.

$$d\xi = -\left(\frac{\sqrt{a^2 + b^2}}{a}\right) \sin \phi \, d\phi$$

Hence we have

$$\begin{aligned} W &= -\frac{2\pi p a b^2}{\sqrt{a^2 + b^2}} \int_{\xi_o}^{\xi} \frac{d\xi}{(1 - \xi^2)^2} \\ &= -\frac{\pi p a b^2}{2\sqrt{a^2 + b^2}} \left[\frac{2\xi}{1 - \xi^2} + \ln \frac{1 + \xi}{1 - \xi}\right]_{\xi_o}^{\xi} \end{aligned}$$

where ξ_o is the value of ξ at $\phi = \phi_o$. Denoting the quantity inside the brackets as $f(\xi)$, the expression above, after substitution of the limits of integration, can be written as a function of $[f(\xi) - f(\xi_o)]$. We now equate W to the upward vertical component of N_ϕ (given by $-2\pi N_\phi r_2 \sin^2 \phi$), to obtain

$$N_\phi = \frac{pab^2}{4\sqrt{a^2 + b^2}\, r_2 \sin^2 \phi} [f(\xi) - f(\xi_o)]$$

Substituting for r_2 from Eq. (e4), we have

$$N_\phi = \frac{pb^2}{4a\sqrt{a^2 + b^2}} \frac{(a^2 \sin^2 \phi - b^2 \cos^2 \phi)^{1/2}}{\sin^2 \phi} [f(\xi) - f(\xi_o)]$$

A simple computer program can be used to generate N_ϕ, from the expression above, for various values of ϕ. Alternatively, Soare has given tables of $f(\xi)$ for various values of ξ, and these have been reproduced in Reference [5]. Then N_θ is given by Eq. (6.20), wherein we substitute $p_r = -p \cos \phi$ and the values of $|r_1|$ and r_2 in terms of ϕ.

In case we have a nonzero value for $N_\phi = N_{\phi_o}$ at $\phi = \phi_o$, due to a line load, the vertical component of this resultant is given by $2\pi r_o N_{\phi_o} \sin \phi_o$ (acting upward since N_ϕ is positive in tension). To take into account the effect of this component, we must add to the above solution for N_ϕ a single term, given by Eq. (6.10) as

$$\frac{N_{\phi_o} r_o \sin \phi_o}{r_2 \sin^2 \phi}$$

For large cooling tower structures, the shell thickness will usually be designed to vary from top to bottom. Hence the shell can be divided into discrete segments of constant thickness. Then N_ϕ can be obtained in any segment, from the above, by letting N_ϕ at the bottom of an upper segment be equal to N_{ϕ_o} on the next lower segment, and so on. Due to the important practical applications in the design of natural-draft cooling towers for power plants, and also because of previous failures of such towers, several references by Gould, Billington, and others, concerning the analysis of hyperboloidal shells of cooling towers subjected to various loading conditions, have been written and are available in the literature [6,7].

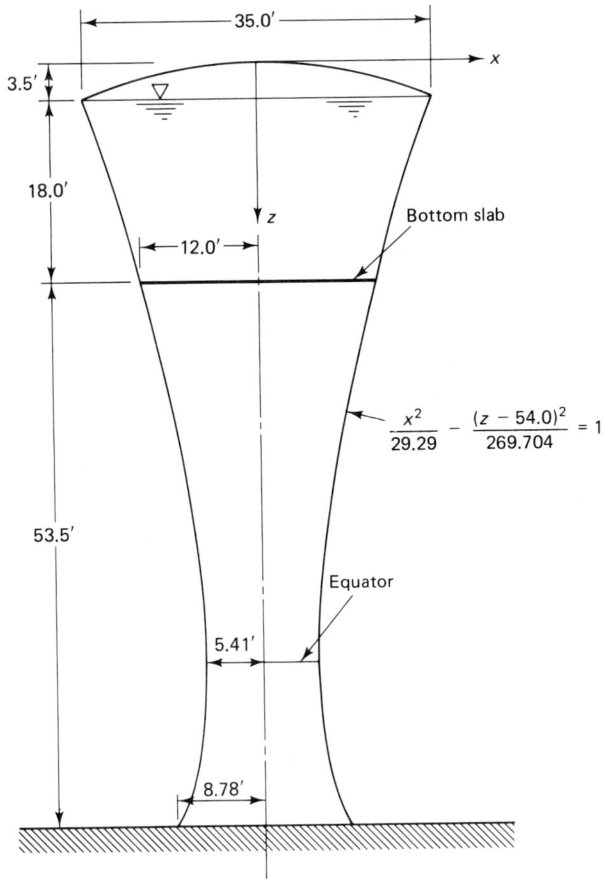

Figure 6.19 Elevated water tank and supporting structure comprised of one continuous hyperboloid of revolution.

Expressions for the membrane stresses under water load for a tank of hyperboloid shape can also be derived. For an elevated water tank, several feet above the ground, the supporting structure can also be a hyperboloid shell in continuation of the hyperboloid shell of the tank wall above. Thus the entire structure, from the ground to the top of the tank, can be one hyperboloid shell with different thicknesses for the tank and supporting shaft portions. The curvature of the meridian for such a shell should be rather low so that a large ground area is not unnecessarily covered by the shaft. Such an elevated water tank is shown in Fig. 6.19. This tank has a shallow roof dome and a circular slab at its bottom. The equation of the hyperbola, with origin of coordinates at the apex of the roof dome, is as indicated. The equator is at $z = 54.0$ ft, where the radius of the parallel circle is 5.41 ft. The curvature of the wall of the tank is small and it looks similar to a cone. Membrane stresses in the tank and shaft walls can be obtained using the expressions just derived.

Again, incompatibility problems arise at the junctions of the wall and dome, and the wall and slab, and these have to be solved by applying edge loads.

Other forms of shells of revolution of negative κ, which are of practical interest, are obtained by revolving an arc of a circle about an axis, as shown in Fig. 6.20. This shell is actually a part of a toroid which is obtained by revolving a circle about an axis, as in a tire tube. Portions of the toroid, such as shown in the figure, are used in pressure vessels for a smoother transition from one shell to another, thereby minimizing the discontinuity stresses at the junction. We shall see an example of such a shell in the next section.

6.14 PRESSURE VESSELS

Metal pressure vessels are mainly designed to resist internal pressure loads, the effect of self-weight usually being small compared to these stresses. Let us consider a pressure vessel with an arbitrary shape of the meridian which, however, intersects the axis of revolution at a right angle (Fig. 6.21). Substituting $p_r = p$ in Eq. (6.7), and integrating from $\phi = 0$ to $\phi = \phi$, we obtain

$$N_\phi = \frac{1}{r_2 \sin^2 \phi} \int_0^\phi p \sin \phi \cos \phi \, r_1 r_2 \, d\phi$$

where the lower integration limit has been taken as zero. Now we note that

$$r = r_2 \sin \phi$$

and

$$dr = ds \cos \phi = r_1 \, d\phi \cos \phi$$

Substituting these relations, we have the following expression:

$$N_\phi = \frac{1}{r_2 \sin^2 \phi} \int_{r=0}^{r=r} pr \, dr = \frac{pr^2}{2r_2 \sin^2 \phi}$$

or upon substituting $r = r_2 \sin \phi$, we obtain the simpler expression

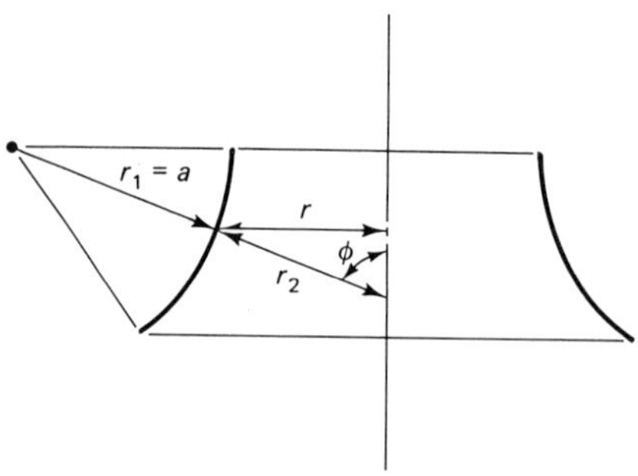

Figure 6.20 Shell of revolution with negative Gaussian curvature, whose meridian is a portion of a circular arc.

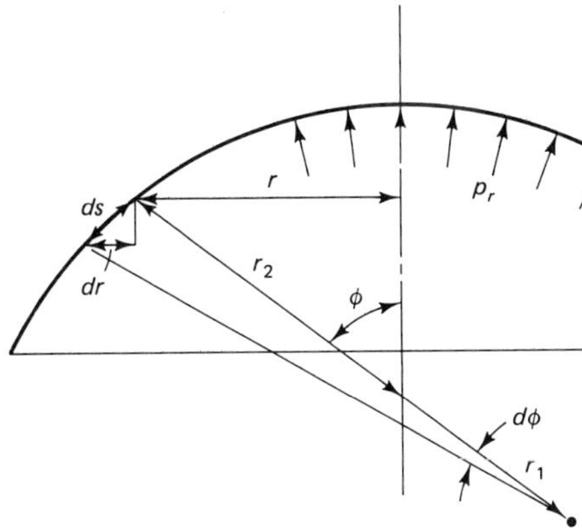

Figure 6.21 Pressure vessel with a meridian of arbitrary shape which meets the axis of revolution at a right angle.

$$N_\phi = \frac{pr_2}{2}$$

which is valid for any shape of meridian. Of course, r_2 must be substituted in terms of ϕ in this expression. This result could have also been obtained easily from the method of sections, by equating the total vertical load $\pi r^2 p$ to $2\pi r N_\phi \sin\phi$, which is the vertical component of N_ϕ. Now, using Eq. (6.2), the expression for N_θ becomes

$$N_\theta = pr_2\left(1 - \frac{r_2}{2r_1}\right)$$

Let us now consider a cylindrical pressure vessel with hemispherical heads, under an internal pressure p, as shown in Fig. 6.22(a). For the spherical head with $r_1 = r_2 = a$, we obtain from the above, $N_\phi = N_\theta = pa/2$. For the cylinder, we have $N_\theta = pa$, while the meridional stress resultant N_x will be equal to N_ϕ from the sphere. Now, at the junction of the cylinder and sphere, the radial deflections of two neighboring points A and B on the cylinder and sphere, respectively, will be given by

$$\Delta_H^W = a\varepsilon_\theta^W = \frac{a}{Et_W}(N_\theta^W - \nu N_x^W) = \frac{pa^2}{2Et_W}(2 - \nu) \tag{6.21a}$$

$$\Delta_H^D = a\varepsilon_\theta^D = \frac{a}{Et_D}(N_\theta^D - \nu N_\phi^D) = \frac{pa^2}{2Et_D}(1 - \nu) \tag{6.21b}$$

It is seen that, for identical thicknesses, the deflection of the cylinder, Δ_H^W, is substantially greater than that of the sphere, Δ_H^D. Hence there will develop edge loads X_1 and X_2, acting on the two shells, such that compatibility of displacements is satisfied [Fig. 6.22(b)]. The situation becomes more critical if we have a shallow spherical cap instead of a hemispherical head. In this case, the radial component

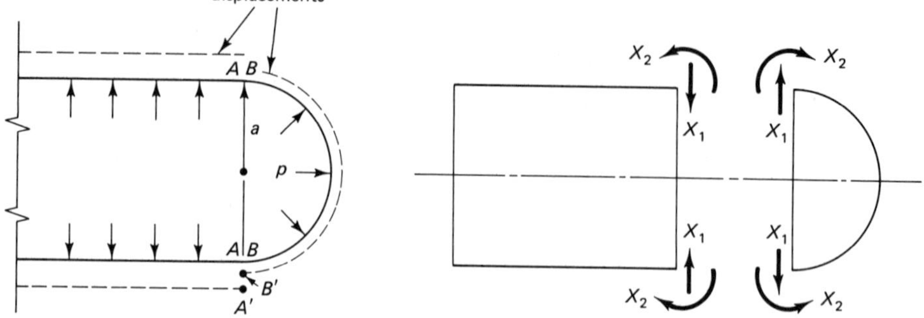

Figure 6.22 Cylindrical pressure vessel with hemispherical heads. (a) Membrane displacements. (b) Corrective edge loads at dome/wall junction.

of N_ϕ in the sphere, at its junction with the cylinder, has to be resisted by the cylinder as an edge load. This makes the differential deflection of the two primary structures much greater. Normally, a ring beam would be provided at the junction to resist the radial component of N_ϕ.

When the diameter of a cylindrical pressure vessel is to be reduced, a conical shell may be used as a transition from the larger to the smaller shell [Fig. 6.23(a)]. As discussed above for the shallow spherical cap, the discontinuity stresses at the junctions of the cone and cylinders will be high without the presence of a ring beam. Hence, to have a smoother transition, pieces of shells called knuckles are inserted between the cone and the cylinders, as shown in Fig. 6.23(b). These knuckles can be considered as rings cut from a toroidal shell having a small radius of curvature r_1 for the meridian. The re-entrant knuckle at the junction of the cone and the smaller cylinder is seen to have a negative κ. The ASME Boiler and Pressure Vessel Code gives guidelines for minimum radii of knuckles, based on expected stresses [4].

Example 6.2

Figure 6.24 shows a metal containment vessel of a nuclear reactor, which is mounted on a skirt during testing. Although the vessel is later covered by thick concrete shielding, it is designed to independently withstand large pressures that may develop due to a malfunction. The vessel is bottle shaped, with a spherical bottom connected to a cylindrical tube that has a hemispherical cap. The knuckle has negative Gaussian curvature κ. The values of N_ϕ and N_θ at various levels, under self-weight and an internal test pressure of 45 psi, can be easily calculated and the stresses at various levels can be checked. We shall calculate stresses in the knuckle at $\phi = 45°$. It will be easiest to use the method of sections.

Self-weight. The weight of the dome is $W_1 = 2\pi h^2 \gamma t$, where h is the height/radius of the dome and γ is the density of steel (490 lb/ft^3). Substituting numerical values, we get

$$W_1 = 2\pi(11.5)^2(490)\left(\frac{0.875}{12}\right) = 29{,}689.2 \text{ lb}$$

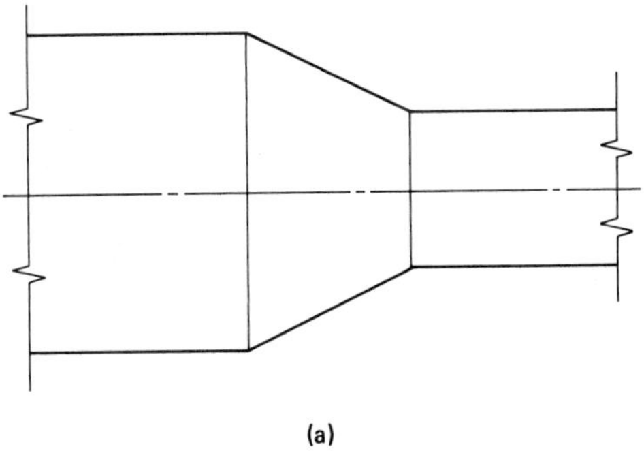

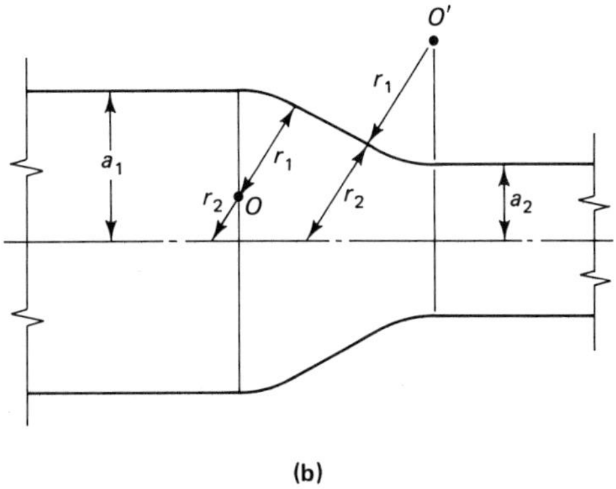

Figure 6.23 Transitions in pressure vessels. (a) Conical shell transition between cylinders. (b) Toroidal knuckles used as transitions between cylinders and cone.

The weight of the cylinder, W_2, is given by

$$W_2 = 2\pi(11.5)(18.604)\left(\frac{0.377}{12}\right)(490) = 20{,}693.8 \text{ lb}$$

Hence N_x at point A, which is also N_{ϕ_o} for the knuckle, is given by

$$N_x = N_{\phi_o} = -\frac{29{,}689.2 + 20{,}693.8}{2\pi(11.5)} = -697.28 \text{ lb/ft}$$

For the knuckle we can write

$$(|r_1| + r_2) \sin \phi = 11.5 + 5.0 = 16.5$$

Hence

$$r_2 = \frac{16.5}{\sin \phi} - 5.0$$

Also,

$$r = r_2 \sin \phi = 16.5 - 5.0 \sin \phi$$

Thus we can find N_ϕ as

$$-N_\phi(2\pi r \sin \phi) = W_1 + W_2 + \int_\phi^{\pi/2} (2\pi r)(\gamma t) r_1 \, d\phi$$

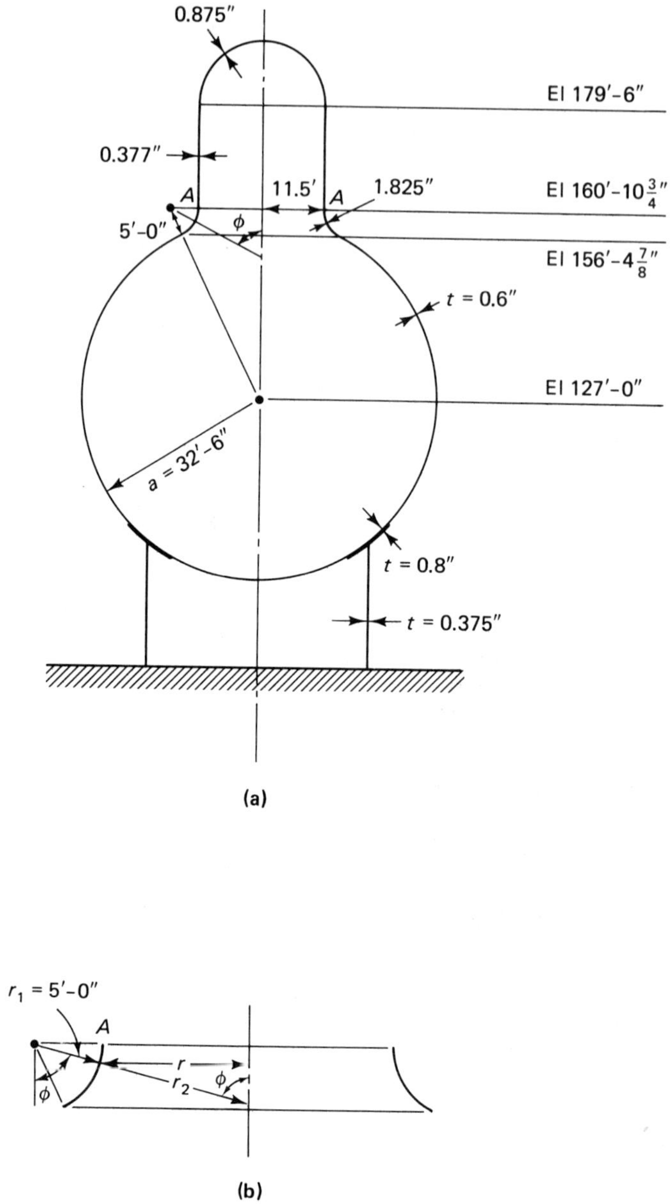

Figure 6.24 Nuclear reactor containment vessel. (a) Geometry of vessel. (b) Detail of toroidal knuckle transition.

$$= 50{,}383.0 + 2\pi(5.0)\left(\frac{1.825}{12}\right)(490.0)\int_{\phi}^{\pi/2} (16.5 - 5.0 \sin \phi)\, d\phi$$

$$= 50{,}383.0 + 2341.14(25.92 - 16.5\phi - 5.0 \cos \phi)$$

or

$$N_\phi = \frac{-17{,}676.6 + 372.6(16.5\phi + 5.0 \cos \phi)}{(16.5 - 5.0 \sin \phi) \sin \phi}$$

Substituting $\phi = \pi/4$, we obtain

$$N_\phi = -1257.81 \text{ lb/ft}$$

From Eq. (6.20), N_θ can also be obtained. We substitute $p_r = -\gamma t \cos \phi$, and obtain

$$N_\theta = r_2\left(p_r + \frac{N_\phi}{|r_1|}\right)$$

$$= \left(\frac{16.5}{\sin \phi} - |r_1|\right)\left(-\gamma t \cos \phi + \frac{N_\phi}{|r_1|}\right)$$

Substituting in the appropriate numerical values, along with $\phi = \pi/4$, we have

$$N_\theta = -5578.39 \text{ lb/ft}$$

Although both N_ϕ and N_θ are compressive, the magnitude of the larger compressive stress, given by $\sigma = -5578.39/1.825(12) = -254.72$ psi, is relatively small and should not create any buckling problems.

Test pressure load. We now have $p_\phi = 0$ and $p_r = 45$ psi (6480 lb/ft^2). The boundary stress resultant N_{ϕ_o} from the cylinder can be obtained by the relation

$$N_{\phi_o}(2\pi)(11.5) = \pi(11.5)^2(6480)$$

Hence

$$N_{\phi_o} = 37{,}260 \text{ lb/ft}$$

Now from Eq. (6.7), we have

$$[N_\phi r_2 \sin^2 \phi]_{\phi_o}^{\phi} = \int_{\phi_o}^{\phi} p_r \cos \phi \; r_1 r_2 \sin \phi \, d\phi$$

where r_1 will have to be taken as (-5.0), so that

$$[N_\phi r_2 \sin^2 \phi]_{\phi_o}^{\phi} = p_r \cdot r_1 \int_{\phi_o}^{\phi} (16.5 \cos \phi - 5.0 \sin \phi \cos \phi)\, d\phi$$

$$= p_r \cdot r_1 \left[16.5 \sin \phi + \frac{5}{4} \cos 2\phi\right]_{\phi_o}^{\phi}$$

or

$$N_\phi = \frac{p_r r_1}{r_2 \sin^2 \phi}\left[16.5 \sin \phi + \frac{5}{4} \cos 2\phi\right]_{\phi_o}^{\phi} + \frac{[N_\phi r_2 \sin^2 \phi]_{\phi = \phi_o}}{r_2 \sin^2 \phi}$$

Now, substituting the values $r_1 = -5.0$, $\phi_o = \pi/2$, $[r_2]_{\phi=\phi_o} = 11.5$, $N_{\phi_o} = 37{,}260$, and $\phi = \pi/4$, we obtain

$$N_\phi = 59{,}403.86 \text{ lb/ft}$$

Hence

$$N_\theta = \left(\frac{16.5}{\sin\phi} - 5.0\right)\left(6480.0 + \frac{N_\phi}{|r_1|}\right)$$

or

$$N_\theta = 336{,}636.01 \text{ lb/ft}$$

Both N_ϕ and N_θ are in tension under internal pressure, and the hoop stress will be

$$\frac{336{,}636.01}{12.0 \times 1.825} = 15{,}371.51 \text{ psi}$$

This stress, when added to the self-weight stress, gives a net tensile stress less than 20,000 psi at $\phi = \pi/4$, which should be acceptable for A36 steel. From the general expressions for N_ϕ and N_θ above, their variation with ϕ could be plotted to obtain the magnitude and location of the maximum stress in the knuckle.

6.15 DESIGN CONSIDERATIONS

We have already discussed in Chapter 4 general considerations and code recommendations for the design of civil engineering shell structures, which can also be applied to doubly curved shells. Membrane analysis gives a quick solution for preliminary design. Indeed, many shell structures and water tanks have been built on designs based on membrane theory alone, keeping in view, of course, the likely effects at the discontinuities.

The doubly curved shell roofs are very efficient in resisting loads, and hence large areas can be covered with minimum concrete thickness and reinforcement. We have seen for the spherical roof of Example 6.1, that the maximum resultant N_ϕ or N_θ is only -1768.46 lb/ft. Therefore, the maximum compressive stress is only $1768.46/(12.0 \times 3.0) = 49.12$ psi, which is very small for a 3000-psi cylinder-strength concrete. Such low stresses will also require only minimal temperature and shrinkage reinforcement. Thus in this example we can cover an area having a 60-ft plan diameter with only a 3-in. concrete thickness. This can be compared with the roof slab of the example in Section 5.7, which was 9 in. thick, to cover an area with a diameter of 24 ft. Such a roof slab would also require much more reinforcement than does a domed roof. Therefore, although preparation of formwork for such domes is difficult due to their double curvature, they are commonly used as roofs over tanks, coliseums, planetariums, and other structures. Very large diameter domes, such as Nervi's sports stadium in Rome and the Kingdome in Seattle, have been constructed to cover large civic facilities. Dome roofs in timber, covering plan diameters of 500 ft for stadiums, have also recently been constructed in the United States [8,9].

We shall next illustrate the design of a water tank by means of membrane theory.

6.16 DESIGN OF AN INTZE TANK

Water towers form an essential part of the infrastructure for the development of any city or town. Even a town with a population of only 2000 may need a water tower for its water supply. Such water towers become the landmarks of these towns, and hence they should be constructed keeping in view their aesthetic appeal as well as their economy. Tanks made from different forms of shells of revolution serve both of these purposes. In the United States most such tanks are made of steel shells supported on steel columns. In Europe and Asian countries, however, concrete is commonly used for the construction of such tanks [10]. Some of the tanks house restaurants and viewing galleries on top. An economical form of these concrete tanks, which is commonly constructed, is the Intze tank.

Figure 6.25 shows a typical Intze tank. It may be supported on a cylindrical shaft, as shown, or on a deep circular ring beam that spans over six or eight columns. It is interesting to note that on this support, the horizontal components of N_ϕ from the cone and the bottom dome oppose each other, and hence the support will carry primarily vertical loads with a minimum of horizontal thrust. This desirable aspect, as well as having a cone and dome at the bottom, instead of an expensive circular bottom slab, make these tanks economical in material consumption for a given storage capacity. Of course, the labor involved for forming the bottom dome and cone, compared to that for a circular slab, can offset this economy. For this reason, these tanks are not commonly constructed in the United States.

The dimensions shown are assumed to be centerline (midsurface) dimensions. Although uniform thicknesses for individual members have been assumed here for simplicity, it is quite likely that in practice the cylindrical tank wall will be only about 6 in. thick, flared to about 8 in. in the lower 6 ft or so near the ring beam. The bottom dome can also be tapered down to a 6-in. thickness at its crown, as this is considered worthwhile to save some concrete. No ring beam is provided at the junction of the roof dome and the wall, since usually the wall is sufficiently stiff to resist the horizontal thrust of the dome (which carries mainly self-weight only). At the most, a small ring beam may be provided. At the junction of the wall and the cone, however, a good-sized ring beam, as shown, is required. By reducing the thickness of this ring and increasing its width to about 3 ft, it can also serve as an inspection gallery walkway. The slope of the bottom cone is kept high (to minimize bending moments by "plate action"), usually close to 45 degrees, but not greater, to avoid double formwork during concreting. In the present example, this

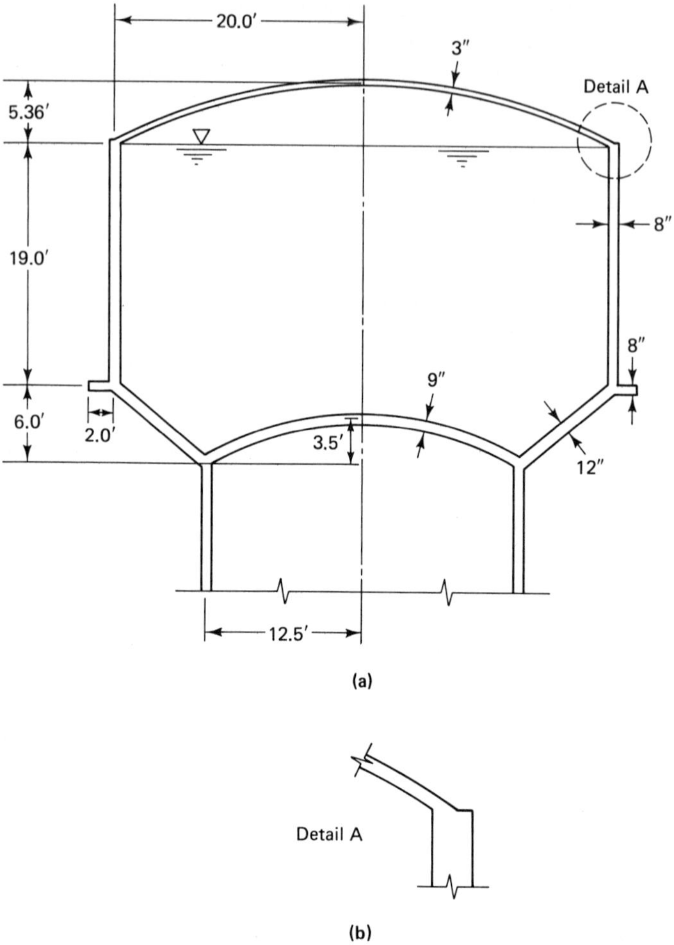

Figure 6.25 Intze water tank. (a) Geometry of tank. (b) Detail of dome/wall junction.

angle works out to be 38.66°. The thickness of the cone is usually kept constant due to its short length.

In the present example, the radius of the roof dome is 40.0 ft, and the angle ϕ at the base is 30° (which is within the normal range of 25° to 35° for such domes). The radius of the bottom dome is 24.071 ft; its base angle ϕ is calculated to be 31.28°. This angle may have been designed steeper to aid the dome in carrying the heavy water load; also, not much extra concrete would be consumed by increasing this angle, since the dome area is not particularly large.

For simplicity, the water level has been assumed to be at the top of the cylindrical wall, although in practice, a freeboard of 6 to 12 in. should be maintained. We shall also assume Poisson's ratio as zero.

We shall now calculate the membrane forces, under dead load (DL) and water load (WL), in the various elements of the tank of Fig. 6.25, and also simultaneously design each element, one by one, based on membrane theory.

Roof Dome: Membrane Analysis

The total load on the roof dome is taken as $p = 55$ lb/ft^2 over its surface area, as in Example 6.1. By simple geometry (Fig. 6.26), we have $\phi_1 = 30°$ and $a = 40.0$ ft. Substituting these numerical values into Eqs. (6.6a,b), we calculate N_ϕ and N_θ for ϕ varying at 5° intervals. These values are given in Table 6.1.

Under water load only, there are no further membrane forces developed in the dome.

Roof Dome: Design

As expected, both N_ϕ and N_θ are compressive in the dome, and the maximum compressive stress will be

$$\sigma_{\text{max}} = -\frac{1178.98}{12(3.0)} = -32.75 \text{ psi}$$

This stress is very small,* and only temperature and shrinkage steel of $A_s = 0.3\%$ in each direction needs to be provided. Hence we have

$$A_s = 0.003(12)(3.0) = 0.108 \text{ in.}^2\text{/ft}$$

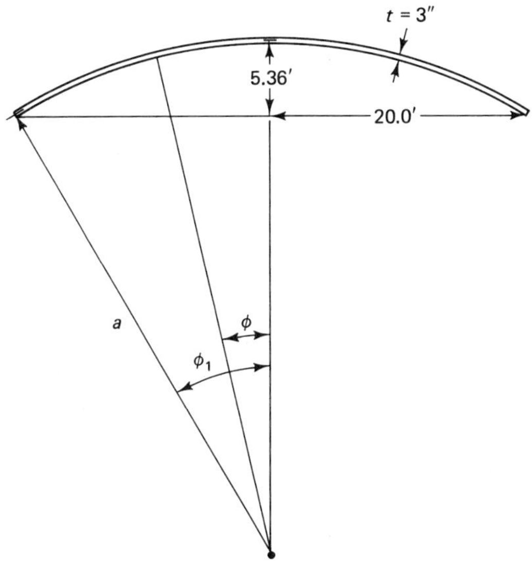

Figure 6.26 Detail of Intze tank dome roof.

*When designing very large domes, one would also check the allowable buckling load [11]. In the present problem, however, buckling stresses are not expected to be critical.

TABLE 6.1 Membrane Forces in Roof Dome

ϕ (deg)	N_ϕ (lb/ft)	N_θ (lb/ft)
0	−1100.00	−1100.00
5	−1102.10	−1089.53
10	−1108.42	−1058.16
15	−1119.07	−1005.97
20	−1134.20	− 933.12
25	−1154.06	− 839.81
30	−1178.98	− 726.28

We can provide a single layer of No. 3 bars at 12.0 in. center-to-center spacing in the meridional direction, and the same in the form of circular rings in a perpendicular direction. A typical reinforcement arrangement is shown in Fig. 6.27.

Cylindrical Wall: Analysis

Dead load. The loads at the top of the wall, from the dome, are as shown in Fig. 6.28(a). If N_{ϕ_1} is the meridional force at the bottom of the dome, t_w is the thickness of the cylindrical wall, and γ_c is the density of concrete, then N_x and N_θ in the wall are given by

$$N_x = -N_{\phi_1} \sin \phi_1 - \gamma_c t_w x$$

$$= -1178.98 \sin 30° - 150.0\left(\frac{8.0}{12.0}\right)x$$

or

$$N_x = -589.49 - 100.0x$$

and

$$N_\theta = p_r r_w = 0.0$$

Since we have not provided a ring beam, the horizontal component of N_{ϕ_1} from the dome will be resisted jointly by the edge zones of the dome and the wall (acting themselves as rings) near the junction. We shall see the effect of this on the design from the bending theory derived in the next chapter. Presently, we shall restrict our design to be based on membrane theory alone.

Water load. From Fig. 6.28(b), we can easily obtain

$$N_x = 0$$

and

$$N_\theta = 62.5(20.0)(x) = 1250.0x \text{ lb/ft}$$

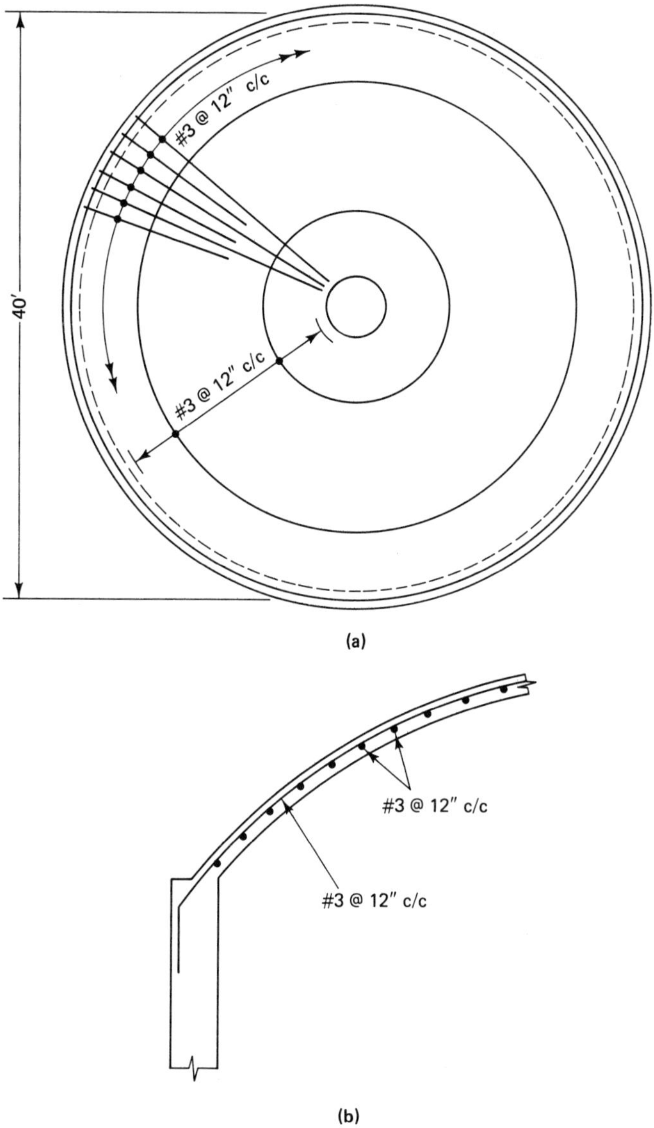

Figure 6.27 Typical reinforcement detail for Intze tank dome roof. (a) Plan view of reinforcement. (b) Meridional section showing reinforcement.

Cylindrical Wall: Design

The maximum value of N_x occurs at the base of the cylinder ($x = 19.0$), this value being $N_x = -2489.49$ lb/ft. Therefore, the maximum vertical compressive stress is

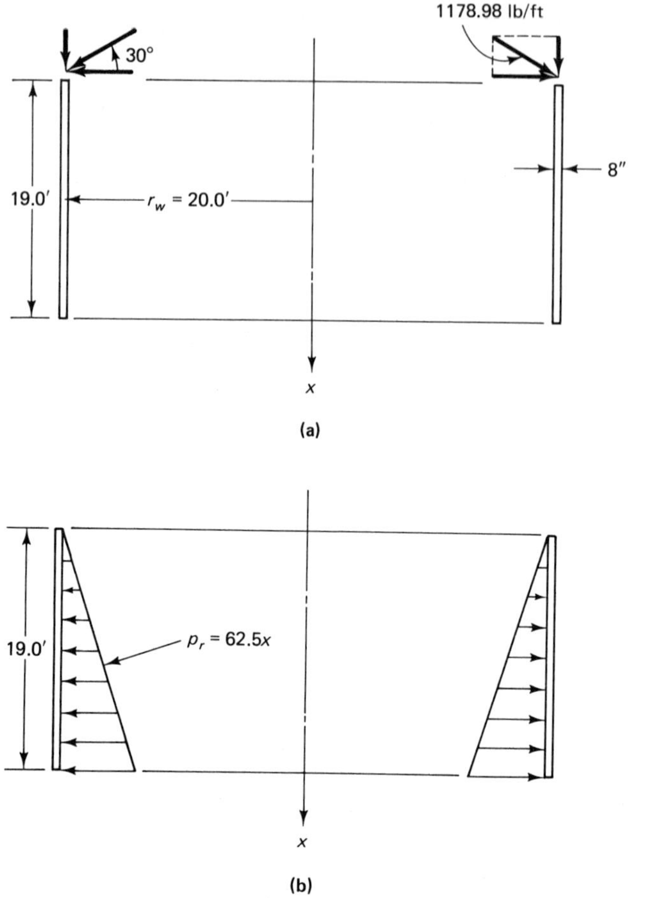

Figure 6.28 Cylindrical wall portion of Intze tank. (a) Dead load. (b) Water load.

$$\sigma^c_{x_{\max}} = \frac{2489.49}{12.0(8.0)} = 25.93 \text{ psi}$$

Since this stress is small, only the minimum percentage of steel ($A_s = 0.3\%$) is provided in the vertical direction and is divided equally on each wall face:

$$A_s = 0.003(12.0)(8.0) = 0.288 \text{ in.}^2/\text{ft}$$

We shall provide No. 3 bars at 9 in. center-to-center vertically on each face. Now values of N_θ are calculated at suitable intervals, and rebar in the corresponding zones is designed. The results are given in Table 6.2. The zonal length, taken here as 4.0 ft, could be reduced for more refinement in actual design. The maximum tensile stress in the concrete, neglecting the effect of steel reinforcement, will be

$$\frac{23{,}750.0}{12.0(8.0)} = 247.4 \text{ psi}$$

TABLE 6.2 Design of Hoop Reinforcement in Cylindrical Wall[a]

x (ft)	N_θ (lb/ft)	$(N_\theta)_{avg}$ = Average N_θ for Zone Δx (lb/ft)	Δx Zonal Length (ft)	Tension in Zone = $(N_\theta)_{avg}\,\Delta x$, T (lb)	A_s in Zone = T/ 20,000.0 (in.^{2})	Bars on Each Face (Min. No. 3 at 9 in. c-c)
0.0	0.0					
		2,500.0	4.0	10,000.0	0.5	6 No. 3 (min.)
4.0	5,000.0					
		7,500.0	4.0	30,000.0	1.5	7 No. 3
8.0	10,000.0					
		12,500.0	4.0	50,000.0	2.5	7 No. 4
12.0	15,000.0					
		17,500.0	4.0	70,000.0	3.5	9 No. 4
16.0	20,000.0					
		21,875.0	3.0	65,625.0	3.28	6 No. 5
19.0	23,750.0					

[a]It should be noted that near the junction of the wall with the roof dome, the actual hoop reinforcement required will be greater than the minimum indicated in this table. This is because the junction must resist the hoop tension caused by the thrust from the dome. We shall consider this issue further in Chapter 7 involving bending theory.

This stress is below 400 psi and is considered acceptable. Hence the assumed wall thickness of 8 in. is satisfactory. The arrangement of reinforcement in the wall will be as shown in Fig. 6.29. The splices in the circular rings should be well staggered.

Ring Beam: Analysis

At the top of the cone, N_x from the shell is -2489.49 lb/ft under dead load. To this we add the self-weight of the ring beam = $(2.0 \times 0.67 \times 150) = 200$ lb/ft. Hence the total line load at the top of the cone is 2689.49 lb/ft. As shown in Fig. 6.30, this load can be resolved into a component along the cone, of 2689.49/ sin α, and a horizontal component of 2689.49 cot α. We have the relation that $\tan\alpha = 6.0/7.5$, which gives $\alpha = 38.66°$. Hence the horizontal component to be resisted by the ring beam is 3361.86 lb/ft. The hoop tension in the ring beam is given by

$$T_R = 3361.86(20.0 + 1.0) = 70{,}599.06 \text{ lb}$$

Under water pressure, there is no unbalanced load at the joint, and hence the ring beam will not carry any additional forces under membrane conditions.

Ring Beam: Design

The tensile stress in the ring beam is

$$\sigma_t = \frac{70{,}599.06}{24.0(8.0)} = 367.7 \text{ psi}$$

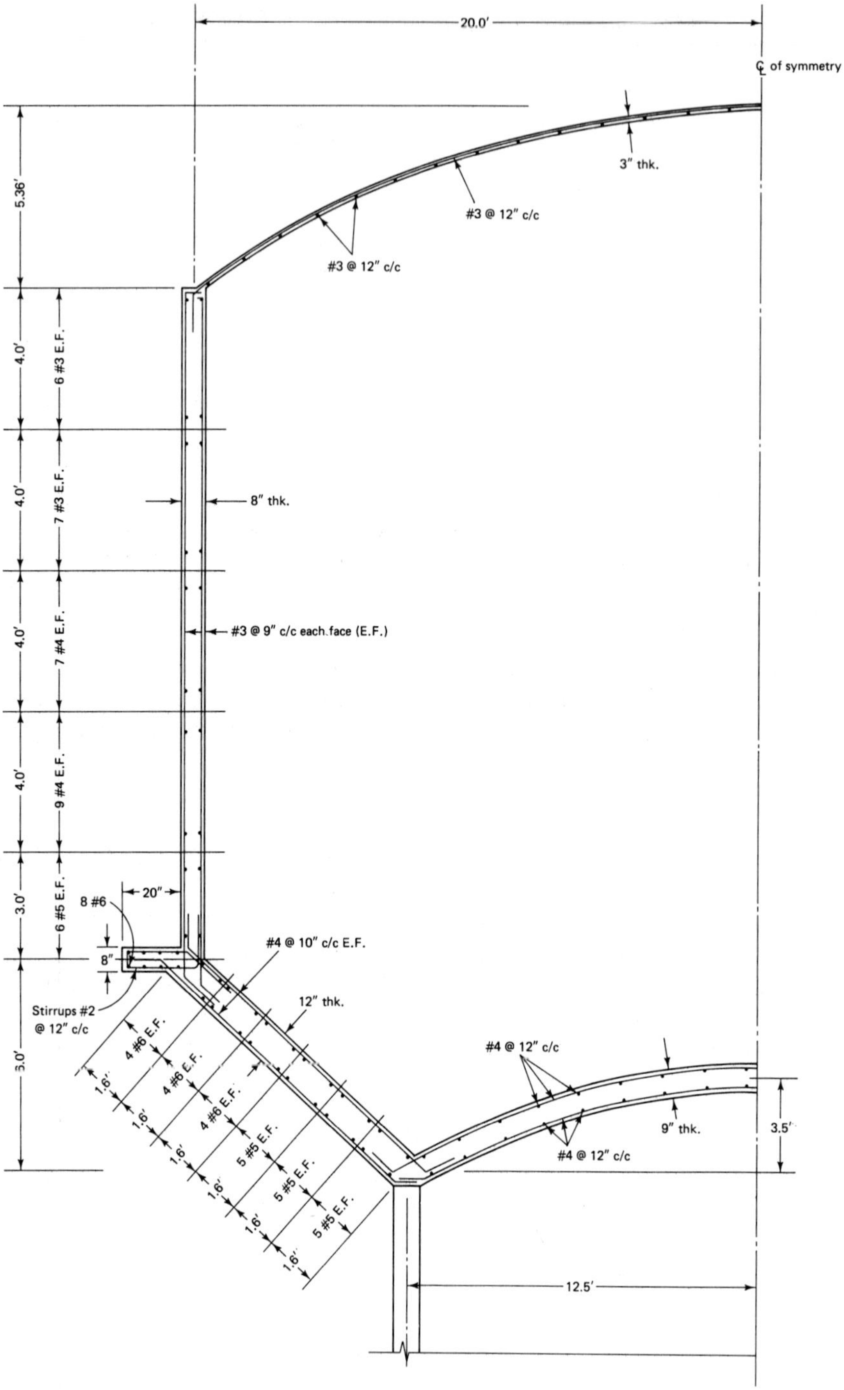

Figure 6.29 Details of reinforcement for Intze tank.

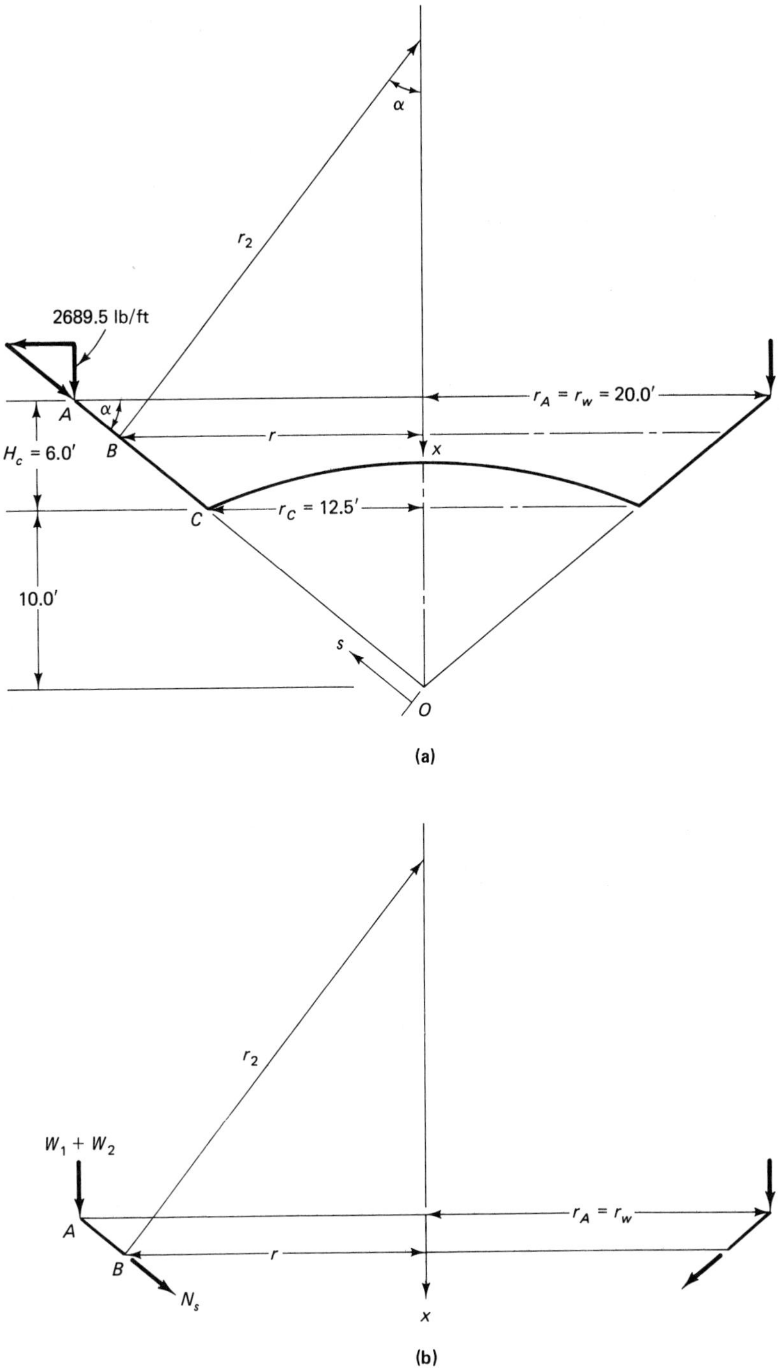

Figure 6.30 Conical shell portion of Intze tank. (a) Coordinate system. (b) Freebody of section *AB*.

This stress is less than the allowable value of 400 psi. The steel required is determined as

$$A_s = \frac{70{,}599.06}{20{,}000.0} = 3.53 \text{ in.}^2$$

Hence eight No. 6 bars as rings, with stirrups of No. 2 bars at 12 in. center-to-center, may be provided in the ring beam. The laps in the ring bars should, of course, be well staggered.

Conical Shell: Analysis

Dead load. We shall denote by r_A and r_C, the radii of the parallel circle at the top and bottom of the cone, respectively, and the length $\overline{OA}$ we shall denote by L [Fig. 6.30(a)]. We measure the coordinate s along the line OA starting from the point O. By simple geometry, we have

$$\tan \alpha = \frac{H_c}{r_A - r_C}$$

$$L = \frac{r_A}{\cos \alpha}$$

and

$$L - s = \frac{x}{\sin \alpha}$$

To find N_s, we take a section x = *constant,* and consider equilibrium of the freebody above [Fig. 6.30(b)]. Let W_1 be the total vertical load carried on top of the cone which comes from the wall and the roof dome, and let W_2 be the weight of the ring beam. The self-weight of the conical frustum, W_3, is given by

$$W_3 = \pi(r_A + r)[x^2 + (r_A - r)^2]^{1/2}(t_c\gamma_c)$$

where t_c is the thickness of the cone and γ_c is the density of concrete. Then we have

$$N_s(2\pi r \sin \alpha) = -(W_1 + W_2 + W_3)$$

Hence

$$N_s = -\frac{W_1 + W_2 + W_3}{2\pi r \sin \alpha} \tag{f1}$$

For our present problem, we have

$$\tan \alpha = \frac{6.0}{7.5} = 0.8$$

and

$$L = \frac{20.0}{\cos \alpha} = 25.612 \text{ ft}$$

Also, at the base of the cylindrical wall, we had previously found $N_x = -2489.49$ lb/ft. Hence we can obtain the following:

$$W_1 = 2489.49(2\pi)(20.0) = 312{,}838.54 \text{ lb}$$

$$W_2 = 2\pi(21.0)(200.0) = 26{,}389.38 \text{ lb}$$

and

$$W_3 = \pi(20.0 + r)[x^2 + (20.0 - r)^2]^{1/2}(1.0)(150.00)$$

Substituting

$$r = \left(\frac{16.0 - x}{16.0}\right)(20.0) = (20.0 - 1.25x)$$

into the above, we obtain

$$W_3 = 754.35(x)(40.0 - 1.25x)$$

Hence from Eq. (f1) we have

$$N_s = -\left[\frac{312{,}838.54 + 26{,}389.38 + 754.35(x)(40.0 - 1.25x)}{2\pi(20.0 - 1.25x)(16.0/25.612)}\right]$$

or

$$N_s = -\left[\frac{69{,}139.34 + 153.75(x)(40.0 - 1.25x)}{16 - x}\right] \tag{g1}$$

To obtain N_θ we use Eq. (6.2) and note that $r_1 = \infty$ for the cone. Hence

From Fig. 6.30(a), we have

$$\overline{OB} = (L - \overline{AB}) = L - \frac{x}{\sin \alpha}$$

and hence

$$r_2 = \overline{OB} \cot \alpha = \left(L - \frac{x}{\sin \alpha}\right) \cot \alpha$$

Also,

$$p_r = \gamma_c t_c \cos \alpha$$

Thus we have

$$N_\theta = \gamma_c t_c \cos \alpha \cot \alpha \left(L - \frac{x}{\sin \alpha}\right)$$

or

$$N_\theta = \gamma_c t_c \cos \alpha \left(L \cot \alpha - \frac{x \cot \alpha}{\sin \alpha} \right) \tag{f2}$$

We can now substitute in the numerical values for our problem, to obtain

$$N_\theta = 117.13(32.015 - 2.0x) \tag{g2}$$

We shall obtain numerical values of N_s and N_θ for various values of x (or s). But first we derive expressions for N_s and N_θ in the cone under water load.

Water load. For the case of water load on the conical shell, we have $p_s = 0$ and $p_r = (H_W + x)(\gamma_w)$, where H_W is the height of water above the top of the cone and γ_w is the density of water. In terms of coordinate s, we have

$$p_r = \gamma_w[H_W + (L - s) \sin \alpha] = \gamma_w(H_W + L \sin \alpha - s \sin \alpha)$$

It is now easy to obtain N_s from the integral of Eq. (4.7a). This equation is rewritten here as

$$N_s = -\frac{1}{s} \int (p_s - p_r \cot \alpha) s \, ds + \frac{C}{s}$$

Substituting for p_s and p_r, we have

$$N_s = \frac{\gamma_W \cot \alpha}{s} \int (H_W + L \sin \alpha - s \sin \alpha) s \, ds + \frac{C}{s}$$

Upon integrating, we obtain

$$N_s = \gamma_w \cot \alpha \left[(H_W + L \sin \alpha) \frac{s}{2} - \frac{s^2}{3} \sin \alpha \right] + \frac{C}{s}$$

For the case of water load, we will have $N_s = 0$ at the top of the cone (i.e., at $s = L$). Substituting this boundary condition into the equation above, we can determine C. We thus find

$$C = -\gamma_w \cot \alpha \left(\frac{H_W L^2}{2} + \frac{L^3}{6} \sin \alpha \right)$$

Hence

$$N_s = \gamma_w \cot \alpha \left[-\frac{H_W L^2}{2s} - \frac{L^3}{6} \frac{\sin \alpha}{s} + (H_W + L \sin \alpha) \frac{s}{2} - \frac{s^2}{3} \sin \alpha \right] \tag{f3}$$

For our problem we substitute $\gamma_w = 62.5$ lb/ft^3, $L = 25.612$ ft, $\cot \alpha = 1.25$, and $H_W = 19.0$ ft, etc., into the above and simplify, to obtain

$$N_s = 1367.19s - 16.268s^2 - \frac{623{,}517.47}{s} \tag{g3}$$

Also,

$$N_\theta = p_r \cdot r_2$$

or

$$N_\theta = \gamma_w(H_W + L \sin\alpha - s \sin\alpha)s \cot\alpha \tag{f4}$$

For our problem we substitute $x = (L - s)\sin\alpha$ and $s = (25.612 - x/\sin\alpha)$. Thus we obtain, in terms of x,

$$N_\theta = 62.5(19.0 + x)(32.015 - 2.0x) \tag{g4}$$

For various values of s or x, forces N_s and N_θ were calculated under dead and water load, from Eqs. (g1–g4). These values are summarized in Table 6.3.

Conical Shell: Design

From Table 6.3 we see that N_s is compressive throughout, and the maximum compressive stress will occur at the bottom of the cone, this stress being $\sigma = (31{,}132.11)/(12.0)^2 = 216.20$ psi. Since this stress is small, we shall provide only 0.3% steel in the meridional direction, or

$$A_s = 0.003(12.0)^2 = 0.432 \text{ in.}^2/\text{ft}$$

Hence No. 4 bars at 10-in. center-to-center spacing on each face should be provided in the meridional direction. The 10-in. center-to-center spacing should, of course, be measured at the top of the cone, and it will become closer toward the bottom due to the reduction in the circumference. Now, in the circumferential direction, N_θ reduces as we go toward the bottom of the cone, because although p_r increases (for the water load case), r_2 decreases rapidly. The maximum tensile stress in the concrete (neglecting steel reinforcement) will be $\sigma = 41{,}767.73/(12.0)^2 = 290.05$ psi, which is acceptable, and hence the thickness assumed for the cone is satisfactory. The hoop reinforcement can be calculated in tabular form, as seen in Table 6.4.

The spacing of the rings comes to be about 4 in., which should be adequate

TABLE 6.3 Membrane Forces in Conical Shell

			DL {Eqs. (gl), (g2)}		WL {Eqs. (g3), (g4)}		DL + WL	
$(25.612 - s)$ (ft)	s (ft)	x (ft)	N_s	N_θ	N_s	N_θ	N_s (lb/ft)	N_θ (lb/ft)
0.0	25.612	0.0	−4,321.21	3,749.92	0.00	38,017.81	−4,321.21	41,767.73
1.6	24.012	1.0	−5,006.48	3,515.66	−2,517.69	37,518.75	−7,524.17	41,034.41
3.2	22.412	2.0	−5,762.19	3,281.40	−5,350.61	36,769.69	−11,112.80	40,051.09
4.8	20.812	3.0	−6,604.59	3,047.14	−8,551.87	35,770.63	−15,156.46	38,817.77
6.4	19.212	4.0	−7,555.36	2,812.88	−12,192.66	34,521.56	−19,748.02	37,334.44
8.0	17.612	5.0	−8,644.06	2,578.62	−16,370.09	33,022.50	−25,014.15	35,601.12
9.6	16.012	6.0	−9,912.06	2,344.36	−21,220.05	31,273.44	−31,132.11	33,617.80

TABLE 6.4 Design of Hoop Reinforcement for Conical Shell

$(25.612 - s)$ (ft)	N_θ (lb/ft)	Average N_θ for Zone, $(N_\theta)_{avg}$ (lb/ft)	Zone Length, ds (ft)	Tension in Zone $(N_\theta)_{avg}\,ds$, T (lb)	A_s $(T/20{,}000)$ (in.^{2})	Bars on Each Face (Min. No. 4 at 10 in. c-c)
0.0	41,767.73					
		41,401.07	1.6	66,241.71	3.31	4 No. 6
1.6	41,034.41					
		40,542.75	1.6	64,868.40	3.24	4 No. 6
3.2	40,051.09					
		39,434.43	1.6	63,095.09	3.15	4 No. 6
4.8	38,817.77					
		38,076.11	1.6	60,921.78	3.05	5 No. 5
6.4	37,334.44					
		36,467.78	1.6	58,348.45	2.92	5 No. 5
8.0	35,601.12					
		34,609.46	1.6	55,375.14	2.77	5 No. 5
9.6	33,617.80					

for the actual placing of reinforcement. The arrangement of reinforcement is shown in Fig. 6.29.

Bottom Dome: Analysis

Dead load. The load due to self-weight of the bottom dome is $p = 150.0(0.75) = 112.5$ lb/ft^2. N_ϕ and N_θ values can be obtained readily from Eqs. (6.6a,b) as

$$N_\phi = -\frac{2707.99}{1 + \cos\phi}$$

and

$$N_\theta = -2707.99\left(\cos\phi - \frac{1}{1 + \cos\phi}\right)$$

Water load. Referring to Fig. 6.31, we have $p_\phi = 0$, and p_r is given by

$$p_r = -\gamma_w\,[H_D + a(1 - \cos\phi)] \tag{h1}$$

where

$$\gamma_w = 62.5 \text{ lb/ft}^3$$

Thus from Eq. (6.7), we have

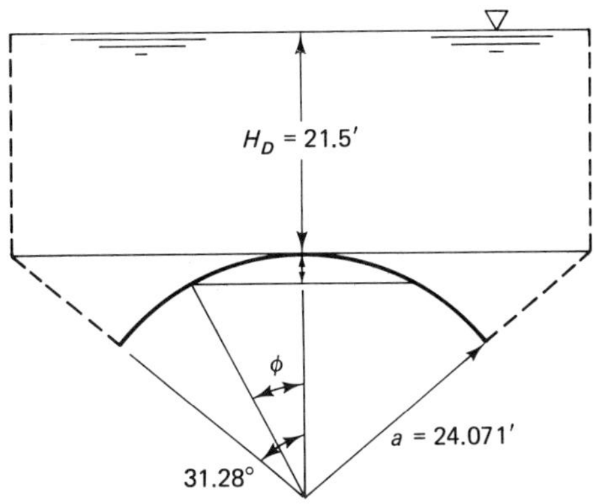

Figure 6.31 Bottom dome geometry of Intze tank.

$$\begin{aligned} N_\phi &= \frac{1}{a \sin^2 \phi} \int_0^\phi p_r \cos \phi \sin \phi \ \ a^2 \, d\phi \\ &= -\frac{\gamma_w a}{\sin^2 \phi} \int_0^\phi [H_D + a(1 - \cos \phi)] \cos \phi \sin \phi \, d\phi \\ &= \frac{\gamma_w a}{\sin^2 \phi} \left[H_D \frac{\cos^2 \phi}{2} + a\left(\frac{\cos^2 \phi}{2} - \frac{\cos^3 \phi}{3} \right) \right]_0^\phi \end{aligned}$$

Finally,

$$N_\phi = \frac{\gamma_w a}{\sin^2 \phi} \left[\frac{H_D + a}{2} \cos^2 \phi - \frac{a}{3} \cos^3 \phi - \frac{H_D}{2} - \frac{a}{6} \right] \tag{h2}$$

Substituting in the appropriate numerical values, we obtain

$$N_\phi = \frac{1504.438}{\sin^2 \phi} (22.786 \cos^2 \phi - 8.024 \cos^3 \phi - 14.762) \tag{h3}$$

Also,

$$N_\theta = p_r \cdot a - N_\phi \tag{h4}$$

We can calculate numerical values of N_ϕ and N_θ, for dead and water load, from Eqs. (6.6a,b) and Eqs. (h1)–(h4) above; these values are summarized in Table 6.5. It may be noted that the value of N_ϕ at $\phi = 0$ under water load, has to be obtained by taking the limit as $\phi \to 0$ in Eq. (h3).

Bottom Dome: Design

The bottom dome has compressive stresses throughout. The maximum stress is equal to 21,009.36/12.0(9.0) = 194.53 psi, which is very small. Since this dome is also fairly thick compared to its radius, buckling stresses are not expected to be critical. We shall provide only the minimum steel area, 0.3% of the cross-sectional area, in each direction. Thus,

TABLE 6.5 Membrane Forces in Bottom Dome[a]

ϕ (deg)	DL {Eqs. (6.6a,b)} N_ϕ	DL N_θ	WL {Eqs. (h_1),(h_3),(h_4)} p_r	WL N_ϕ	WL N_θ	DL + WL N_ϕ	DL + WL N_θ
0	−1,354.00	−1,354.00	−1,343.75	−16,172.71	−16,172.71	−17,526.71	−17,526.71
5	−1,356.58	−1,341.11	−1,349.47	−16,207.14	−16,276.08	−17,563.72	−17,617.19
10	−1,364.36	−1,302.49	−1,366.61	−16,309.90	−16,585.77	−17,674.26	−17,888.26
15	−1,377.46	−1,238.25	−1,395.01	−16,479.42	−17,099.86	−17,856.88	−18,338.11
20	−1,396.09	−1,148.59	−1,434.48	−16,713.06	−17,816.31	−18,109.15	−18,964.90
25	−1,420.54	−1,033.73	−1,484.70	−17,007.07	−18,731.14	−18,427.61	−19,764.87
31.28	−1,460.11	− 854.24	−1,562.44	−17,454.37	−20,155.12	−18,914.49	−21,009.36

[a] N_ϕ and N_θ are in lb/ft and p_r in lb/ft^2.

$$A_s = 0.003(12.0)(9.0) = 0.324 \text{ in.}^2/\text{ft}$$

We can provide No. 4 bars at 12-in. center-to-center spacing as meridional and circumferential rings on each face.

Reactions on the Supports

The supporting cylindrical structure will have to resist the resultants N_s and N_ϕ from the cone and bottom dome, respectively. The forces acting on the support are as shown in Fig. 6.32. It is seen that under DL and WL conditions, the total

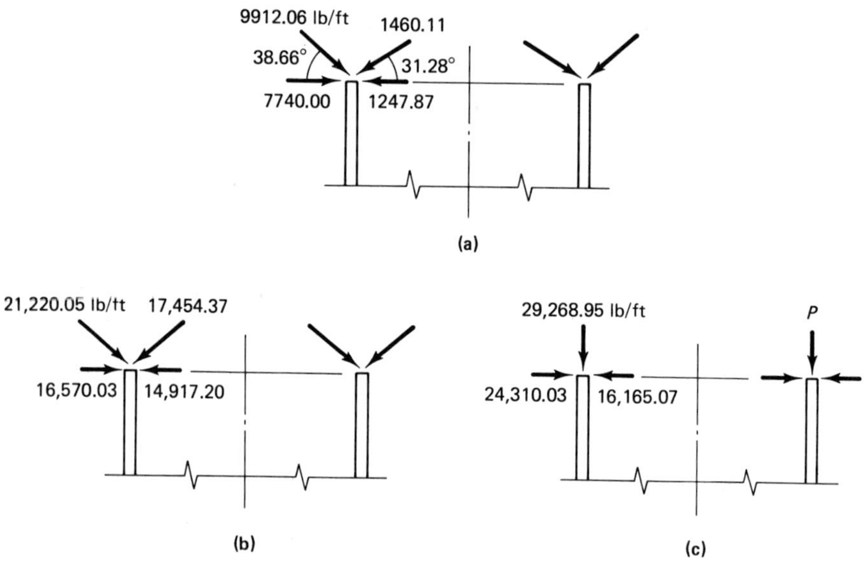

Figure 6.32 Support loads for the Intze tank. (a) Dead load forces. (b) Water load forces. (c) Dead plus water load forces.

horizontal component of the meridional forces from the cone exceeds that from the dome, thereby giving an unbalanced horizontal load. This load must be resisted by the supporting structure. It is clearly seen that if the slope of the cone were increased to about 45°, then the net horizontal force from the dome and the cone acting on the support would reduce further. The vertical load on the support, P, is obtained by addition of the vertical components of the meridional forces.

This completes the design of the Intze water tank, based on membrane theory alone. It should now be an elementary task to design any other Intze tank. Forces in the roof and bottom domes under DL are given by Eqs. (6.6a,b). Forces in the cylindrical wall are also readily obtainable. Equations (h1), (h2), and (h4) for forces in the bottom dome under water load, and Eqs. (f1)–(f4) for the conical shell under DL and WL are quite general and can be used for any other tank by substituting the proper numerical values. We may, therefore, summarize the formulas for membrane forces in various elements of the Intze tank. Using the notations of this example, these expressions are as follows:

Summary of Formulas for Membrane Forces in Intze Tank

Roof dome

1. *Dead Load*

$$N_\phi = \frac{-pa}{1 + \cos\phi} \qquad N_\theta = -pa\left(\cos\phi - \frac{1}{1 + \cos\phi}\right) \tag{6.6a,b}$$

p = total uniformly distributed load (over surface area)

2. *Water Load*

$$N_\phi = N_\theta = 0$$

Cylindrical wall

1. *Dead Load*

$$N_x = -N_{\phi_1} \sin\phi_1 - (\gamma_c t_w)(x)$$

$$N_\theta = 0$$

ϕ_1 = angle of the bottom edge of the roof dome
N_{ϕ_1} = N_ϕ at $\phi = \phi_1$ from the roof dome
γ_c = density of concrete
t_w = thickness of wall

2. *Water Load*

$$N_x = 0$$

$$N_\theta = \gamma_w r_w x$$

r_w = radius of cylindrical wall = radius at top of cone

γ_w = density of water/fluid

Ring beam

1. *Dead Load*

$$\text{Hoop tension} = N_{x_o}(\cot \alpha)R$$

R = mean radius of the ring

α = angle of cone to the horizontal

N_{x_o} = N_x in the cylindrical wall at its junction with the ring beam and cone

2. *Water Load*

$$\text{Hoop tension} = 0$$

Conical shell

1. *Dead Load*

$$N_s = -\frac{W_1 + W_2 + W_3}{2\pi r(\sin \alpha)} \tag{f1}$$

$$N_\theta = \gamma_c t_c \cos \alpha \left(L \cot \alpha - x \frac{\cot \alpha}{\sin \alpha} \right) \tag{f2}$$

W_1 = total load from cylindrical wall (including roof dome)

W_2 = weight of ring beam

$W_3 = \pi(r_A + r)\, t_c \gamma_c [x^2 + (r_A - r)^2]^{1/2}$

t_c = thickness of cone

L = length of conical shell along meridian

x = vertical coordinate, measured downward from the junction of the cone with the wall

r = radius of the parallel circle at x.

r_A = radius of the parallel circle at the top of the cone ($x = 0$).

2. *Water Load*

$$N_s = \gamma_w \cot\alpha \left[-\frac{1}{s}\left(\frac{H_W L^2}{2} + \frac{L^3 \sin\alpha}{6}\right) + \frac{H_W + L\sin\alpha}{2}\, s - \frac{s^2}{3}\sin\alpha \right] \tag{f3}$$

$$N_\theta = \gamma_w \cot\alpha\, [H_W + (L - s)\sin\alpha]s \tag{f4}$$

H_W = height of water above the top of the cone

s = coordinate along the meridian of the cone from its apex

$= (L - x/\sin\alpha)$

Bottom dome

1. *Dead Load*
Equations for N_ϕ and N_θ are the same as those for the roof dome [Eqs (6.6a,b)]

2. *Water Load*

$$N_\phi = \frac{\gamma_w a}{\sin^2\phi}\left(\frac{H_D + a}{2}\cos^2\phi - \frac{a}{3}\cos^3\phi - \frac{H_D}{2} - \frac{a}{6}\right) \tag{h2}$$

$$N_\theta = p_r \cdot a - N_\phi \tag{h4}$$

$$p_r = -\gamma_w[H_D + a(1 - \cos\phi)] \tag{h1}$$

H_D = height of water above the apex of the dome

As stated previously, discontinuity problems will arise at the junctions of shells and ring beams. In Chapter 5 we discussed how to solve such edge-load problems using the bending theory for cylindrical shells. In the following chapter we consider a similar bending theory derivation for any shell of revolution subjected to axisymmetric loads. Such theory will allow us to solve for redundant forces acting at the juncture of these doubly curved shells.

REFERENCES

[1] "Design of Circular Domes," Portland Cement Association, Publication No. IS076.01D, Skokie, Ill.

[2] Hass, A. M., *Design of Thin Concrete Shells,* Vol. 1, John Wiley & Sons, Inc., New York, 1967.

[3] Flügge, W., *Stresses in Shells,* 2nd ed., Springer-Verlag, New York, 1973.

[4] Bednar, H. H., *Pressure Vessel Design Handbook,* Van Nostrand Reinhold Company, Inc., New York, 1981.

[5] Ramaswamy, G. S., *Design and Construction of Concrete Shell Roofs,* Revised ed., R.E. Krieger, Malabar, Florida, 1984.

[6] Billington, D. P., Gould, P. L., et al., "Reinforced Concrete Cooling Tower Shells—Practice and Commentary," Report of ACI-ASCE Joint Committee 334, *Am. Concr. Inst. J.*, Jan. 1977, pp. 22–31.

[7] Billington, D. P., *Thin Shell Concrete Structures,* 2nd ed., McGraw-Hill Book Company, New York, 1982.

[8] Turner, M. R., "Timber Dome Roof over Stadium Spans Record 502 ft," *Civ. Eng. (N.Y.),* Aug. 1977.

[9] "Tacoma Dome," *Civ. Eng. (N.Y.),* July 1984.

[10] Klus, J. P., and Wortley, C. A., "Finns Emphasize Esthetics in Water Towers," *Civ. Eng. (N.Y.),* Sept. 1973, pp. 84–87.

[11] Fischer, L., *Theory and Practice of Shell Structures,* Wilhelm Ernst & Sohn, Berlin, 1968.

[12] Behr, R. A., Mehta, K. C., and Kiesling, E. W., "Strength and Stability of Earth Covered Dome Shells," *ASCE, J. Struct. Div.*, Vol. 110, No. 1, 1984.

PROBLEMS

6.1. A cylindrical water tank with a hemispherical bottom is supported as shown in Fig. P6.1. Plot the variations of meridional and circumferential stress resultants in the cylinder and the dome under water load.

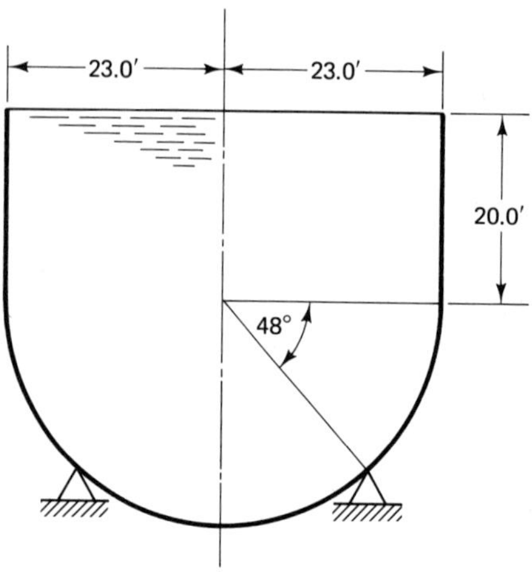

Figure P6.1 Cylindrical water tank with a hemispherical bottom.

6.2. For the dome with a skylight discussed in Section 6.8, obtain expressions for N_ϕ and N_θ under a snow load of p_o, distributed uniformly on the plan area. Use the definite integral approach.

6.3. For the tank and its supporting tower, in the shape of a hyperboloid of revolution, shown in Fig. 6.19 (Section 6.13), obtain membrane stress resultants under water load, in the tank wall, as well as in the supporting shell wall. Discuss the problems that will arise at the junction of the shell with the bottom slab.

6.4. At the junction of the cone and cylinder of a pressure vessel, a toroidal knuckle $AB'C$ of radius $r_1 = 1.0$ ft is introduced, as shown in Fig. P6.4. The thicknesses of the cylinder and the cone are 0.375 in. each, while the knuckle is 1.25 in. thick. Poisson's ratio is equal to 0.3. Find the membrane stress resultants in the cone and the knuckle under an internal pressure of 50 psi. Obtain outward differential deflections between the cylinder and the knuckle at their junction C. Assume that $E = 30 \times 10^6$ psi.

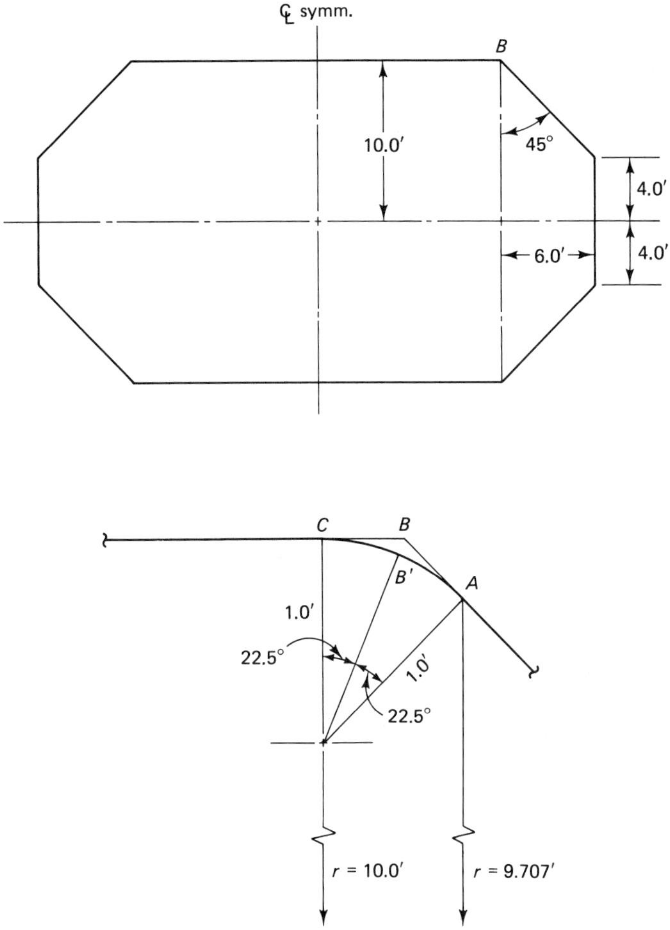

Figure P6.4 Toroidal knuckle used as a transition between conical and cylindrical portions of a pressure vessel.

6.5. The meridian of a dome roof of a mosque, conforming to the style of Islamic architecture, consists of a toroidal shell and a conoidal/spherical shell (Fig. P6.5). The toroidal shell AB has a meridian with a radius of curvature equal to 10 ft. The radius of curvature of the meridian BC of the spherical shell is 40 ft. Both meridians have a common tangent at their junction B. Obtain expressions for the membrane stress resultants in the dome, and design the reinforcement in the roof. Assume a total load of 65 lb/ft^2 over the surface area. Discuss the problems that arise in the membrane analysis at point B.

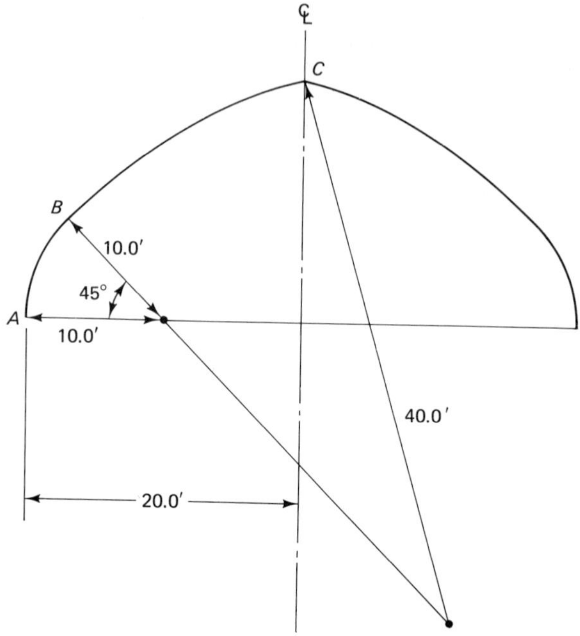

Figure P6.5 Meridian of a shell used as a mosque roof.

6.6. Figure P6.6 shows a "greenhouse" design of an energy-efficient (earth-sheltered) house [12]. It consists of a dome roof covered with 3 ft. of earth. The 36 in.-wide ring beam also serves as a footing. The loading for which this structure has to be designed is summarized as follows:

1. 500 lb/ft^2 over the *surface* area of the roof dome and ring beam, for self-weight, earth load, and live load.
2. An upward soil reaction on the ring beam which is assumed as 3000 lb/ft^2, uniformly distributed.
3. There is a lateral triangular pressure on the cylindrical wall, having a value of 480 lb/ft^2 at the base, plus a uniformly distributed lateral pressure of 900 lb/ft^2 due to "surcharge" pressure below the ring beam.

Obtain the membrane stresses in the dome, ring beam, and cylindrical wall. Design the reinforcement in these elements under membrane conditions. Work with the mid-surface dimensions of the elemental members.

6.7. Figure P6.7 shows the cross section of a structure that somewhat resembles the structure of the departure hall of the international airport at Abu Dhabi, United Arab Emirates. Its meridian consists partly of a circular arc of radius $a = 80$ ft, and partly of a parabolic curve whose equation is $(r - 20)^2 = 7.225z$. Both the curves meet smoothly at A, where they have a common horizontal tangent. Show, step by step, how to obtain the membrane stress resultants in this structure for a total load of 75 lb/ft^2 of surface area. Design the thickness and reinforcement for the structure.

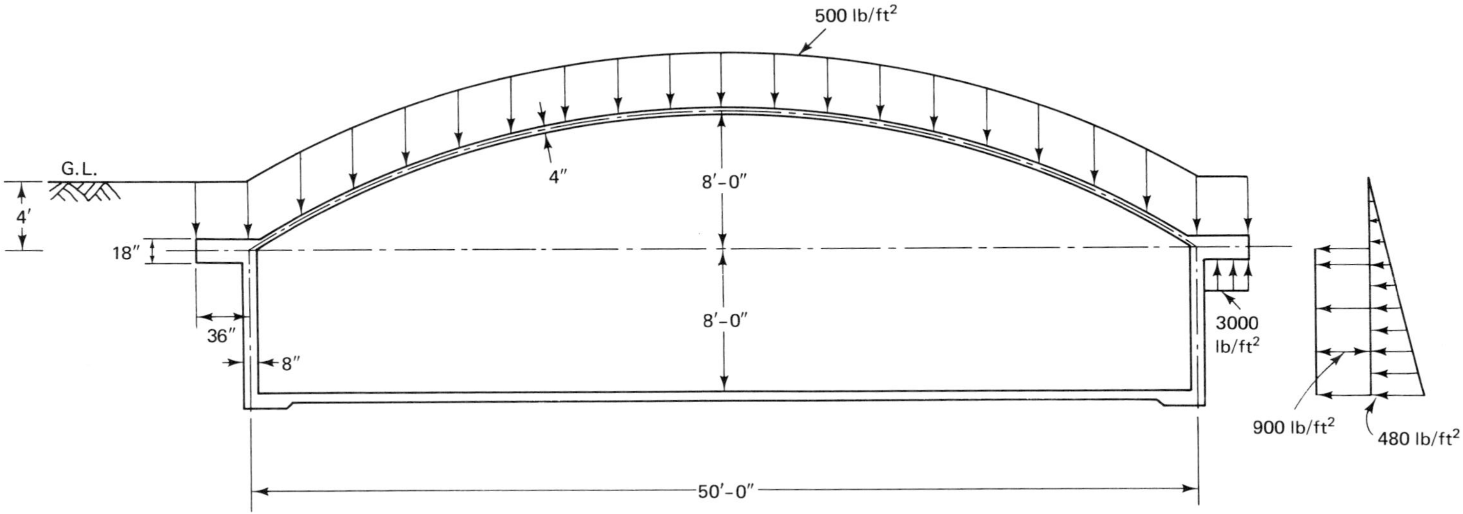

Figure P6.6 Shell structure used in an earth-sheltered home.

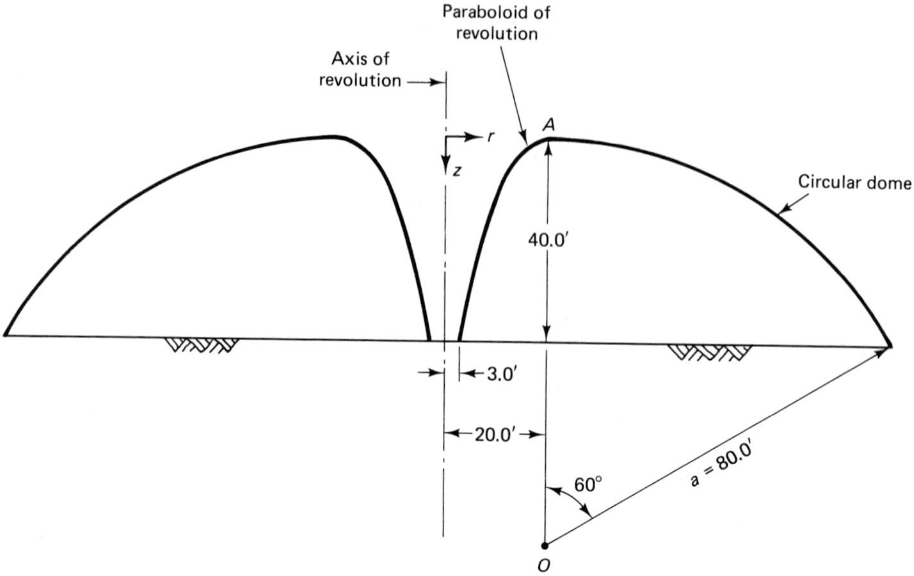

Figure P6.7 Cross section of a structure that is idealized as a surface of revolution with parabolic and circular portions.

6.8. For the hyperboloid of revolution discussed in Section 6.13, write a computer program for obtaining membrane stress resultants N_ϕ and N_θ under self-weight plus a line load of P per unit length of circumference at the top.

7

Bending Theory for Shells of Revolution: Axisymmetric Loading

7.1 INTRODUCTION

In the membrane analysis of shells of revolution considered in Chapter 6, we had seen that the solutions obtained by the membrane theory failed to satisfy continuity of displacements at the junctions of two shells, shells and ring beams, etc. We have to apply edge loads, consisting of shears and bending moments, in order to correct the discrepancies between the membrane displacements. We shall therefore derive in this chapter the bending theory for shells of revolution. As for cylindrical shells (Chapter 5), we shall assume that no distributed loads act on the shell, and that only axisymmetric edge loads are present. For the case of distributed loading, the membrane solution will be satisfactory away from the edges. Even under axisymmetric edge loading only, the solution of the shell equations involves many complexities for a general shell of revolution, and hence we shall solve the equations making a further assumption that the shell is spherical. We shall then consider how this solution can be applied approximately in solving problems of other shells of revolution as well.

7.2 DERIVATION OF EQUATIONS

For deriving the equations of the spherical shell, we make the same assumptions as those made in the bending theory of circular cylindrical shells (Chapter 5). We shall, as before, write the equations of equilibrium, kinematics, and Hooke's law, and then reduce these to a single governing differential equation.

Equilibrium Equations

Let us consider equilibrium of an element of the shell obtained by taking sections ϕ, $\phi + d\phi$, and θ, $\theta + d\theta$, as shown in Fig. 7.1(a). Due to axisymmetry, the only nonzero internal stress resultants will be N_ϕ, N_θ, M_ϕ, M_θ, and Q_ϕ, and these will all be constant in θ.

From Fig. 7.1(b), we see that N_ϕ and $N_\phi + (dN_\phi/d\phi)\,d\phi$, acting on opposite edges, have a resultant along the inward normal equal to $N_\phi\, rd\theta\, d\phi$ if we neglect terms of higher order (i.e., terms containing the product of more than two differential quantities). Similarly, the shears Q_ϕ and $Q_\phi + (dQ_\phi/d\phi)\,d\phi$, acting on opposite

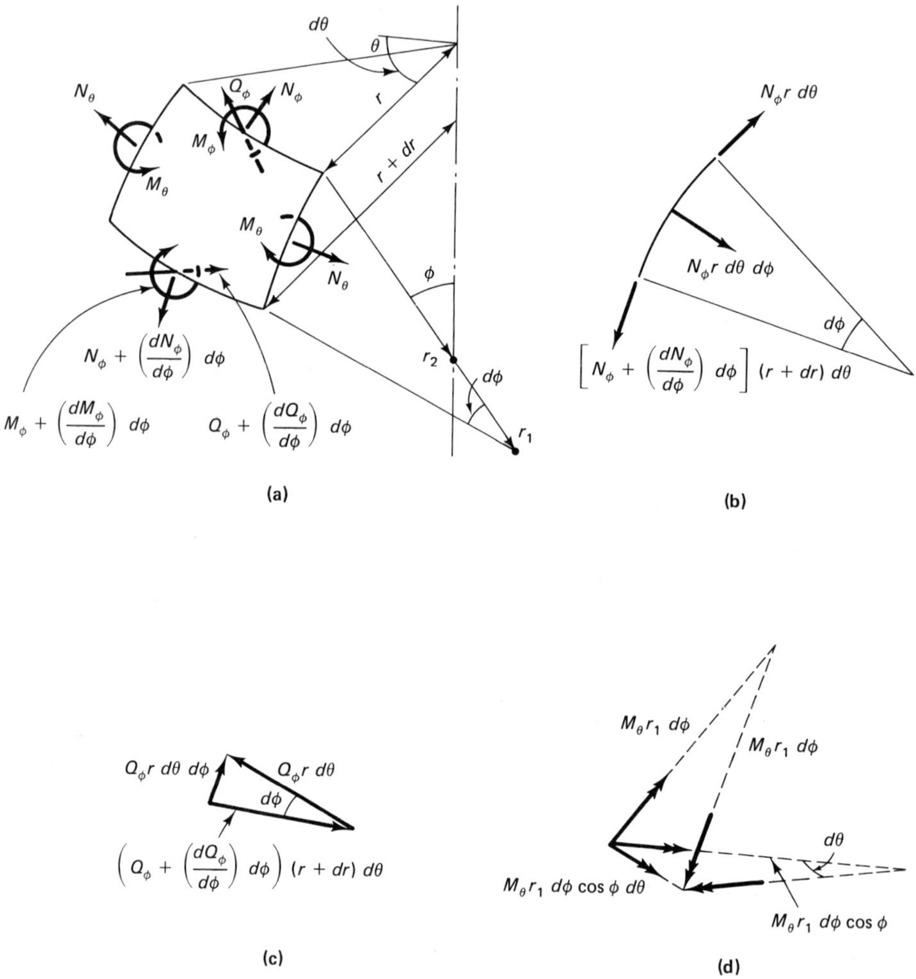

Figure 7.1 Element of a shell of revolution under bending loads. (a) Element geometry and loading. (b) Normal component of meridional forces. (c) Tangential component of shear forces. (d) Circumferential resultant of moments M_θ.

edges, will have a resultant force of $Q_\phi\, r d\theta\, d\phi$ acting along the tangent to the meridian, as shown in Fig. 7.1(c). Due to N_θ acting on the two edges of length $r_1 d\phi$, we shall have resultant components along the tangent and normal to the meridian given by $N_\theta\, r_1 d\phi\, d\theta \cos\phi$ and $N_\theta\, r_1 d\phi\, d\theta \sin\phi$, respectively, as obtained earlier for membrane behavior. We can now write the equations for equilibrium of forces in the directions tangential and normal to the meridian, as

$$\frac{d}{d\phi}(N_\phi \cdot r\, d\theta)\, d\phi - N_\theta \cdot r_1\, d\phi\, d\theta \cos\phi - Q_\phi \cdot r\, d\theta\, d\phi = 0$$

and

$$-N_\phi \cdot r\, d\theta\, d\phi - N_\theta \cdot r_1\, d\phi\, d\theta \sin\phi - \frac{d}{d\phi}(Q_\phi \cdot r\, d\theta)\, d\phi = 0$$

A third equilibrium equation can be obtained by taking moments about the tangent to the parallel circle, $\phi = constant$. Figure 7.1(d) shows a vectorial representation of moments $(M_\theta\, r_1 d\phi)$ acting on edges θ and $\theta + d\theta = constant$. The components of these vectors in the plane of the parallel circle $\theta = constant$ are of magnitude $M_\theta\, r_1 d\phi \cos\phi$, and they subtend an angle $d\theta$, as shown. These components have a resultant $(M_\theta\, r_1 d\phi \cos\phi\, d\theta)$ which acts in the opposite direction of the moment $M_\phi + (dM_\phi/d\phi)\, d\phi$. We can thus write our third equilibrium equation as

$$\frac{d}{d\phi}(M_\phi \cdot r\, d\theta)\, d\phi - M_\theta \cdot r_1\, d\phi \cos\phi\, d\theta - Q_\phi \cdot r\, d\theta\, r_1\, d\phi = 0$$

Dropping the term $(d\theta\, d\phi)$ from the three equilibrium equations above, and denoting $(\cdot)^{\bullet} = d(\cdot)/d\phi$, we can rewrite these equations as

$$(N_\phi \cdot r)^{\bullet} - N_\theta \cdot r_1 \cos\phi = Q_\phi \cdot r \tag{a1}$$

$$(N_\phi \cdot r) + N_\theta \cdot r_1 \sin\phi = -(Q_\phi \cdot r)^{\bullet} \tag{a2}$$

and

$$(M_\phi \cdot r)^{\bullet} - M_\theta \cdot r_1 \cos\phi = (Q_\phi \cdot r) r_1 \tag{a3}$$

Multiplying Eqs. (a1) and (a2) by $\sin\phi$ and $\cos\phi$, respectively, and adding, we obtain

$$(N_\phi \cdot r)^{\bullet} \sin\phi + (N_\phi \cdot r)\cos\phi = -(Q_\phi \cdot r)^{\bullet}\cos\phi + (Q_\phi \cdot r)\sin\phi$$

or

$$(N_\phi \cdot r \sin\phi)^{\bullet} = -(Q_\phi \cdot r\cos\phi)^{\bullet}$$

Integrating both sides with respect to ϕ gives us

$$N_\phi r \sin\phi + Q_\phi r \cos\phi = C \tag{a4}$$

where C is a constant.

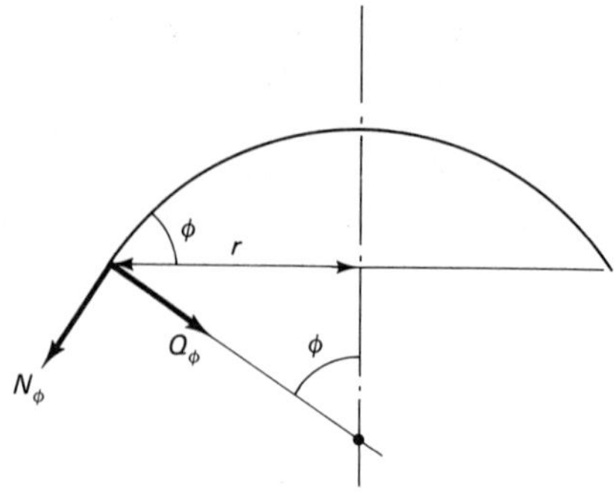

Figure 7.2 Freebody diagram of shell section above coordinate ϕ = *constant*, showing internal forces N_ϕ and Q_ϕ.

To understand the physical interpretation of this equation, let us consider equilibrium of the freebody above a section ϕ = *constant*, as shown in Fig. 7.2. Since there are no loads acting on the shell, for equilibrium of forces in the vertical direction we must have

$$2\pi r(N_\phi \sin\phi) + 2\pi r(Q_\phi \cos\phi) = 0$$

or

$$N_\phi r \sin\phi + Q_\phi r \cos\phi = 0$$

This equation is the same as Eq. (a4) with $C = 0$. If we had a concentrated upward load at the apex of the shell, or a line load with a resultant equal to V acting on the freebody, the equilibrium equation above would be

$$2\pi r(N_\phi \sin\phi) + 2\pi r(Q_\phi \cos\phi) = V$$

or

$$N_\phi r \sin\phi + Q_\phi r \cos\phi = \frac{V}{2\pi}$$

where the right-hand side of this equation is constant in ϕ. This equation is the same as Eq. (a4) with $C = V/2\pi$. Thus we may say that the constant C in Eq. (a4) merely represents the effect of a concentrated load or a line load in the equation for equilibrium of vertical forces. For our edge-load problems, since we shall not consider any concentrated or line loads, we take $C = 0$. Hence Eq. (a4) gives us

$$N_\phi = -Q_\phi \cot\phi$$

We can use this simple equation [which is obtained by a combination of the first two equations (a1) and (a2)], any one of the original first two equations, and the third equation (a3) as the equilibrium relationships for our problem. These relationships are summarized below.

$$N_\phi = -Q_\phi \cot\phi \tag{7.1a}$$

$$N_\phi r + N_\theta r_1 \sin\phi = -(Q_\phi \cdot r)^{\cdot} \tag{7.1b}$$

$$(M_\phi \cdot r)^{\cdot} - M_\theta r_1 \cos\phi = (Q_\phi \cdot r) r_1 \tag{7.1c}$$

Kinematics

We now consider the deformations of the shell, in order to establish additional equations to be used in solving the edge-load problem. We shall denote by v and w the displacements of a point on the shell along the tangent to the meridian and along the outward normal, respectively. We shall also call ε_ϕ and ε_θ the strains in the ϕ- and θ-directions caused by the in-plane stress resultants N_ϕ and N_θ, respectively. These are the values of the strains that are constant over the thickness t. On the other hand, we shall denote by $\bar{\varepsilon}_\phi$ and $\bar{\varepsilon}_\theta$ the strains (in the ϕ- and θ-directions, respectively) which vary over the thickness and are caused by the moments M_ϕ and M_θ. We have already derived in Chapter 6 (Eqs. 6.12a,b) equations for ε_ϕ and ε_θ as follows:

$$\varepsilon_\phi = \frac{(v)^{\cdot} + w}{r_1} \qquad \varepsilon_\theta = \frac{v \cot\phi + w}{r_2} \tag{7.2a,b}$$

If we denote by χ the "clockwise" rotation of the tangent to the (left-side) meridian, we had also derived (in Eq. 6.15b) the expression

$$\chi = \frac{(w)^{\cdot} - v}{r_1} \tag{7.2c}$$

The changes of curvature κ_ϕ and κ_θ caused by moments M_ϕ and M_θ, respectively, shall be considered positive when they have the same sense as that for positive moments. Thus for a shell of positive Gaussian curvature, as in Fig. 6.1, a positive κ_ϕ or κ_θ will mean a decrease in the original curvature. Now let us consider the deflections of two neighboring points, A and B, at a distance $r_1 d\phi$ on the shell (Fig. 7.3). After deflections, the points move to new positions A' and B', respectively. If χ and $\chi + (d\chi/d\phi)\, d\phi$ are the rotations of the tangents at A and B, respectively, the change of angle between the tangents at A and B, due to the deformations, will be $(d\chi/d\phi)\, d\phi$. This term divided by the arc length $AB = r_1 d\phi$ gives us the change of curvature κ_ϕ, and hence we can write

$$\kappa_\phi = \frac{(\chi)^{\cdot}}{r_1} \tag{7.2d}$$

It may be noted from Fig. 7.3, that a positive $(\chi)^{\cdot}$ will cause a change of curvature as would be caused by a positive M_ϕ (i.e., a positive κ_ϕ), and this is reflected by the equation above.

It may also be seen from Fig. 7.3 that as point A deflects to A', the normal ACO will take the new position $A'C'O'$ and will make an angle of $(\phi - \chi)$ with the vertical, instead of simply ϕ. The radius of curvature r_2 (AC) of the second principal curve will become $\bar{r}_2$ $(A'C')$ after deflection. Hence the change of curvature κ_θ will be given by

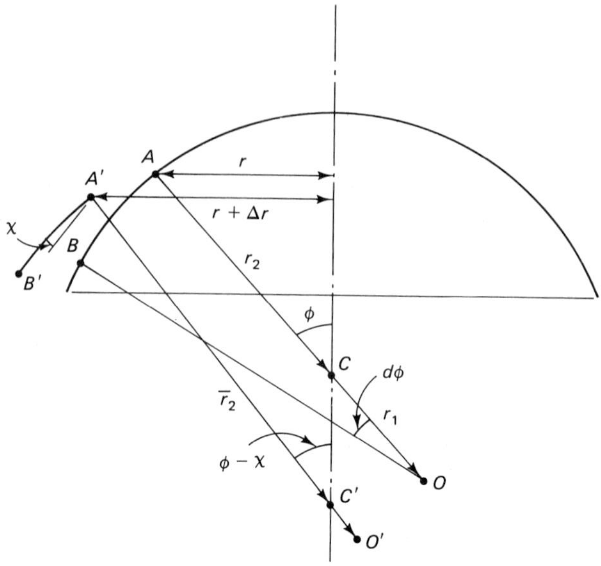

Figure 7.3 Deformations of shell meridian, showing changes in lengths and angles.

$$\kappa_\theta = -\left(\frac{1}{\bar{r}_2} - \frac{1}{r_2}\right)$$

The negative sign outside the parentheses in the equation above is obviously needed to be consistent with our notation. By our definition, a positive κ_θ corresponds to a positive M_θ. A positive M_θ tends to reduce the initial curvature, whereas a positive value of the quantity inside the parentheses above would mean an increase in the initial curvature. We can now write the equation above as

$$\begin{aligned}\kappa_\theta &= \frac{1}{r_2} - \frac{1}{\bar{r}_2} \\ &= \frac{\sin\phi}{r} - \frac{\sin(\phi - \chi)}{r + \Delta r} \\ &= \frac{1}{r}\left[\sin\phi - \sin(\phi - \chi)\left(1 + \frac{\Delta r}{r}\right)^{-1}\right]\end{aligned}$$

or

$$\kappa_\theta \approx \frac{1}{r}\left[\sin\phi - \sin(\phi - \chi)\left(1 - \frac{\Delta r}{r}\right)\right]^*$$

Neglecting the term $\Delta r/r$ compared to unity, and expanding the term $\sin(\phi - \chi)$, we have

*Using Maclaurin's series, we have $(1 + \Delta r/r)^{-1} \approx 1 - \Delta r/r$, neglecting higher-order terms.

$$\kappa_\theta \approx \frac{1}{r}(\sin\phi - \sin\phi\cos\chi + \cos\phi\sin\chi)$$

For small values of χ, we can take $\cos\chi \approx 1$ and $\sin\chi \approx \chi$. Then, from the above, we finally obtain

$$\kappa_\theta \approx \frac{\chi\cos\phi}{r} \qquad \text{or} \qquad \kappa_\theta \approx \frac{\chi\cot\phi}{r_2} \tag{7.2e}$$

Hooke's Law

We now derive Hooke's law equations by performing similar steps as those followed earlier for beams, plates, and cylindrical shells. The strains ε_ϕ and ε_θ, which are constant over the thickness, are related to the in-plane stress resultants N_ϕ and N_θ, by Hooke's law, according to

$$\varepsilon_\phi = \frac{1}{Et}(N_\phi - \nu N_\theta) \qquad \varepsilon_\theta = \frac{1}{Et}(N_\theta - \nu N_\phi) \tag{7.3a,b}$$

Conversely, solving for N_ϕ and N_θ, we obtain

$$N_\phi = D(\varepsilon_\phi + \nu\varepsilon_\theta) \qquad N_\theta = D(\varepsilon_\theta + \nu\varepsilon_\phi) \tag{7.4a,b}$$

where $D = Et/(1 - \nu^2)$ is the extensional rigidity of the shell.

Next, we must obtain relationships between the moments and the changes of curvature. We shall measure z from the middle surface, positive along the inward normal. Let us consider the elongation of a fiber EF, of length $(r_1\,d\phi)$* and thickness dz, at a distance z from the middle surface [Fig. 7.4(a)]. If χ and $\chi + (d\chi/d\phi)\,d\phi$ are the rotations of the tangents to points A and B, respectively, the relative rotation of the normals at A and B is $(d\chi/d\phi)d\phi = (\kappa_\phi \cdot r_1 d\phi)$. This conclusion also

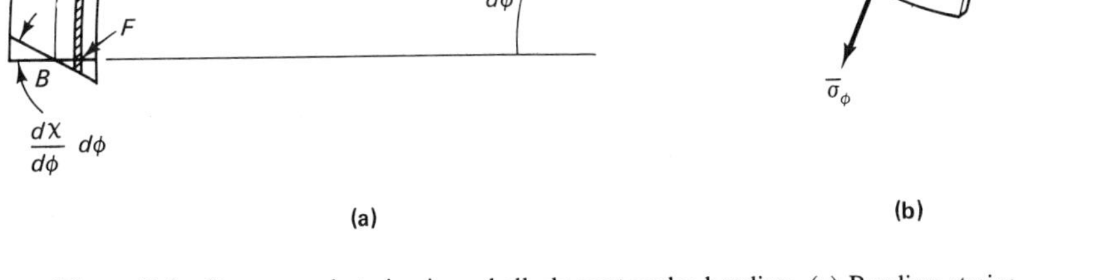

Figure 7.4 Stresses and strains in a shell element under bending. (a) Bending strains over shell thickness. (b) Bending stresses acting on an elemental slice of thickness dz.

*The actual length of EF is $(r_1 - z)\,d\phi$, as opposed to $r_1\,d\phi$ of the middle surface fiber AB. However, this difference is neglected for reasons explained in Chapter 5 concerning cylindrical shells.

follows from the definition of κ_ϕ, which is the change of angle between the tangents (or normals) at A and B divided by the arc length $r_1 d\phi$. Hence the elongation of EF is given by $z\kappa_\phi r_1 d\phi$, and thus the strain $\bar{\varepsilon}_\phi$, at a distance z from the middle surface becomes

$$\bar{\varepsilon}_\phi(z) = \frac{z\kappa_\phi r_1 d\phi}{r_1 d\phi} = z\kappa_\phi$$

In a similar manner, from the definition of κ_θ, we can derive the relation

$$\bar{\varepsilon}_\theta(z) = z\kappa_\theta$$

If we consider a slice of thickness dz from the element [Fig. 7.4(b)], stresses $\bar{\sigma}_\phi$ and $\bar{\sigma}_\theta$ acting on the slice, due to M_ϕ and M_θ, will be related to the strains by Hooke's law according to

$$\bar{\varepsilon}_\phi(z) = \frac{1}{E}[\bar{\sigma}_\phi(z) - \nu\bar{\sigma}_\theta(z)]$$

$$\bar{\varepsilon}_\theta(z) = \frac{1}{E}[\bar{\sigma}_\theta(z) - \nu\bar{\sigma}_\phi(z)]$$

Or, upon solving for $\bar{\sigma}_\phi$ and $\bar{\sigma}_\theta$, we obtain

$$\bar{\sigma}_\phi(z) = \frac{E}{1-\nu^2}[\bar{\varepsilon}_\phi(z) + \nu\bar{\varepsilon}_\theta(z)]$$

$$\bar{\sigma}_\theta(z) = \frac{E}{1-\nu^2}[\bar{\varepsilon}_\theta(z) + \nu\bar{\varepsilon}_\phi(z)]$$

The moment about the midsurface, due to $\bar{\sigma}_\phi(z)$ acting along the fiber EF (of thickness dz and of unit width in the θ-direction), is given by $(\bar{\sigma}_\phi(z)\cdot 1\cdot dz\cdot z)$. Integrating over the thickness t, for all such fibers, we obtain

$$M_\phi = \int_{-t/2}^{+t/2} \bar{\sigma}_\phi(z) z\, dz$$

$$= \frac{E}{1-\nu^2}\int_{-t/2}^{+t/2} z[\bar{\varepsilon}_\phi(z) + \nu\bar{\varepsilon}_\theta(z)]\, dz$$

Substituting for $\bar{\varepsilon}_\phi(z)$ and $\bar{\varepsilon}_\theta(z)$, we obtain

$$M_\phi = \frac{E}{1-\nu^2}(\kappa_\phi + \nu\kappa_\theta)\int_{-t/2}^{+t/2} z^2 dz = K(\kappa_\phi + \nu\kappa_\theta) \tag{7.4c}$$

where $K = Et^3/12(1-\nu^2)$ is the flexural rigidity of the shell. Proceeding in a similar manner, we can derive the relation

$$M_\theta = K(\kappa_\theta + \nu\kappa_\phi) \tag{7.4d}$$

We can now summarize the various equations needed for deriving the final governing differential equation for edge-load behavior.

Summary of Equations: General Shells of Revolution—Axisymmetric Edge Loads Only

Equilibrium

$$N_\phi = -Q_\phi \cot \phi \tag{7.1a}$$

$$(N_\phi \cdot r) + N_\theta r_1 \sin \phi = -(Q_\phi \cdot r)^{\cdot} \tag{7.1b}$$

$$(M_\phi \cdot r)^{\cdot} - M_\theta r_1 \cos \phi = (Q_\phi \cdot r) r_1 \tag{7.1c}$$

Kinematics

$$\varepsilon_\phi = \frac{(v)^{\cdot} + w}{r_1} \qquad \varepsilon_\theta = \frac{v \cot \phi + w}{r_2}$$
$$\chi = \frac{(w)^{\cdot} - v}{r_1} \qquad \kappa_\phi = \frac{(\chi)^{\cdot}}{r_1} \qquad \kappa_\theta = \frac{\chi \cot \phi}{r_2} \tag{7.2a–e}$$

Hooke's Law

$$N_\phi = D(\varepsilon_\phi + \nu \varepsilon_\theta)$$
$$N_\theta = D(\varepsilon_\theta + \nu \varepsilon_\phi) \qquad D = \frac{Et}{1 - \nu^2}$$
$$M_\phi = K(\kappa_\phi + \nu \kappa_\theta) \tag{7.4a–d}$$
$$M_\theta = K(\kappa_\theta + \nu \kappa_\phi) \qquad K = \frac{Et^3}{12(1 - \nu^2)}$$

These 12 equations contain 12 unknowns, and thus we should be able to obtain a solution for a given edge-load problem. However, it is not easy to solve these equations for a general shell of revolution unless some approximations are introduced [1]; even then the solution is rather complicated. We shall therefore consider solving these equations only for the special case of a spherical shell. We can use this solution for solving many practical problems. We shall also see how it can be used to obtain approximate solutions for problems involving nonspherical shells.

7.3 EQUATIONS FOR SPHERICAL SHELLS: AXISYMMETRIC EDGE LOADS

Substituting $r_1 = r_2 = a$ and $r = a \sin \phi$ into Eqs. (7.1) and (7.2) and simplifying, we can rewrite these expressions as follows:

Equilibrium Equations

$$N_\phi = -Q_\phi \cot \phi \tag{7.5a}$$

$$N_\phi \sin \phi + N_\theta \sin \phi = -(Q_\phi \sin \phi)^{\cdot} \tag{7.5b}$$

$$(M_\phi \sin\phi)^\bullet - M_\theta \cos\phi = Q_\phi\, a \sin\phi \tag{7.5c}$$

Kinematics

$$a\varepsilon_\phi = (v)^\bullet + w \qquad a\varepsilon_\theta = v\cot\phi + w$$
$$a\chi = (w)^\bullet - v \qquad a\kappa_\phi = (\chi)^\bullet \qquad a\kappa_\theta = \chi\cot\phi \tag{7.6a–e}$$

The Hooke's law relations remain unchanged, as given by Eqs. (7.4a–d).

Having obtained all these equations, our procedure for previous structures, such as beams, plates or cylindrical shells, had been to eliminate all other quantities except w, and thus obtain a governing differential equation in w. In the present case, however, it is more convenient to obtain a differential equation in Q_ϕ or χ. To achieve this, we first substitute Eqs. (7.6d,e) into Eqs. (7.4c,d), to obtain

$$M_\phi = \frac{K}{a}\left[(\chi)^\bullet + \nu\chi\cot\phi\right] \qquad \text{and} \qquad M_\theta = \frac{K}{a}\left[\chi\cot\phi + \nu(\chi)^\bullet\right]$$

We can now eliminate M_ϕ and M_θ by substituting these relations into Eq. (7.5c). We thus obtain the following equation after performing the differentiation [assuming E and t (i.e., K) to be constant in ϕ] and simplifying:

$$(\chi)^{\bullet\bullet} + (\chi)^\bullet\cot\phi - \chi(\cot^2\phi + \nu) = \left(\frac{a^2}{K}\right)Q_\phi \tag{7.7}$$

For obtaining another equation involving Q_ϕ and χ, we first substitute Eq. (7.5a) into (7.5b) and simplify, to obtain the following relation between N_θ and Q_ϕ:

$$N_\theta = -(Q_\phi)^\bullet \tag{7.8}$$

Since now both N_ϕ and N_θ are expressed in terms of Q_ϕ by Eqs. (7.5a) and (7.8), we can substitute these equations into Eq. (6.16b) derived in Chapter 6 (and also valid here), to obtain an equation in terms of χ and Q_ϕ. Equation (6.16b) can also be derived directly here from Eqs. (7.6a–c) and (7.3a,b), as follows. We can write, from Eqs. (7.6a,b),

$$a\varepsilon_\phi = (v)^\bullet + w$$
$$a\varepsilon_\theta^\bullet = (v)^\bullet\cot\phi - v\csc^2\phi + (w)^\bullet$$

From these two equations, we obtain

$$a[(\varepsilon_\theta)^\bullet - \varepsilon_\phi\cot\phi] = -v\csc^2\phi + (w)^\bullet - w\cot\phi$$

Again, from Eq. (7.6b), we have

$$a\varepsilon_\theta\cot\phi = v\cot^2\phi + w\cot\phi$$

Adding together these last two equations and simplifying, we obtain

$$a[(\varepsilon_\theta)^\bullet + (\varepsilon_\theta - \varepsilon_\phi)\cot\phi] = -v + w^\bullet$$

In this equation we can substitute, in the left-hand side, relations from Eqs. (7.3a,b), and on the right-hand side Eq. (7.6c), to obtain the following equation [which is the same as Eq. (6.16b)]:

$$(N_\theta)^{\cdot} - \nu(N_\phi)^{\cdot} + (1 + \nu)(N_\theta - N_\phi)\cot\phi = Et\chi \tag{7.9}$$

On substituting for N_ϕ and N_θ from Eqs. (7.5a) and (7.8) into the equation above and simplifying, we have the following second equation in terms of Q_ϕ and χ:

$$(Q_\phi)^{\cdot\cdot} + (Q_\phi)^{\cdot}\cot\phi - (\cot^2\phi - \nu)Q_\phi = -Et\chi \tag{7.10}$$

Both Eqs. (7.7) and (7.10) are second-order ordinary differential equations with variable coefficients, and both have similar form. If we use a linear differential operator $\mathscr{L}$ such that

$$\mathscr{L}(\cdot) = (\cdot)^{\cdot\cdot} + (\cdot)^{\cdot}\cot\phi - (\cdot)\cot^2\phi$$

the two equations can be written as

$$\mathscr{L}[\chi] - \nu\chi = \left(\frac{a^2}{K}\right)Q_\phi \tag{7.7}$$

and

$$\mathscr{L}[Q_\phi] + \nu Q_\phi = -Et\chi \tag{7.10}$$

To obtain a single equation in Q_ϕ, we use the differential operator $\mathscr{L}$ on Eq. (7.10) once again, to obtain

$$\mathscr{L}\{\mathscr{L}[Q_\phi]\} + \nu\mathscr{L}[Q_\phi] = -Et\mathscr{L}[\chi]$$

Again, from Eq. (7.10), we have

$$\nu\mathscr{L}[Q_\phi] + \nu^2 Q_\phi = -\nu Et\chi$$

Subtracting the second equation above from the first and using Eq. (7.7), we have

$$\mathscr{L}\{\mathscr{L}[Q_\phi]\} - \nu^2 Q_\phi = -Et\{\mathscr{L}[\chi] - \nu\chi\} = -\frac{Eta^2}{K}Q_\phi$$

and hence

$$\mathscr{L}\{\mathscr{L}[Q_\phi]\} + 4\lambda^4 Q_\phi = 0 \tag{7.11}$$

where

$$4\lambda^4 = \left(\frac{Eta^2}{K} - \nu^2\right) = 12\left(\frac{a}{t}\right)^2(1 - \nu^2) - \nu^2$$

Instead of eliminating χ, we could have eliminated Q_ϕ from Eqs. (7.7) and (7.10), in order to obtain an equation for χ identical to Eq. (7.11) for Q_ϕ.

Equation (7.11) is the final equation of our bending theory derivation. It is a fourth-order ordinary differential equation with variable coefficients. It can be split into two second-order equations by writing it, in complex form, as

$$(\mathcal{L} - 2i\lambda^2)(\mathcal{L}[Q_\phi] + 2i\lambda^2 Q_\phi) = 0 \tag{7.12a}$$

or

$$(\mathcal{L} + 2i\lambda^2)(\mathcal{L}[Q_\phi] - 2i\lambda^2 Q_\phi) = 0 \tag{7.12b}$$

Thus the solutions of the two second-order equations

$$\mathcal{L}[Q_\phi] \pm 2i\lambda^2 Q_\phi = 0$$

will satisfy the fourth-order equation also, and hence four independent solutions for forming the general solution of Eq. (7.11) can be obtained. The first second-order equation has the form

$$(Q_\phi)^{\cdot\cdot} + \cot\phi\,(Q_\phi)^{\cdot} - \cot^2\phi\,(Q_\phi) + 2i\lambda^2 Q_\phi = 0 \tag{7.13}$$

Since this equation has variable coefficients, the solution is not easily obtained. By making suitable transformations, it can be put into a standard form of the *hypergeometric equation,* the solutions of which can be obtained in the form of power series called *hypergeometric series.*

7.4 APPROXIMATE SOLUTION OF EQUATION FOR SPHERICAL SHELLS

The solution procedure discussed in the preceding section becomes quite complex. To simplify the solution, we make some approximations introduced first by Geckeler, and obtain a simpler equation, the solutions of which may still be fairly accurately applied to many problems of practical interest.

For our self-equilibrating edge-load problem, we expect Q_ϕ to die out rapidly away from the edge according to Saint-Venant's principal, as discussed in Chapter 5. It is likely, therefore, that $(Q_\phi)^{\cdot}$ will be much greater than Q_ϕ, and $(Q_\phi)^{\cdot\cdot}$ will be much greater than $(Q_\phi)^{\cdot}$. Hence if we assume that $(Q_\phi)^{\cdot\cdot} \gg (Q_\phi)^{\cdot} \gg Q_\phi$, we can neglect the terms $(Q_\phi)^{\cdot}\cot\phi$ and $(Q_\phi)\cot^2\phi$ compared to $(Q_\phi)^{\cdot\cdot}$ in Eq. (7.13), provided that $\cot\phi$ itself is not very large. The value of $\cot\phi$ increases from zero at $\phi = 90°$ to 2.747 at $\phi = 20°$. This being the range of ϕ values of practical interest, we can simplify Eq. (7.13) to take the following form:

$$(Q_\phi)^{\cdot\cdot} + 2i\lambda^2 Q_\phi = 0$$

The fourth-order equation (7.12a) will then become

$$(\mathcal{L} - 2i\lambda^2)[(Q_\phi)^{\cdot\cdot} + 2i\lambda^2 Q_\phi] = 0$$

where the operator $\mathcal{L}(\cdot)$ now takes the simple form $\mathcal{L}(\cdot) \approx (\cdot)^{\cdot\cdot}$, the remaining two terms being neglected. Upon expanding the equation above, we obtain the following simple fourth-order equation for Q_ϕ, instead of Eq. (7.11):

$$(Q_\phi)^{\cdot\cdot\cdot\cdot} + 4\lambda^4 Q_\phi = 0 \tag{7.14a}$$

where, as before,

$$4\lambda^4 = \left(\frac{Eta^2}{K} - \nu^2\right) \approx \frac{Eta^2}{K} = 12\left(\frac{a}{t}\right)^2(1 - \nu^2) \tag{7.14b}$$

Since Eq. (7.14a) now has constant coefficients, we can assume its solution as $Q_\phi = Ae^{m\phi}$. Substituting this expression into the differential equation of Eq. (7.14a), we obtain for m the equation $m^4 + 4\lambda^4 = 0$. As done for cylindrical shells, we can use deMoivre's theorem to obtain the four roots of m, as $m = \pm(1 \pm i)\lambda$. Hence the general solution for Q_ϕ can be written as

$$Q_\phi = A_1e^{-(1+i)\lambda\phi} + A_2e^{-(1-i)\lambda\phi} + A_3e^{(1+i)\lambda\phi} + A_4e^{(1-i)\lambda\phi}$$

where A_1, A_2, A_3, and A_4 are constants to be determined from the boundary conditions. Rearranging terms, we can also write Q_ϕ in the following form, with a new set of constants, $B_1, \ldots, B_4$:

$$Q_\phi = e^{-\lambda\phi}(B_1 \cos \lambda\phi + B_2 \sin \lambda\phi) + e^{\lambda\phi}(B_3 \cos \lambda\phi + B_4 \sin \lambda\phi) \tag{7.15}$$

Figure 7.5 shows a shell structure with boundaries at $\phi = \phi_o$ and $\phi = \phi_1$, where edge loads Q_ϕ, M_ϕ, or N_ϕ* can be prescribed. If we measure angles β_1 and β_2 from the lower and upper edges, respectively, as shown, we can write the relations

$$\beta_1 = \phi_1 - \phi \qquad \text{and} \qquad \beta_2 = \phi - \phi_o$$

Substituting these expressions into Eq. (7.15) and rearranging, we can write the solution for Q_ϕ in the form

$$Q_\phi = e^{-\lambda\beta_1}(C_1 \cos \lambda\beta_1 + C_2 \sin \lambda\beta_1) + e^{-\lambda\beta_2}(C_3 \cos \lambda\beta_2 + C_4 \sin \lambda\beta_2) \tag{7.16}$$

The quantity λ given by Eq. (7.14b) is greater than unity, and in fact, is expected to be much larger for thin shells. Hence, because of the presence of the multiplier $(e^{-\lambda\beta_1})$, we can say that the first two terms in the equation above will die out as

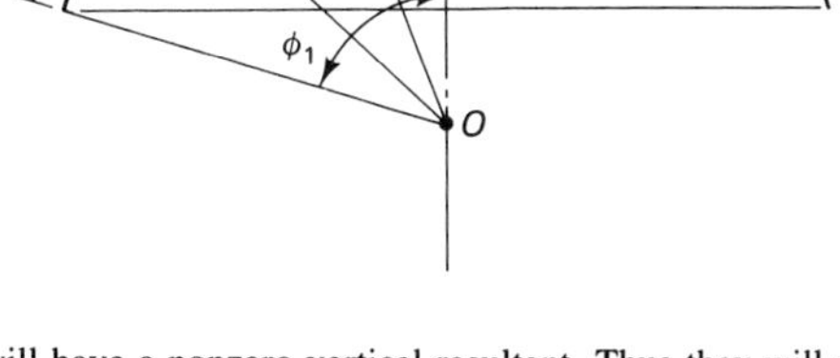

Figure 7.5 Shell structure with boundaries at $\phi = \phi_o$ and $\phi = \phi_1$, showing angular coordinates measured from the edges.

*By themselves, edge loads Q_ϕ or N_ϕ will have a nonzero vertical resultant. Thus they will not be self-equilibrating, which is incorrect, since we have no other loading on the shell. Q_ϕ and N_ϕ should, therefore, be applied in combination such that the resultant forces are zero. N_ϕ and Q_ϕ must therefore satisfy, at the edges, the relation given by Eq. (7.1a).

we go away from the bottom edge ($\beta_1 = 0$). Similarly, the last two terms will die out as we go away from the top edge ($\beta_2 = 0$). Hence if we are solving the problem of edge loads at one boundary only, say $\beta_1 = 0$, we can neglect the last two terms in the solution, and consider only the first two. Henceforth, we shall consider edge loads at the edge $\beta_1 = 0$ only. The solution for Q_ϕ can then be written in the following convenient form:

$$Q_\phi = C\, e^{-\lambda\beta_1} \sin(\lambda\beta_1 + \psi) \tag{7.17a}$$

where C and ψ are new constants to be determined from the boundary conditions. We shall now obtain expressions for the other quantities (N_ϕ, M_ϕ, etc.) in terms of C and ψ by using Eqs. (7.4) to (7.10). From Eq. (7.5a) we have

$$N_\phi = -Q_\phi \cot\phi = -Ce^{-\lambda\beta_1} \sin(\lambda\beta_1 + \psi) \cot\phi \tag{7.17b}$$

From Eq. (7.8), noting that $d(\cdot)/d\phi = -d(\cdot)/d\beta_1$, we obtain

$$N_\theta = -(Q_\phi)^{\boldsymbol{\cdot}} = \frac{dQ_\phi}{d\beta_1} = C\lambda e^{-\lambda\beta_1}[-\sin(\lambda\beta_1 + \psi) + \cos(\lambda\beta_1 + \psi)]$$

or

$$N_\theta = -C\lambda\sqrt{2}\, e^{-\lambda\beta_1} \sin\left(\lambda\beta_1 + \psi - \frac{\pi}{4}\right) \tag{7.17c}$$

Now, from Eq. (7.10), neglecting Q_ϕ and $(Q_\phi)^{\boldsymbol{\cdot}}$ in comparison with $(Q_\phi)^{\boldsymbol{\cdot\cdot}}$, we have

$$\chi = -\frac{1}{Et}(Q_\phi)^{\boldsymbol{\cdot\cdot}} = \frac{2C\lambda^2}{Et}\, e^{-\lambda\beta_1} \cos(\lambda\beta_1 + \psi) \tag{7.17d}$$

Earlier, we had

$$M_\phi = \frac{K}{a}[(\chi)^{\boldsymbol{\cdot}} + \nu\chi\cot\phi]$$

By similar arguments as used for Q_ϕ, we can neglect $\chi \cot\phi$ compared to $(\chi)^{\boldsymbol{\cdot}}$ in the equation above, to obtain

$$M_\phi \approx \frac{K}{a}(\chi)^{\boldsymbol{\cdot}} = -\frac{K}{Eta}(Q_\phi)^{\boldsymbol{\cdot\cdot\cdot}}$$

Noting from Eq. (7.14b) that $4\lambda^4 \approx Eta^2/K$, we substitute this relation into the equation above, to find

$$M_\phi = -\left(\frac{a}{4\lambda^4}\right)(Q_\phi)^{\boldsymbol{\cdot\cdot\cdot}}$$

or

$$M_\phi = \frac{a}{\sqrt{2}\,\lambda}\, Ce^{-\lambda\beta_1} \sin\left(\lambda\beta_1 + \psi + \frac{\pi}{4}\right) \tag{7.17e}$$

We also had

$$M_\theta = \frac{K}{a}\,[\chi \cot \phi + \nu(\chi)^{\cdot}]$$

which we can approximate as

$$M_\theta \approx \nu\left(\frac{K}{a}\right)(\chi)^{\cdot} = \nu M_\phi \tag{7.17f}$$

We shall also require an expression for the horizontal deflection, which we shall denote by Δ_H; positive when radially outward in the plane of the parallel circle. Thus we derive

$$\begin{aligned}\Delta_H &= a \sin \phi \; \varepsilon_\theta \\ &= \frac{a \sin \phi}{Et}(N_\theta - \nu N_\phi) \\ &= \frac{a \sin \phi}{Et}[-(Q_\phi)^{\cdot} + \nu Q_\phi \cot \phi]\end{aligned}$$

We can again neglect the second term inside the brackets (which means neglecting the effect of the meridional forces on Δ_H), to obtain

$$\Delta_H \approx \frac{a \sin \phi}{Et}(-Q_\phi)^{\cdot} = \frac{a \sin \phi \; N_\theta}{Et} \tag{7.17g}$$

For future reference, we shall summarize below, the expressions derived for the various quantities of importance.

Summary of Equations: Spherical Shell, Edge Loads at $\beta_1 = 0$ ($\phi = \phi_1$)

$$Q_\phi = Ce^{-\lambda\beta_1} \sin (\lambda\beta_1 + \psi) \tag{7.17a}$$

$$N_\phi = -Ce^{-\lambda\beta_1} \sin (\lambda\beta_1 + \psi) \cot \phi \tag{7.17b}$$

$$N_\theta = -C\lambda\sqrt{2}\, e^{-\lambda\beta_1} \sin\left(\lambda\beta_1 + \psi - \frac{\pi}{4}\right) \tag{7.17c}$$

$$\chi = C\frac{2\lambda^2}{Et} e^{-\lambda\beta_1} \cos (\lambda\beta_1 + \psi) \tag{7.17d}$$

$$M_\phi = C\frac{a}{\sqrt{2}\,\lambda} e^{-\lambda\beta_1} \sin\left(\lambda\beta_1 + \psi + \frac{\pi}{4}\right) \tag{7.17e}$$

$$M_\theta = \nu M_\phi \tag{7.17f}$$

$$\Delta_H = \frac{a \sin \phi}{Et} N_\theta \tag{7.17g}$$

We have mixed the variables β_1 and ϕ in the expression for N_ϕ above. However, these are related simply by $\beta_1 = \phi_1 - \phi$. It must be remembered that the equations above have been obtained for spherical shells only, with additional simplifications, and hence are approximate so as not to be valid for problems of other shells or also for shallow spherical domes, where ϕ is less than 20° or so. Keeping these restrictions in mind, we shall use these equations to obtain solutions of two basic problems, which can then be used repeatedly for solving other problems.

7.5 SOLUTIONS OF TWO BASIC EDGE-LOAD PROBLEMS

First Basic Problem: Edge Shear Load

We consider horizontal edge loads H per unit length, applied to the bottom edge of a spherical shell, as shown in Fig. 7.6. We have the boundary conditions at $\beta_1 = 0$, $M_\phi = 0$ and $Q_\phi = -H \sin \phi_1$. These conditions, when substituted into Eqs. (7.17e) and (7.17a), give us

$$\psi = -\frac{\pi}{4} \qquad \text{and} \qquad C = \sqrt{2}\, H \sin \phi_1$$

Substituting these values into Eqs. (7.17a–g), we obtain

$$Q_\phi = \sqrt{2}\, H \sin \phi_1\, e^{-\lambda\beta_1} \sin\left(\lambda\beta_1 - \frac{\pi}{4}\right) \tag{7.18a}$$

$$N_\phi = -\sqrt{2}\, H \sin \phi_1 \cot \phi\, e^{-\lambda\beta_1} \sin\left(\lambda\beta_1 - \frac{\pi}{4}\right) \tag{7.18b}$$

$$N_\theta = -2\lambda H \sin \phi_1\, e^{-\lambda\beta_1} \sin\left(\lambda\beta_1 - \frac{\pi}{2}\right) \tag{7.18c}$$

$$\chi = \frac{2\sqrt{2}\, \lambda^2 H \sin \phi_1}{Et}\, e^{-\lambda\beta_1} \cos\left(\lambda\beta_1 - \frac{\pi}{4}\right) \tag{7.18d}$$

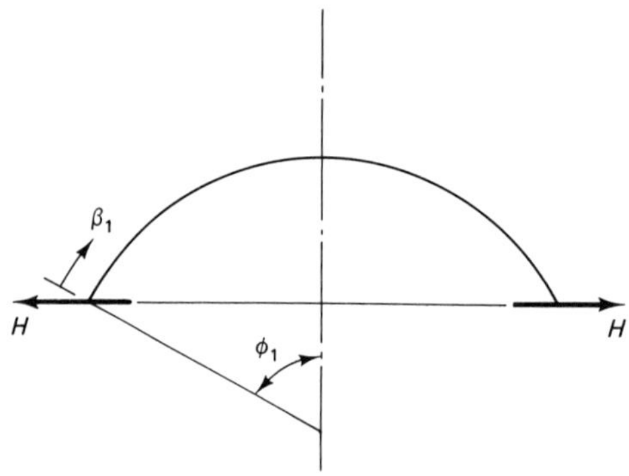

Figure 7.6 Spherical dome subjected to horizontal edge shears at its bottom edge.

$$M_\phi = \frac{aH \sin \phi_1}{\lambda} e^{-\lambda\beta_1} \sin \lambda\beta_1 \tag{7.18e}$$

$$M_\theta = \nu M_\phi \tag{7.18f}$$

$$\Delta_H = \frac{a \sin \phi}{Et} N_\theta \tag{7.18g}$$

We shall also need the values of χ and Δ_H at the edge $\beta_1 = 0$. Hence we write

$$(\chi)_{\beta_1=0} = \frac{2\lambda^2 H \sin \phi_1}{Et} \tag{7.18h}$$

$$(\Delta_H)_{\beta_1=0} = \frac{2a\lambda H \sin^2 \phi_1}{Et} \tag{7.18i}$$

Second Basic Problem: Edge Moment Load

We consider next edge moments M per unit length applied along the bottom of a spherical shell, as shown in Fig. 7.7. Now the boundary conditions at $\beta_1 = 0$ are $Q_\phi = 0$ and $M_\phi = M$. These conditions give, from Eqs. (7.17a,e), $\psi = 0$ and $C = 2\lambda M/a$. Substituting these values into Eqs. (7.17a–g), we have

$$Q_\phi = \frac{2\lambda M}{a} e^{-\lambda\beta_1} \sin \lambda\beta_1 \tag{7.19a}$$

$$N_\phi = -\frac{2\lambda M}{a} \cot \phi \, e^{-\lambda\beta_1} \sin \lambda\beta_1 \tag{7.19b}$$

$$N_\theta = -\frac{2\sqrt{2}\,\lambda^2}{a} M e^{-\lambda\beta_1} \sin \left(\lambda\beta_1 - \frac{\pi}{4}\right) \tag{7.19c}$$

$$\chi = \frac{4\lambda^3}{Eta} M e^{-\lambda\beta_1} \cos \lambda\beta_1 \tag{7.19d}$$

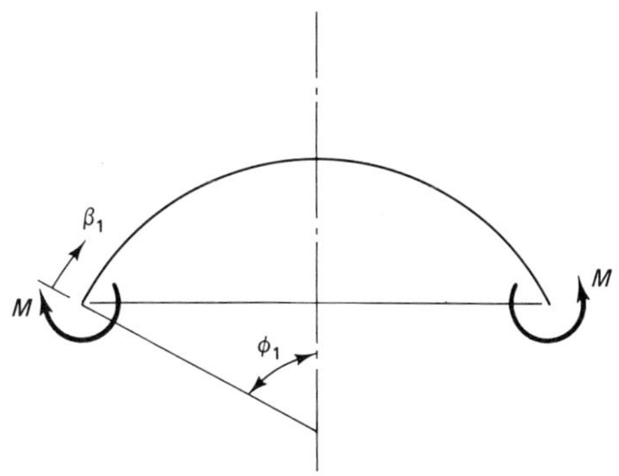

Figure 7.7 Spherical dome subjected to edge moments at its bottom edge.

$$M_\phi = \sqrt{2}\, Me^{-\lambda\beta_1} \sin\left(\lambda\beta_1 + \frac{\pi}{4}\right) \tag{7.19e}$$

$$M_\theta = \nu M_\phi \tag{7.19f}$$

$$\Delta_H = \frac{a \sin\phi}{Et} N_\theta \tag{7.19g}$$

Also,

$$(\chi)_{\beta_1=0} = \frac{4\lambda^3 M}{Eta} \quad \text{and} \quad (\Delta_H)_{\beta_1=0} = \frac{2\lambda^2 M}{Et}\sin\phi_1 \tag{7.19h,i}$$

Again, we note

$$4\lambda^4 = \left(\frac{Eta^2}{K} - \nu^2\right) \approx \frac{Eta^2}{K} = 12\left(\frac{a}{t}\right)^2(1 - \nu^2)$$

7.6 SOLUTIONS OF PROBLEMS BY THE FLEXIBILITY METHOD

Using the solutions obtained for the two basic edge-load problems in Section 7.5, we can now solve several problems of practical interest by satisfying the compatibility equations [Eqs. (6.19a,b)], i.e., by the flexibility method. Our procedure will be to divide the shell suitably, into two or more "primary structures," by cutting it at points where discontinuities of displacements under membrane conditions occur. Each primary structure is so chosen that it is stable under the actions of loads and membrane forces, and also, that it can have unrestrained membrane displacements. As an example, we consider the dome and ring problem of Example 6.1. We have seen there that at the junction of the dome and ring beam, there is a discontinuity of displacements. Hence we separate the dome and ring in order to obtain two primary structures, as shown in Fig. 7.8(b). The dome is now assumed to be supported on rollers which move on inclined planes and give a reaction $R = N_{\phi_1}$ to keep the shell in equilibrium under self-weight loading. The rollers also allow the displacement w of the dome to take place freely.* On the second primary structure, which here consists only of the ring beam on rollers (moving in a horizontal plane), opposite forces R will act. Under these forces, the ring is in equilibrium and is also free to deform horizontally.

Having chosen the primary structures, we apply to them redundant forces X_1 and X_2, the values of which are determined from the compatibility equations. These equations state that the net relative displacements of points A and A_1 [Fig. 7.8(c)], which are connected in the original structure, must be zero [2]. We can write, as we did for cylindrical shells,

*Note that under membrane conditions, we had seen in Chapter 6 that one displacement could be specified at one boundary. Hence if the meridional displacement v is zero at the edge, as in this case, w must be freely allowed.

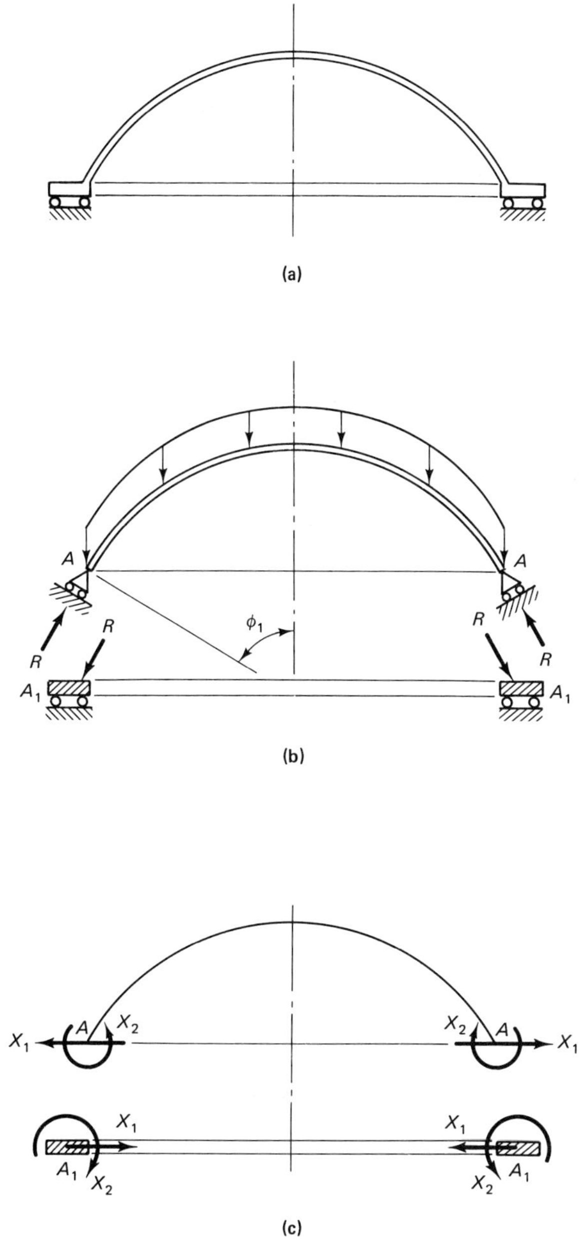

Figure 7.8 Spherical-dome/ring-beam interaction. (a) Dome and ring beam. (b) Primary structures under self-weight. (c) Corrective edge loads.

$$\delta_{11}X_1 + \delta_{12}X_2 + \delta_{10} = 0 \tag{7.20a}$$

$$\delta_{21}X_1 + \delta_{22}X_2 + \delta_{20} = 0 \tag{7.20b}$$

where δ_{ij} $(i, j = 1, 2)$ represent the relative displacements of A and A_1 under unit values of X_1 and X_2, and δ_{i0} $(i = 1, 2)$ represent the relative displacements of A and A_1 for the primary structures under membrane loadings. Obviously,

$$\delta_{ij} = \delta_{ij}^1 + \delta_{ij}^2 \qquad i = 1, 2 \qquad j = 0, 1, 2 \tag{7.20c}$$

where superscripts 1 and 2 represent displacements of the primary structures 1 and 2, (i.e., dome and ring), respectively, positive in the positive senses of X_1 and X_2 acting on them. For ready reference in the future, we can summarize the flexibility coefficients δ_{ij} $(i, j = 1, 2)$ for the dome and the ring beam, when the assumed directions of X_1 and X_2 are as shown in Fig. 7.9.

Dome Flexibility Coefficients

When $X_1 = 1$ we obtain, by substituting $H = 1$ into Eqs. (7.18i,h), the following flexibility coefficients for the dome:

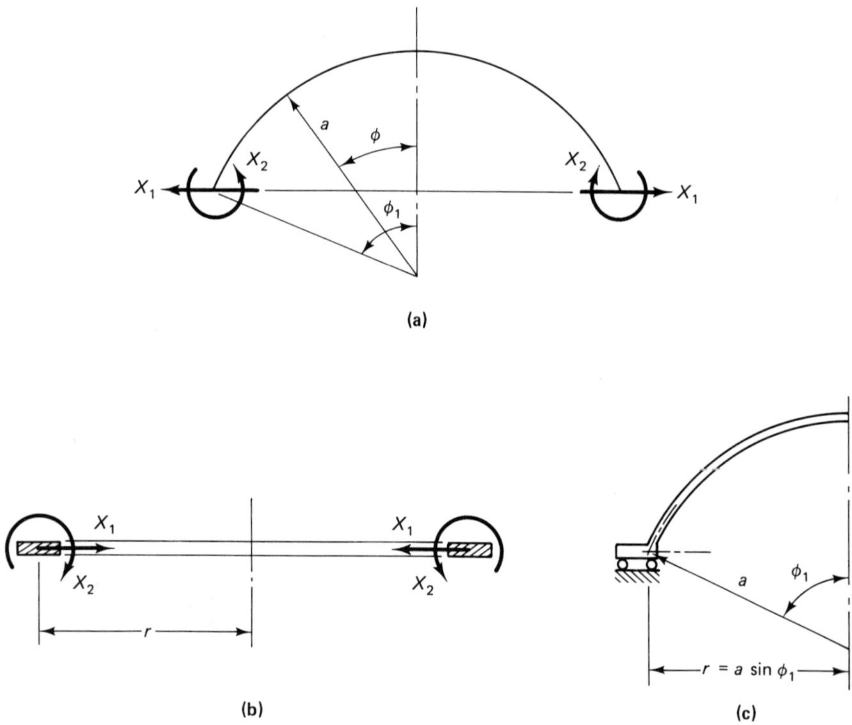

Figure 7.9 Sign conventions for flexibility coefficients. (a) Positive directions assumed for dome edge loads. (b) Positive directions for edge loads on the ring beam. (c) Geometry of structure, showing intersection of dome middle surface with ring beam centroid.

$$\delta_{11}^D = (\Delta_H)_{\beta_1=0} = \frac{2a\lambda \sin^2\phi_1}{Et}$$

and

$$\delta_{21}^D = -(\chi)_{\beta_1=0} = -\frac{2\lambda^2 \sin \phi_1}{Et}$$

Also, when $X_2 = 1$, we obtain, by substituting $M = -1$ into Eqs. (7.19i,h), the coefficients

$$\delta_{12}^D = (\Delta_H)_{\beta_1=0} = -\frac{2\lambda^2 \sin \phi_1}{Et}$$

and

$$\delta_{22}^D = -(\chi)_{\beta_1=0} = \frac{4\lambda^3}{Eta}$$

Hence we can summarize the flexibility matrix for the dome as

$$\boldsymbol{\delta}^D = \begin{bmatrix} \delta_{11}^D & \delta_{12}^D \\ \delta_{21}^D & \delta_{22}^D \end{bmatrix} = \begin{bmatrix} \dfrac{2a\lambda \sin^2\phi_1}{Et} & \dfrac{-2\lambda^2 \sin \phi_1}{Et} \\ \dfrac{-2\lambda^2 \sin \phi_1}{Et} & \dfrac{4\lambda^3}{Eta} \end{bmatrix} \tag{7.21a–d}$$

Although obviously δ_{11}^D and δ_{22}^D will always be positive, the signs of δ_{21}^D and δ_{12}^D will have to be suitably modified depending on the assumed directions of X_1 and X_2 in a given problem.

Ring Beam Flexibility Coefficients

We shall assume that the force X_1 is applied at the centroid of the cross section of the ring. Although further refinements of the analysis by considering X_1 eccentric to the centroidal axis of the ring could be made when necessary [3], it is felt that they are not warranted for the solutions of normal practical problems. Hence we shall generally assume that the dome and the ring are always so connected that the meridian of the dome passes through the centroid of the ring's cross section [Fig. 7.9(c)]; i.e., X_1 always passes through the centroid of the ring.

Now, for $X_1 = 1$, we have

$$\delta_{11}^R = r\varepsilon_\theta = \frac{r(X_1 r)}{AE} = \frac{r(1)(r)}{AE} = \frac{r^2}{AE}$$

where A is the cross-sectional area of the ring. Since X_1 passes through the centroid of the cross section of the ring, it will not cause any rotation, and hence $\delta_{21}^R = 0$. Now, under the action of X_2, the bending moment in the ring, M^R, is

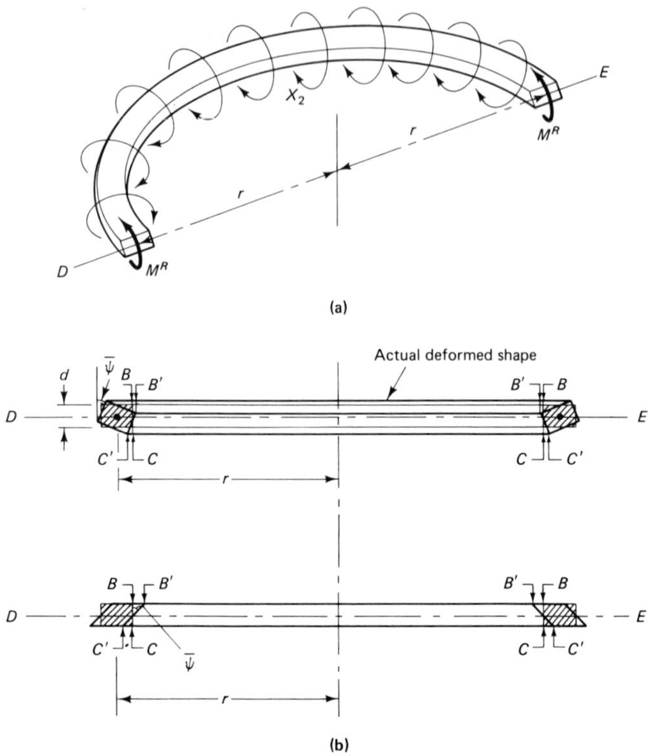

Figure 7.10 Axisymmetric loading and deformation of ring beam. (a) Section of the ring showing internal and external loads. (b) Bending deformation of ring cross section.

obtained by taking a section along any diameter DE and considering the equilibrium of all moments to one side of this section [Fig. 7.10a]. We thus obtain $2M^R = (2rX_2)$ or $M^R = (X_2 r)$. This moment causes hoop compression at the top of the ring and hoop tension at the bottom of the ring. Maximum hoop stresses, at the top and bottom fibers, will be given by

$$\sigma_{\text{top}} = -\frac{M^R}{I}\frac{d}{2} \qquad \sigma_{\text{bottom}} = +\frac{M^R}{I}\frac{d}{2}$$

where I is the moment of inertia (about the line DE) and d is the depth of the ring cross section, assuming it to be rectangular. These stresses cause hoop strains which, when multiplied by the radius of the ring, will give the radial deflections. These radial deflections will be inward above and outward below the neutral axis, and will vary linearly from top to bottom. Thus the rotation of the ring will be as shown in Fig. 7.10(b). Radial deflections BB' and CC' at the top and bottom, respectively, are given by*

*For simplicity, we also assume that E for the ring and the dome (or cylinder) are always the same.

$$CC' = |BB'| = \frac{r(\sigma_{\text{bottom}})}{E} = \frac{rM^R}{EI}\left(\frac{d}{2}\right)$$

where we have assumed that the width of the ring beam cross section is small compared to r. Rotation of the ring, $\bar{\psi}$, will be given by

$$\bar{\psi} = \frac{|BB'| + CC'}{d} = \frac{rM^R}{EI} = \frac{X_2 r^2}{EI}$$

Hence, substituting $X_2 = 1$, we obtain

$$\delta_{22}^R = \frac{r^2}{EI}$$

Again, due to X_2, there will be no horizontal deflection of the centroid of the ring, and hence $\delta_{12}^R = 0$. We can now summarize the flexibility matrix for the ring, as follows:

$$\boldsymbol{\delta}^R = \begin{bmatrix} \dfrac{r^2}{AE} & 0 \\ 0 & \dfrac{r^2}{EI} \end{bmatrix} \qquad (7.22\text{a–d})$$

Of course, these coefficients are valid only for a ring of rectangular cross section, and for the case where X_1 passes through the centroid of the ring. However, the coefficients can easily be derived for other conditions. We next consider the solutions of some practical examples, using the flexibility approach.

Example 7.1

As a first example, let us consider a complete spherical shell subjected to self-equilibrating radial loads F per unit length acting along the equator [Fig. 7.11(a)]. Considering the freebody of the upper half of the shell, we shall have edge loads of $F/2$ acting as shown in Fig. 7.11(b). Also, by symmetry, we can see that the deformation of the shell will be such that it will have a zero slope (to the vertical) at the equator. Hence there will be an unknown moment X_2 per unit length acting along the shell's equator. The rotation of the shell must be zero under the action of $F/2$ and X_2. Denoting $-\delta_{21}(F/2)$ as the rotation at the equator (in the direction of X_2 [Fig. 7.11(b)]) under the action of forces $X_1 = -F/2$, and δ_{22} as the rotation caused by $X_2 = 1$, we obtain a single compatibility equation as*

$$-\delta_{21}\left(\frac{F}{2}\right) + X_2\delta_{22} = 0$$

Flexibility δ_{21} is obtained by substituting $\phi_1 = 90°$ into Eq. (7.21c), and δ_{22} is obtained directly from Eq. (7.21d). Thus we find that

$$\delta_{21} = \frac{-2\lambda^2}{Et} \qquad \delta_{22} = \frac{4\lambda^3}{Eta}$$

*There are no distributed loads on the shell, and hence no membrane deformations of the structure. As a result, the term δ_{20} is equal to zero and does not enter into this equation.

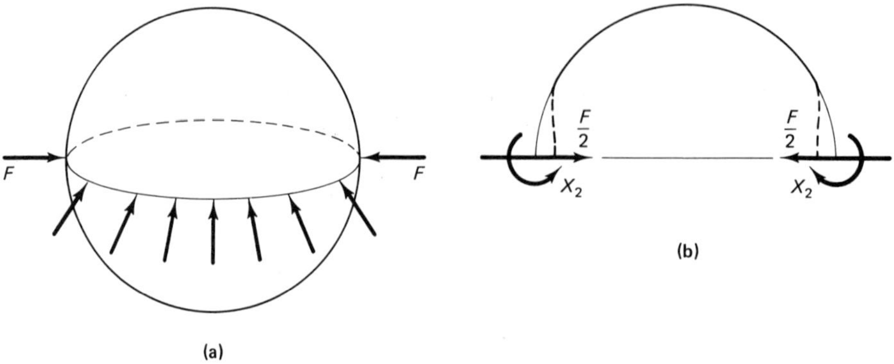

Figure 7.11 Spherical shell with line load along equator. (a) Actual structure. (b) Structure analyzed using symmetry.

Hence

$$X_2 = \frac{\delta_{21}}{\delta_{22}}\left(\frac{F}{2}\right) = -\frac{Fa}{4\lambda}$$

The final values for stress resultants or displacements at any point in the shell are given by the superposition of the two solutions obtained by substituting $H = -(F/2)$ into Eqs. (7.18 a–i), and $M = Fa/4\lambda$ into Eqs. (7.19 a–i).

7.7 ROOF DOME AND RING JUNCTION: FLEXIBILITY ANALYSIS

We shall now consider the problem of the junction of a roof dome with a ring beam, for which we had earlier obtained membrane stresses and displacements in Example 6.1. We had seen that due to discontinuity of displacements, unknown redundant forces X_1 and X_2, as shown in Fig. 7.12, must be applied so that compatibility of displacements is achieved. We shall determine these redundants from Eqs. (7.20 a–c).

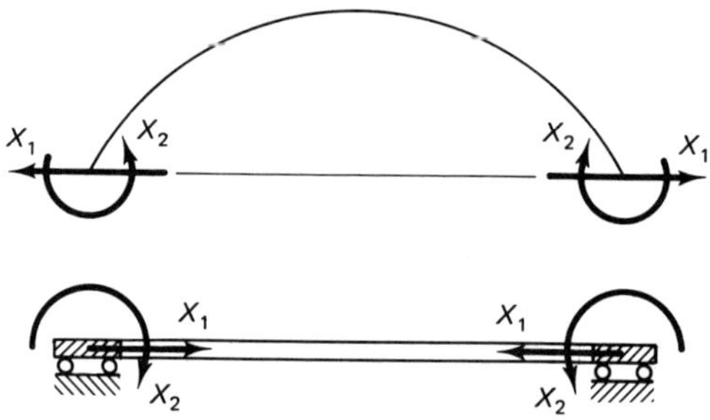

Figure 7.12 Corrective edge redundants X_1 and X_2 for roof-dome/ring-beam analysis.

Displacements of the Primary Structures

We had already obtained the membrane deflection and rotation at the bottom of the shell [Eqs. (6.18 a–c)]. Hence, taking displacements as positive in the positive directions of X_1 and X_2 (assumed as in Fig. 7.12), we can write

$$E\delta_{10}^{D} = -130{,}730.3 \text{ lb/ft}$$

$$E\delta_{20}^{D} = -13{,}200.0 \text{ lb/ft}^2$$

Equation (6.18c) gives us the horizontal displacement of the ring, while the rotation of the ring is zero. Hence we find

$$E\delta_{10}^{R} = -918{,}918.80 \text{ lb/ft}$$

$$E\delta_{20}^{R} = 0.0$$

Flexibility δ_{10}^{R} is negative because the deflection is outward, whereas X_1 is assumed inward on the ring. Now the total displacements become

$$E\delta_{10} = -1{,}049{,}649.10 \text{ lb/ft}$$

$$E\delta_{20} = -13{,}200.0 \text{ lb/ft}^2$$

Bending Analysis

Shell Constants. We shall assume that $\nu = 0$; hence

$$\lambda^4 = 3\left(\frac{a}{t}\right)^2(1 - \nu^2) = \frac{3(60.0)^2}{(0.25)^2} = 172{,}800.0$$

which gives $\lambda = 20.389$. Substituting $\phi_1 = 30°$ and the appropriate values of λ, a, and t into Eqs. (7.21 a–d), we obtain

$$E\boldsymbol{\delta}^{D} = \begin{bmatrix} 2446.56 & -1662.68 \\ -1662.68 & 2259.92 \end{bmatrix}$$

Ring Constants. The ring size assumed in Example 6.1 for obtaining its displacements was 1.5 ft × 1.0 ft. Assuming 1.5 ft as its width, the moment of inertia of the ring is $I = 0.125$ ft^4 and its area is $A = 1.5$ ft^2. Substituting these values, together with $r \approx 30.0$ ft, into Eqs. (7.22 a–d), we obtain

$$E\boldsymbol{\delta}^{R} = \begin{bmatrix} 600.0 & 0.0 \\ 0.0 & 7200.0 \end{bmatrix}$$

Hence the combined flexibility matrix can be written as

$$E\boldsymbol{\delta} = E\boldsymbol{\delta}^D + E\boldsymbol{\delta}^R = \begin{bmatrix} 3046.56 & -1662.68 \\ -1662.68 & 9459.92 \end{bmatrix}$$

Thus Eqs. (7.20a,b) become

$$3046.56X_1 - 1662.68X_2 = +1{,}049{,}649.10 \tag{7.23a}$$

$$-1662.68X_1 + 9459.92X_2 = +13{,}200.00 \tag{7.23b}$$

Upon solving, we have

$$X_1 = 381.93 \text{ lb/ft} \qquad \text{and} \qquad X_2 = 68.52 \text{ ft-lb/ft}$$

Both X_1 and X_2 are positive, indicating that their assumed directions were correct. Their values are also small and will not affect the membrane stresses significantly. The final forces in the dome are obtained by superimposing on the membrane values the effects of X_1 and X_2 from Eqs. (7.18 a–i) and (7.19 a–i). The final tension in the ring beam is given by

$$T = 45{,}945.94 - 381.93(30.0) = 34{,}488.04 \text{ lb}$$

It may be noted that after considering bending deformations, the hoop tension in the ring beam is much reduced as compared to that obtained in the membrane analysis. This fact can be used for economizing in the reinforcement detail of the ring beam.

It is thus very easy to carry out the bending analysis at junctions of a dome and ring beam. For any other loading, such as live load or snow load, the left-hand sides of Eqs. (7.23a,b) will remain the same, and we have only to calculate the membrane displacements of the primary structures, δ_{10} and δ_{20}, under the new loading, in order to obtain the unknown redundants.

7.8 INTZE TANK: ROOF DOME AND WALL ANALYSIS

For the Intze tank in Section 6.16, we designed the roof dome and the cylindrical wall by membrane theory alone, ignoring the incompatibility at their junction. We shall now satisfy the compatibility of displacements at this junction, and see the effect on the design of bending stresses thus developed. For simplicity, we shall assume that the meridian of the dome and cylinder meet at a common point, without any eccentricity, and we shall also assume that $\nu = 0$.

We shall consider the case of dead load only. Under water load, both the roof dome and wall primary structures will have zero horizontal deflections at their junction. Also, the roof dome will not have any rotation, but the wall will have a nonzero rotation. Hence, under water load, only incompatibility of rotations will occur. The bending stresses due to matching of the rotations for water load are small compared to those under the dead load of the roof dome, and hence neglecting them would not affect the design considerably.

Shell Constants

Dome. We have the following for the dome:

$$4\lambda_D^4 = 12(1 - \nu^2)\frac{a^2}{t^2} - \nu^2 = 12\frac{a^2}{t^2}$$

Hence

$$\lambda_D^4 = \frac{3.0(40.0)^2}{(0.25)^2} = 76{,}800.0$$

so

$$\lambda_D = 16.65$$

Cylindrical Wall. (see Chapter 5). Again, we find that

$$\lambda_w^4 = \frac{3r_w^2}{t_w^2}$$

where r_w and t_w are the radius and thickness of the wall. Hence

$$\lambda_w^4 = \frac{3.0(20.0)^2}{(0.67)^2} = 2673.2$$

so

$$\lambda_w = 7.19$$

Also, we find that

$$K_w = \frac{Et_w^3}{12(1 - \nu^2)} = \frac{Et_w^3}{12}$$

or

$$\frac{K_w}{E} = \frac{t_w^3}{12} = \frac{(0.67)^3}{12.0} = 0.02506 \text{ ft}^3$$

Displacements of the Primary Structures

Figure 7.13 (a) and (b) show the primary structures of the dome and the wall. Since there is no ring beam at their junction, the horizontal and vertical components of N_ϕ at the bottom of the dome must be resisted by the primary structure of the cylinder.

Dome. For the dome we had, from Eqs. (6.6a,b),

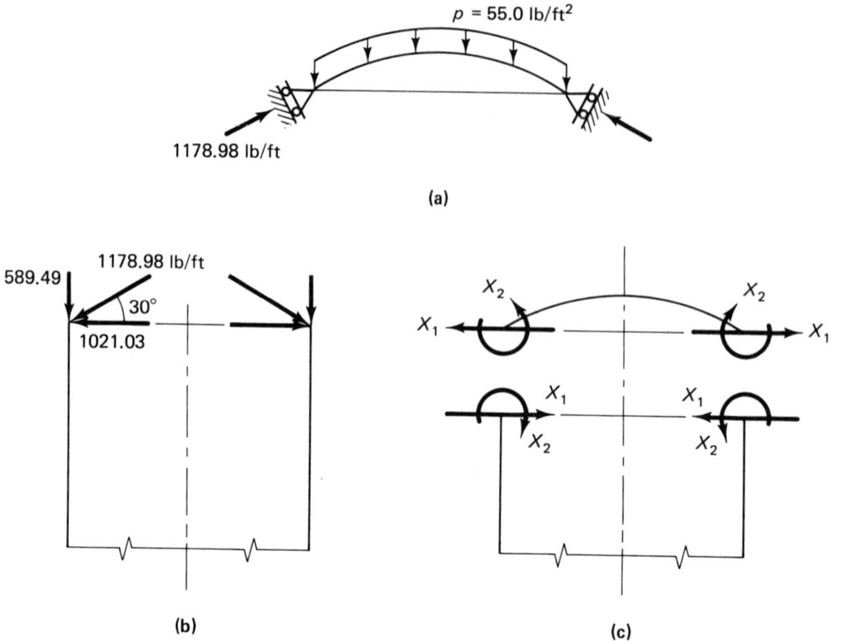

Figure 7.13 Primary structures for the dome and wall of the Intze tank. (a) Primary structure of dome. (b) Primary structure of cylindrical wall. (c) Edge redundants for dome/wall interaction.

$$N_\phi = -\frac{pa}{1 + \cos\phi} \qquad N_\theta = -pa\left(\cos\phi - \frac{1}{1 + \cos\phi}\right)$$

From Eq. (6.15a), we have for the dome

$$E\Delta_H = \frac{a \sin\phi}{t}(N_\theta - \nu N_\phi)$$

Hence at the bottom of the dome ($\phi = 30°$), we have

$$E\Delta_H = \left[\frac{40.0(\sin 30°)}{0.25}\right](-726.28) = -58{,}102.4 \text{ lb/ft}$$

where the value of N_θ is taken from Example 6.2, Table 6.1. Thus we have

$$E\delta_{10}^D = +E\Delta_H = -58{,}102.4 \text{ lb/ft}$$

Also, from Eq. (6.16b), we have

$$Et\chi = (N_\theta)^\cdot - \nu(N_\phi)^\cdot + \cot\phi\,(1 + \nu)(N_\theta - N_\phi)$$

Or, with $\nu = 0$,

$$Et\chi = (N_\theta)^\cdot + \cot\phi\,(N_\theta - N_\phi)$$

or

$$Et\chi = pa \sin \phi \left[1 + \frac{1}{(1 + \cos \phi)^2} \right] + \cot \phi \, (N_\theta - N_\phi)$$

Substituting $p = 55$ lb/ft^2, $a = 40.0$ ft, $t = 0.25$ ft, and $\phi = 30°$, along with the appropriate values for N_θ and N_ϕ from Table 6.1, we obtain

$$E\delta_{20}^D = -E\chi = -8800.0 \text{ lb/ft}^2$$

Wall. Under membrane conditions, the wall carries the following resultants:

$$N_\theta = 0 \quad N_x = -589.49 - \gamma_c t_w x$$

The horizontal deflection is given by

$$w = r_w \varepsilon_\theta = \frac{r_w}{Et_w} (N_\theta - \nu N_x)$$

For our case, since $N_\theta = 0$ and $\nu = 0$, we shall get w and dw/dx as zero under membrane conditions. But the primary structure of the cylinder [Fig. 7.13(b)] must also resist a horizontal edge load of $H = 1021.03$ lb/ft. We must use Eqs. (5.11g,h) to obtain the deflection and rotation at the top of the wall under this load. We thus have

$$\Delta_H = \frac{Hr_w^3}{2K_w \lambda_w^3}$$

or

$$E\Delta_H = \frac{Hr_w^3}{2(K_w/E)\lambda_w^3}$$

Also, we had

$$\frac{dw}{dx} = -\frac{Hr_w^2}{2K_w \lambda_w^2}$$

Hence,

$$E\frac{dw}{dx} = -\frac{Hr_w^2}{2(K_w/E)\lambda_w^2}$$

Substituting the appropriate numerical values, we obtain

$$E\delta_{10}^W = -E\Delta_H = \frac{-1021.03(20.0)^3}{2.0(0.02506)(7.19)^3} = -438{,}460.78$$

$$E\delta_{20}^W = +E\frac{dw}{dx} = \frac{-1021.03(20.0)^2}{2.0(0.02506)(7.19)^2} = -157{,}626.65$$

Hence we have for the displacements of the primary structures,

$$E\delta_{10} = E\delta_{10}^{D} + E\delta_{10}^{W} = -58{,}102.4 - 438{,}460.78 = -496{,}563.18 \text{ lb/ft}$$

$$E\delta_{20} = E\delta_{20}^{D} + E\delta_{20}^{W} = -8800.0 - 157{,}626.65 = -166{,}426.65 \text{ lb/ft}^2$$

Displacements under Unit Edge Loads

Dome. From Eqs. (7.21a–d), substituting in the appropriate numerical values, we have

$$E\delta_{11}^{D} = \frac{2a\lambda_D \sin^2 \phi_1}{t_D} = \frac{2.0(40.0)(16.65)(\sin^2 30°)}{0.25} = 1332.00 \text{ lb/ft}$$

$$E\delta_{21}^{D} = E\delta_{12}^{D} = -\frac{2\lambda_D^2 \sin \phi_1}{t_D} = -\frac{2.0(16.65)^2(\sin 30°)}{0.25} = -1108.89 \text{ lb/ft}^2$$

$$E\delta_{22}^{D} = \frac{4\lambda_D^3}{t_D a} = \frac{4.0(16.65)^3}{0.25(40.0)} = 1846.30 \text{ lb/ft}^2$$

Wall. From Eqs. (5.19a–d), we have

$$E\delta_{11}^{W} = \frac{r_w^3}{2(K_w/E)\lambda_w^3} = \frac{(20.0)^3}{2.0(0.02506)(7.19)^3} = 429.43 \text{ lb/ft}$$

$$E\delta_{21}^{W} = E\delta_{12}^{W} = \frac{r_w^2}{2(K_w/E)\lambda_w^2} = \frac{(20.0)^2}{2.0(0.02506)(7.19)^2} = 154.38 \text{ lb/ft}^2$$

$$E\delta_{22}^{W} = \frac{r_w}{(K_w/E)\lambda_w} = \frac{20.0}{0.02506(7.19)} = 111.00 \text{ lb/ft}^2$$

Hence for the combined displacements under unit edge loads, we have

$$E\delta_{11} = E\delta_{11}^{D} + E\delta_{11}^{W} = 1332.0 + 429.43 = 1761.43 \text{ lb/ft}$$

$$E\delta_{12} = E\delta_{21} = E\delta_{21}^{D} + E\delta_{21}^{W} = -1108.89 + 154.38 = -954.51 \text{ lb/ft}^2$$

$$E\delta_{22} = E\delta_{22}^{D} + E\delta_{22}^{W} = 1846.30 + 111.0 = 1957.3 \text{ lb/ft}^2$$

It is seen from the above that the wall is much stiffer than the dome, due to its smaller r/t ratio. Substituting the above into Eqs. (7.20a,b), we obtain

$$1761.43X_1 - 954.51X_2 = 496{,}563.18$$

$$-954.51X_1 + 1957.30X_2 = 166{,}426.65$$

Upon solving these equations, we get

$$X_1 = 445.79 \text{ lb/ft}$$

$$X_2 = 302.43 \text{ ft-lb/ft}$$

It is thus seen that a rather small moment and small shear are developed at the junction of the dome and wall. The final stresses will now be obtained by superimposing the effects of the edge loads onto the stresses of the primary structures.

Final Stresses in the Dome

The meridional component of X_1 will yield a tension resultant N_ϕ of $X_1 \cos 30° = 386.07$ lb/ft at the edge. The final value of N_ϕ at the edge is obtained by adding to this the membrane force $N_\phi = -1178.98$ lb/ft, from Table 6.1. Hence the final value for N_ϕ will be compressive at the edge as well as at all other locations. Therefore, this change in N_ϕ will not significantly influence the design of the dome performed previously. We shall next compute the final values of N_θ. Using Eqs. (7.18c) and (7.19c), we have

$$N_\theta = -2\lambda_D X_1 \sin \phi_1 \, e^{-\lambda_D \beta_1} \sin\left(\lambda_D \beta_1 - \frac{\pi}{2}\right) + \frac{2\sqrt{2}\,\lambda_D^2 X_2}{a} e^{-\lambda_D \beta_1} \sin\left(\lambda_D \beta_1 - \frac{\pi}{4}\right)$$

Substituting in the proper numerical values, we obtain

$$N_\theta = -7422.40 e^{-16.65\beta_1} \sin\left(16.65\beta_1 - \frac{\pi}{2}\right) + 5928.41 e^{-16.65\beta_1} \sin\left(16.65\beta_1 - \frac{\pi}{4}\right)$$

or

$$N_\theta = A_1 + B_1$$

where A_1 and B_1 are the first and second terms in the equation above, due to an edge shear and moment, respectively. This resultant N_θ can be calculated for various values of β_1 (or ϕ), and then superimposed onto the membrane stresses of Table 6.1, in order to obtain the final results. This is done in Table 7.1. It is seen that

TABLE 7.1 Resultant N_θ in Dome due to Bending and Membrane Effects

ϕ (deg)	β_1 (deg)	β_1 (rad)	A_1	B_1	N_θ (Bending) $= A_1 + B_1$	N_θ (Membrane) (See Table 6.1)	Final N_θ (lb/ft)
30	0	0.0000	7422.40	−4192.02	3230.38	−726.28	2504.10
25	5	0.0873	202.96	858.49	1061.45	−839.81	221.64
20	10	0.1745	−394.92	276.72	−118.20	−933.12	−1051.32
15	15	0.2618	−32.86	−31.75	−64.61	−1005.97	−1070.58
10	20	0.3491	19.78	−16.85	2.93	−1058.16	−1055.23

only near the edge for about 6° or so, N_θ is a tensile force, due to the predominant bending stresses in this region. In the rest of the dome, N_θ is still compressive and almost the same as for the membrane case. Assuming, for simplicity, a linear variation of N_θ with ϕ, we find that tension occurs up to a value of β_1 given by

$$\beta_1 = 5° + \left(\frac{221.64}{221.64 + 1051.32}\right)(5°) = 5.87°$$

i.e., over a length equal to $5.87(\pi/180.0)(40.0) = 4.1$ ft from the edge along the meridian. We must provide reinforcing steel for these tensile forces. The total tension in this zone will be approximately

$$T = 4.1\left(\frac{2504.10 + 0}{2}\right) = 5133.41 \text{ lb}$$

Hence the amount of steel required for this zone is

$$A_s = \frac{5133.41}{20{,}000} = 0.257 \text{ in.}^2$$

In our design based on membrane theory, we have already provided the minimum hoop steel as No. 3 bars at 12-in. center-to-center spacing. This minimum reinforcing is more than that required above, and hence our preliminary design is satisfactory.

The variation in the bending moment with β_1, due to X_1 and X_2, can also be calculated. However, since this moment is small, we can simply assume that the maximum moment is equal to the edge moment X_2. The edge section is subjected to this moment, which causes tension on the outer face, in addition to a direct compression N_ϕ. Neglecting the small compression, we obtain the area of steel required for resisting the moment, as

$$A_s = \frac{X_2}{f_a c} = \frac{302.43(12)}{20{,}000(0.87)(2.25)} = 0.093 \text{ in.}^2/\text{ft}$$

where f_a is the allowable design strength of steel and c is the effective moment arm of the concrete compression and steel tension forces (taken as 87 percent of the section depth minus rebar cover). This amount of steel is very small, and thus the bars anchored from the cylindrical wall into the dome, on its outer face, would be more than sufficient. The moment will also die out in less than 5° away from the edge. Sometimes the meridional steel provided in the dome may be taken to its outer face, by bending the rebar near the edge, in order to cater for this moment.

In this example we demonstrated that very little change is required in the design of the dome, from the membrane theory design, to account for the bending effects.

Final Stresses in the Wall

Resultant N_θ in the wall can be obtained by superimposing the effects of X_1 and X_2 onto the forces in the primary structure. Returning to Fig. 7.13(b,c), we see that the net horizontal edge load on the cylinder will be $H = (1021.03 - X_1) = 575.24$ lb/ft, outward. N_θ, due to this force plus X_2, is found by adding Eqs. (5.11c) and (5.12c), as

$$N_\theta = 2H\lambda_w e^{-(\lambda_w x/r_w)} \cos\left(\lambda_w \frac{x}{r_w}\right) + \frac{2\sqrt{2}\, X_2 \lambda_w^2}{r_w} e^{-(\lambda_w x/r_w)} \sin\left(\lambda_w \frac{x}{r_w} - \frac{\pi}{4}\right)$$

Substituting in the proper numerical values, we have

$$N_\theta = 8271.95e^{-0.3595x} \cos 0.3595x + 2211.05e^{-0.3595x} \sin\left(0.3595x - \frac{\pi}{4}\right)$$

or

$$N_\theta = A_2 + B_2$$

where A_2 and B_2 are the contributions to N_θ coming from edge shear and edge bending, respectively. The values of N_θ for different values of x are given in Table 7.2.

Due to the small value of r_w/t_w (i.e., small value of λ_w), it is seen that the edge effects do not die out as rapidly as they do in the dome. We get tension in the top 5 ft or so of the wall. When these values are added to the membrane hoop tension N_θ, due to water load (Table 6.2), there will only be a small increase in tension in the top 5 ft of the wall. However, it can easily be checked that the minimum hoop steel provided in Table 6.2 will still be sufficient to resist the resultant forces. Also, the moments due to edge shear and moment will be so small that the minimum vertical steel provided earlier will be sufficient. Thus we can again say that even though we have not provided a ring beam at the dome-wall junction, the bending stresses generated in the wall are so small that the membrane theory design remains almost unaffected.

TABLE 7.2 Final N_θ Forces in Wall (Dead Load)

x (ft)	A_2	B_2	$N_\theta = A_2 + B_2$ (lb/ft)
0.0	8271.95	−1563.45	6708.50
2.0	3032.75	−71.48	2961.27
4.0	260.02	318.75	578.77
6.0	−529.32	250.70	−278.62
8.0	−449.86	108.15	−341.71
10.0	−204.20	19.79	−184.41

In solving this problem we had applied the horizontal component of the membrane force N_ϕ, of the dome, to the primary structure of the wall [Fig. 7.13(b)]. Alternatively, we could have applied this force to the dome itself, with only the vertical component of N_ϕ being applied to the wall. The primary structures would then be as shown in Fig. 7.14. Displacements δ_{10}^D and δ_{20}^D will now have to be calculated as membrane displacements, plus displacements due to the edge load [which can be obtained by using Eqs. (7.18h,i)]. On the other hand, for the wall, now both δ_{10}^W and δ_{20}^W will be zero (assuming that $\nu = 0$). Again, new values of X_1 and X_2 can be obtained. X_2 will come out to be the same as before. The final net edge loads on the dome and wall will, of course, be the same as those obtained above. The reader may verify this by performing the revised numerical calculations.

7.9 ANALYSIS OF JUNCTIONS WITH MORE THAN TWO INTERSECTING MEMBERS

In the example just considered in Section 7.8, if we had a ring beam at the junction of the dome and wall, the bending theory solutions would have become more complex. The same situation will arise when we want to solve the problem at the junction of the wall, ring beam, and the cone for the Intze tank in Section 6.16.

For the dome/ring/wall junction of Fig. 7.15(a), unknown edge forces X_1, X_2 between the dome and the ring beam, and X_3, X_4 between the ring beam and the wall, will develop. As discussed earlier, for simplicity we assume that the dome and wall meet at one point on the horizontal axis through the centroid of the ring cross section [Fig. 7.15(a)]. This can always be closely achieved in practice when forming the geometry of the structure. Hence X_1 and X_3 may be assumed to pass through the centroid of the ring. Considering the small effect of bending on the

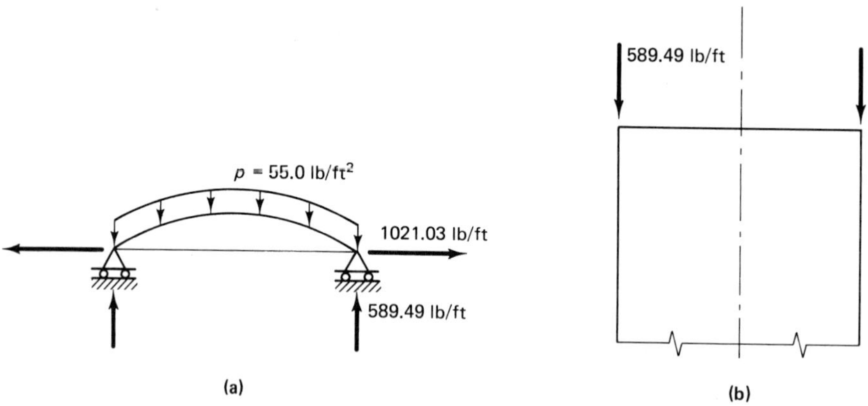

Figure 7.14 Alternative primary structures for dome and wall of the Intze tank. (a) Primary structure of dome. (b) Primary structure of wall.

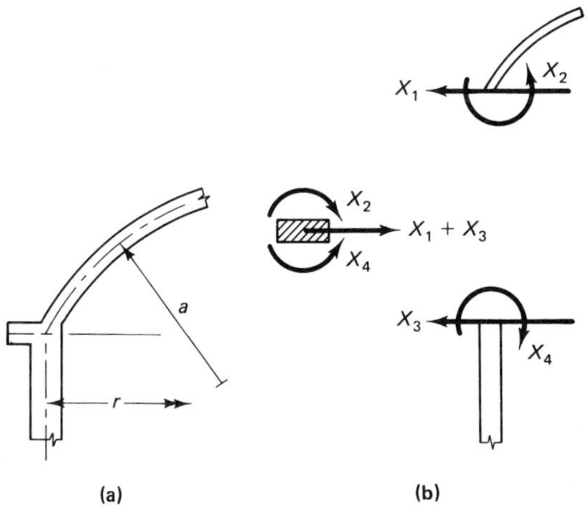

Figure 7.15 Dome/wall/ring-beam junction of Intze tank. (a) Detail of junction. (b) Edge redundants.

shell design, as illustrated in Section 7.8, it is felt that further refinements are not warranted for common practical designs.

We can now write four equations stating that the relative deflections and rotations between the dome and ring, and the ring and wall, are zero. These equations will be of the form

$$\delta_{i0} + \delta_{i1}X_1 + \delta_{i2}X_2 + \delta_{i3}X_3 + \delta_{i4}X_4 = 0 \qquad i = 1, 2, 3, 4$$

We shall have, as before, $\delta_{11} = \delta_{11}^D + \delta_{11}^R$ and $\delta_{44} = \delta_{44}^R + \delta_{44}^W$, and so on. From Fig. 7.15(b), we also see that we shall obtain relations of the type $\delta_{11}^R = \delta_{13}^R = \delta_{31}^R = \delta_{33}^R$, etc.

Although such problems can be solved, in principle, in the manner described above, in practice it becomes cumbersome to solve these problems by hand when four unknowns are involved. Instead, such problems can be more easily solved by the "stiffness method." In this method, whether or not there is a ring beam, we shall have only two equations for the two unknown joint displacements: the horizontal deflection and the rotation. We discuss this method in Section 7.12.

7.10 PRESSURE VESSELS

Let us now consider the cylindrical vessel with hemispherical heads, shown in Fig. 7.16, under an internal pressure p. We had seen in Section 6.14 that the membrane displacements of the cylindrical and spherical shells are different at their junction, and thus edge loads X_1 and X_2, as shown in Fig. 7.16(b), will be developed to satisfy compatibility at the junction.* We had already obtained in Eqs. (6.21a,b) the displacements of the primary structures. We can therefore write

$$\delta_{10} = \delta_{10}^{D} + \delta_{10}^{W}$$

$$\delta_{10} = \frac{pa^2}{2Et_D}(1 - \nu) - \frac{pa^2}{2Et_W}(2 - \nu)$$

or, if $t_D = t_W = t$,

$$E\delta_{10} = -\frac{pa^2}{2t}$$

where the subscripts D and W refer to the dome and the cylindrical shell wall, respectively. Also, since under internal pressure both the cylinder and the dome will not have any rotation, we find that $\delta_{20} = 0$. Using Eqs. (7.21a–d) with the proper signs and $\phi_1 = 90°$, we have for the dome

$$E\delta_{11}^{D} = \frac{2a\lambda_D}{t_D} \tag{7.24a}$$

$$E\delta_{21}^{D} = E\delta_{12}^{D} = \frac{2\lambda_D^2}{t_D} \tag{7.24b,c}$$

and

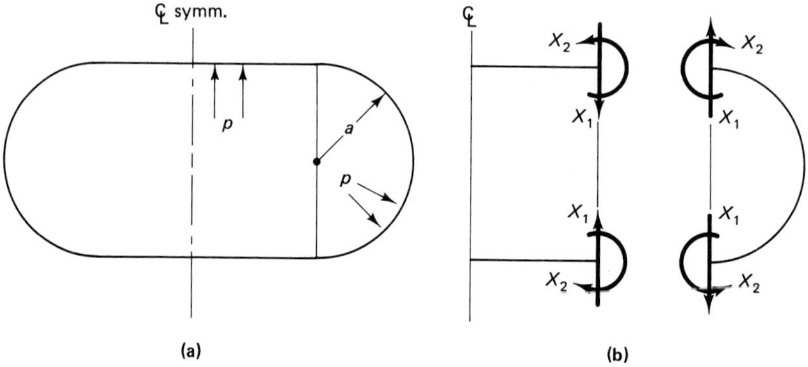

Figure 7.16 Interaction of cylindrical pressure vessel and hemispherical cap. (a) Configuration of structure. (b) Edge redundants that satisfy displacement compatibility.

*It may be noted that N_ϕ from the dome at $\phi = 90°$ will be taken up as N_x by the cylinder. Also, for both the cylinder and the sphere, we have the freedom of prescribing one membrane displacement without affecting the membrane forces. We assume this displacement to be the axial displacement of the cylinder, which is taken as zero at its center. Also, whatever axial displacement the cylinder has at its junction with the dome is taken as the prescribed displacement for the dome. Hence there is no incompatibility of displacements along the axis of the cylinder at its junction with the dome.

$$E\delta_{22}^{D} = \frac{4\lambda_D^3}{at_D} \tag{7.24d}$$

For the cylindrical shell, referring to Eqs. (5.19a–d), we have

$$\delta_{11}^{W} = \frac{a^3}{2K_W\lambda_W^3}$$

$$\delta_{21}^{W} = \delta_{12}^{W} = -\frac{a^2}{2K_W\lambda_W^2}$$

$$\delta_{22}^{W} = \frac{a}{K_W\lambda_W}$$

Recalling that $4\lambda_W^4 = (Et_Wa^2/K_W)$, we can rewrite these equations as

$$\delta_{11}^{W} = \frac{a^3\lambda_W}{2K_W\lambda_W^4} = \frac{2a\lambda_W}{Et_W} \tag{7.25a}$$

Similarly,

$$\delta_{21}^{W} = \delta_{12}^{W} = -\frac{2\lambda_W^2}{Et_W} \tag{7.25b,c}$$

and

$$\delta_{22}^{W} = \frac{4\lambda_W^3}{Et_Wa} \tag{7.25d}$$

Adding the flexibility coefficients for the dome and the cylindrical shell, we finally obtain

$$E\delta_{11} = \frac{2a\lambda_D}{t_D} + \frac{2a\lambda_W}{t_W}$$

$$E\delta_{12} = E\delta_{21} = \frac{2\lambda_D^2}{t_D} - \frac{2\lambda_W^2}{t_W}$$

$$E\delta_{22} = \frac{4\lambda_D^3}{at_D} + \frac{4\lambda_W^3}{at_W}$$

Substituting these terms into the compatibility equations (7.20a,b), values of X_1 and X_2 can be readily obtained. It is interesting to consider the special case when the thicknesses of the dome and cylinder are the same, i.e., when $t_D = t_W$. We recall that

$$4\lambda_D^4 = 12(1 - \nu^2)\frac{a^2}{t_D^2} - \nu^2 \qquad \text{and} \qquad 4\lambda_W^4 = 12(1 - \nu^2)\frac{a^2}{t_W^2}$$

Neglecting the term (ν^2), we shall have $\lambda_D = \lambda_W = \lambda$ when $t_D = t_W = t$. Then, from the above, we obtain

$$E\delta_{11} = \frac{4a\lambda}{t}$$

$$E\delta_{12} = E\delta_{21} = 0$$

and

$$E\delta_{22} = \frac{8\lambda^3}{at}$$

The second compatibility equation (7.20b) will now simply be $(\delta_{22})X_2 = 0$, which gives $X_2 = 0$. This means that there is no moment at the junction of the dome and the cylinder. Also, the first equation will become

$$\delta_{10} + \delta_{11}X_1 = 0$$

or

$$X_1 = -\frac{\delta_{10}}{\delta_{11}} = \frac{pa^2}{2t}\frac{t}{4a\lambda} = \frac{pa}{8\lambda}$$

Having obtained X_1, we can superimpose the stresses it induces onto the membrane stresses, in order to obtain the final values. Initially, in the primary structure, the dome and the cylinder both had different deflections, but both had zero rotations. The absence of moment at their junction ($X_2 = 0$) indicates that the rotations of the cylinder and dome, caused by X_1, are exactly the same; thus when their deflections are matched, their slopes will also automatically be matched. We shall use this interesting fact in order to apply the bending theory solutions developed for the spherical shell, in an approximate manner, to other shells considered in the next section.

For the case when the pressure vessel head is not a hemispherical dome, but instead, only a shallower spherical cap, the primary structure of the cylinder (or dome) will have to carry the horizontal component of the membrane resultant N_ϕ at the edge of the sphere. This problem will be similar to the Intze tank roof-dome/wall problem solved in Section 7.8.

7.11 APPLICATIONS TO CONICAL AND OTHER SHELLS

Due to the mathematical complexities involved, we derived in this chapter equations only for an approximate bending theory of axisymmetric edge loads on spherical shells. For conical shells, Flügge [1] has derived equations of bending theory that involve evaluations of Bessel functions. By making some further simplifications, approximate equations which are simpler and more familiar to those derived previously for spherical shells can be obtained for conical shells. Instead of using such equations, however, we can approximately apply the equations of spherical shells to solve problems of conical shells as well.

In Section 7.10 we have seen that the flexibility coefficients of the hemispherical dome in Eqs. (7.24a–d) are the same as those for the cylinder in Eqs. (7.25a–d) if both have the same thicknesses. This means that the edge displacements of a cylindrical shell, under edge loads X_1 or X_2, can be approximately obtained by finding the edge displacements of an equivalent inscribed spherical shell of the same thickness, or vice versa [Fig. 7.17(a)]. Also, since the effects of these edge forces generally die out rapidly, the bending stresses or displacements near the edges of one type of shell (e.g., cylinder) could also be approximately obtained by using corresponding solutions for another similar shell (e.g., sphere).

In a similar manner, for the conical shell of Fig. 7.17(b), we can approximately obtain bending theory solutions for edge loads by using the solutions for an inscribed sphere having the same thickness, as shown. The radius of this equivalent spherical shell would be $a = r/\sin\alpha$, where r is the base radius of the conical shell. This approach can be used to solve the incompatibility problems arising at the junction of the cone, cylinder, and ring beam of the Intze tank considered earlier.

7.12 STIFFNESS METHOD FOR SOLVING DISCONTINUITY PROBLEMS

When more than two members, such as a dome, ring beam, and cylinder, meet at a joint,* we had seen that the flexibility method of analysis becomes too complicated for hand solution. In such cases, the stiffness method, similar to that discussed for cylindrical shells earlier, is simpler to use. Figure 7.18 shows a typical joint where a dome, ring beam, and cylinder meet. Let u_1 and u_2, as shown, denote the horizontal

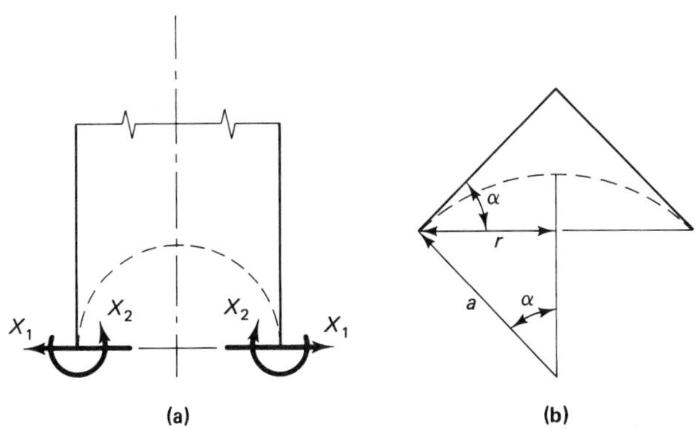

Figure 7.17 Approximation of flexibility coefficients for various shells using an equivalent inscribed spherical shell. (a) Approximation for cylinders. (b) Approximation for cones.

*It is understood that by "joint" we mean here the junction in the form of a parallel circle.

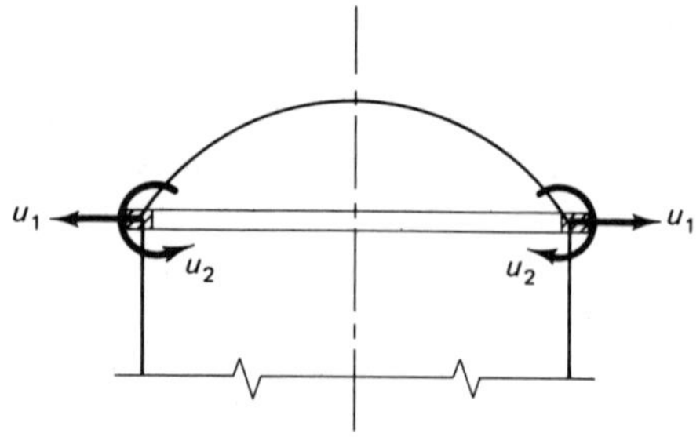

Figure 7.18 Dome/wall/ring-beam junction, showing joint displacements to be determined in the stiffness analysis.

deflection and rotation of the joint. Then, as for cylindrical shells, we can write the equilibrium equations for the forces acting on the joint in the following manner:

$$S_{11}u_1 + S_{12}u_2 + \text{FER}_1 = 0 \tag{7.26a}$$

$$S_{21}u_1 + S_{22}u_2 + \text{FER}_2 = 0 \tag{7.26b}$$

where S_{ij} $(i, j = 1, 2)$ represent the total stiffnesses of the members meeting at the joint, and corresponding to u_1 and u_2. Thus

$$S_{ij} = S_{ij}^D + S_{ij}^R + S_{ij}^W \qquad (i, j = 1, 2) \tag{7.26c}$$

where the superscripts D, R, and W refer to the dome, ring beam, and cylinder (wall), respectively. FER_1 and FER_2 are the unbalanced fixed-edge reactions acting on the members (or the sum of the fixed-edge reactions, corresponding to u_1 and u_2, acting on the different members). Hence again we can write

$$\text{FER}_i = \text{FER}_i^D + \text{FER}_i^R + \text{FER}_i^W \qquad (i = 1, 2) \tag{7.26d}$$

Thus, in matrix form, Eqs. (7.26a–d) can be written as

$$\mathbf{SU} + \mathbf{FER} = \mathbf{0}$$

where

$$\mathbf{S} = \mathbf{S}^D + \mathbf{S}^R + \mathbf{S}^W \qquad \text{and} \qquad \mathbf{FER} = \mathbf{FER}^D + \mathbf{FER}^R + \mathbf{FER}^W$$

Solving Eqs. (7.26a–d) we can obtain u_1 and u_2. Then the edge loads acting on the dome are given by

$$X_1^D = S_{11}^D u_1 + S_{12}^D u_2 + \text{FER}_1^D \tag{7.27a}$$

$$X_2^D = S_{21}^D u_1 + S_{22}^D u_2 + \text{FER}_2^D \tag{7.27b}$$

where the positive directions of X_1 and X_2 correspond to the positive directions of u_1 and u_2. In matrix form we can write these equations as

$$\mathbf{X}^D = \mathbf{S}^D\mathbf{U} + \mathbf{FER}^D$$

Similar equations can be written for the ring beam and the cylinder by changing the superscripts. The final stresses in the dome will be obtained by superimposing on the "fixed-edge" solution the effect of $(\mathbf{X}^D - \mathbf{FER}^D)$, or more easily, by superimposing on the results of a "primary" dome structure the effects of edge loads $\mathbf{X}^D$.

7.13 STIFFNESS COEFFICIENTS FOR SPHERICAL SHELLS AND FOR RING BEAMS

Spherical Shells

It is easy to obtain stiffness coefficients for a spherical shell. We proceed as we did for the cylindrical shell, by considering two separate cases of deformation, (Fig. 7.19).

Case I. We apply to the shell a unit displacement u_1, while keeping $u_2 = 0$. The values of H and M developed can be calculated as follows. From Eqs. (7.18h,i) and (7.19h,i) we obtain, due to H and M in the assumed directions,

$$u_2 = -(\chi)_{\beta_1=0} = -\frac{2\lambda^2 H \sin\phi_1}{Et} + \frac{4\lambda^3 M}{Eta} = 0$$

Hence

$$M = \frac{a \sin\phi_1}{2\lambda} H$$

Also, deflection u_1, due to H and M, is then obtained as

$$u_1 = \frac{2a\lambda \sin^2\phi_1}{Et} H - \frac{2\lambda^2 \sin\phi_1}{Et} M$$

Or by substituting for M in terms of H from above and adding, we obtain

$$u_1 = \frac{a\lambda \sin^2\phi_1}{Et} H$$

Hence for $u_1 = 1$, we have

$$S_{11}^D = H = \frac{Et}{a\lambda \sin^2\phi_1} \tag{7.28a}$$

and then

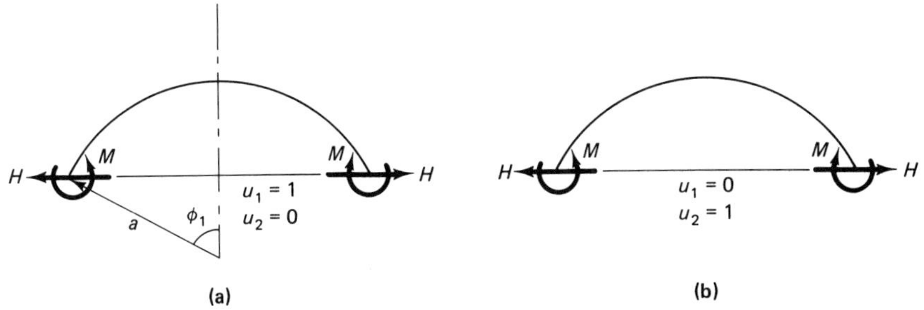

Figure 7.19 Shell edge-load configurations for determining stiffness coefficients. (a) Case I, spherical dome with $u_1 = 1$ and $u_2 = 0$. (b) Case II, spherical dome with $u_1 = 0$ and $u_2 = 1$.

$$S_{21}^D = M = \frac{Et}{2\lambda^2 \sin \phi_1} \tag{7.28b}$$

Case II. We now apply to the shell a unit rotation $u_2 = 1$, while keeping $u_1 = 0$. The values of H and M thus developed can be obtained again by using Eqs. (7.18h,i) and (7.19h,i). We have

$$u_1 = (\Delta_H)_{\beta_1=0} = \frac{2a\lambda \sin^2 \phi_1}{Et} H - \frac{2\lambda^2 \sin \phi_1}{Et} M = 0$$

Hence

$$H = \frac{\lambda}{a \sin \phi_1} M$$

Now, rotation u_2, due to H and M, is given by

$$u_2 = -(\chi)_{\beta_1=0} = -\frac{2\lambda^2 \sin \phi_1}{Et} H + \frac{4\lambda^3}{Eta} M$$

Upon substituting for H in terms of M from above and adding, we obtain

$$u_2 = \frac{2\lambda^3}{Eta} M$$

Hence for $u_2 = 1$, we have

$$S_{22}^D = M = \frac{Eta}{2\lambda^3} \tag{7.28d}$$

and

$$S_{12}^D = H = \frac{Et}{2\lambda^2 \sin \phi_1} \tag{7.28c}$$

We can summarize the four stiffness coefficients, in matrix form, as

$$\mathbf{S}^D = \begin{bmatrix} S_{11}^D & S_{12}^D \\ S_{21}^D & S_{22}^D \end{bmatrix} = \begin{bmatrix} \dfrac{Et}{a\lambda \sin^2 \phi_1} & \dfrac{Et}{2\lambda^2 \sin \phi_1} \\ \dfrac{Et}{2\lambda^2 \sin \phi_1} & \dfrac{Eta}{2\lambda^3} \end{bmatrix} \tag{7.28a–d}$$

We could have also obtained the stiffness matrix above by simply inverting the flexibility matrix $\boldsymbol{\delta}^D$, obtained earlier in Eqs. (7.21a–d) for the spherical shell. Indeed, the reader can easily verify that $\boldsymbol{\delta}^D \mathbf{S}^D = \mathbf{I}$, where $\mathbf{I}$ is the identity matrix.

Ring Beams

We assume that u_1 is the horizontal deflection of the centroid of the ring cross section and that u_2 is its rotation. We can then obtain stiffness coefficients for the ring, in the manner above, by considering two cases. However, because of our previous simplifications, the ring flexibility matrix was a diagonal matrix [Eqs. (7.22a–d)]. It is therefore easier simply to invert this matrix in order to obtain the stiffness matrix for the ring. We shall have

$$\mathbf{S}^R = \begin{bmatrix} S_{11}^R & S_{12}^R \\ S_{21}^R & S_{22}^R \end{bmatrix} = [\boldsymbol{\delta}^R]^{-1} = \begin{bmatrix} \dfrac{AE}{r^2} & 0 \\ 0 & \dfrac{EI}{r^2} \end{bmatrix} \tag{7.29a–d}$$

where A and I are the area and the moment of inertia, respectively, of the cross section of the ring. We may recall here that we have assumed a rectangular ring cross section in our derivation.

7.14 FIXED-EDGE REACTIONS FOR A SPHERICAL SHELL

Figure 7.20 shows the positive directions of FER_1 and FER_2, which are in the same directions as u_1 and u_2, respectively. We replace FER_1^D and FER_2^D by X_1 and X_2, respectively, for simplicity. Then, using the flexibility coefficients of Eqs. (7.21a–d), we can write the compatibility equations (7.20a,b) for finding X_1 and X_2, as follows:

$$(2a\lambda \sin^2 \phi_1)X_1 - (2\lambda^2 \sin \phi_1)X_2 + Et\delta_{10}^D = 0$$

$$-(2\lambda^2 \sin \phi_1)X_1 + \frac{4\lambda^3}{a}X_2 + Et\delta_{20}^D = 0$$

where δ_{10}^D and δ_{20}^D are the displacements of the primary structure of the dome for any given loading, taken positive in the positive senses of X_1 and X_2 (or u_1 and u_2), respectively. On solving the equations above, we obtain the following expressions:

$$\text{FER}_1^D = X_1 = \frac{-t}{a\lambda \sin^2 \phi_1}\left[E\delta_{10}^D + \left(\frac{a}{2\lambda} \sin \phi_1\right)E\delta_{20}^D\right] \tag{7.30a}$$

and

$$\text{FER}_2^D = X_2 = \frac{-t}{2\lambda^2 \sin \phi_1}\left[E\delta_{10}^D + \frac{a \sin \phi_1}{\lambda} E\delta_{20}^D\right] \tag{7.30b}$$

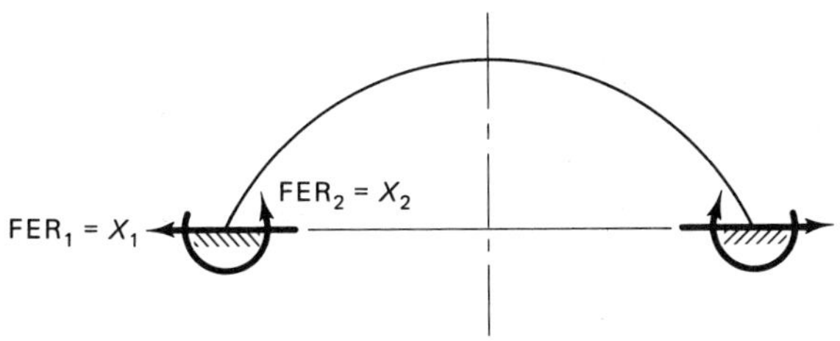

Figure 7.20 Fixed-edge reactions for a spherical dome.

It is thus seen, as is intuitively obvious, that

$$\text{FER}_1^D = X_1 = -\left[S_{11}^D \delta_{10}^D + S_{12}^D \delta_{20}^D \right] \tag{7.31a}$$

and

$$\text{FER}_2^D = X_2 = -\left[S_{21}^D \delta_{10}^D + S_{22}^D \delta_{20}^D \right] \tag{7.31b}$$

Or, in matrix form, we have

$$\mathbf{FER}^D = -\mathbf{S}^D \boldsymbol{\delta}_0^D$$

where $\boldsymbol{\delta}_0$ represents the vector of membrane displacements for a given loading. This equation can be understood or interpreted clearly as follows. $\boldsymbol{\delta}_0^D$ is the vector of displacements of the primary structure. Under fixed conditions, these primary-structure displacements must be suppressed; i.e., we must apply displacements equal to $-\boldsymbol{\delta}_0^D$. Then, by definition, the edge loads induced will be $\mathbf{S}^D(-\boldsymbol{\delta}_0^D)$, which are the fixed-edge reactions $\mathbf{FER}^D$. Proceeding similarly, we can also obtain similar expressions for the ring beam. Hence

$$\text{FER}_1^R = -\left[S_{11}^R \delta_{10}^R + S_{12}^R \delta_{20}^R \right] \tag{7.32a}$$

$$\text{FER}_2^R = -\left[S_{21}^R \delta_{10}^R + S_{22}^R \delta_{20}^R \right] \tag{7.32b}$$

or

$$\mathbf{FER}^R = -\mathbf{S}^R \boldsymbol{\delta}_0^R$$

Values of $\delta_{10}^{D,R}$ and $\delta_{20}^{D,R}$ can be calculated for given loadings on the primary structures, and hence $\text{FER}_1^{D,R}$ and $\text{FER}_2^{D,R}$ can be obtained. Similar expressions will hold for FER_1^W and FER_2^W in a cylindrical shell wall.

7.15 SOLUTION OF PROBLEMS BY THE STIFFNESS METHOD

Roof Dome and Ring Junction

As a first example of the stiffness method, we shall solve the problem of a roof-dome/ring junction, under self-weight of the dome, which was solved in Section 7.7 by the flexibility method. Figure 7.21(a) shows the primary structures for the dome and ring beam, and Fig. 7.21(b) shows the fixed-edge reactions after the structures are restrained against horizontal deflection and rotation. We had already obtained for the primary structures (refer to Example 6.1 and Section 7.7), the following results:

$$E\delta_{10}^D = -130{,}730.3 \text{ lb/ft}$$

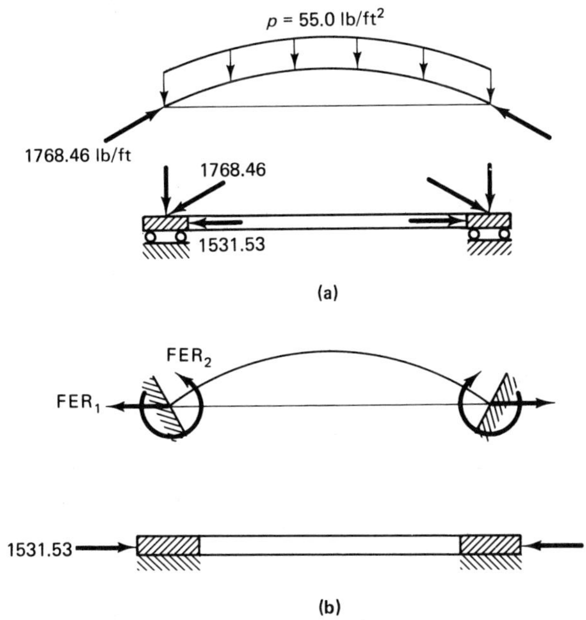

Figure 7.21 Primary structures for the dome and ring beam of the example in Section 7.7. (a) Configurations of primary structures. (b) Fixed-edge reactions.

$$E\delta_{20}^{D} = -13{,}200.0 \text{ lb/ft}^2$$

$$E\delta_{10}^{R} = +918{,}918.80 \text{ lb/ft}$$

and

$$E\delta_{20}^{R} = 0.0$$

It may be noted that we have changed the sign of δ_{10}^{R} here because δ_{10}^{R} is now assumed positive in the positive direction of u_1. Also, we had $\lambda = 20.389$, $a = 60.0$ ft, $\phi_1 = 30°$, $t = 0.25$ ft, and for the ring, $A = 1.5$ ft^2 and $I = 0.125$ ft^4. Substituting these values into Eqs. (7.28a–d) and (7.29a–d), we obtain

$$\mathbf{S}^{D} = \begin{bmatrix} 8.175 & 6.014 \\ 6.014 & 8.849 \end{bmatrix} (E \times 10^{-4})$$

and

$$\mathbf{S}^{R} = \begin{bmatrix} 16.667 & 0.0 \\ 0.0 & 1.389 \end{bmatrix} (E \times 10^{-4})$$

Substituting these values and δ_{i0} ($i = 1, 2$) into Eqs. (7.31a,b), we obtain

$$\text{FER}_1^D = -\left[8.175(E \times 10^{-4})\frac{-130{,}730.3}{E} + 6.014(E \times 10^{-4})\frac{-13{,}200.0}{E}\right] = 114.81 \text{ lb/ft}$$

Similarly, we find $\text{FER}_2^D = 90.30$ ft-lb/ft.

For the ring beam, we have, from Eq. (7.32a),*

$$\text{FER}_1^R = -16.667(E \times 10^{-4})(918{,}918.80)\frac{1}{E}$$

or

$$\text{FER}_1^R = -1531.53 \text{ lb/ft}$$

This result should be obvious. The ring primary structure carries a horizontal force equal to 1531.53 lb/ft (acting outward) which comes from the horizontal component of N_ϕ in the dome. This force will cause an outward deflection of the ring. To suppress this deflection under fixed conditions, we must apply an equal and opposite force of 1531.53 lb/ft on the ring. By definition, this force is FER_1^R, and being opposite in direction to u_1, it has a negative value.

Also, from Eq. (7.32b), we find that $\text{FER}_2^R = 0$. Adding for the dome and ring beam, we have

$$\mathbf{S} = \begin{bmatrix} 24.842 & 6.014 \\ 6.014 & 10.238 \end{bmatrix}(E \times 10^{-4})$$

and

$$\mathbf{FER} = \begin{bmatrix} -1416.72 \\ 90.30 \end{bmatrix}$$

Our equations, $\mathbf{SU} + \mathbf{FER} = \mathbf{0}$, can now be solved by inverting $\mathbf{S}$, or we can solve them explicitly. We then obtain,

$$u_1 = \frac{68.973}{E} \times 10^4 \qquad \text{and} \qquad u_2 = -\frac{49.336}{E} \times 10^4$$

Final edge loads on the shell are now given by Eqs. (7.27a,b). Thus

$$X_1^D = 8.175(68.973) + 6.014(-49.336) + 114.81 = 381.96 \text{ lb/ft}$$

and

$$X_2^D = 6.014(68.973) + 8.849(-49.336) + 90.30 = 68.53 \text{ ft-lb/ft}$$

*If we assume the primary structure for the dome as similar to that shown in Fig. 7.14(a), there will be no load on the ring beam, and hence $\text{FER}_1^R = \text{FER}_2^R = 0$. Then FER_1^D will be 114.81 (from above) minus 1531.53, which equals -1416.72 lb/ft.

These values match very closely (within numerical accuracy) with those obtained earlier by the flexibility method in the example of Section 7.7.

The forces on the ring beam are also obtained as

$$X_1^R = 16.667(68.973) - 1531.53 = -381.96 \text{ lb/ft}$$

$$X_2^R = 1.389(-49.336) = -68.53 \text{ ft-lb/ft}$$

Force X_1^R obtained above is equal to the sum of the edge load due to the displacements u_1 (and $u_2 = 0$) and the fixed-end reaction. To X_1^R we must add the load of 1531.53 lb/ft due to N_ϕ from the dome, in order to obtain the net outward force on the ring beam as $1531.53 - 381.96 = 1149.57$ lb/ft. As expected, X_1^R and X_2^R are equal and opposite to the corresponding forces acting on the dome.

If we assume that the primary structure for the dome is similar to that shown in Fig. 7.14(a), there will be no load on the ring beam. Hence we will have $\text{FER}_1^R = \text{FER}_2^R = 0$. In this case there will be an outward edge load of 1531.53 lb/ft acting on the dome. This will yield a dome fixed-edge reaction of

$$\text{FER}_1^D = 114.81 - 1531.53 = -1416.72 \text{ lb/ft}$$

Reaction FER_2^D will be the same as before. Hence values of u_1 and u_2 will be obtained the same as before. The net horizontal force on the dome will again be due to the sum of the effects of u_1 and u_2, plus FER_1^D and the horizontal load due to N_ϕ. Thus we shall have

$$X_1^D(\text{total}) = [8.175(68.973) + 6.014(-49.336)] - 1416.72 + 1531.53$$

$$= 381.96 \text{ lb/ft} \quad \text{(as obtained previously)}$$

Dome-Wall Problem

We shall now solve the example of Section 7.8 by the stiffness method; this problem having been solved earlier by the flexibility method. The primary structures for the dome and cylinder will be as shown in Fig. 7.13(a) and (b). Assuming u_1 and u_2 as in Fig. 7.18, we had obtained:

$$E\delta_{10}^D = -58{,}102.4 \qquad E\delta_{20}^D = -8800.0$$

and

$$E\delta_{10}^W = +438{,}460.78 \qquad E\delta_{20}^W = +157{,}626.65$$

Again, the signs of δ_{10}^W and δ_{20}^W are different here from those used in the flexibility method, since they are now positive in the positive senses of u_1 and u_2. For the dome we have $a = 40.0$ ft, $t_D = 0.25$ ft, $\phi_1 = 30°$, and $\lambda_D = 16.65$. Substituting these values into Eqs. (7.28a–d), we obtain

$$\mathbf{S}^D = \begin{bmatrix} 15.015 & 9.018 \\ 9.018 & 10.832 \end{bmatrix} (E \times 10^{-4})$$

Also, from Eqs. (7.31a,b), we obtain

$$\mathbf{FER}^D = \begin{bmatrix} 95.177 \\ \\ 61.929 \end{bmatrix}$$

For the cylindrical wall, we have $r_W = 20.0$ ft, $t_W = 0.67$ ft, and $\lambda_W = 7.19$. Substituting these values into Eqs. (5.28a–d), we can obtain

$$\mathbf{S}^W = \begin{bmatrix} 46.59 & -64.80 \\ \\ -64.80 & 180.26 \end{bmatrix} (E \times 10^{-4})$$

Also, using equations similar to those in Section 7.14 for the dome, we find

$$\mathrm{FER}_1^W = -[S_{11}^W \delta_{10}^W + S_{12}^W \delta_{20}^W]$$

$$\mathrm{FER}_2^W = -[S_{21}^W \delta_{10}^W + S_{22}^W \delta_{20}^W]$$

Substituting in the proper numerical values, we obtain

$$\mathrm{FER}_1^W = -1021.37 \qquad \text{and} \qquad \mathrm{FER}_2^W \approx 0.0$$

Adding the stiffness and FER matrices of the dome and the cylinder, we obtain

$$\mathbf{S} = \mathbf{S}^D + \mathbf{S}^W = \begin{bmatrix} 61.605 & -55.782 \\ \\ -55.782 & 191.092 \end{bmatrix} (E \times 10^{-4}) \tag{7.33}$$

$$\mathbf{FER} = \mathbf{FER}^D + \mathbf{FER}^W = \begin{bmatrix} -926.193 \\ \\ 61.929 \end{bmatrix} \tag{7.34}$$

Using these matrices to solve Eqs. (7.26a,b), we obtain the joint displacements as

$$u_1 = \frac{20.037 \times 10^4}{E} \qquad \text{and} \qquad u_2 = \frac{5.525 \times 10^4}{E}$$

Final edge loads for the dome are obtained from Eqs. (7.27a,b) as

$$X_1^D = \frac{20.037 \times 10^4}{E}(15.015 \times 10^{-4})E + \frac{5.525 \times 10^4}{E}(9.018 \times 10^{-4})E + 95.177$$

or

$$X_1^D = 445.86 \text{ lb/ft}$$

Similarly, $X_2^D = 302.47$ ft-lb/ft.

These values match with those obtained in Section 7.8. For the cylinder we obtain

$$X_1^W = 46.59(20.037) - 64.8(5.525) - 1021.37 = -445.87 \text{ lb/ft}$$

and

$$X_2^W = -302.46 \text{ ft-lb/ft}$$

The net horizontal edge load on the wall will be obtained by adding X_1^W to the horizontal load (1021.37 lb) on the primary structure, which then becomes $1021.37 - 445.87 = 575.5$ lb/ft, outward, as obtained in Section 7.8.

7.16 PROBLEMS INVOLVING MORE THAN TWO MEMBERS AT A JOINT

It can be seen that using the stiffness method, junction problems can be solved easily. The major advantage of the stiffness method lies in the fact that it can handle the problem of more than two members meeting at a joint with relative ease, while such problems become too complicated in the flexibility method. We shall illustrate this by solving the same problem of the dome and cylindrical wall solved in Section 7.15, but now assuming that a ring beam also exists at the junction.

Example 7.2: Dome/Ring-Beam/Wall Junction

Let us assume that in the preceding dome-wall problem we now have also a ring beam of cross-sectional dimensions: 12-in. width and 8-in. thickness. To avoid changes in numerical work already done, we shall assume that the primary structures of the dome and wall are the same as before; i.e., we shall assume that the horizontal component of the membrane force N_ϕ from the dome still acts on the primary structure of the wall. Hence the primary structure of the ring will carry no load. Thus we find

$$\delta_{10}^R = \delta_{20}^R = 0.0$$

Taking the radius to the center of the ring as 20.5 ft, and using Eqs. (7.29a–d), we obtain the stiffness matrix of the ring beam as

$$\mathbf{S}^R = \begin{bmatrix} 15.943 & 0 \\ 0 & 0.596 \end{bmatrix} (E \times 10^{-4})$$

Also, since δ_{10}^R and δ_{20}^R are zero, we shall have $\mathbf{FER}^R = \mathbf{0}$. For the dome and the wall, the stiffness and FER matrices remain the same as in the preceding dome-wall problem. Adding for the three members, we obtain

$$\mathbf{S} = \mathbf{S}^D + \mathbf{S}^R + \mathbf{S}^W = \begin{bmatrix} 77.548 & -55.782 \\ -55.782 & 191.688 \end{bmatrix} (E \times 10^{-4})$$

while **FER** remains the same as before [Eq. (7.34)]. Upon solving these new equations, we obtain

$$u_1 = \frac{14.812 \times 10^4}{E} \text{ ft} \qquad \text{and} \qquad u_2 = \frac{3.987 \times 10^4}{E}$$

It is seen that with the presence of the ring beam (even though small in size), the displacements have become substantially reduced in comparison to those obtained earlier. Edge loads on the dome are now obtained from Eqs. (7.27a,b) as

$$X_1^D = 353.53 \text{ lb/ft} \qquad \text{and} \qquad X_2^D = 238.69 \text{ ft-lb/ft}$$

Also, proceeding as before, we obtain

$$X_1^W = -589.64 \text{ lb/ft} \qquad \text{and} \qquad X_2^W = -241.12 \text{ ft-lb/ft}$$

The forces on the ring beam will be

$$X_1^R = 15.943(14.812) = 236.15 \text{ lb/ft}$$

and

$$X_2^R = 0.596(3.987) = 2.38 \text{ ft-lb/ft}$$

As expected, edge forces on the dome and wall are reduced due to the addition of a ring beam. The reduction could be even greater if a larger edge beam was assumed. The edge forces at the joint are as shown in Fig. 7.22. It may be checked that the summations of all internal edge forces and moments are zero (within numerical accuracy). Also, the forces N_{ϕ_1} of the primary structure are equal and opposite on the dome and the cylinder. The final horizontal edge load on the cylinder will, of course, be obtained by superimposing the component of N_{ϕ_1} onto the edge force. This net horizontal load on the wall will thus be $1021.37 - 589.64 = 431.73$ lb/ft, outward, as opposed to 575.5 lb/ft obtained without the ring beam in the example of the previous Section 7.15.

From this example we see that multimember joint problems can be easily handled by the stiffness method. The stiffness method has also some other inherent advantages over the flexibility method, as discussed in texts on matrix analysis of structures [2].

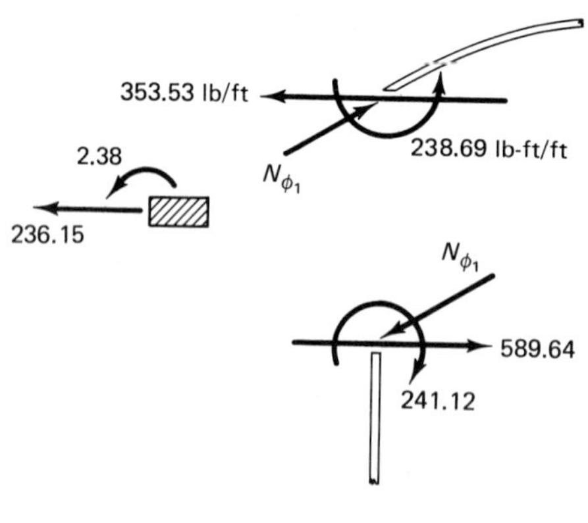

Figure 7.22 Edge loads determined for dome/wall/ring-beam interaction problem.

Using the stiffness method, the incompatibility problems arising at the junctions of the wall, ring beam, and cone or the cone, bottom dome, and supporting shaft, for the Intze tank problem, can now readily be solved. For the stiffness of the cone, we can use an equivalent inscribed spherical shell of the same thickness, as discussed in Section 7.11.

7.17 COMPUTER ANALYSIS OF SHELLS OF REVOLUTION

The theory and examples discussed so far should give the reader a very good insight into the behavior and methods of analysis for shells of revolution. For solutions of various problems in practice, it is desirable that the designer should use a computer program for the analysis of shells. The computer program should be such that the same program can be used for solving problems of various types of shells of revolution.

If we consider an axisymmetric ring element of the spherical shell, bounded by $\phi = \phi_o$ and $\phi = \phi_1$, there will be three stress resultants: N_ϕ, M_ϕ, and Q_ϕ acting at each edge of the element under axisymmetric loading. There will be correspondingly three displacements: v, w, and χ at each edge of the element. Earlier in this chapter we considered only one edge of the element, assuming that the effects of forces at one edge die out before the other edge is reached. We also considered only two displacements (horizontal movement and rotation) at one edge, and derived a 2×2 stiffness matrix. (The third displacement was a known quantity and was not considered in deriving the stiffness matrix.) In a general shell ring element, however, we must consider all three displacements at each edge, and since $(\phi_1 - \phi_o)$ will be a small angle for such an element, the effect of loads at one edge will not die out at the other edge. Hence we shall have a 6×6 stiffness matrix for this element. Such a stiffness matrix can be developed by using a solution similar to Eq. (7.16). The stiffness matrix then can be used to write a standard computer program, using methods of structural analysis [2], for analyzing spherical shells with any distributed or line loads and any boundary conditions. The shell can be divided into a number of ring elements. By taking smaller elements of different thicknesses, problems of shells with variable thickness can also be solved.

The program discussed above could be applied only to spherical shells, and hence would have limited use. For wider applications using a single program, we can consider the basic element to be a frustum of a conical shell and derive its stiffness matrix using solutions similar to those given by Flügge [1] for a conical shell. Popov et al. [4] have used such a conical ring element and derived its stiffness matrix from the differential equations of bending theory. Using such conical shell elements, we can idealize a shell of revolution of any shape, i.e., spherical, ellipsoidal, or also a ring beam or even a circular plate, by a series of such elements. The results obtained will be sufficiently accurate for most practical problems in civil engineering.

The derivation of the stiffness matrix of a conical frustum element, from the differential equations, becomes a complex task. Instead, Grafton and Strome [5] have derived a stiffness matrix for the conical frustum element explicitly by assuming a displacement function, and then using the standard procedures of finite element analysis [6]. Although this stiffness matrix is approximate, good accuracy can be obtained if a large number of elements is considered in a given problem. The authors have analyzed and designed a variety of tanks, domes, etc., using a computer program based on a stiffness matrix similar to that given in Reference [5].

Analysis of the Intze tank of Sections 6.16 and 7.8, idealized into a number of conical ring elements, as shown in Fig. 7.23, was performed by using the stiffness matrix of Grafton and Strome. The shell has been divided into a large number of elements to obtain good results with this simple program. Small spherical caps, near the axis of revolution in the top and bottom domes, are not considered so as

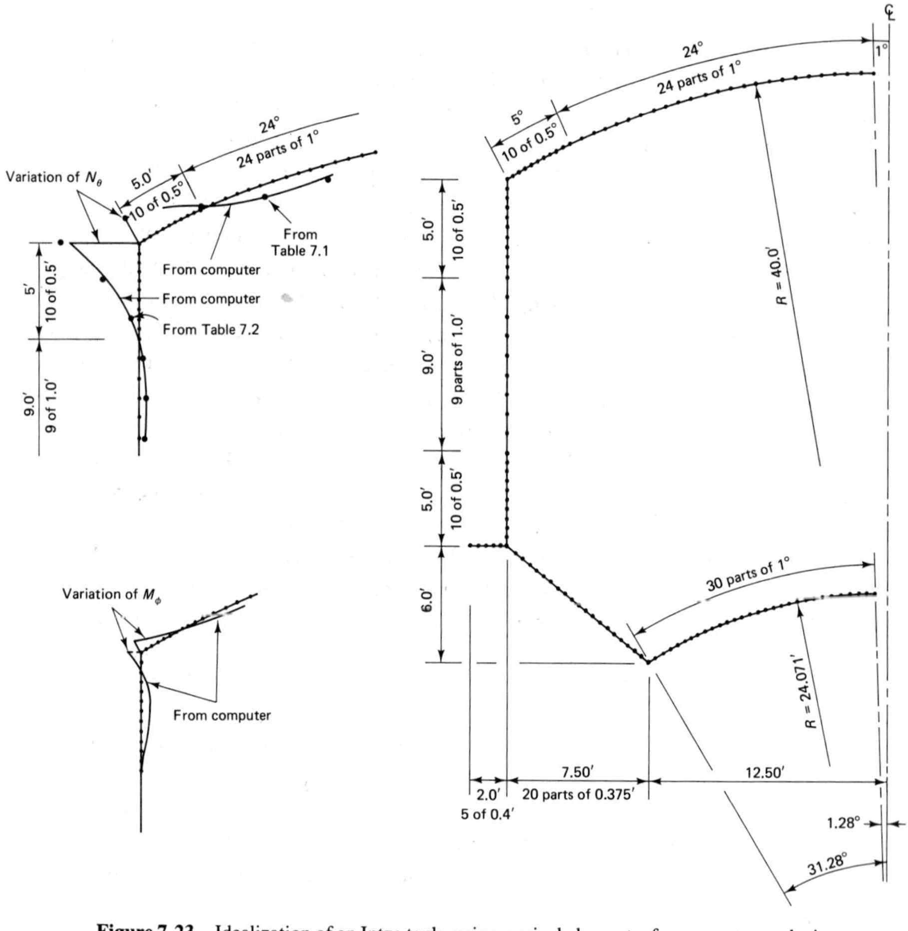

Figure 7.23 Idealization of an Intze tank, using conical elements, for computer analysis.

to avoid errors, since the stiffness matrix of an element contains division by r. The values of all stress resultants in the entire structure could thus be obtained in one analysis. The computer analysis was performed separately for dead load and water load. The loads were idealized within the program as statically equivalent concentrated loads at the elemental nodal points. The program also gives results for the sum of the stress resultants under the two loadings. Results obtained for forces in the roof dome and cylinder compare well with those obtained using X_1 and X_2 values at the junction of the cylinder and roof dome, as found in Section 7.8.

Ross [7] has also discussed a program for the analysis of axisymmetric shells. He has used a general axisymmetric curved element. His program can also be applied to conical shells, or shells of revolution of negative Gaussian curvature (such as for cooling towers), by giving the radius of curvature an infinite or negative value, respectively.

There are, of course, a number of other articles in the literature which use different elements or displacement functions for the analysis of axisymmetric shells. The reader can find these references in standard books on finite element analysis [6] and in the various engineering journals. In a civil engineering design office, a program with a basic conical frustum element can be used to solve, with sufficient accuracy, nearly all problems of axisymmetric shells.

REFERENCES

[1] Flügge, W., *Stresses in Shells,* 2nd ed., Springer-Verlag, New York, 1973.

[2] Weaver, W., Jr., and Gere, J. M., *Matrix Analysis of Framed Structures,* 2nd ed., Van Nostrand Reinhold Company, Inc., New York, 1980.

[3] Billington, D. P., *Thin Shell Concrete Structures,* 2nd ed., McGraw-Hill Book Company, New York, 1982.

[4] Popov, E. P., Penzien, J., and Lu, Z. A., "Finite Element Solution for Axisymmetric Shells," *Proc. Am. Soc. Civ. Eng., EM Div.,* 1964, pp. 119–145.

[5] Grafton, P. E., and Strome, D. R., "Analysis of Axisymmetric Shells by the Direct Stiffness Method," *J.A.I.A.A.,* Vol. 1, 1963, pp. 2342–2347.

[6] Zienkiewicz, O. C., *The Finite Element Method,* 3rd ed., McGraw-Hill Book Company, New York, 1977.

[7] Ross, C. T. F., *Finite Element Programs for Axisymmetric Problems in Engineering,* Ellis Horwood Ltd., West Sussex, England, 1984.

PROBLEMS

7.1. For the example in Section 7.7, assume that the primary structure of the dome itself carries the edge horizontal thrust. Thus the primary structure of the ring beam will not carry any horizontal load. Using these primary structures, obtain δ_{10} and δ_{20} and solve the problem. Check that the results agree with those of Section 7.7.

7.2. Re-solve the example in Section 7.8 by taking the primary structures as shown in Fig. 7.14, and check that the results match with those in Section 7.8.

7.3. A 138,000 (U.S.)-gallon elevated storage reservoir (ESR) of Intze type is to be designed (see Fig. P7.3). There is no ring beam at the junction of the top dome and the wall. Find membrane and bending stresses in the dome and wall under

(a) DL + LL on the dome of 80 lb/ft² of surface area.

(b) Water load (assume that freeboard = 0 ft, for simplicity). Neglect the effect of the boundary conditions at the lower edge of the wall. Assume that $E_{\text{dome}} = E_{\text{wall}}$ and $\nu = 0.1$. Plot N_θ, M_ϕ (or M_x) values in the dome and wall near the junction (up to $\phi = 20°$ for the dome and $x/a = 0.3$ for the wall). Design the reinforcement in the dome. Also design the reinforcement in the wall near the dome/wall junction.

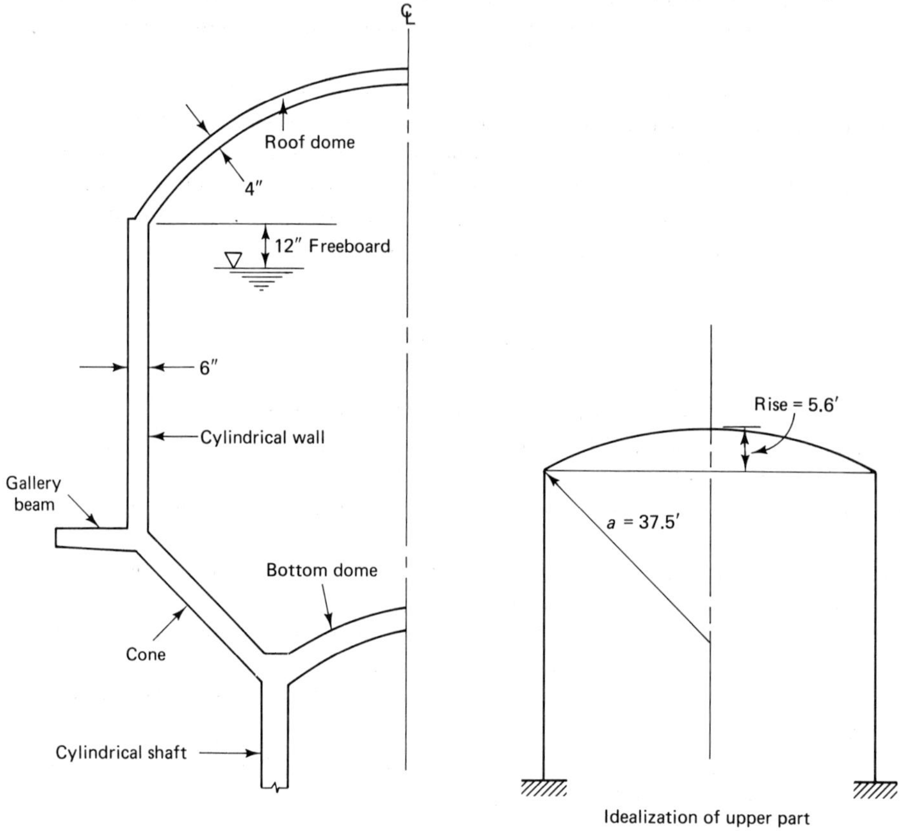

Figure P7.3 Configuration of Intze tank and idealization of its upper portion.

7.4. An Intze-type water tank is open at its top and is filled with water (see Fig. P7.4). Assume that the thickness of the tank is 12 in. throughout, Poisson's ratio is zero, and the concrete density is 150 lb/ft³.

(a) Assume that there are ring beams (not shown) provided at B and C. Under membrane conditions, find the values of hoop forces you will consider for designing the ring beams under (i) dead load only, and (ii) water load only.

(b) Assume that there is no ring beam at B. Find the final value of N_θ in the cone at B using bending theory. The loading to be considered is the dead load of the cylinder only. (For simplicity, assume the cone to be weightless.) Neglect the effects of discontinuities at C on the stresses at B. (*Hint:* Bending stresses in the cone may be taken as those in an inscribed equivalent spherical shell.)

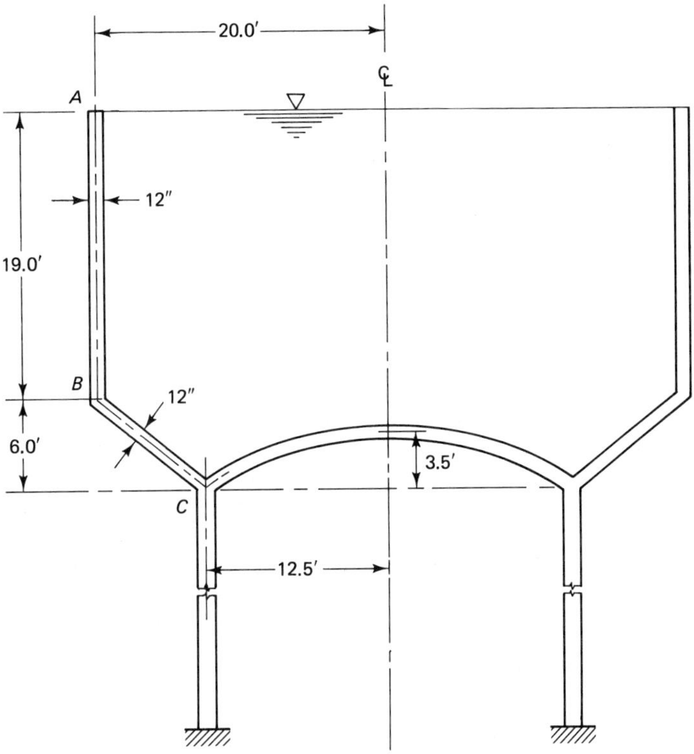

Figure P7.4 Configuration of an open Intze tank.

7.5. In Problem 7.3, assume that there is a gallery ring beam at the junction of the top dome and the wall. The horizontal width of this ring beam is 3 ft and its thickness is 4 in. Find, by the stiffness method, the moments and horizontal forces in the dome, wall, and ring beam at their junction, under a load of 80 lb/ft^2 on the surface area of the dome. Assume that $\nu = 0$. For finding the rotational stiffness of the ring beam, treat it as an annular plate and use the stiffness derived in Problem 5.2. For finding its stiffness against horizontal displacement, the ring beam may be simply treated as a thin ring with a diameter that is an average of its inner and outer diameters.

7.6. A steel water tank, as shown in Fig. P7.6, is composed of a spherical shell supported on a cylindrical wall. Assume the thickness of both shells as 0.25 in., $\nu = 0.3$, and the unit weight of steel as 490 lb/ft^3. Obtain the membrane stress resultants in the primary structures under water load. Solve the problem of incompatibility of displacements at the junction of the two shells. Check the maximum stresses in the shells near their junction.

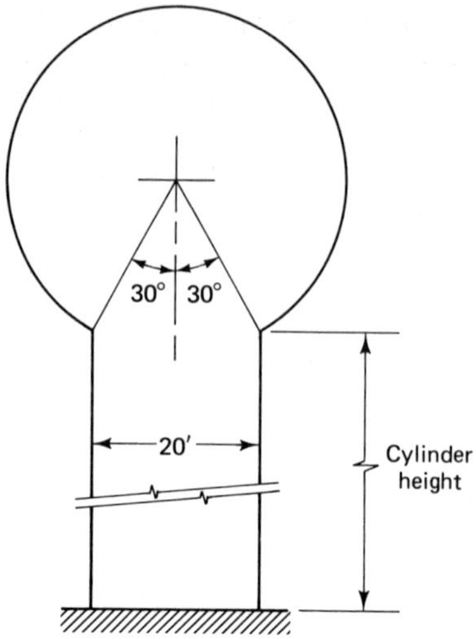

Figure P7.6 A steel "standpipe" water tank consisting of a spherical shell supported on a cylindrical shaft.

7.7. A structure consisting of a cylindrical shell connected to a spherical dome is shown in Fig. P7.7. The structure is restrained against vertical displacement at points A and C. In all the examples considered thus far, we have restrained axial deformations at one location only. In this case, under self-weight, the vertical deformation of point B is now given by the deformation of the cylinder. Hence vertical deformation is specified at two locations, B and C, for the spherical shell. The membrane analysis of the primary structure gives us the freedom of prescribing only one vertical displacement. Hence there will now be a third redundant, X_3, in the vertical direction at B, in addition to X_1 and X_2. Discuss in detail, step by step, the procedure for solving this problem under self-weight and internal pressure loadings.

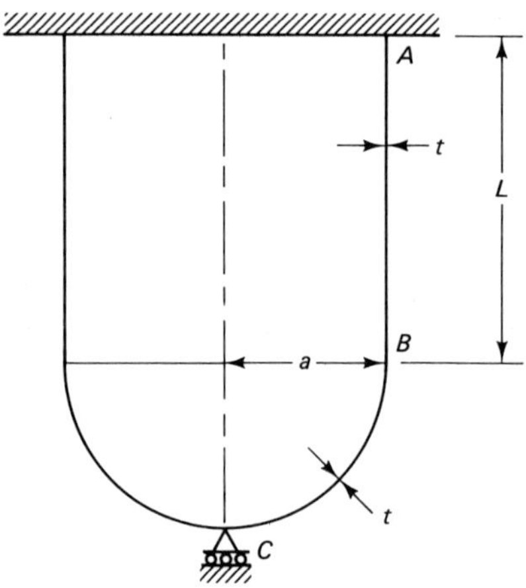

Figure P7.7 Structure, consisting of a cylindrical wall and hemispherical base, which is restrained against vertical translation.

7.8. For the energy-efficient earth-sheltered house structure of Problem 6.6, obtain the final bending moments at the dome/ring/wall junction for the given loading. Use the stiffness method to obtain edge redundants.

7.9. The dome roof of a hall is supported on a cylindrical wall (or on beams over columns) around its circumference A, as shown in Fig. P7.9. An 8-ft-wide annular slab surrounds the dome, acting as an overhead canopy. The loading on the dome is 55 lb/ft^2 over the surface area, and the loading on the slab is 100 lb/ft^2. Dimensions for the dome and canopy are shown, and Poisson's ratio is assumed to be zero.

Obtain the redundant edge loads X_1 and X_2 acting at the dome/slab junction. Neglect in-plane deformations of the slab. Use the results of Problem 2.8 in solving this problem.

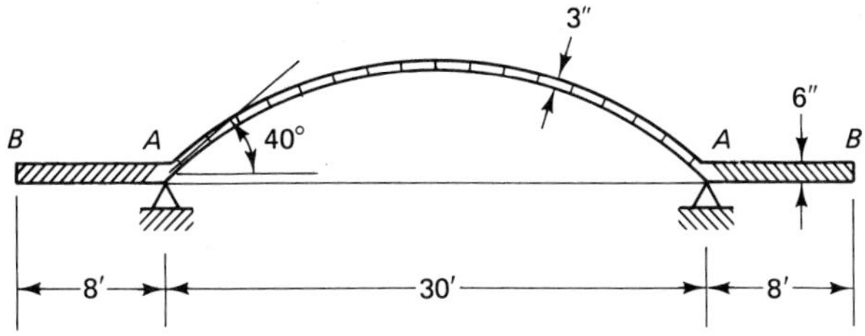

Figure P7.9 A dome roof supported around its circumference A, with an annular slab acting as a canopy.

7.10. Assume that the conical tank of Section 4.8 is covered by a dome roof. The radius of the dome is 40 ft, and its thickness is 4 in. Assume that the total load on the surface of the dome is 65 lb/ft^2, and also assume Poisson's ratio as zero.

Obtain the edge loads at the junction of the dome and the cone under the following conditions: (i) dead load, and (ii) water load. Design reinforcement for the dome and the cone.

8

Membrane Theory of General Shells of Double Curvature

8.1 INTRODUCTION

In earlier chapters we studied the membrane theory for specific types of shells with single or double curvature, i.e., circular cylinders, cones, domes, hyperboloids of revolution, etc. These were all shells of revolution with zero, positive, or negative Gaussian curvatures. Now we shall consider the membrane theory for shells of double curvature which may have any arbitrary shape. We shall then apply this theory to solve examples of hyperbolic paraboloids, elliptic paraboloids, and conoids, as well as to other shells.

8.2 DERIVATION OF DIFFERENTIAL EQUATION

In the rectangular Cartesian coordinate system, x, y, and z, we consider an element $ABCD$ of the shell obtained by making cuts at sections x, $x + dx$, and y, $y + dy$ (Fig. 8.1). These sections are not the "normal sections" of the shell; i.e., the planes $x = constant$ and $y = constant$ do not necessarily contain the normals to the shell. Hence angle ω between sides AB and AD is not necessarily a right angle, and element $ABCD$ is a skew quadrilateral in space. Its projection in the x-y plane is a rectangle $\bar{A}\bar{B}\bar{C}\bar{D}$ with sides dx and dy.

Denoting by ϕ and θ, respectively, the angles which the sides AD and AB make with the x-y plane, the lengths $\overline{AD}$ and $\overline{AB}$ of the shell element can be written as $(dx/\cos\phi)$ and $(dy/\cos\theta)$, respectively. The midsurface area of the element then becomes $(dx/\cos\phi)(dy/\cos\theta)\sin\omega$.

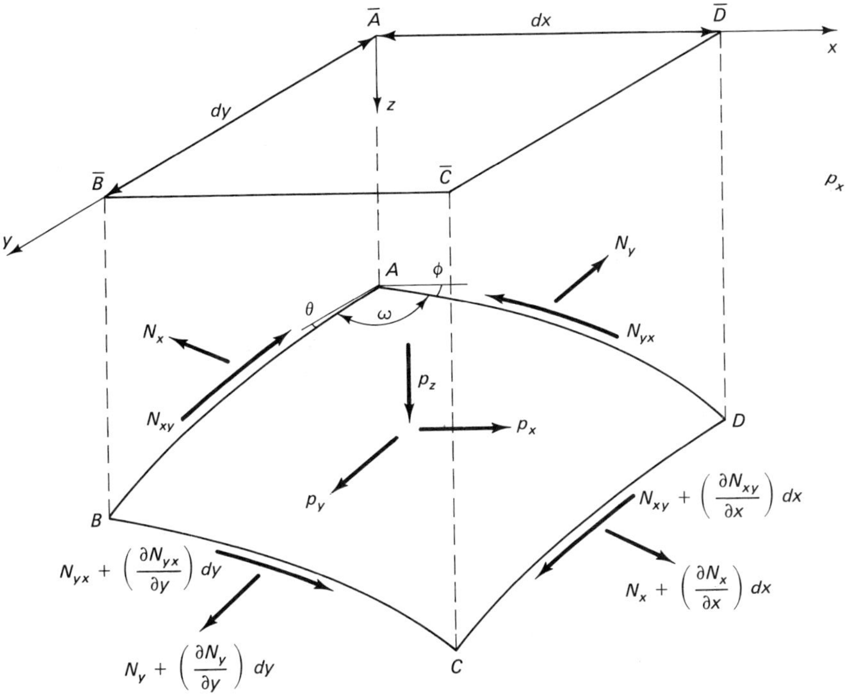

Figure 8.1 Element of a general shell of double curvature.

The curves $x = constant$ and $y = constant$ on the shell surface can be used as coordinate curves, and we denote by N_x, N_y, and N_{xy} the membrane stress resultants at any point on the shell, defined per unit length along the coordinate curves of the shell. The stress resultants, acting on the opposite faces of the element, are shown in Fig. 8.1. Also shown are the loads on the shell, p_x and p_y (which are in the x-y plane parallel to the x- and y-axes, respectively), and the load p_z which acts in the z-direction. These distributed loads, p_x, p_y, and p_z, are defined per unit area of the shell surface.

To begin our derivation of the governing differential equation, we consider equilibrium of forces in the x- and y-directions, in the x-y plane. The force acting on side AB of the element is $N_x \cdot (dy/\cos\theta)$, and its component in the x-y plane is $N_x \cdot (dy/\cos\theta)\cos\phi$. The increment of this force in the x-direction, from AB to CD, will then be

$$\frac{\partial}{\partial x}\left(N_x \cdot \frac{dy}{\cos\theta}\cos\phi\right) dx$$

Similar expressions can be written for the differential forces due to N_{xy}, N_{yx}, and N_y. Also, the force in the x-direction due to the loading p_x, is obtained by multiplying p_x by the area of the element. Hence, the equations of equilibrium in the x- and y-directions can be written as

$$\frac{\partial}{\partial x}\left(N_x \frac{dy}{\cos\theta}\cos\phi\right) dx + \frac{\partial}{\partial y}\left(N_{yx}\frac{dx}{\cos\phi}\cos\phi\right) dy + p_x \frac{dx}{\cos\phi}\frac{dy}{\cos\theta}\sin\omega = 0$$

and

$$\frac{\partial}{\partial x}\left(N_{xy} \frac{dy}{\cos\theta}\cos\theta\right) dx + \frac{\partial}{\partial y}\left(N_y\frac{dx}{\cos\phi}\cos\theta\right) dy + p_y \frac{dx}{\cos\phi}\frac{dy}{\cos\theta}\sin\omega = 0$$

Now let us define

$$\overline{N}_x = N_x \frac{\cos\phi}{\cos\theta} \tag{8.1a}$$

$$\overline{N}_y = N_y \frac{\cos\theta}{\cos\phi} \tag{8.1b}$$

To understand the physical meaning of these definitions, we multiply Eq. (8.1a) by dy, to obtain

$$\overline{N}_x\, dy = N_x \frac{dy}{\cos\theta}\cos\phi$$

The right-hand side represents the horizontal component of the force along the y-curve,* which acts on the shell element. Hence $\overline{N}_x$ is the horizontal component of the force acting along the y-curve on the shell, and is defined per unit length in the horizontal (x-y) plane. Thus while N_x is the stress resultant acting on the shell element along the y-curve per unit length of shell surface, $\overline{N}_x$ is its horizontal component defined per unit length in the horizontal plane. Similarly, $\overline{N}_y$ is the horizontal component of the stress resultant N_y, and is defined per unit length in the horizontal plane. We also note that the horizontal component of shear N_{xy}, acting on side AB, is $N_{xy}(dy/\cos\theta)\cos\theta = N_{xy}\,dy$. Hence the horizontal component of N_{xy} per unit length in the horizontal plane is equal to N_{xy} itself. For uniformity of notations, however, we define

$$\overline{N}_{xy} = N_{xy} = N_{yx} \tag{8.1c}$$

where we recognize, from equilibrium of moments about the z-axis, that N_{xy} and N_{yx} are equal.

Substituting Eqs. (8.1a–c) into the two equilibrium equations derived earlier, and dropping $(dx)(dy)$, we obtain

*For simplicity, we refer to the x-y plane as the horizontal plane. By y-curve, we mean the curve created by intersection of the shell with a plane $x = constant$.

$$\frac{\partial \overline{N}_x}{\partial x} + \frac{\partial \overline{N}_{xy}}{\partial y} + \overline{p}_x = 0 \tag{8.2a}$$

$$\frac{\partial \overline{N}_{xy}}{\partial x} + \frac{\partial \overline{N}_y}{\partial y} + \overline{p}_y = 0 \tag{8.2b}$$

where

$$\overline{p}_x = p_x \frac{\sin \omega}{\cos \phi \cos \theta} \tag{8.3a}$$

and

$$\overline{p}_y = p_y \frac{\sin \omega}{\cos \phi \cos \theta} \tag{8.3b}$$

Multiplying both sides of Eqs. (8.3a,b) by $(dx)(dy)$, we can readily see that while p_x and p_y are the loads in the x- and y-directions, respectively, defined per unit area of the shell surface, $\overline{p}_x$ and $\overline{p}_y$ are the corresponding loads defined per unit area projected in the horizontal (x-y) plane.

Equations (8.2a,b) are similar to those obtained for stresses in plane stress or plane strain problems in the theory of elasticity [1]. Indeed, we have considered here the equilibrium of a rectangular element (dx, dy) in the x-y plane with "stresses" $\overline{N}_x$, $\overline{N}_y$, $\overline{N}_{xy}$ acting on its sides and with "body forces" $\overline{p}_x$ and $\overline{p}_y$.

In plane stress problems, the two equations obtained from statics contain three unknowns. Hence the problem is "statically indeterminate" and we use a condition of compatibility to obtain the third equation (in order to find the three unknowns). Here the third equation can be obtained by considering equilibrium of forces in the z-direction. Since the vertical component of N_x acting on side AB is $(N_x(dy/\cos \theta) \sin \phi)$, the difference in vertical components of N_x acting on opposite faces is then

$$\frac{\partial}{\partial x}\left(N_x \frac{dy}{\cos \theta} \sin \phi\right) dx$$

Thus, from equilibrium in the vertical direction, accounting for similar terms as above, we have

$$\frac{\partial}{\partial x}\left(N_x \frac{dy}{\cos \theta} \sin \phi\right) dx + \frac{\partial}{\partial y}\left(N_y \frac{dx}{\cos \phi} \sin \theta\right) dy + \frac{\partial}{\partial x}\left(N_{xy} \frac{dy}{\cos \theta} \sin \theta\right) dx$$
$$+ \frac{\partial}{\partial y}\left(N_{xy} \frac{dx}{\cos \phi} \sin \phi\right) dy + p_z \frac{dx}{\cos \phi} \frac{dy}{\cos \theta} \sin \omega = 0$$

or

$$\frac{\partial}{\partial x}(\overline{N}_x \tan \phi) + \frac{\partial}{\partial y}(\overline{N}_y \tan \theta) + \frac{\partial}{\partial x}(\overline{N}_{xy} \tan \theta) + \frac{\partial}{\partial y}(\overline{N}_{xy} \tan \phi) = -\overline{p}_z \tag{8.4}$$

where

$$\overline{p}_z = p_z \frac{\sin \omega}{\cos \phi \cos \theta} \tag{8.3c}$$

Here $\overline{p}_z$ is the load in the z-direction per unit area projected in the horizontal plane. Substituting $\tan \phi = \partial z/\partial x$ and $\tan \theta = \partial z/\partial y$ into Eq. (8.4) and simplifying, we can write

$$\overline{N}_x \frac{\partial^2 z}{\partial x^2} + 2\overline{N}_{xy} \frac{\partial^2 z}{\partial x \, \partial y} + \overline{N}_y \frac{\partial^2 z}{\partial y^2} + \left(\frac{\partial \overline{N}_x}{\partial x} + \frac{\partial \overline{N}_{xy}}{\partial y} \right) \frac{\partial z}{\partial x} + \left(\frac{\partial \overline{N}_{xy}}{\partial x} + \frac{\partial \overline{N}_y}{\partial y} \right) \frac{\partial z}{\partial y} = -\overline{p}_z \tag{8.5}$$

For the expressions in the parentheses, we can readily substitute from Eqs. (8.2a,b), in order to obtain

$$\overline{N}_x \frac{\partial^2 z}{\partial x^2} + 2\overline{N}_{xy} \frac{\partial^2 z}{\partial x \, \partial y} + \overline{N}_y \frac{\partial^2 z}{\partial y^2} = -\overline{p}_z + \overline{p}_x \frac{\partial z}{\partial x} + \overline{p}_y \frac{\partial z}{\partial y} \tag{8.6}$$

Thus for a given shell surface $z = f(x, y)$ and known loading, Eq. (8.6) together with Eqs. (8.2a,b) govern the three internal stress resultants in the shell. By introducing a stress function, Φ, as in plane stress problems, we can convert these equations into a single differential equation containing a single unknown, which can then be solved. But before doing this, we shall consider some simple cases to gain more insight into the interpretation of Eq. (8.6).

8.3 SIMPLE CASES

For most practical problems, loadings $\overline{p}_x$ and $\overline{p}_y$ can be taken as zero. Substituting $\overline{p}_x = \overline{p}_y = 0$ into Eqs. (8.2a,b) and (8.6), we can summarize the resulting equations as

$$\frac{\partial \overline{N}_x}{\partial x} + \frac{\partial \overline{N}_{xy}}{\partial y} = 0 \tag{8.7a}$$

$$\frac{\partial \overline{N}_{xy}}{\partial x} + \frac{\partial \overline{N}_y}{\partial y} = 0 \tag{8.7b}$$

$$\overline{N}_x \frac{\partial^2 z}{\partial x^2} + 2\overline{N}_{xy} \frac{\partial^2 z}{\partial x \, \partial y} + \overline{N}_y \frac{\partial^2 z}{\partial y^2} = -\overline{p}_z \tag{8.7c}$$

Plane Problems: Cables and Arches

Let us consider a cable (or an arch) structure that lies in the x-z plane. We can then take $\overline{N}_y$ and $\overline{N}_{xy}$ as zero in Eqs. (8.7a,b), to obtain

$$\frac{d\overline{N}_x}{dx} = 0 \qquad \text{or} \qquad \overline{N}_x = H \tag{8.8a}$$

where H is a constant. This means that $\overline{N}_x$, the horizontal component of the axial force in the cable, is constant along the cable length.*

Also, we find from Eq. (8.7c) that

$$\overline{N}_x \frac{d^2z}{dx^2} = -\overline{p}_z$$

or

$$H \frac{d^2z}{dx^2} = -\overline{p}_z \tag{8.8b}$$

where the derivative is no longer a partial derivative. Knowing the geometry of a cable and the loading on it, we can substitute for d^2z/dx^2 and $\overline{p}_z$ into Eq. (8.8b) in order to obtain H, the horizontal component of the cable force. Let us consider two cases of loading on the cable.

For the first case, we assume that the cable is loaded by a "live load" p_o which is uniformly distributed on its plan length. Since we have defined $\overline{p}_z$ as the distributed load per unit length along the x-axis, we take $\overline{p}_z = p_o$. Then, from Eq. (8.8b), we obtain

$$\frac{d^2z}{dx^2} = -\frac{p_o}{H}$$

or

$$z = C_1 + C_2 x - \frac{p_o}{H}\frac{x^2}{2}$$

where C_1 and C_2 are two constants to be determined from the boundary conditions.

It is seen that the shape of the cable in this case is parabolic. This was expected, since it is well known that a parabola is the "funicular" curve which can sustain a vertical load distributed uniformly over its plan length under "membrane conditions," i.e., by development of in-plane forces only, without any bending moments or shears.

As a second case, we assume that the cable is loaded by its own weight (i.e., by a load p_o which is uniformly distributed along the length of the cable). In this case we have $p_z = p_o$, and from Eq. (8.3c)

$$\overline{p}_z = \frac{p_o}{\cos\phi}$$

We can further write

*In undergraduate statics, we obtain the same result by considering equilibrium of a freebody of the cable.

$$\bar{p}_z = \frac{p_o}{\cos\phi} = p_o \sec\phi = p_o \sqrt{1 + \left(\frac{dz}{dx}\right)^2}$$

Substituting this into Eq. (8.8b), we get

$$H \frac{d^2z}{dx^2} = -p_o \sqrt{1 + \left(\frac{dz}{dx}\right)^2}$$

It can easily be verified that the solution of this equation is

$$z = -\frac{H}{p_o} \cosh \frac{p_o x}{H} + C_3$$

where C_3 is a constant. One can recognize this as the equation of a catenary, which is well known as the funicular curve for a cable with self-weight uniformly distributed over its curved length.

Equation Governing the Response of a Membrane

A thin membrane such as a soap bubble is capable of developing only in-plane tensile forces. Therefore, we substitute $\bar{N}_{xy} = 0$ into Eqs. (8.7a–c), to obtain

$$\bar{N}_x = f(y) \qquad \bar{N}_y = g(x)$$

and

$$\bar{N}_x \frac{\partial^2 z}{\partial x^2} + \bar{N}_y \frac{\partial^2 z}{\partial y^2} = -\bar{p}_z$$

where $f(y)$ is some function of y, and $g(x)$ is some function of x. Assuming that the tension in the membrane is constant all around and that for small slopes, $\bar{N}_x$ and $\bar{N}_y$ can be approximately taken as equal to this tension T, we can write the equations above as

$$\bar{N}_x = \bar{N}_y = T$$

and

$$\frac{\partial^2 z}{\partial x^2} + \frac{\partial^2 z}{\partial y^2} = -\frac{\bar{p}_z}{T}$$

This last equation is the well-known Laplace equation for thin membranes [1].

Equation for Large Deflections of a Plate and Plate Buckling

In deriving equations for bending of plates [2], it is usually assumed that the deflection w is small compared to the length and width of the plate, so that the slopes can be neglected compared to unity in the expressions for curvatures.

The plate deflects into a nondevelopable surface. Such deformation gives rise to in-plane forces in the plate. It is therefore also assumed that w is small compared to the thickness of the plate, so that the effects of in-plane forces can be neglected in the derivation of the plate equations (refer to Chapter 2).

In large deflection* theory of plates, as well as in deriving equations for buckling of plates, we must include the effects of in-plane forces acting on the plate. In small deflection theory, the equation governing deflection w of the plate, in rectangular or polar coordinates, is given by (refer to Chapter 2)

$$\nabla^2\nabla^2 w = \frac{p_z}{K}$$

where p_z is the vertically downward load on the plate. Using Eq. (8.7c), to include the contribution of the in-plane forces, we can replace p_z by

$$p_z - \overline{N}_x \frac{\partial^2 w}{\partial x^2} - 2\overline{N}_{xy} \frac{\partial^2 w}{\partial x\, \partial y} - \overline{N}_y \frac{\partial^2 w}{\partial y^2}$$

to obtain

$$K\nabla^2\nabla^2 w + \overline{N}_x \frac{\partial^2 w}{\partial x^2} + 2\overline{N}_{xy} \frac{\partial^2 w}{\partial x\, \partial y} + \overline{N}_y \frac{\partial^2 w}{\partial y^2} = p_z$$

as the differential equation for w. This is the well-known *von Kármán* equation for large deflections of plates [2]. Unlike shells, where the geometry of the surface $z = f(x, y)$ is known, here the deflected surface of the plate $w = f(x, y)$ is not known. Hence even if we replace $\overline{N}_x$, $\overline{N}_y$, and $\overline{N}_{xy}$ by a single stress function Φ, this equation still contains two unknowns w and Φ. By considering deformations and Hooke's law, another equation in terms of w and Φ was derived by von Kármán to finally obtain two equations for w and Φ.

By taking $p_z = 0$ in the equation above, we obtain an equation for the buckling of plates, similar to the fourth-order differential equation for beam-columns (see Problem 1.5). Assuming w in the form of a harmonic function satisfying the given boundary conditions, simple solutions for buckling loads of plates can also be obtained from this equation [3].

Shell Problems

Equation (8.7c) shows that the vertical load $\overline{p}_z$ on the shell is supported by vertical components of the three in-plane stress resultants. The relative contribution of each resultant in resisting $\overline{p}_z$ depends on the respective second derivative of z with which it is multiplied [i.e., on the geometry of the shell surface given by $z = f(x, y)$]. Partial derivatives $\partial^2 z/\partial x^2$ and $\partial^2 z/\partial y^2$ represent the curvatures κ_x and κ_y of the x- and y-curves, respectively, on the shell surface. The derivative

*Deflection w is still assumed to be small compared to the lateral dimensions of the plate, but may not be small compared to the plate thickness.

$(\partial^2 z/\partial x \partial y)$ represents the "twist" γ_{xy} of the surface. We can thus write Eq. (8.7c) as

$$\kappa_x \overline{N}_x + \kappa_y \overline{N}_y + 2\gamma_{xy} \overline{N}_{xy} = -\overline{p}_z$$

or

$$\frac{\overline{N}_x}{r_x} + \frac{\overline{N}_y}{r_y} + 2\gamma_{xy} \overline{N}_{xy} = -\overline{p}_z$$

The first two terms in the equation above represent the familiar contribution of the hoop forces in the "rings" along the x- and y-curves of the shell. It may be noted that these curves are not necessarily the orthogonal principal curves of the surface.

The twist $(\partial^2 z/\partial x \partial y)$ is the change in the slope $\theta = \partial z/\partial y$ as we move from AB to DC (Fig. 8.1). Similarly, the twist is also equal to the change in slope $\phi = \partial z/\partial x$ as we move from AD to BC. When this twist is zero, $ABCD$ becomes a parallelogram, having its opposite edges AB and DC parallel. Then the vertical components of shears on these edges will cancel each other. When the twist is nonzero, however, the vertical components of the shears on AB and DC will be different due to the difference in the slopes of AB and DC. Therefore, the differential vertical component can support a vertical load. The same can be said about shears on AD and BC. Surfaces that are defined by an equation

$$z = f(x) + g(y)$$

have $(\partial^2 z/\partial x \partial y) = 0$ (i.e., the "twist" at any point is zero). Hence for these shells, the differential vertical component of $\overline{N}_{xy}$ will be zero. Shells of translation, such as elliptic paraboloids and hyperbolic paraboloids with generating parabolas as their boundaries, are such surfaces which are generated by translation of one curve over another at right angles. The contribution of shears N_{xy} to resist vertical loads in these shells will be zero.

Hyperbolic paraboloid surfaces can also be formed by straight-line generators. If the x- and y-axes are so chosen that the x- and y-curves on the shell surface are along the straight-line generators, then obviously κ_x and κ_y will be zero. Hence only N_{xy} will resist the vertical loads, and the resistance due to resultants N_x and N_y will be zero except due to their values applied at the boundaries, under this coordinate system. We shall see this in more detail later.

8.4 SHELL EQUATION USING A STRESS FUNCTION

We now return to Eqs. (8.2a,b) and (8.6), which govern the membrane stress resultants in a shell. To obtain from them a single equation containing one unknown, we introduce a stress function, Φ, such that

$$\overline{N}_{xy} = -\frac{\partial^2 \Phi}{\partial x\, \partial y} \tag{8.9a}$$

Substituting this into Eqs. (8.2a,b) and integrating, we obtain

$$\overline{N}_x = \frac{\partial^2 \Phi}{\partial y^2} - \int \overline{p}_x \, dx \tag{8.9b}$$

and

$$\overline{N}_y = \frac{\partial^2 \Phi}{\partial x^2} - \int \overline{p}_y \, dy \tag{8.9c}$$

Upon substituting these expressions into Eq. (8.6) and simplifying, we obtain the following equation for Φ:

$$\left(\frac{\partial^2 z}{\partial y^2}\right)\left(\frac{\partial^2 \Phi}{\partial x^2}\right) - 2\left(\frac{\partial^2 z}{\partial x \, \partial y}\right)\left(\frac{\partial^2 \Phi}{\partial x \, \partial y}\right) + \left(\frac{\partial^2 z}{\partial x^2}\right)\left(\frac{\partial^2 \Phi}{\partial y^2}\right)$$
$$= -\overline{p}_z + \overline{p}_x \frac{\partial z}{\partial x} + \overline{p}_y \frac{\partial z}{\partial y} + \frac{\partial^2 z}{\partial x^2} \int \overline{p}_x \, dx + \frac{\partial^2 z}{\partial y^2} \int \overline{p}_y \, dy \tag{8.10}$$

Thus we have a second-order partial differential equation in Φ. Knowing the surface geometry and the loading, this equation can be solved. Let us denote $\partial^2 z/\partial y^2$, $\partial^2 z/\partial x \partial y$, and $\partial^2 z/\partial x^2$ by a, b, and c, respectively. Then, in the theory of partial differential equations [4], depending on whether $(ac - b^2)$ is equal to, greater than, or less than zero, the equation is called, respectively, parabolic, elliptic, or hyperbolic. It can be shown from the theory of surfaces that for synclastic shells, $(ac - b^2)$ is greater than zero, and hence the differential equation is of the elliptic type. Also, for shells with anticlastic and developable surfaces, the equations are of the hyperbolic and parabolic types, respectively.

The methods of solution for these partial differential equations are different for the different shell classifications [4]. For the elliptic equations, separation of the variables x and y can generally be achieved by adopting in one direction, say x, a solution for Φ in the form of a Fourier series expansion which also satisfies the boundary conditions in that direction. Then the resulting ordinary differential equation in y may be solved for each harmonic. The solution for Φ is obtained by summation over a sufficient number of harmonics. Thus, knowing Φ, expressions for the stress resultants can also be obtained in series form. We shall use this procedure for solving problems of elliptic paraboloids and also for hyperbolic paraboloids (with generating parabolas in the x-z and y-z planes) in Chapter 11, but before that we shall review the theory of Fourier series in Chapter 10.

For hypar shells (i.e., hyperbolic paraboloids), if we choose the coordinate axes along the straight-line generators, Eq. (8.10) or (8.6) reduces to a very simple form which can be directly integrated. Hence simple "closed-form" solutions can be obtained for many practical problems. Therefore, we shall first study solutions of simple hypar shell problems of practical interest in the next chapter.

REFERENCES

[1] Timoshenko, S. P., and Goodier, J. N., *Theory of Elasticity*, 3rd ed., McGraw-Hill Book Company, New York, 1970.

[2] Timoshenko, S. P., and Woinowsky-Krieger, S., *Theory of Plates and Shells*, 2nd ed., McGraw-Hill Book Company, New York, 1959.

[3] Timoshenko, S. P., and Gere, J. M., *Theory of Elastic Stability*, McGraw-Hill Book Company, New York, 1961.

[4] Carrier, G. F., and Pearson, C. E., *Partial Differential Equations*, Academic Press, Inc., New York, 1976.

PROBLEMS

8.1. For the design of skewed hypar shells, we shall later require the shell equilibrium equations to be derived when the coordinate axes x and y are not at right angles, but say, at an angle 2ψ (Fig. P8.1). Assuming loads $p_x = p_y = 0$, show that the stress resultants are still governed by equations similar to Eqs. (8.7a–c), except that $\bar{p}_z$ now has to be replaced by $(\bar{p}_z \cdot \sin 2\psi)$. Hence show that for skew hypars, with $p_x = p_y = 0$, the governing equations are

$$\frac{\partial \bar{N}_x}{\partial x} + \frac{\partial \bar{N}_{xy}}{\partial y} = 0 \tag{8.11a}$$

$$\frac{\partial \bar{N}_{xy}}{\partial x} + \frac{\partial \bar{N}_y}{\partial y} = 0 \tag{8.11b}$$

$$\bar{N}_x \frac{\partial^2 z}{\partial x^2} + 2\bar{N}_{xy} \frac{\partial^2 z}{\partial x\, \partial y} + \bar{N}_y \frac{\partial^2 z}{\partial y^2} = -\bar{p}_z \sin 2\psi \tag{8.11c}$$

[*Hint:* When writing equilibrium of forces, say in the x-direction, the contribution of

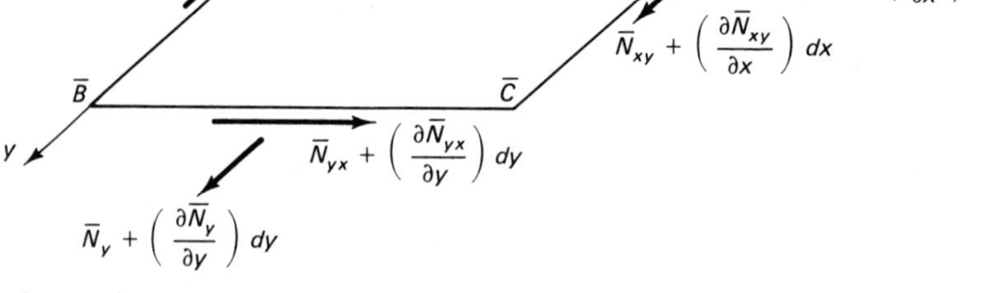

Figure P8.1 Horizontal projection of skew shell element when coordinate axes x and y are not necessarily orthogonal.

$\overline{N}_{xy}$ acting on $\overline{A}\overline{B}$ and $\overline{C}\overline{D}$ and the contribution of $\overline{N}_y$ acting on $\overline{B}\overline{C}$ and $\overline{A}\overline{D}$ must also be included. From the resulting equations, the relations of Eqs. (8.7a,b) can readily be deduced.]

8.2. For each of the shells with surfaces defined by the following equations, check whether the differential equation governing the membrane stress resultants will be elliptic, hyperbolic, or parabolic:

(a) Dome: $z^2 + x^2 + y^2 = c$

(b) Cone: $z^2 = (x^2 + y^2) \tan^2 \alpha$

(c) Elliptic paraboloid: $z = \dfrac{x^2}{2h_1} + \dfrac{y^2}{2h_2}$

(d) Hyperbolic paraboloid: $z = \dfrac{x^2}{2h_1} - \dfrac{y^2}{2h_2}$

(e) Hypar with coordinate axes along straight-line generators: $z = cxy$

9

Membrane Theory of Hyperbolic Paraboloid Shells

9.1 INTRODUCTION

Hypar shells (hyperbolic paraboloids) are shells of translation generated by moving a parabola along another parabola which has an opposite curvature (the planes containing the generating parabolas being at a right angle). These saddle-shaped shells have negative Gaussian curvatures. They can also be generated by the translation of straight-line generators [Fig. 3.5(c)]. It is this property of hypars which makes them attractive in design practice, because different elegant shapes of the doubly curved surfaces can be constructed using only straight-line forms. Candela [1] has demonstrated the versatility of this shell by designing a number of aesthetically pleasing hypar shell structures of different forms which have been constructed in Mexico.

For a better understanding of the membrane analysis of hypar shells, we shall begin by reviewing the properties of hyperbolas, as studied in coordinate geometry [2].

9.2 HYPERBOLA: DEFINITIONS

A hyperbola is defined as the locus of a point that moves so that the ratio of its distance from a fixed point S (or S') [see Fig. 9.1(a)] to its perpendicular distance from a fixed straight line AA (or $A'A'$) is a constant greater than 1. The point S (or S') is called the *focus*, and the line AA (or $A'A'$) is called the *directrix*. The ratio

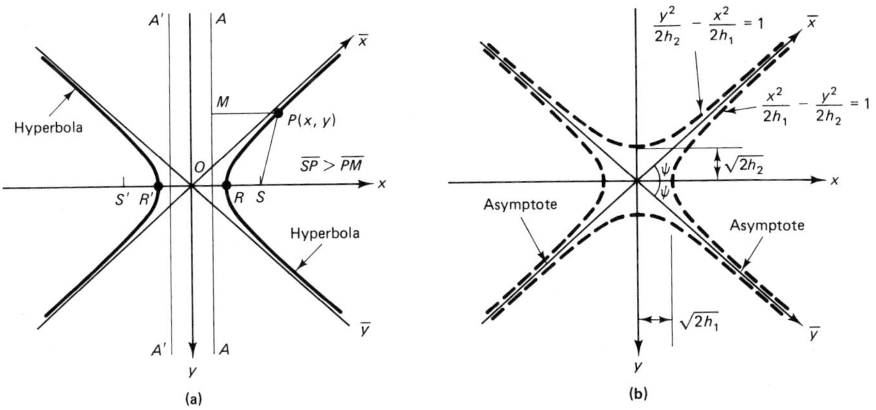

Figure 9.1 Coordinate geometry of the hyperbola. (a) Generation of the hyperbola. (b) Definition of asymptotes and equations of conjugate hyperbolas.

$e = \overline{SP}/\overline{PM}$ is constant and is greater than 1. If $\overline{OR} = \overline{OR'} = \sqrt{2h_1}$, the coordinates of S are $(e\sqrt{2h_1}, 0)$, and the equation of the hyperbola is given as

$$\frac{x^2}{2h_1} - \frac{y^2}{2h_1(e^2 - 1)} = 1$$

or

$$\frac{x^2}{2h_1} - \frac{y^2}{2h_2} = 1 \tag{9.1a}$$

where $2h_2 = 2h_1(e^2 - 1)$.

The equation of the *conjugate hyperbolas* [Fig. 9.1(b)] is given by

$$\frac{y^2}{2h_2} - \frac{x^2}{2h_1} = 1 \tag{9.1b}$$

Asymptotes

Asymptotes to the hyperbolas are defined as straight lines that do not cross the hyperbolas, but rather meet them at infinity. To determine the equations of the asymptotes, we assume them in the form $y = mx + c$, where m and c are constants to be determined. Substituting this equation into Eq. (9.1a), we obtain the following equation for the points of intersection of the hyperbola and the straight-line asymptote.

$$\frac{x^2}{2h_1} - \frac{(mx + c)^2}{2h_2} = 1$$

We shall transform the variable x to a function of the variable ξ, by substituting $x = 1/\xi$. Hence the equation above becomes

$$\frac{1}{2h_1\xi^2} - \frac{(m + c\xi)^2}{2h_2\xi^2} = 1$$

or, on expanding, we have

$$(4h_1h_2 + 2h_1c^2)\xi^2 + 4mch_1\xi + (2h_1m^2 - 2h_2) = 0$$

Since, by definition, the hyperbola and the asymptote intersect at infinity, then $x = \pm\infty$; i.e., $\xi = \pm 0$ must be the two solutions of the equation above. This requires that the last term, as well as the coefficient of ξ in the equation above, must vanish. Thus, we have

$$2h_1m^2 - 2h_2 = 0 \qquad \text{and} \qquad 4mch_1 = 0$$

This gives us $c = 0$ and $m = \pm\sqrt{h_2/h_1}$. These two roots for m give two asymptotes whose equations are

$$y = \pm\sqrt{\frac{h_2}{h_1}}\,x \tag{9.2a}$$

We see that the asymptotes pass through the origin. They are shown in Fig. 9.1(b). These lines make an angle ψ with the x-axis such that

$$\tan\psi = \sqrt{\frac{h_2}{h_1}} \tag{9.2b}$$

It may easily be verified that the asymptotes [Eq. (9.2a)] of the hyperbolas [Eq. (9.1a)] are also asymptotes to the conjugate hyperbolas of Eq. (9.1b).

Rectangular Hyperbolas

When the conjugate hyperbolas are identical, they are called *rectangular hyperbolas*. In such cases, $h_1 = h_2$, and the equations of the asymptotes become $y = \pm x$. The asymptotes thus make an angle of 45° with the x-axis and are at right angles to each other. Otherwise, in general, the asymptotes will make an angle of 2ψ with each other.

Equation of Hyperbola with Asymptotes as Coordinate Axes

Let us consider a set of new coordinate axes $\bar{x}$, $\bar{y}$ taken along the asymptotes to the hyperbola (Fig. 9.2). These axes are not necessarily at right angles to each other (except in the case of rectangular hyperbolas). By transforming coordinates, we can rewrite Eq. (9.1a) for the hyperbola in terms of the new coordinates $\bar{x}$ and $\bar{y}$. Both $\bar{x}$ and $\bar{y}$ make an angle of ψ with the x-axis, and we can write from Eq. (9.2b)

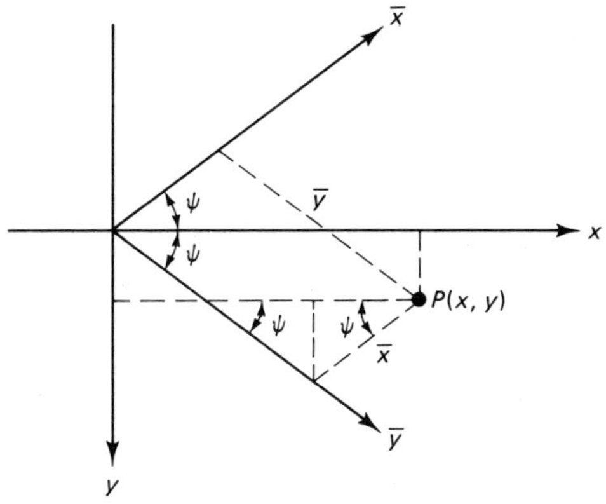

Figure 9.2 Transformation of coordinate axes from x and y to axes $\bar{x}$ and $\bar{y}$ which are taken along the asymptotes of the hyperbola.

$$\tan^2 \psi = \frac{h_2}{h_1} \tag{9.3a}$$

and hence

$$\cos^2 \psi = \frac{h_1}{h_1 + h_2} \tag{9.3b}$$

and also

$$\sin^2 \psi = \frac{h_2}{h_1 + h_2} \tag{9.3c}$$

From Fig. 9.2 we can also readily deduce the following relations:

$$x = \bar{x} \cos \psi + \bar{y} \cos \psi$$

and

$$y = -\bar{x} \sin \psi + \bar{y} \sin \psi$$

Substituting these relations into Eq. (9.1a), we obtain

$$(\bar{x} + \bar{y})^2 \frac{\cos^2 \psi}{2h_1} - (\bar{x} - \bar{y})^2 \frac{\sin^2 \psi}{2h_2} = 1$$

Using Eqs. (9.3b,c) and simplifying, we obtain the equation of the hyperbola as

$$\frac{2\bar{x}\bar{y}}{h_1 + h_2} = 1 \tag{9.4a}$$

or, by virtue of Eq. (9.3c), this can also be written as

$$\frac{2\bar{x}\bar{y} \sin^2 \psi}{h_2} = 1 \tag{9.4b}$$

For rectangular hyperbolas, we have $h_1 = h_2 = h_o$ and also $\psi = 45°$. Then Eq. (9.4a or b) becomes

$$\frac{\bar{x}\bar{y}}{h_o} = 1 \tag{9.4c}$$

This equation has a very simple form. For hypar shell surfaces we can also derive such simplified equations, as we shall see in the following section.

9.3 GEOMETRY OF HYPERBOLIC PARABOLOID SHELLS

We had defined the hypar shell as a surface generated by translation of one parabola on another parabola of opposite curvature and at right angles to the first parabola [Fig. 9.3(a)]. Such a surface will be given by the equation

$$z = \frac{x^2}{2h_1} - \frac{y^2}{2h_2} \tag{9.5}$$

We note that by taking x = *constant* and y = *constant* in the preceding equation, we obtain equations for parabolas of opposite curvatures. When $z = 0$, we obtain two straight lines $y = \pm\sqrt{h_2/h_1}\, x$. These straight lines, indicated by axes $\bar{x}$ and $\bar{y}$ in Fig. 9.3(a), make angles of ψ with the x-axis such that $\tan\psi = \sqrt{h_2/h_1}$. Now, when we substitute $z = c$, where c is a constant, into Eq. (9.5), we obtain

$$\frac{x^2}{2h_1} - \frac{y^2}{2h_2} = c$$

or

$$\frac{x^2}{2h_1c} - \frac{y^2}{2h_2c} = 1$$

For different values of c, the equation above gives us a series of hyperbolas on the hypar surface, as shown in Fig. 9.3(a). For negative values of c we obtain the conjugate hyperbolas. By comparing the equation above with Eq. (9.1a), and by examining Eq. (9.1b), we find that all these hyperbolas will have asymptotes that make an angle ψ with the x-axis, such that

$$\tan\psi = \sqrt{\frac{h_2c}{h_1c}} = \sqrt{\frac{h_2}{h_1}}$$

Thus the asymptotes of the hyperbolas on the surface of the shell will be in the vertical planes formed by $\bar{x}$-z and $\bar{y}$-z.

We can now write the equation of the shell using $\bar{x}$, $\bar{y}$, and z as the coordinate axes. If we substitute for x and y in terms of $\bar{x}$ and $\bar{y}$ as in the preceding section, the right-hand side of Eq. (9.5) will become the same as the left-hand side of Eq. (9.4a), and hence the equation of the shell becomes

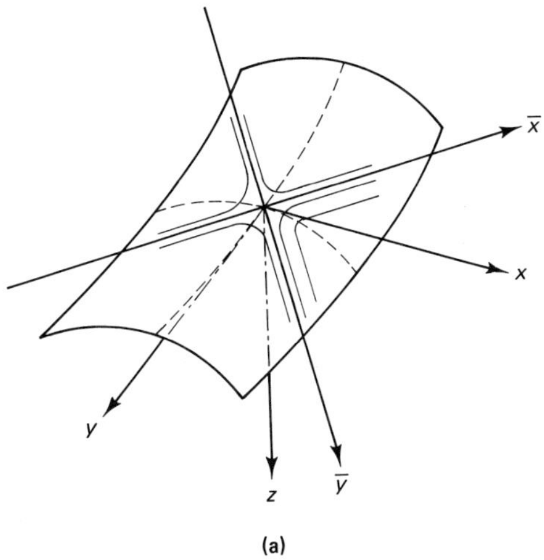

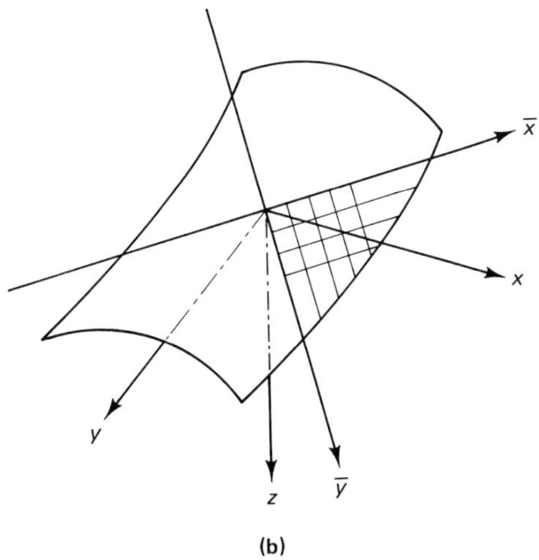

Figure 9.3 Hypar, showing (a) hyperbolic contours and (b) straight-line generators.

$$z = \frac{2\bar{x}\bar{y}}{h_1 + h_2} \tag{9.6a}$$

or, by using Eq. (9.4b), we can also write

$$z = \frac{2\bar{x}\bar{y}\sin^2\psi}{h_2} \tag{9.6b}$$

For "rectangular" hypar shells, we substitute $h_1 = h_2 = h_o$ to obtain the equation

$$z = \frac{\bar{x}\bar{y}}{h_o} \tag{9.6c}$$

Such an equation of the shell surface is of very simple form, and hence the solution of the membrane stress problem as described in Chapter 8 becomes greatly simplified.

Noting from Eqs. (9.6a–c) that intersections of the shell surface with planes $\bar{x}$ = *constant* or $\bar{y}$ = *constant* are straight lines, we see that the shell surface can also be generated by translation of straight lines in the $\bar{x}$ and $\bar{y}$ directions. These straight-line generators of the shell are shown in Fig. 9.3(b).

Now, let us consider in Fig. 9.4(a) a parallelogram $OAB'C$ in the $\bar{x}$-$\bar{y}$ plane, where $\bar{x}$ and $\bar{y}$ are not necessarily at right angles, but form an angle 2ψ. The lengths of sides OC and OA are assumed to be a and b, respectively. By dipping point B' to B, for a distance h along the z-direction, we get a skew quadrilateral $OABC$ in space. The quadrilateral will be a doubly curved surface. Line $DP'E'$ in the plane $OAB'C$ will become DPE after "dipping." If we assume the coordinates of P as $(\bar{x}, \bar{y}, z)$, then from triangle DEE' we can write

$$z = \overline{EE'}\,\frac{\bar{y}}{b}$$

Now, from triangle ABB', we can substitute into the equation above the relation

$$\overline{EE'} = h\,\frac{\bar{x}}{a}$$

Hence we obtain

$$z = h\,\frac{\bar{x}\bar{y}}{ab} \tag{9.7a}$$

as the relation governing the coordinates of a point P on the surface $OABC$. Comparing this equation with Eqs. (9.6a,b) or (9.6c), we find that they are exactly the same if we take

$$\frac{h_1 + h_2}{2} = \frac{ab}{h} \tag{9.7b}$$

or

$$\frac{h_2}{2\sin^2\psi} = \frac{ab}{h} \tag{9.7c}$$

or, for rectangular hypars,

$$h_o = \frac{ab}{h} \tag{9.7d}$$

It may be noted above that a and b represent the plan projections of the sides of the quadrilateral along straight-line generators (in $\bar{x}$, $\bar{y}$ coordinate axes), while h_1, h_2, or h_o are the parameters relating to the hyperbolas in the x, y coordinate system.

From the discussion above, we conclude that the skew quadrilateral surface generated by dipping B' to B is indeed a hyperbolic paraboloid with straight-line

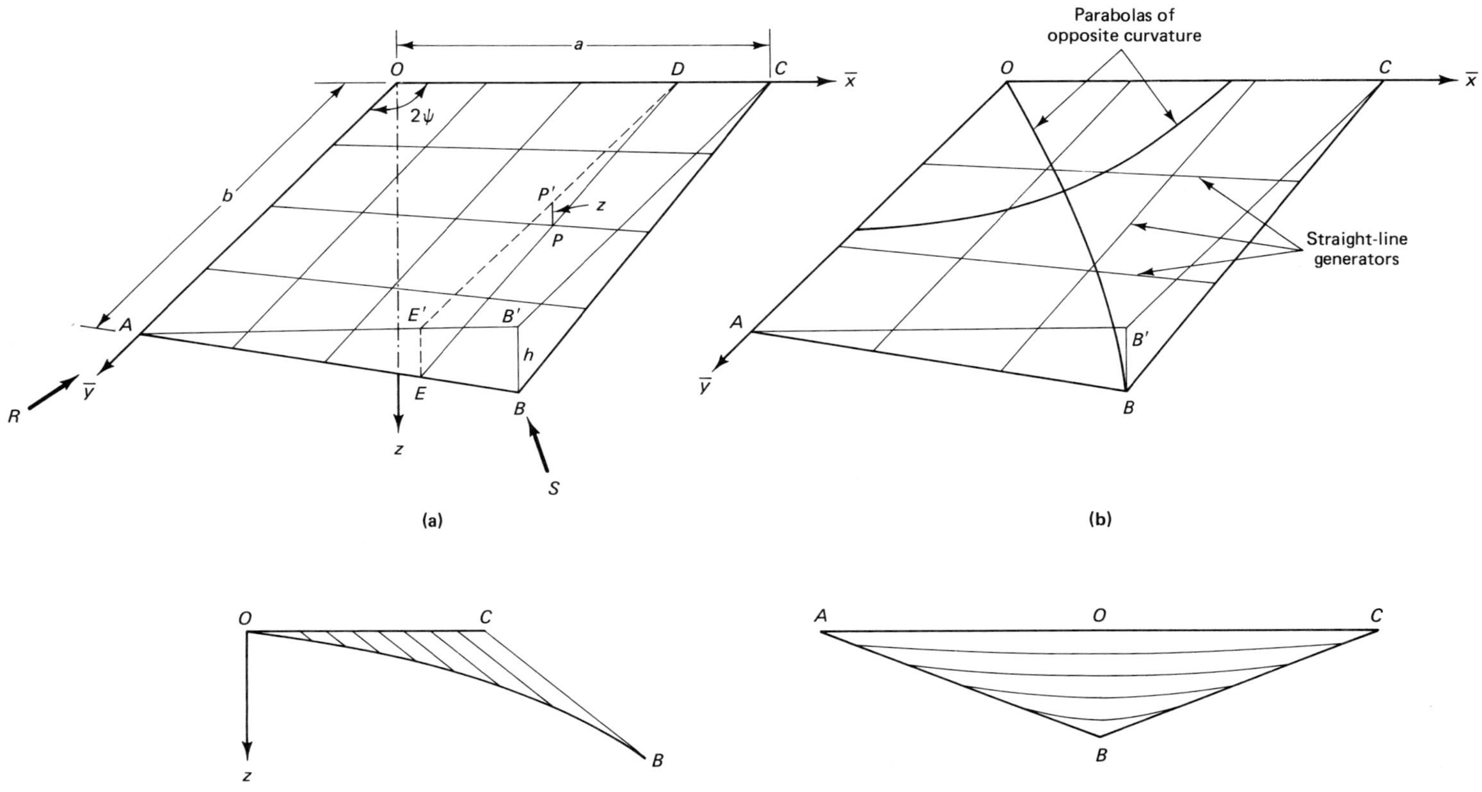

Figure 9.4 Skew quadrilateral formed by dipping one end of a parallelogram. (a) Configuration of the surface. (b) Straight-line generators. (c) View R. (d) View S.

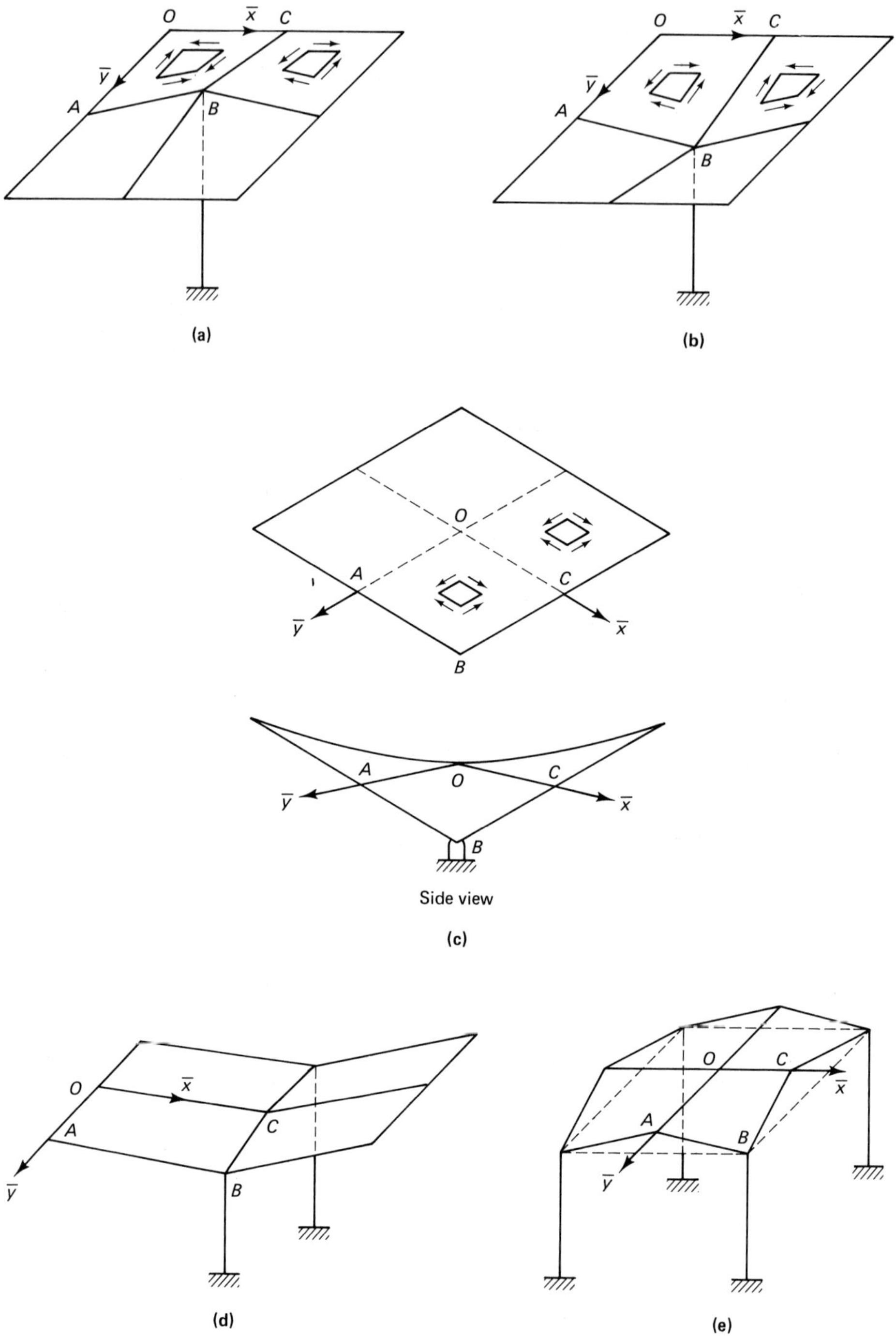

Figure 9.5 Hypar roofs constructed from skew quadrilateral panels. (a) Normal umbrella. (b) Inverted umbrella. (c,d) Two forms of "butterfly" shells. (e) "Hipped" roof.

generators $\bar{x}$, $\bar{y}$ as the coordinate axes. Figure 9.4(b–d) shows, for the same surface, the generating parabolas of opposite curvature which are obtained by intersection of the surface with the vertical planes at angles ψ with plane $\bar{x}$-z. Thus we see that a basic panel of a hypar shell can be generated by dipping sides AB' and CB' of a plane parallelogram to AB and CB. Such a panel has straight lines $\bar{x}$ = *constant* and $\bar{y}$ = *constant* as generators. It may be noted that Eq. (9.7a) for the shell surface is valid whether the coordinate axes $\bar{x}$, $\bar{y}$ are at right angles or at angle 2ψ (i.e., it is valid for rectangular as well as skew hypars).

By combining different basic quadrilateral panels similar to $OABC$, we can create various interesting forms of hypar shells for use as roof structures. Figure 9.5 shows some forms of roof structures obtained by combining four basic panels $OABC$ in different ways. In each case, O, A and C are in one plane, whereas B is either dipped down or up with respect to this plane. Roofs in Fig. 9.5(a) and (b) may be called "umbrella" roofs—normal and inverted, respectively. We may call the roofs in Fig. 9.5(c) and (d) "butterfly" types, and that in Fig. 9.5(e) a "hipped" type. All of these doubly curved structures can easily be constructed in practice by using straight forms along the straight-line generators. We shall study the membrane stresses in such shells in the next section.

9.4 MEMBRANE STRESSES IN HYPAR SHELLS WITH BOUNDARIES ALONG STRAIGHT-LINE GENERATORS

We shall consider membrane stresses in hypar shells which have their boundaries along straight-line generators. We consider the basic quadrilateral $OABC$ (Fig. 9.6), and we now denote the coordinate axes along the straight-line generators as x and

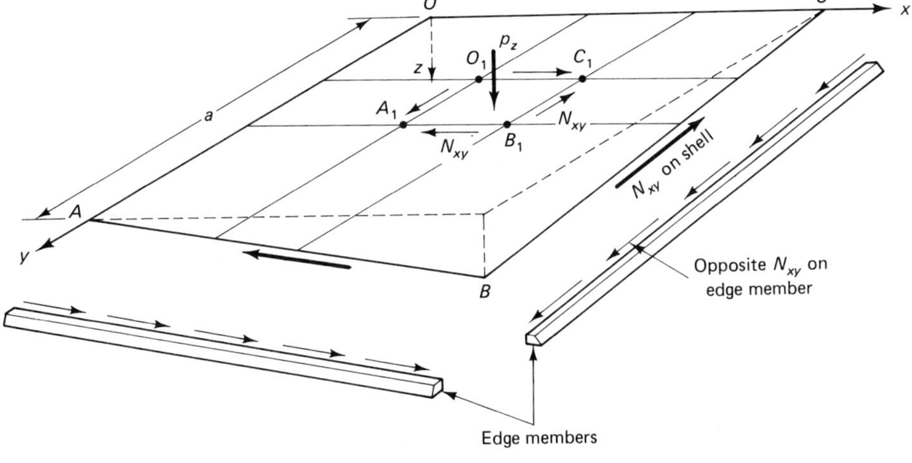

Figure 9.6 Stress resultants in a basic skew quadrilateral panel.

y, dropping the bars used in the preceding section, for simplicity. As derived in Section 9.3, the equation of the surface is

$$z = \frac{h}{ab} xy \tag{9.7a}$$

where a and b are the lengths of the sides of the quadrilateral and h is the "drop" of B from the plane OAC. In Chapter 8 we had seen that the membrane stress resultants in shells of double curvature are governed by Eq. (8.6) or (8.10). We assume that the loads p_x and p_y are zero and that the vertical load p_z is uniformly distributed over the plan area (i.e., $\bar{p}_z$ is constant).* Substituting Eq. (9.7a) in Eq. (8.10), we obtain the following equation for Φ:

$$\frac{\partial^2 \Phi}{\partial x\, \partial y} = \frac{ab}{2h} \bar{p}_z \tag{9.8a}$$

and hence, from Eq. (8.9a) or directly from Eq. (8.6), we have

$$\bar{N}_{xy} = -\frac{ab}{2h} \bar{p}_z \tag{9.8b}$$

Although Eq. (9.7a) is valid for either rectangular or skew hypar surfaces, Eqs. (8.6) and (8.10) were derived with rectangular coordinate axes. Hence Eqs. (9.8a,b) are valid for rectangular hypars only. We shall later see how to modify them for skew hypar shells. Integrating Eq. (9.8a) twice, we obtain

$$\Phi = \frac{ab}{2h} \bar{p}_z\, xy + f_1(x) + f_2(y) \tag{9.8c}$$

where f_1 and f_2 are unknown functions of x and y, respectively. From Eqs. (8.9b,c), we then obtain

$$\bar{N}_x = f_2^{\cdot\cdot}(y) \tag{9.8d}$$

and

$$\bar{N}_y = f_1''(x) \tag{9.8e}$$

where $(\cdot)'$ and $(\cdot)^{\cdot}$ denote derivatives with respect to x and y, respectively. Thus $\bar{N}_x$ and $\bar{N}_y$, respectively, are functions of y and x only. This fact can also be readily derived by substituting Eq. (9.8b) into Eqs. (8.7a,b).

We note from the above that $\bar{N}_x$ and $\bar{N}_y$, the plan projections of N_x and N_y, are constant along x and y, respectively. If values of N_x and N_y are defined on boundaries BC and AB (Fig. 9.6), respectively, then from Eqs. (8.1a,b) $\bar{N}_x$ and $\bar{N}_y$ are known on these boundaries. Hence since $\bar{N}_x$ and $\bar{N}_y$ are constant along x and

*In practical problems, self-weight of the shell is constant over its surface area and not over the plan area. However, for shells with shallow slopes, this assumption can be made since it simplifies the solution without significant loss in accuracy.

y, respectively, they are also known everywhere else in the shell. N_x and N_y can then be found at any point from Eqs. (8.1a,b).

If we prescribe N_x and N_y on the two boundaries BC and AB, we do not have the additional freedom of prescribing them on boundaries OA and OC; rather, we must accept the values given by Eqs. (8.1a,b). If we prescribe N_x and N_y as zero at the two boundaries, they will be zero throughout, and the shell will have only one nonzero stress resultant, $N_{xy} = \overline{N}_{xy}$. This resultant is also constant throughout and is given by Eq. (9.8b). The shell thus has a very simple stress system.

In Eq. (9.7a), h is positive if the "drop" of B is in the positive z-direction. Thus for positive h, $\overline{N}_{xy} = N_{xy}$ is negative throughout the shell, as found from Eq. (9.8b). For the shell with positive value of h shown in Fig. 9.6, the directions of N_{xy} acting on an element $O_1A_1B_1C_1$ are shown. It is clear that the differences of the vertical components of resultants N_{xy} acting on the opposite edges of the element balance p_z. Since N_{xy} is constant in this case, the nonzero difference of vertical components of N_{xy} is obtained only due to the difference in slopes of the opposite edges (i.e., the "skewness" of the element provides the means for load resistance, as discussed in Chapter 8).

At the boundaries, there are nonzero values of N_{xy} in the shell. These must be resisted by supports, such as walls, on which N_{xy} will act as loads in the opposite direction. Usually, such supports are in the form of edge beams that carry these loads in direct tension or compression to the column supports. Roof structures are generally formed by combining two or more basic hypar quadrilateral elements. At their junction, shears from adjacent quadrilaterals may cancel each other or they may add up, in which case additional internal edge members may also be required at such junctions.

Thus for the umbrella shell in Fig. 9.5(a), the shears in two adjacent panels will be as shown.* We can easily see that the outer edge members, such as those along OA and OC, have to carry shears in directions opposite to those of N_{xy}, and will thus be in direct compression. At the internal junction BC, shears from adjacent panels add up and an edge member along BC will have to be provided to resist these forces in direct tension. For the inverted umbrella in Fig. 9.5(b), the stresses are in opposite directions; hence edge members OC and BC will now be in tension and compression, respectively. For the butterfly shell in Fig. 9.5(c), we can see that shears from adjacent panels will cancel each other, thereby requiring no edge beams along OA and OC, whereas edge beams along CB and AB will be in direct compression. Readers can readily visualize for themselves the state of stress in shell panels and edge beams for the other examples in Fig. 9.5.

*It is easy to obtain the directions of shears acting on the element by physical interpretation. First, notice which two sides of the element are in one plane. The other two "skew" sides must be dipped either up or down. Fix the directions of shears along these skew sides such that their vertical components will oppose p_z. The shears on the other two sides will be complementary shears. Alternatively, one can always obtain the directions from the sign of $\overline{N}_{xy} = N_{xy}$ found in Eq. (9.8b), by using the correct signs for h and p_z.

Summarized below are the important equations for the basic quadrilateral of the shell, to be used for ready reference when solving practical problems.

Hypar Shell Summary: Coordinate Axes *x, y* along Straight-Line Generators (Refer to Fig. 9.6.)

The equation of the surface is

$$z = \frac{h}{ab} xy \tag{9.7a}$$

where h is positive in the positive z-direction. For *rectangular* hypars,

$$N_{xy} = \overline{N}_{xy} = -\frac{ab}{2h} \overline{p}_z \tag{9.8b}$$

where $\overline{p}_z$ is the load intensity over the plan area. If $\overline{p}_z$ = *constant,* then $\overline{N}_x$ and $\overline{N}_y$ are constant along the generators and are the same as those forces applied at the boundaries.

We shall now consider a few examples of the design of hypar shell roofs.

9.5 DESIGN OF AN INVERTED UMBRELLA SHELL

Inverted umbrella hypar shells are commonly used as roofs over industrial plants, canteen buildings, and gate offices of industries. We shall design one such typical shell which has plan dimensions of 30 × 30 ft (Fig. 9.7). The rise of such units is kept around one-tenth of the maximum plan dimension, and we assume it here as 3 ft.

The shell is composed of four basic quadrilateral panels, at the junctions of which edge beams *AB, CB,* etc., are provided. Also, at the boundaries, edge beams *OAH, OCD,* etc., are also provided.

Panel: Analysis and Design

Since the edge beams *OAH, OCD,* etc., cannot resist any normal loads, we assume N_x and N_y as zero at the boundaries. Assuming a 3-inch shell thickness, the self-weight load will be 37.5 lb/ft^2. To this, we add a load for finishes (10 lb/ft^2), the weight of edge beams (which we consider as an equivalent distributed load of 15 lb/ft^2), and a live load of 15 lb/ft^2. The total load on the shell thus becomes 77.5 lb/ft^2. Although only the live load is constant over the plan area, for simplicity, we assume that $\overline{p}_z$ = 77.5 lb/ft^2. With $\overline{p}_z$ assumed as constant and N_x, N_y taken as zero at the boundaries, we shall have N_x and N_y equal to zero throughout the shell. Substituting $a = b = 15$ ft and $h = 3$ ft into Eq. (9.8b), we obtain

$$N_{xy} = -\frac{15(15)}{2(3)}(77.5) = -2906 \text{ lb/ft}$$

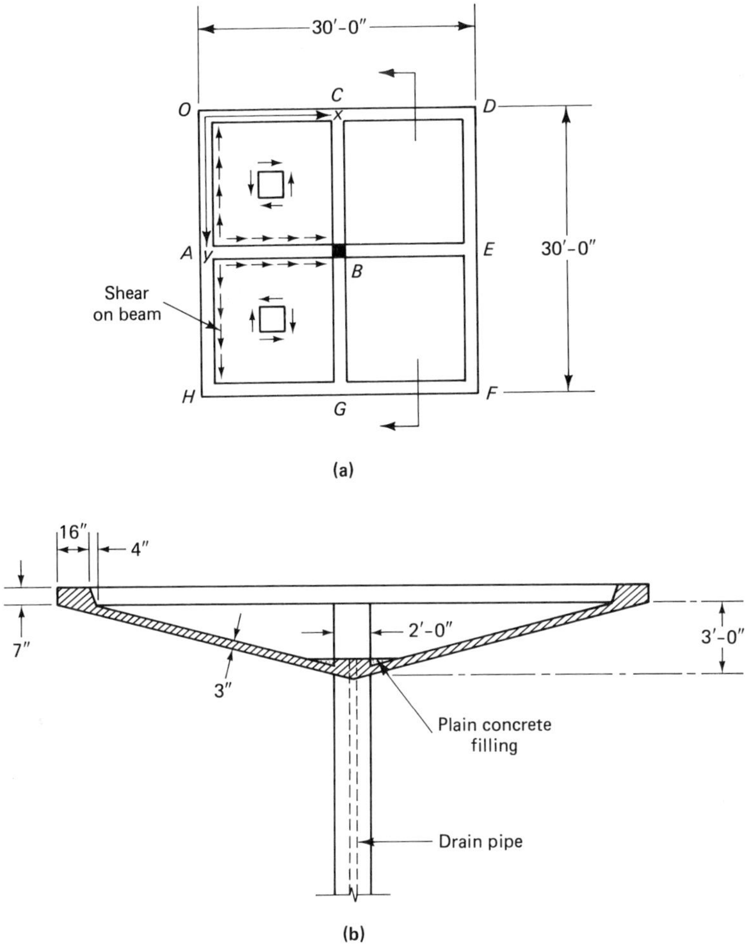

Figure 9.7 Inverted umbrella hypar shell roof. (a) Plan view. (b) Section view.

Since the shell is in pure shear, from Mohr's circle we can see that the diagonal tension will have the same value as N_{xy}, so that the tensile stress will equal $2906/12(3) = 80.73$ psi. This stress is small for concrete with a cylinder strength of 3000 psi. For the reinforcement, we have

$$\text{steel required} = \frac{2906}{20{,}000} = 0.1453 \text{ in.}^2/\text{ft}$$

Taking the minimum steel as 0.3% in each direction (refer to Section 4.9), we have

$$\text{minimum steel} = \frac{0.3(12)(3)}{100} = 0.108 \text{ in.}^2/\text{ft} \quad \text{(in each direction)}$$

We can provide No. 3 bars at 12 in. center-to-center spacing in both directions. This will provide sufficient reinforcement along the tension diagonal, as well.

Beam *OAH*. The force $N_{xy} = 2906$ lb/ft, which comes from the shell, acts on the beam, as shown in Fig. 9.7(a). Since ends O and H are free, the tensile force in the beam will be zero at O and H, and will increase linearly to a maximum at A. Thus we have

$$\text{maximum tension} = 2906(15) = 43{,}590 \text{ lb}$$

Assuming the edge beam to be 16 in. × 7 in., we have

$$\text{tensile stress} = 43{,}590/16(7) = 389 \text{ psi}$$

$$\text{steel required} = 43{,}590/20{,}000 = 2.18 \text{ in.}^2$$

We can provide four No. 5 bars at the top and bottom of the beam, and also provide nominal stirrups of No. 2 bars at 8 in. center-to-center spacing. The design of beams OD, DF, and HF will be similar.

Edge Beams *AB, CB, EB,* and *GB*. These edge beams carry shears N_{xy}, from two adjacent panels, in direct compression. This compression will vary from zero at A to a maximum at B (for beam AB). We have

$$\text{inclined length of } AB = \sqrt{15^2 + 3^2} = 15.3 \text{ ft}$$

$$\text{maximum compressive force} = 15.3(2906)(2) = 88{,}924 \text{ lb}$$

We assume a beam width of 20 in. and a thickness varying from 6 in. at A to 9 in. at B. So we find

$$\text{maximum compressive stress} = \frac{88{,}924}{20(9)} = 494 \text{ psi}$$

Although this member acts like a cantilever compression member, the stress above is low and can easily be resisted by the concrete; only nominal steel needs to be provided. Some extra steel is usually provided, however, to resist moments that can develop due to unsymmetrical wind or other live loads. Therefore, we use four No. 5 bars at the top and bottom of the beam, with No. 2 bars as stirrups at 8 in. center-to-center spacing.

Reinforcement Details

The panel reinforcement may be placed at the shell's midsurface along the straight-line generators (i.e., parallel to the edge beams). Alternatively, it may be placed along directions of principal tension and compression at 45°.

The main reinforcement of the tension beams (OH) should be anchored into the orthogonal beams (OD and HF) at right angles. The column bars should be anchored in beams BA, BC, etc., so that any moments generated due to unbalanced loading can be resisted.

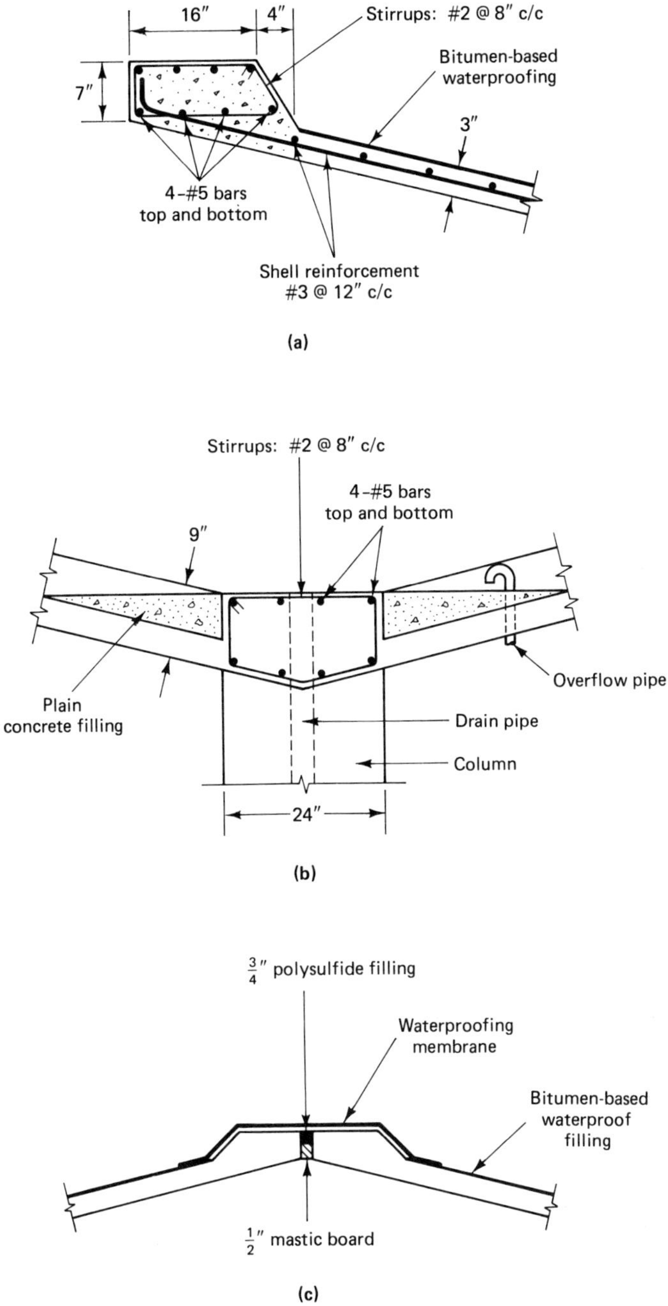

Figure 9.8 Reinforcement and design details for the inverted umbrella hypar shell. (a) Detail of panel and outer edge beam. (b) Detail of central edge beam. (c) Detail at junction of outer edge beams.

Figure 9.8(a and b) show the reinforcement arrangement in the outer and inner edge beams. It may be noted that to facilitate drainage, a small portion of the panel adjoining the column is filled with ordinary concrete up to the top of the central beams. The water is allowed to drain through a pipe embedded in the column. Adequate care should be taken to ensure that this pipe never gets clogged; otherwise, the shell may be subjected to heavy water loads. A few overflow pipes, located a short distance away from the column, as shown in Fig. 9.8(b), could be provided as a means to provide emergency runoff of excessive water, in case the central pipe clogs.

To span large areas, a number of such inverted umbrellas may be constructed adjoining one another. A typical detail at the junction of two umbrellas could be as shown in Fig. 9.8(c). To allow natural lighting inside industrial plants, alternate rows of umbrellas are commonly constructed at higher and lower levels, and glazing is then provided at the junctions to allow the entrance of light.

The edge beams of the roof will have axial displacements, due to tensile or compressive forces, which will not match the corresponding displacements of the shell. Also, the edge beams, due to their own weight, transmit vertical loads locally on the shell, thus causing larger deflections at the edges. The incompatibilities of displacements that arise will cause bending moments in the shell. It is thus desirable to provide additional reinforcement in the shell at its junction with the edge beam. This reinforcement is arranged so that steel is available at both the top and bottom of the shell near the junctions.

In this example, if the shell were loaded by a live load only in the half portion *OCGH* [Fig. 9.7(a)], tension in beams *OC* and *HG* would not be balanced by corresponding tensions in *CD* and *GF*. The unbalanced tensions can be considered to act as horizontal forces applied at *C* and *G* on the shell, and these forces will cause bending moments in beams *BE* and *BA*. It is desirable that a provision be made in the design of edge beams and columns, for resisting such moments which arise due to unbalanced live load or wind load. For such temporary loadings, many codes permit a 33% increase in the allowable stresses for steel and concrete.

The membrane analysis of the umbrella shell can also be performed by considering the shell as a beam. This treatment of the problem also helps us to easily take into consideration, approximately, the additional stress in the edge members under unbalanced live loads and wind loads, as we shall see in the next section.

9.6 ANALYSIS OF UMBRELLA SHELL AS A BEAM

Referring to Fig. 9.7, we can consider the shell as four beams cantilevering from the column in four directions. Taking a section parallel to *CBG*, the cross section of one of the beams will be as seen in Fig. 9.7(b). If we cut down the middle of the top flange and the web of an I-beam, and spread the two portions apart, the resulting cross section will look like that seen in Fig. 9.7(b). We can consider beams *OD* and *HF* as the split top flange of the I-section, beam *BE* as the bottom flange, and the shell itself as the split inclined web. It must be realized that the depth of this I-section reduces from 3 ft to zero as we move to the free end of the cantilever (i.e., as we move from B to E).

For the example of Section 9.5 we can obtain the stresses in the shell and the beams by using this beam analogy, as follows. From elementary statics, the bending moment at section CBG is

$$\text{moment at } CBG = 15(30)(77.5)\left(\frac{15}{2}\right) = 261{,}562.5 \text{ lb-ft}$$

The lever arm is taken as 3 ft; hence

$$\text{tensile force in top edge beam} = \frac{261{,}562.5}{2(3)} = 43{,}594 \text{ lb}$$

Also, the horizontal component of the compressive force in BE is

$$\frac{261{,}562.5}{3} = 87{,}187.5 \text{ lb}$$

Hence

$$\text{compressive force in } BE = \frac{87{,}187.5(15.3)}{15} = 88{,}931 \text{ lb}$$

These values match well with those obtained previously. To obtain the net shear to be resisted by the webs at section CBG, we subtract from the shear force due to vertical loads, the vertical component of the compressive force in BE. Thus we find

$$\text{net shear on webs} = 15(30)(77.5) - \frac{88{,}931(3)}{15.3}$$

$$= 34{,}875 - 17{,}437.5 = 17{,}437.5 \text{ lb}$$

If N_{xy} is the shear flow in the webs, its total vertical component must equal the net shear load. Hence

$$N_{xy}(15.3)\left(\frac{3}{15.3}\right)(2) = 17{,}437.5$$

or

$$N_{xy} = 2906.25 \text{ lb/ft}$$

as obtained before. It is thus seen that the membrane stresses in the umbrella shell and its edge beams could be obtained correctly from simple statics and beam theory.

Earlier we had stated that the rise of the umbrella shell is usually kept to within about $\frac{1}{10}$ of the plan dimensions. This gives a depth of $\frac{1}{5}$ of the span for the cantilever beam, which seems to be appropriate for reinforced concrete structures.

If only one-half of the shell *BCDEFG* is loaded by a wind pressure (or other live load) of 15 lb/ft^2, the moment on the column will be

$$15(15)(30)\left(\frac{15}{2}\right) = 50{,}625 \text{ ft-lb}$$

This moment is transmitted to the column by the two beams *AB* and *BE*, assuming that beams *BC* and *BG* cannot resist torsion. Hence beams *AB* and *BE* each have to be designed for a moment of 25.3 k-ft in addition to the compressive forces under uniform dead load. The steel and size of beams already adopted in the preceding section will be sufficient.

9.7 SKEW HYPARS

For aesthetic reasons, or otherwise, we may construct hypar shells by combining basic quadrilateral panels whose sides are not at right angles in plan, but form an angle 2ψ [Fig. 9.4(a)]. In such cases the straight-line generators of the shell are also not at right angles in plan. The equation of the surface of such a shell is still governed by Eq. (9.7a), which was derived for a general case. So, dropping the bar notation, we have

$$z = \frac{hxy}{ab} \tag{9.7a}$$

For obtaining membrane stress resultants, we note that Eqs. (8.6) and (8.10) are no longer valid, since they were derived for rectangular coordinate axes. However, we have seen in Problem 8.1 that with p_x and p_y as zero, the membrane stresses in shells with skew coordinate axes are governed by Eqs. (8.11a–c). We can also readily check that Eqs. (8.9a–c) and (8.10) will still be valid with $p_x = p_y = 0$ and $\bar{p}_z$ replaced by $\bar{p}_z \sin 2\psi$. Hence substituting Eq. (9.7a) into Eq. (8.11c), we have the shear stress as

$$\bar{N}_{xy} = -\frac{ab}{2h}\bar{p}_z \sin 2\psi \tag{9.9}$$

Proceeding as before, we can also see that Eqs. (9.8c–e) will still be valid for skew hypars. Hence we conclude that the membrane stresses in a skew hypar under vertical loads are similar to those in a rectangular hypar. Later in Section 9.10, we shall derive this equation [Eq. (9.9)] in a different way called "arch analogy."

Design of a Skew Hypar Shell: Butterfly Type

Figure 9.9 shows a butterfly-type shell which is supported on two buttresses at *B* and *F*. Such roofs have a pleasing appearance and can be used to cover gas stations, farmhouses, as well as other facilities. The heights of the buttresses from the ground level can be suitably chosen so as to obtain sufficient headroom.

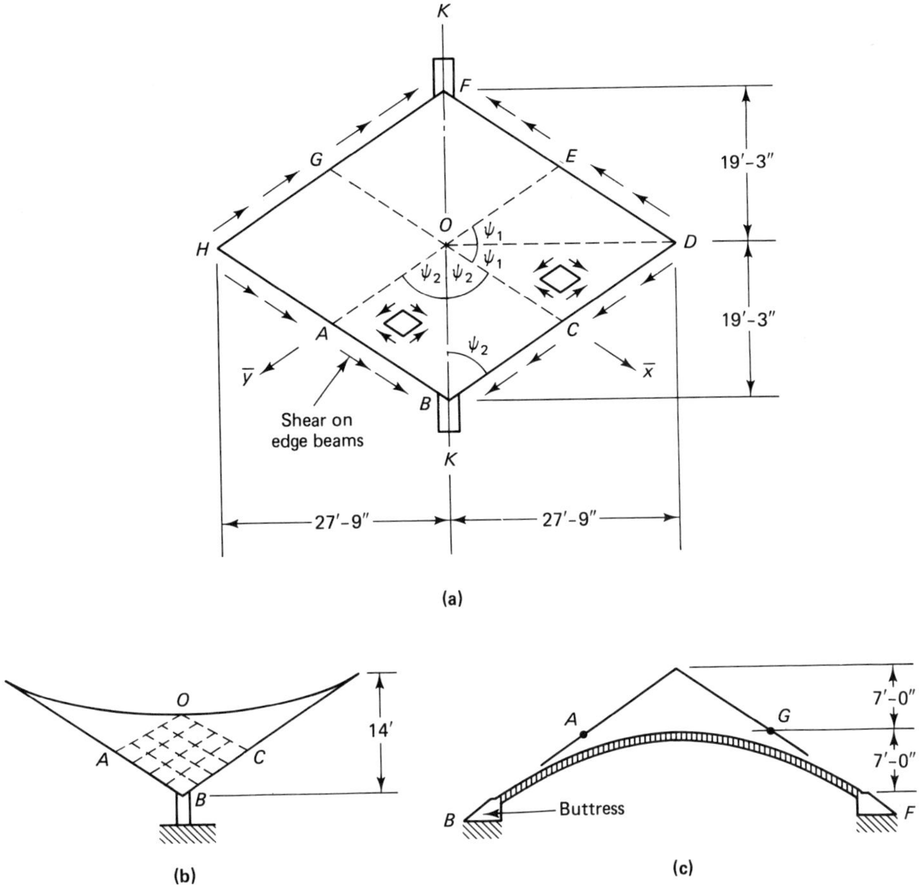

Figure 9.9 Skew hypar shell roof of butterfly type. (a) Plan view. (b) Side view. (c) View of section *KK*.

The shell shown in Fig. 9.9 is composed of four basic quadrilateral panels. Panels *OABC*, *OEFG* and panels *OCDE*, *OGHA* are similar to each other. Points *O*, *A*, *C*, *E*, and *G* will all lie in one plane. Points *B* and *F* are dipped down 7 ft, whereas points *D* and *H* are raised up 7 ft from this plane. Considering the directions of shear forces on elements within each quadrant, as shown in Fig. 9.9(a), we can readily conclude that shears from adjacent panels will cancel each other along *GOC* and *AOE*. Therefore, no beams are necessary there, whereas along the external edges, beams will be required. These edge beams will be in compression and will transmit the load to the buttresses.

For a 3-in. shell thickness, we assume the total load as 75 lb/ft^2 constant over the plan area. Thus we have $\overline{p}_z = 75$ lb/ft^2. Since the edges are free, and because $\overline{p}_z$ is constant, we conclude that N_x and N_y must be zero throughout the shell. From geometry we find

$$\text{plan length } BD = \sqrt{27.75^2 + 19.25^2} = 33.77 \text{ ft}$$

We now proceed with the analysis and design of each roof element.

Roof Elements: Analysis

Quadrant *OABC*. We have $a = b = 16.89$ ft and $h = 7.0$ ft. In the plan view of triangle BOD we find

$$\tan \psi_2 = \frac{27.75}{19.25} = 1.442$$

or

$$\psi_2 = 55.25°$$

Hence

$$\psi_1 = 90° - 55.25° = 34.75°$$

From Eq. (9.9) we find N_{xy} as

$$N_{xy} = -\frac{ab\bar{p}_z \sin 2\psi_2}{2h}$$

$$= -\frac{16.89(16.89)}{2(7.0)}(75)[\sin (110.5°)] = -1431.5 \text{ lb/ft}$$

Quadrant *OCDE*. Here also we have $a = b = 16.89$ ft and $h = -7.0$ ft.* Hence, we find

$$N_{xy} = \frac{-16.89(16.89)(75)}{2(-7.0)} \sin 69.5° = +1431.5 \text{ lb/ft}$$

It may be noted that although ψ_1 and ψ_2 are different for the two quadrants, values of $\sin 2\psi_1$ and $\sin 2\psi_2$ are the same, since $2\psi_1 + 2\psi_2 = 180°$. Hence N_{xy} will have the same numerical value in both quadrants. This is necessary, as otherwise edge beams would have been required along GC and AE to resist unbalanced shears.

Edge Beams *BD, FD, BH,* and *FH*. We determine the inclined length of BD as

$$\overline{BD} = \sqrt{33.77^2 + 14^2} = 36.56 \text{ ft}$$

*For this quadrant, if axes x and y are taken along OE and OC, respectively, and z is taken downward, then referring to Fig. (9.4), h will be negative and $\bar{p}_z$ will be positive downward. Hence N_{xy} will be positive. In any case, it is easiest to fix the directions of shears on elements in each quadrant by direct visualization, as explained previously.

Shears on beam *BD* from the shell panels are directed as shown in Fig. 9.9(a). Since *D* is a free end, the compression in the edge beam will increase linearly from zero at *D* to a maximum at *B*. This maximum compression is given as

$$\text{maximum compression} = 36.56(1431.5) = 52{,}335.6 \text{ lb}$$

Roof Elements: Design

For the shell, the diagonal tensile stress is 1431.5/12(3) = 39.76 psi. This stress is small, and the minimum steel of 0.3% in each direction will be sufficient. We can provide No. 3 bars at 12-in. center-to-center spacing in two directions, either along the straight-line generators or along the two opposing parabolic generators.

Assuming the cross section of the edge beam to be 12 in. × 12 in., the compressive stress in the beam is 52,335.6/12(12) = 363.4 psi. This stress is also small, and hence nominal steel of 1% of the beam's cross-sectional area may be provided. However, the actual steel provided is four No. 8 bars at the top and bottom of the beam, as shown in Fig. 9.10(a). This larger amount of steel is provided to resist any moments that will develop in the beams due to unbalanced live load or wind loads. For example, if 15 lb/ft^2 is the unbalanced load, due to live or wind loading on portion *FBD* (Fig. 9.9) of the shell, the unbalanced moment about *FB* is

$$\frac{1}{1000}(15)(27.75)\left(\frac{38.5}{2}\right)\left(\frac{27.75}{3}\right) = 74.12 \text{ k-ft}$$

This moment will have to be resisted by the bending moments in edge beams *FD*, *BD*, *FH*, and *BH*.* If M_o is the moment in each of these beams, then

$$4M_o \sin \psi_2 = 74.12$$

or

$$4M_o\left(\frac{27.75}{33.77}\right) = 74.12$$

which gives

$$M_o = 22.55 \text{ k-ft}$$

Neglecting direct compression in the beams, for simplicity, and assuming a lever arm of 0.87 times the effective beam depth, the steel required to resist M_o is

*The shell will rotate as a whole, in order to resist the external moment. If we consider a freebody of the right half of the shell [Fig. 9.9(a)], the external moment about *FB* will be resisted by moments in beams *BD*, *FD*, and by resisting couples from stresses in the shell arch, which are transmitted to the other half of the shell, to cause moments in beams *FH* and *BH*.

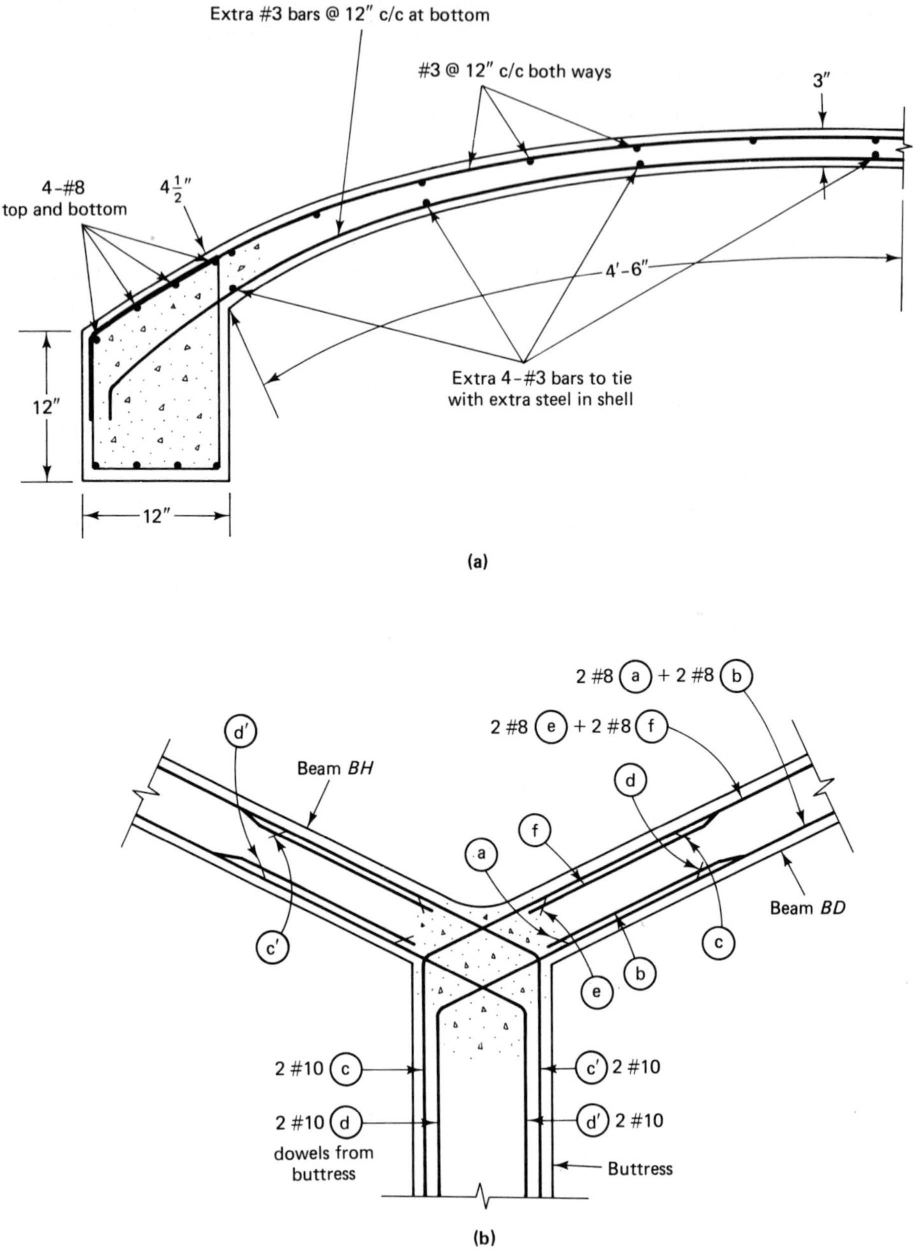

Figure 9.10 Design of butterfly hypar shell. (a) Detail at outer edge beam. (b) Detail at junction of edge beams and buttress.

$$A_s = \frac{22.55(12)}{20(0.87)(10.5)} = 1.48 \text{ in.}^2$$

where 20 k/in^2. is the allowable stress in steel and 10.5 in. (12 in. minus cover) is the beam's effective depth. Thus four No. 8 bars provided at the top and bottom of the beams are sufficiently conservative. These bars must be well anchored in the buttresses, which themselves must be designed to resist the unbalanced moments. To avoid congestion of steel, it is desirable to take out dowels of larger-diameter bars from the buttresses and lap these with the beam bars, as shown in Fig. 9.10(b). It may be noticed from Fig. 9.10(a) that in a distance of 4 ft-6 in. near the beam, the shell thickness is gradually increased from 3 in. to $4\frac{1}{2}$ in. Also, additional steel at the bottom is provided in the shell in this portion to cater for any bending moments developing near the junction.

Since the compressive forces in the edge beams reduce to zero toward D and H, the depths of the edge beams could be tapered to, say 6 in., at D and H. The section of the beams could be further reduced since a portion of the shell can be considered to act with the edge beam (as an L-section) in resisting compression. Such refinements could be made by a designer in practical designs to obtain a more elegant structure.

Buttress: Analysis and Design

The thrust on the buttress is obtained by taking the component, in the horizontal plane, of the compression in beams BD and BH, and then resolving this force along BF (Fig. 9.9). Hence we can write

$$\text{thrust on buttress} = 52{,}335.6\left(\frac{33.77}{36.56}\right)(\cos 55.25°)(2) = 55{,}109.2 \text{ lb}$$

The buttresses can be designed to resist moments and shears caused by this thrust. Alternatively, tie beams connecting the two buttresses can be provided; however, these will obviously cause a hindrance when they are situated above the ground. The buttresses will also have to resist the vertical load from the shell and the moments due to wind or unbalanced loads, as discussed earlier.

The concept of beam behavior could also be applied to analyze this type of shell. The shell can be considered as a beam cantilevering out from BF on both sides. The cross section of the beam at BF will consist of a parabolic arch terminating into flanges formed by the edge beams. The width and depth of this cross section continues to reduce as we move to the ends D or H of the cantilevers. By finding the moment of inertia of the cross section, the compression in the edge beams and the tensile stresses in the shell could be obtained [3].

9.8 EXAMPLE OF HYPAR WITH BOUNDARIES SKEW TO STRAIGHT-LINE GENERATORS

A "hipped"-type hypar shell is shown in Fig. 9.11. It is formed by combining four similar quadrants, such as *OBCE*. Lines *JOE* and *GOB* are at one level, while *F*, *C*, *L*, and *I*, the positions of column supports, are 4 ft below. Normally, the shell

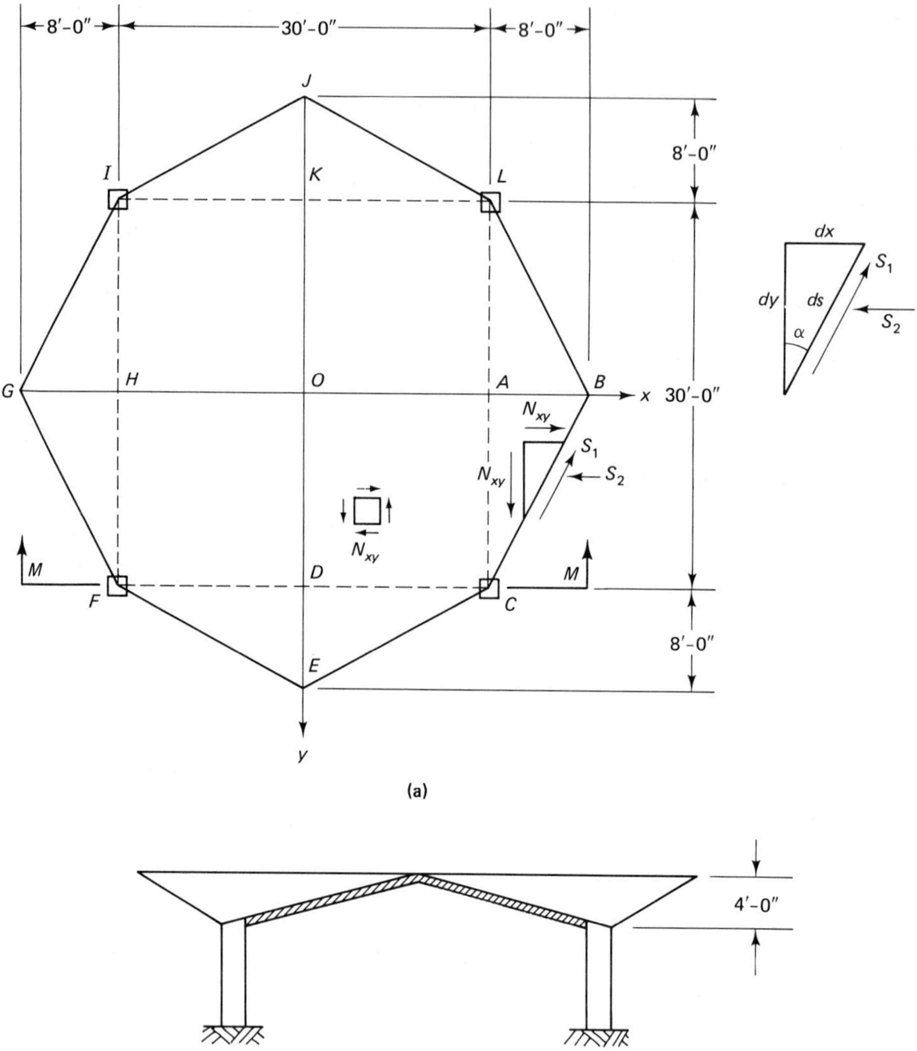

Figure 9.11 Hipped hypar shell roof. (a) Plan view. (b) View of section *MM*.

would have boundaries along the straight-line generators, such as along CA and CD, in which case the solution would have become simple. However, for architectural reasons, here the boundaries are not along the straight-line generators, and hence the membrane solution becomes more complex. The generators of the shell will be "parallel" to the coordinate axes OX and OY which are at right angles. Thus we have a rectangular hypar (and not a skew hypar as in the preceding section) but with its boundaries not along its generators. Under vertical load (assuming it as uniformly distributed over the plan area), the membrane shear in the shell is then

$$N_{xy} = -\frac{ab\,\bar{p}_z}{2h}$$

where $a = b = 15$ ft and $h = 4$ ft.

On the edge BC, the shell will have unbalanced forces which we can resolve into: (1) S_1, a shear along the inclined edge BC; and (2) S_2, a force along the straight-line generator in the negative x-direction. Both S_1 and S_2 are defined per unit length along the inclined edge BC. We assume that the generators in the x- and y-directions make angles ϕ and θ, respectively, with the x- and y-axes (refer to Fig. 8.1). Also, edge BC is assumed to make an angle β with the horizontal. Measuring s along the plan projection of BC, for the triangular edge element shown, we have

$$dx = ds\,(\sin\alpha) \tag{a}$$

and

$$dy = ds\,(\cos\alpha) \tag{b}$$

Writing equations of equilibrium in the x- and y-directions for this element, we obtain

$$-S_2\frac{ds}{\cos\beta}(\cos\phi) + S_1\frac{ds}{\cos\beta}(\cos\beta)\sin\alpha + N_{xy}\frac{dx}{\cos\phi}\cos\phi = 0$$

and

$$S_1\frac{ds}{\cos\beta}(\cos\beta)\cos\alpha - N_{xy}\frac{dy}{\cos\theta}\cos\theta = 0$$

Substituting from Eqs. (a) and (b) and simplifying, we obtain

$$S_1 = N_{xy} \tag{c}$$

and

$$S_2 = 2N_{xy}\sin\alpha\,\frac{\cos\beta}{\cos\phi} \tag{d}$$

Forces opposite to S_1 can be applied on an edge beam along BC. This beam will resist these forces in compression and transfer the load to the column at C. Since the edge BC is free, we must apply back on the shell forces opposite to S_2, so that equilibrium is maintained on the edge. We have the freedom of prescribing N_x at this edge [refer to Eqs. (9.8 d,e)]. We also know that the value of $\overline{N}_x$, applied at the edge, remains constant along the generators, and also does not affect the shear stresses in the shell (which depend solely on $\overline{p}_z$). Since $\overline{N}_x$ is the plan projection of N_x, measured per unit length along the y-direction, we can write the relation

$$\overline{N}_x \, dy = -S_2 \frac{ds}{\cos \beta} \cos \phi$$

Therefore,

$$\overline{N}_x = -S_2 \frac{\cos \phi}{\cos \beta} \frac{1}{\cos \alpha}$$

Substituting from Eq. (d), we obtain

$$\overline{N}_x = -2\overline{N}_{xy} \tan \alpha \tag{e}$$

So then, N_x at any point in the shell is given by Eq. (8.1a), i.e.,

$$N_x = \overline{N}_x \frac{\cos \theta}{\cos \phi} = -2\overline{N}_{xy} \frac{\cos \theta}{\cos \phi} \tan \alpha \tag{f}$$

Since unbalanced forces S_2 on the shell, along BC, are compressive, N_x applied at the edge to nullify these forces will be tensile. Hence N_x given by Eq. (f) will be tensile throughout. We can obtain a similar expression for N_y due to the balancing loads N_y applied at edge CE.

On the ridge OD, the N_x forces from adjoining quadrants will be inclined with one another. Their horizontal components will balance each other, whereas their vertical components will add up and will act as loads on an edge beam along KOD spanning from K to D. The vertical reactions of this beam at K and D will be transferred to the columns by compression in the shell itself, in edge beams provided along IK, KL and FD, DC.

With this background, we can now solve the numerical example. We assume the shell to be 3 in. thick, and we design for a total load (assumed uniformly distributed over the plan area) of 75 lb/ft^2.

Shell: Analysis and Design

Force N_{xy} in the shell can be obtained as

$$N_{xy} = -\frac{ab}{2h} \overline{p}_z = \frac{-15(15)}{2(4)} (75) = -2109.4 \text{ lb/ft}$$

We also have

$$\tan \alpha = \frac{8}{15} = 0.5333$$

From Eq. (f) we find

$$N_x = 2(2109.4)(0.5333)\left(\frac{\cos\theta}{\cos\phi}\right) = 2250.0\left(\frac{\cos\theta}{\cos\phi}\right) \tag{g}$$

Along OD, $\theta = 0$, and hence $\cos\theta = 1$. Also, at point O, $\phi = 0$, and at point D, $\tan\phi = 4/15 = 0.267$ or $\cos\phi = 0.966$. Hence the maximum value of N_x occurs at D and is equal to

$$\frac{2250}{0.966} = 2329.2 \text{ lb/ft}$$

One can easily see that the value of N_y at point D (which will be the same as the value of N_x at point A) is

$$N_y = 2250.0\left(\frac{0.966}{1.0}\right) = 2173.5 \text{ lb/ft}$$

An element of the shell is now not in pure shear, but also has tensile forces N_x and N_y. At point D we can find the maximum principal tensile force, from Mohr's circle, as

$$N_1 = \frac{2329.2 + 2173.5}{2} + \sqrt{\frac{(2329.2 - 2173.5)^2}{4} + 2109.4^2}$$

or

$$N_1 = 4362.19 \text{ lb/ft}$$

Hence the maximum tensile stress in the shell is

$$\frac{4362.19}{12(3)} = 121.17 \text{ psi}$$

This stress is very small, and we find the required steel as

$$\text{steel required} = \frac{4362.19}{20{,}000} = 0.218 \text{ in.}^2/\text{ft}$$

So No. 3 bars at 8 in. center-to-center spacing, provided in two orthogonal directions, should be sufficient to resist the principal tensile stresses.

Edge Beam *BC:* Analysis and Design

For the edge beam BC, we have

$$\text{plan length } BC = \sqrt{15^2 + 8^2} = 17 \text{ ft}$$

$$\text{inclined length } BC = \sqrt{17^2 + 4^2} = 17.46 \text{ ft}$$

The compressive force in the edge beam will be zero at point B and a maximum at point C. This maximum compressive force is

$$\text{maximum compressive force} = 17.46S_1 = 17.46(2109.4) = 36{,}830 \text{ lb}$$

So we can provide a 16-in.-wide by 7-in.-deep beam, with four No. 5 bars on each face and nominal stirrups of No. 2 bars at 12-in. centers.

Edge Beam *JKODE:* Analysis and Design

For the edge beam *JKODE,* since ends *J* and *E* are free, the compressive force due to N_{xy}, for the two quadrants on each side, will increase linearly from zero at *J* or *E* to a maximum at point *O*. We have

$$\text{maximum compression} = 2109.4(23)(2) = 97{,}032.4 \text{ lb}$$

Also, from Eq. (g) for N_x, substituting $\cos\theta = 1$, we obtain for N_x (along *OD*), the following expression:

$$N_x = \frac{2250.0}{\cos\phi}$$

The unbalanced vertical component, due to N_x acting from the two sides on the beam, is [see Fig. (9.12a)]

$$q = 2N_x \sin\phi = 4500 \tan\phi$$

Now $\tan\phi$ varies linearly from zero at point *O* to $4/15 = 0.267$ at point *D*. So $\tan\phi$ can be written as $4y/225$, where *y* is the distance from point *O* measured along *OD*. Hence the vertical load on the beam is

$$q = 4500\left(\frac{4y}{225}\right) = 80y$$

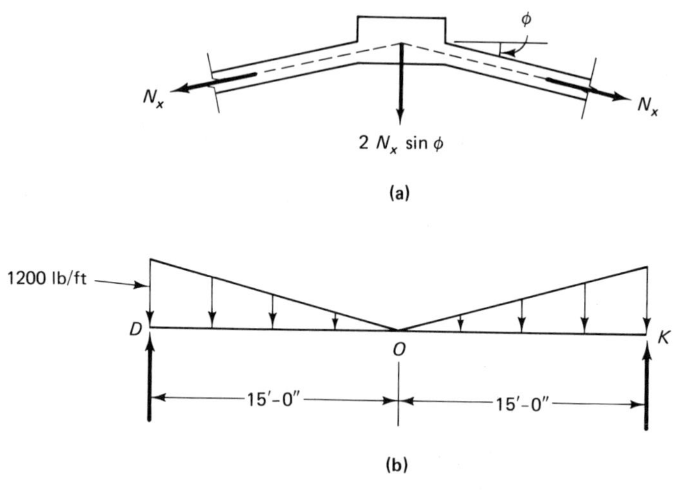

Figure 9.12 Loads acting on edge beam *JKODE*. (a) Unbalanced vertical components acting along the beam. (b) Linear variation of vertical load on member *DOK*.

This vertical load on OD varies linearly from zero at $y = 0$ to a maximum value of 1200 lb/ft at point D, as shown in Fig. 9.12(b). From simple statics of the beam DOK, loaded as shown in the figure, we can calculate

$$\text{maximum internal shear} = \frac{1200(15)}{2} = 9000 \text{ lb}$$

$$\text{maximum internal moment} = \frac{9000(15)}{3} = 45{,}000 \text{ ft-lb}$$

We can provide a beam 18 in. wide by 12 in. deep, with four No. 6 bars at its top and bottom (Fig. 9.13), in order to resist the foregoing bending moment, shear, and the direct compression of 97,032.4 lb.

Beams *DC, DF:* Analysis and Design

The reaction of 9000 lb from beam DOK, at end D, will be transmitted to columns C and F by compression in beams DC and DF. This compression in the beams is found as

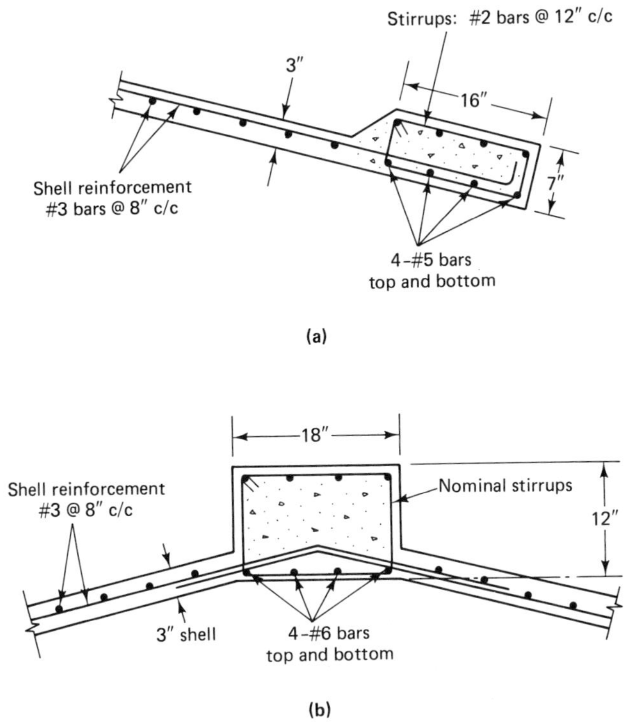

Figure 9.13 Design of edge beams. (a) Detail of beam BC. (b) Detail of beam OD.

$$\text{compression} = \left(\frac{9000}{2}\right)\frac{1}{\sin\phi} = \left(\frac{9000}{2}\right)\frac{\sqrt{225 + 16}}{4} = 17{,}464.7 \text{ lb}$$

This force is small and can be resisted by the shell itself. However, a "hidden" beam 16 in. wide by 3 in. thick, with extra three No. 4 bars provided at its top and bottom, will provide added conservatism.

Forces on the Column

The forces acting on the column are found as shown below. Specifically for column C, we have:

Load from BC

$$\text{vertical load} = (36{,}830)(\sin\beta) = 36{,}830\left(\frac{4}{17.46}\right) = 8438 \text{ lb}$$

$$\text{horizontal load} = (36{,}830)(\cos\beta) = 35{,}860 \text{ lb}$$

Load from DC

$$\text{vertical load} = (17{,}464.7)(\sin\phi_D) = \frac{17{,}464.7(4)}{15.524} = 4500 \text{ lb}$$

$$\text{horizontal load} = (17{,}464.7)(\cos\phi_D) = 16{,}875 \text{ lb}$$

where ϕ_D is the value of ϕ at point D. Similar loads will come from beams EC and AC. The horizontal components of these loads, acting on the column, are shown in Fig. 9.14. The total vertical load on the column will be

$$\text{total vertical load} = (8438 + 4500)(2) = 25{,}876 \text{ lb}$$

As a check on this calculation, using tributary areas, we find

$$\text{plan area } OBCE = 15(15) + 15(8) = 345 \text{ ft}^2$$

Thus the total vertical load carried by the tributary area influencing column C is

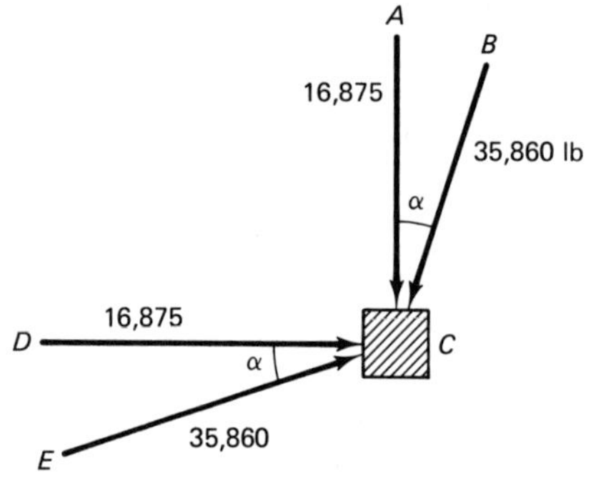

Figure 9.14 Thrusts on column C, coming from edge beams CB, CE and CD, CA.

$$\text{total external load} = 75(345) = 25{,}875 \text{ lb}$$

This matches with the column load calculated above. Now, from Fig. 9.14, the resultant horizontal thrusts acting along *AC* and *DC* will be

$$\text{thrust} = 16{,}875 + 35{,}860 \cos \alpha - 35{,}860 \sin \alpha = 31{,}640.9 \text{ lb}$$

The columns must be sufficiently stiff to resist these thrusts in bending, or if permissible, tie beams may be provided along *CL* and *CF*.

9.9 MEMBRANE STRESSES UNDER LOADS UNIFORMLY DISTRIBUTED OVER THE SURFACE AREA

In all the examples considered thus far, we assumed (approximately) that the vertical loading on the shell was constant over the plan area (i.e., $\bar{p}_z$ = constant). For a shell panel defined by the equation

$$z = \frac{h}{ab} xy \tag{9.7a}$$

this assumption gave us a simple expression for the shear stress:

$$\bar{N}_{xy} = -\frac{ab}{2h} \bar{p}_z \tag{9.8b}$$

Also, $\bar{N}_x$ and $\bar{N}_y$ could be taken as constant along the generators.

Our assumption of the loading being constant over the plan area is valid only for snow or live loads. The dead load, which forms the major load on a shell roof, is constant over the surface area of the shell (assuming the shell thickness as constant). To assess the membrane stresses more correctly under such loads, we proceed as follows.

Referring to Fig. 9.15, we can represent the radius vector of any point on the shell surface by

$$\mathbf{r} = x\mathbf{i} + y\mathbf{j} + z\mathbf{k}$$

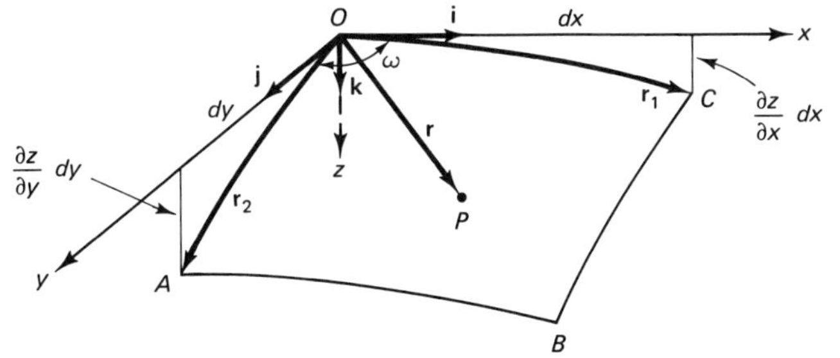

Figure 9.15 Shell element with sides defined by vectors $\mathbf{r}_1$ and $\mathbf{r}_2$.

where **i, j,** and **k** are unit vectors along x, y, and z, respectively. Now vectors $\mathbf{r}_1$ and $\mathbf{r}_2$, along sides OC and OA, respectively, of the shell element, can be written as

$$\mathbf{r}_1 = dx\,\mathbf{i} + \left(\frac{\partial z}{\partial x}\right) dx\,\mathbf{k}$$

$$\mathbf{r}_2 = dy\,\mathbf{j} + \left(\frac{\partial z}{\partial y}\right) dy\,\mathbf{k}$$

The cross product of these two vectors is a vector at right angles to their plane and is given by

$$\mathbf{r}_1 \times \mathbf{r}_2 = \begin{vmatrix} \mathbf{i} & \mathbf{j} & \mathbf{k} \\ dx & 0 & \left(\frac{\partial z}{\partial x}\right)dx \\ 0 & dy & \left(\frac{\partial z}{\partial y}\right)dy \end{vmatrix}$$

$$= -\left(\frac{\partial z}{\partial x}\,dx\,dy\right)\mathbf{i} - \left(\frac{\partial z}{\partial y}\,dx\,dy\right)\mathbf{j} + (dx\,dy)\mathbf{k}$$

The magnitude of this product is given by

$$|\,\mathbf{r}_1 \times \mathbf{r}_2\,| = \sqrt{\left(\frac{\partial z}{\partial x}\right)^2 + \left(\frac{\partial z}{\partial y}\right)^2 + 1}\;(dx\,dy)$$

From the definition of the cross product in *vector algebra,* we know that the magnitude above must be equal to the product of the lengths of the sides OC, OA, and the sine of the angle between the two sides. Hence we have

$$|OC|\,|OA|\sin\omega = |\mathbf{r}_1 \times \mathbf{r}_2| = \left[1 + \left(\frac{\partial z}{\partial x}\right)^2 + \left(\frac{\partial z}{\partial y}\right)^2\right]^{1/2} (dx\,dy)$$

The left-hand side here represents the area of the element $OABC$. Hence the area of the element is

$$\text{area } OABC = \left[1 + \left(\frac{\partial z}{\partial x}\right)^2 + \left(\frac{\partial z}{\partial y}\right)^2\right]^{1/2} (dx\,dy)$$

Now, if p_o is the dead load of the shell, uniformly distributed over its surface area, then load $\overline{p}_z$ (defined on the plan area of the shell) is obtained from

$$\overline{p}_z(dx\,dy) = p_o\left[1 + \left(\frac{\partial z}{\partial x}\right)^2 + \left(\frac{\partial z}{\partial y}\right)^2\right]^{1/2} (dx\,dy)$$

or

$$\overline{p}_z = p_o\left[1 + \left(\frac{\partial z}{\partial x}\right)^2 + \left(\frac{\partial z}{\partial y}\right)^2\right]^{1/2} \tag{9.10}$$

In the examples considered so far, we have been assuming that $\overline{p}_z = p_o = constant$ [i.e., we have been neglecting $(\partial z/\partial x)^2$ and $(\partial z/\partial y)^2$ as compared to 1, which would be true for shallow shells only]. Substituting Eq. (9.10) into Eq. (9.8b), we obtain

$$\overline{N}_{xy} = -\frac{\partial^2\Phi}{\partial x\,\partial y} = -\frac{ab}{2h}\overline{p}_z = -\frac{ab}{2h}p_o\sqrt{1 + \left(\frac{\partial z}{\partial x}\right)^2 + \left(\frac{\partial z}{\partial y}\right)^2} \tag{9.11}$$

We note that N_{xy} is no longer constant throughout the shell, but now varies with x and y. Expanding the square-root term by *Taylor's series*, we obtain

$$\overline{N}_{xy} = -\frac{abp_o}{2h}\left\{1 + \frac{1}{2}\left[\left(\frac{\partial z}{\partial x}\right)^2 + \left(\frac{\partial z}{\partial y}\right)^2\right] - \frac{1}{8}\left[\left(\frac{\partial z}{\partial x}\right)^2 + \left(\frac{\partial z}{\partial y}\right)^2\right]^2 + \cdots\right\}$$

Substituting values of $\partial z/\partial x$ and $\partial z/\partial y$ obtained by taking derivatives in Eq. (9.7a), we have

$$\overline{N}_{xy} = -\frac{abp_o}{2h}\left[1 + \frac{1}{2}\frac{h^2}{a^2b^2}(x^2 + y^2) - \frac{1}{8}\frac{h^4}{a^4b^4}(x^2 + y^2)^2 + \cdots\right] \tag{9.12}$$

For most hypar shell roofs, obtained by combining in various ways the basic shell panel, we generally take h as $\frac{1}{5}$ of the larger side a or b. For a square panel, $a = b$, and if we take $h/a = \frac{1}{5}$, we obtain from Eq. (9.12),

$$\overline{N}_{xy} = -\frac{ab}{2h}p_o\left[1 + \frac{1}{50}\left(\frac{x^2 + y^2}{ab}\right) - \cdots\right]$$

We can readily see that the maximum value of the second term, which occurs for $x = a$ and $y = b$, is only 4% of the first term, and hence the second and higher terms could be neglected compared to the first term for such shallow shells. Now if we take $a = 2b$ and $h = a/5$, we can check from Eq. (9.12) that the second term is about 10% of the first term.

We can therefore conclude that by assuming the total load as constant over the plan area, the error in estimating shear stresses in a hypar shell is not very significant for practical problems of shallow shells. If, however, we consider deep shells, the error can be significant. We can readily see that for $a = b$ and $h/a = \frac{1}{2}$, the second term in Eq. (9.12) is about 25% of the first term at $x = a$ and $y = b$. Hence one must exercise caution when assuming $\overline{p}_z = constant = p_o$ for solving problems of deep hypar shells, based on the simple formulas used so far.

Now to obtain an expression for N_x, we integrate Eq. (9.11) with respect to x. Hence

$$\frac{\partial\Phi}{\partial y} = \frac{ab}{2h}p_o\int\left[1 + \left(\frac{\partial z}{\partial x}\right)^2 + \left(\frac{\partial z}{\partial y}\right)^2\right]^{1/2}dx + f_1(y)$$

Thus

$$\overline{N}_x = \frac{\partial^2 \Phi}{\partial y^2} = \frac{ab}{2h} p_o \frac{\partial}{\partial y} \left\{ \int \left[1 + \left(\frac{\partial z}{\partial x} \right)^2 + \left(\frac{\partial z}{\partial y} \right)^2 \right]^{1/2} dx \right\} + \frac{df_1(y)}{dy}$$

By substituting values of $\partial z/\partial x$ and $\partial z/\partial y$, we obtain

$$\overline{N}_x = \frac{p_o}{2} \frac{\partial}{\partial y} \left\{ \int \left[\left(\frac{ab}{h} \right)^2 + x^2 + y^2 \right]^{1/2} dx \right\} + \frac{df_1(y)}{dy}$$

A similar expression can be obtained for $\overline{N}_y$ also. It may be noted that $\overline{N}_x$ and $\overline{N}_y$ will no longer be constant along the generators.

9.10 MEMBRANE STRESSES IN A HYPAR BY CONSIDERATION OF ARCH ACTION

Let us now consider a hypar shell formed by translation of one parabola along another, as shown in Fig. 9.16(a). We consider x, y, and z as the coordinate axes in the direction of the parabolas, while we denote again $\bar{x}$ and $\bar{y}$ as the axes along the straight-line generators, which make an angle ψ with the x-axis [refer also to Fig. 9.3(b)]. The equation of the shell surface in the x, y, z coordinate system is given by

$$z = \frac{x^2}{2h_1} - \frac{y^2}{2h_2} \tag{9.5}$$

The vertical load at each point P on the shell is shared by two arches at right angles to each other—a tension arch $CAPBD$ and a compression arch $EOPF$ [Fig. 9.16(a)]. Let the coordinates of P be $(x_o, 0, z_o)$, and those of A and B (the points where the tension arch meets the vertical planes $\bar{y}$-z and $\bar{x}$-z) be (x_o, y_1, z_1) and $(x_o, -y_1, z_1)$, respectively. Substituting into Eq. (9.5), we have

$$z_o = \frac{x_o^2}{2h_1} \quad \text{and} \quad z_1 = \frac{x_o^2}{2h_1} - \frac{y_1^2}{2h_2}$$

Now, considering the arch APB [Fig. 9.16(b)], the "drop" of P from AB is equal to

$$z_o - z_1 = \frac{y_1^2}{2h_2}$$

Let $\bar{p}_z$ be the load uniformly distributed over the plan area. We also assume that the load shared by the tension arches is $\alpha \bar{p}_z$. The load carried by the compression arches will then be $(1 - \alpha)\bar{p}_z$. Referring to Fig. 9.16(b), the pull H_1 in arch APB is given from simple statics by

$$H_1(z_o - z_1) = \alpha \bar{p}_z \frac{(2y_1)^2}{8}$$

or

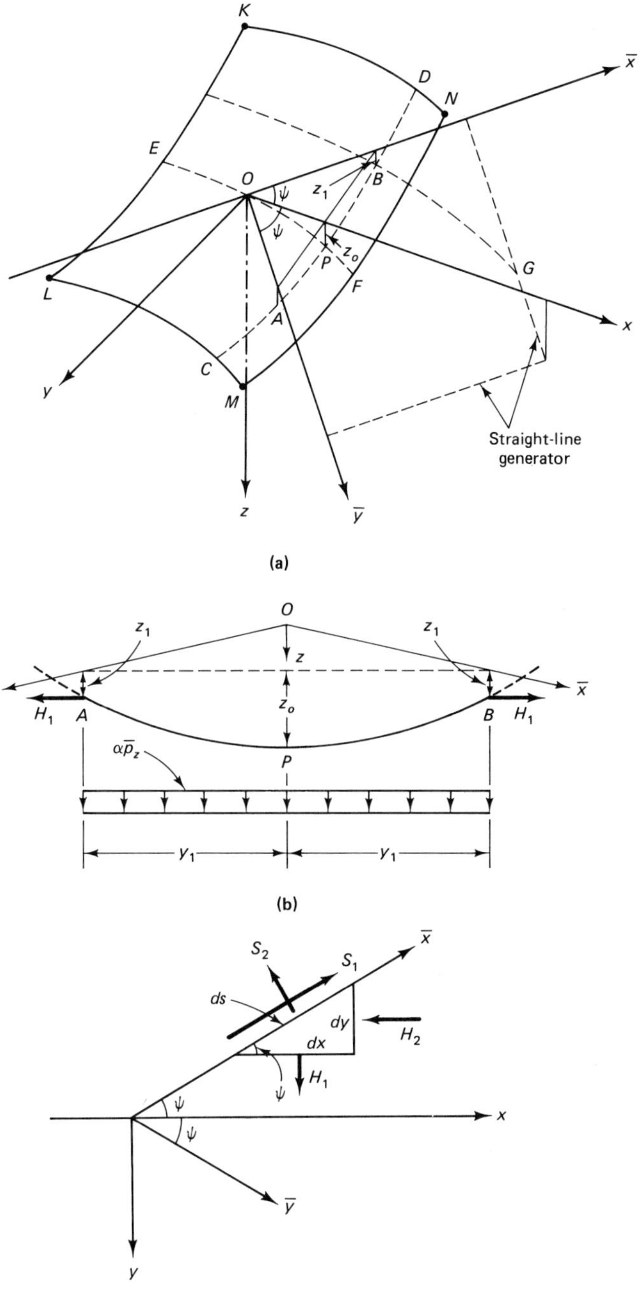

Figure 9.16 Demonstration of "arch action" in a hypar shell. (a) Configuration of hypar and coordinate axes. (b) Illustration of arch *APB*. (c) Triangular element of the shell in plan.

$$H_1 = \alpha \bar{p}_z h_2$$

Similarly, a compression arch BG will give a thrust H_2, at B, equal to

$$H_2 = (1 - \alpha)\bar{p}_z h_1$$

By the symmetry of both arches, we can take $\alpha = \frac{1}{2}$. Hence

$$H_1 = \tfrac{1}{2}\,\bar{p}_z h_2 \qquad \text{and} \qquad H_2 = \tfrac{1}{2}\,\bar{p}_z h_1 \tag{9.13a,b}$$

To find the membrane stresses in the $\bar{x}$, $\bar{y}$, z coordinate system, we consider equilibrium of a triangular element of the shell at B, as shown in Fig. 9.16(c). Resolving the forces acting on the face (ds), we obtain from statics:

$$S_1\,ds - H_1\,dx \sin\psi - H_2\,dy \cos\psi = 0$$

and

$$S_2\,ds - H_1\,dx \cos\psi + H_2\,dy \sin\psi = 0$$

Substituting for H_1 and H_2, with $dx = ds \cos\psi$ and $dy = ds \sin\psi$, the second equation gives

$$S_2 = \tfrac{1}{2}\,\bar{p}_z\,(h_2 \cos^2\psi - h_1 \sin^2\psi)$$

Substituting further for $\cos^2\psi$ and $\sin^2\psi$, from Eqs. (9.3b,c), we obtain $S_2 = 0$. Also, the first equation gives

$$S_1 = \tfrac{1}{2}\,\bar{p}_z\,(h_1 + h_2)\,(\sin\psi)(\cos\psi)$$

For a hypar panel, with $\bar{x}$ and $\bar{y}$ coordinate axes, having sides of length a, b and a drop of h, we can substitute for $(h_1 + h_2)$ from Eq. (9.7b), to obtain

$$S_1 = \frac{ab}{2h}\,\bar{p}_z \sin 2\psi$$

Thus, as before, we find that with $\bar{x}$ and $\bar{y}$ as the coordinate axes, we obtain only shear stresses $S_1 = N_{\bar{x}\bar{y}}$ in the shell, for a vertical load uniformly distributed over the plan area. The value of this shear stress is the same as that from Eq. (9.9) or (9.8b) previously obtained from the differential equations.

9.11 HYPAR SHELL WITH BOUNDARIES ALONG PARABOLIC GENERATORS

The concept of resisting vertical loads by arch action in two directions, as discussed in Section 9.10 can be conveniently used to design groined vaults. Such shells are formed by the intersection of two hypars at right angles, which have their boundaries parallel to parabolic generators (Fig. 9.17).

The arches normal to the free edge will carry very little load since the thrust in them must be zero at the free edges. The load is therefore mainly carried by a series of arches parallel to the free edges. Appropriately, it is seen in the figure

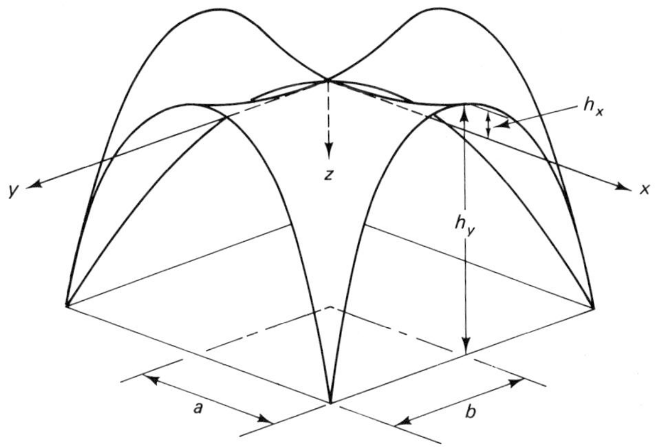

Figure 9.17 Illustration of a groined vault, formed by two hyperbolic paraboloids intersecting at a right angle.

that arches perpendicular to the free edges are made much flatter (i.e., $h_x \ll h_y$). The horizontal components of the thrusts in the arches, for the two directions, will be given by Eqs. (9.13a,b). Substituting $z = -h_x$ when $x = a$ and $y = 0$ into Eq. (9.5), we obtain

$$h_1 = -\frac{a^2}{2h_x}$$

and similarly, substituting $z = (h_y - h_x)$ when $x = a$ and $y = b$, we have

$$h_2 = -\frac{b^2}{2h_y}$$

So the equation of the shell surface becomes

$$z = -\frac{x^2}{a^2/h_x} + \frac{y^2}{b^2/h_y}$$

Generally, the load $\bar{p}_z$ would be shared by forces H_1 and H_2 in the two arches in the two directions. These forces would be obtained by substituting h_1 and h_2, from above, into Eqs. (9.13a,b). However, in the present case, since arches in one direction, the shallower arches, end up in a free edge, the forces in these arches will be zero, and the entire load $\bar{p}_z$ will be resisted by the arches in the other direction, i.e., by the steeper arches. These arches will resist the load in only direct compression, without bending, for a uniform load $\bar{p}_z$ over the plan area (recalling that a parabolic arch is a funicular curve for such loading). We know, however, that the dead load, which forms the major loading on the shell, is not uniformly distributed over the plan area. Also, as noted in previous examples, we cannot assume that the dead load is uniformly distributed over the plan area since, due to steepness of the arches, the error will be significant in this case. Hence the arches, if considered individually, will develop bending moments in addition to axial thrusts.

However, because the parallel arches arc interconnected, shears will develop between adjoining parallel arches to satisfy compatibility of their unequal membrane deflections. We can find the values of these shears such that when they act on an arch together with the vertical loads, only axial forces will be generated in the arch. Such an analysis, although very lengthy, is theoretically possible by starting from the arch at the free edge. Tables have been given in Reference [4] for the arch thrusts and shears that develop under dead load. These values are obtained directly from the shell differential equation, and may readily be used for design.

The thrusts from the arches are transferred to the diagonal arches formed at the intersection of adjacent groins. These diagonal arches will have a varying V-shaped section and can be designed as two-pinned or three-pinned arches to resist the vertical and horizontal components of the arch thrusts. Candela [1] has designed the very impressive restaurant at Xochimilco, as well as other structures in Mexico, using such groined vaults.

Hypar shell roofs may also be constructed using a single hypar supported along the generating parabolas *NK, ML* and/or *KL, MN* [Fig. 9.16(a)]. In this case, as well as in the case of groined vaults just discussed, we have to substitute the derivatives of z from Eq. (9.5) into Eq. (8.10). The second derivative $\partial^2 z/\partial x \partial y$ will now be zero, and we can obtain the following differential equation (assuming that $p_x = p_y = 0$):

$$\frac{1}{h_1}\left(\frac{\partial^2 \Phi}{\partial y^2}\right) - \frac{1}{h_2}\left(\frac{\partial^2 \Phi}{\partial x^2}\right) = -\overline{p}_z$$

This is a second-order partial differential equation of the hyperbolic type and can be solved by using the standard methods for solving partial differential equations. The method usually consists of using a Fourier series expansion in one direction, which also automatically satisfies the boundary conditions in this direction. When substituted into the differential equation, this expansion separates the variables and yields an ordinary differential equation in the other direction. This ordinary differential equation can be solved, and its constants of integration can be determined from the boundary conditions in the second direction.

Similar procedures also have to be followed for solving problems of other shells, such as conoids, elliptic paraboloids, as well as cylindrical barrel shells, as we shall discuss later. It will therefore be appropriate to review briefly the theory of Fourier series. We shall do this in Chapter 10.

9.12 APPLICABILITY OF MEMBRANE THEORY TO PRACTICAL DESIGNS

We have discussed the design of different types of hypar shell structures using membrane theory, which led to a very simple analysis. However, at the junctions of the shell and edge beams, there will actually be incompatibilities of displacements under membrane stresses. The weight of the edge beams will also cause further incompatibilities. A refined analysis, based on bending theory equations for a "shallow" shell [5], becomes quite complex unless a computer program is used. A finite element analysis can also be used [6,7,8] to obtain a detailed picture of actual stresses in the shell. A number of experimental investigations have also been carried out to study the actual behavior of hypar shells as compared to the results of the membrane theory analysis [9]. We shall briefly discuss the conclusions obtained from some of these analytical and experimental studies.

For designing hypar shells in the examples of this chapter, we assumed that the edge shears from the hypar panels were resisted by the beam sections alone. As a matter of fact, the edge beams* were somewhat larger than what one would like to have, since thick edge beams mar the elevational elegance and grace of the thin shell. If the edge beams were deleted altogether, portions of the shell itself would act as edge beams and carry the load. This could easily be accounted for in the design of a shell of medium span. For the inverted umbrella, the valley beams are in compression; however, the compressive loads could easily be resisted by the shell alone. These valley beams, nevertheless, are desirable for better resistance of moments caused by unbalanced loads, such as wind loading.

Even when edge beams are present, a portion of the shell adjoining the beam acts with the edge beam, as would be expected. Thus the stresses in the shell near the edge (or valley) beams are much different than those obtained by a membrane analysis. Also, the stresses in the edge beams are lower than those obtained by membrane theory. The sizes of the edge beams could therefore be reduced after taking into account the contribution of the adjoining shell acting with it. How much effective width of the shell should be considered to act with the beam is a matter of judgment. White [9], in his experiments on inverted umbrella hypar shells, found that the effective width increased as the forces in the edge members increased. Thus for the top edge members, the effective width increased from near zero at the ends to a maximum at the junction of the edge beam with the valley beam. For the valley beam, the width increased toward the column, to a maximum of 33 times the shell thickness.

In the inverted umbrella of Section 9.5, we provided both edge and valley beams above the shell. Therefore, shears from the shell act below the middepth of the beams, and thus cause bending moments in the beams, in addition to the axial forces. The upstand top edge beams are in tension. In these beams, the eccentric shears will cause a sagging moment, which will partially offset the hogging moment due to self-weight of the beam cantilevering on both sides. On the other hand, for the upstand valley beam, the eccentric shears from the shell will cause a hogging moment, which adds to the cantilever hogging moment due to its self-weight. Hence to minimize deflections due to the effects of self-weight of the beams, it is desirable to keep the edge beams upstand and the valley beams downstand (if architecturally

*Here the term "edge beam" will also be used to synonymously describe a valley (or central) beam.

acceptable) in an inverted umbrella shell. For the butterfly shell of Section 9.7, we can similarly conclude that the edge beams should preferably be downstand.

Schnobrich [10,11], in his finite element studies of butterfly and hipped-type hypar shells, has also found that membrane analysis accurately represents the true shell behavior away from the edge beams. For the hipped hypar shell [Fig. 9.5(e)], he found that bending moments in the edge and crown beams were not significantly affected, even when the sizes of the beams were varied. The moments in the crown beams were very small, except when an unusually large beam was considered. It is felt that the bending moments in the edge beams due to self-weight could be conservatively determined in such shells simply by analyzing the beams as a gable frame. It is desirable to have ties connecting the columns rather than designing the columns for the horizontal thrust of the gable, since deflections of the columns increase the moments in the edge beams significantly.

A number of existing hypar shell roofs have been designed on the basis of membrane analysis alone. The designer should, however, keep in mind the effects of unbalanced loads, self-weight of edge beams, etc., for an approximate, yet safe design.

REFERENCES

[1] Faber, Colin, *CANDELA/The Shell Builder,* Reinhold Publishing Corporation, New York, 1963.

[2] Tuma, J. J., *Engineering Mathematics Handbook,* 2nd ed., McGraw-Hill Book Company, New York, 1979.

[3] Leibour, M., "Ultimate Design of Square Hyperbolic Paraboloid Shells," *Indian Concr. J.,* Sept. 1969.

[4] "Elementary Analysis of Hyperbolic Paraboloid Shells," Publication No. IS085.01D, Portland Cement Association, Skokie, Ill., 1960.

[5] Wilby, C. B., and Khwaja, I., *Concrete Shell Roofs,* Applied Science Publishers Ltd., Barking, Essex, England, 1977.

[6] Chu, T. C., and Schnobrich, W. C., "Finite Element Analysis of Translational Shells," *Comput. and Struct.,* Vol. 2, No. 1, 1972.

[7] Pecknold, D. A. W., and Schnobrich, W. C., "Finite Element Analysis of Skewed Shallow Shells," *ASCE J. Struct. Div.,* Vol. 95, Apr. 1969, pp. 715–744.

[8] Tezcan, S. S., Agrawal, K. M., and Kostro, G., "Finite Element Analysis of Hyperbolic Paraboloid Shells," *ASCE J. Struct. Div.,* Vol. 97, Jan. 1971, pp. 407–424.

[9] White, R. N., "Reinforced Concrete Hyperbolic Paraboloid Shells," *ASCE J. Struct. Div.,* Vol. 101, Sept. 1975, pp. 1961–1980.

[10] Schnobrich, W. C., "Analysis of Hyperbolic Paraboloid Shells," Concrete Thin Shells, SP-28, American Concrete Institute, Detroit, 1971, pp. 275–311.

[11] Schnobrich, W. C., "Analysis of Hipped Roof Hyperbolic Paraboloid Structures," *ASCE J. Struct. Div.,* Vol. 98, July 1972, pp. 1575–1583.

[12] Yang, Y. C., "Record Concrete Shell Roof Featured in Coliseum," *Civ. Eng. (N.Y.)*, June 1972.

[13] Nainan, P. K., "Shell Foundations—The New Method," Cement and Concrete, Alok-udyog Cement Series, India, Vol. 12-1, Apr.–June 1971, pp. 42–52.

[14] Banerjee, S. P., "Foundations of Engineering Structures," *J. Inst. Eng. India,* Vol. 44, No. 11, 1964, pp. 633–659.

PROBLEMS

9.1. A hypar shell, with plan as shown in Fig. P9.1, is to be constructed. Points *A, H, G, O, C, D,* and *E* are at elevation (EL) 20′-0″, while points *B* and *F* are at EL 15′-0″. If columns are to be provided at only two points, show where you would locate them.

Using membrane theory, design the shell assuming a thickness of 3 in. and a total uniform load of 60 lb/ft^2. Show where you will provide beams and/or ties. Find forces in all of these, and design them appropriately. Find the vertical and horizontal loads on the columns.

Discuss what additional provisions you would make in design if an additional load of 10 lb/ft^2 is distributed over area *ABFG* only.

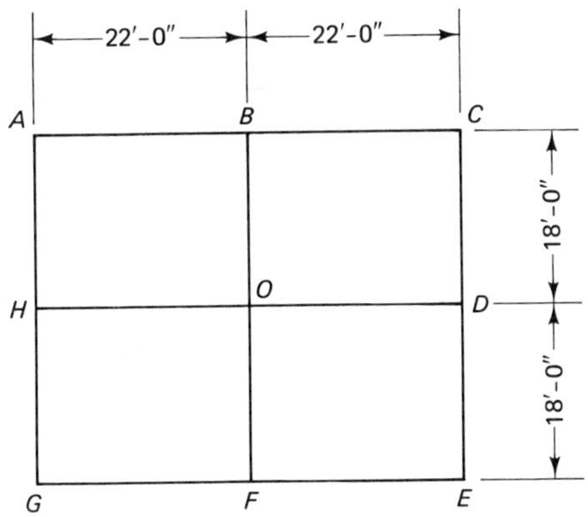

Figure P9.1 Plan view of a hypar shell roof to be constructed.

9.2. For the shell of Section 9.7, obtain properties of the equivalent cantilever beam at section *KK* (Fig. 9.9). Hence obtain approximately the maximum tensile stress in the shell and the compression in the edge beams, by using beam theory. Compare your results with the values obtained earlier.

9.3. A structure is formed from four butterfly-type hypar shells, as shown in Fig. P9.3. Similar structures, covering very large areas, have been designed by Yang [12] (for a coliseum in Puerto Rico), as well as by Candela [1] (in Mexico). Design the shells and edge beams for a load of 75 lb/ft^2 over the plan area. Obtain the forces on the columns, and hence design suitable tie beams connecting the columns.

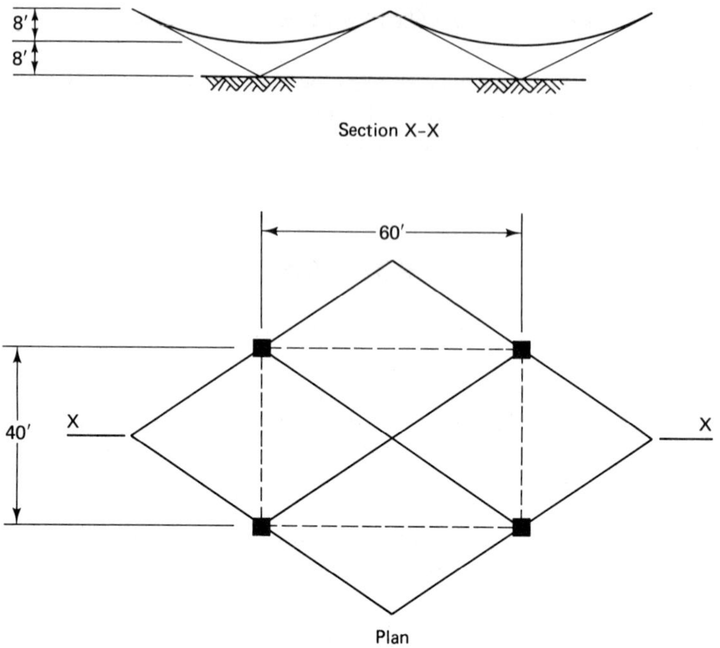

Figure P9.3 Configuration of four adjoining butterfly hypar shells to form a single structure.

9.4. A roof structure for an industrial factory consists of a series of hipped hypar shells (Fig. P9.4). A typical unit *ACIG* has ridges *DF* and *BH* at one level, while points *A*, *C*, *I*, and *G* are column supports located 5 ft below the level of the ridges. The boundaries of the shell, *AD*, *DG*, *CF*, and *FI*, are cut so that they are skew to the straight-line generators. This helps in permitting north light to enter the factory through glazing provided in the inclined triangular portions *ADG* and *CFI*. Bottom tie beams *AG* and *A′G′* of adjoining units are placed apart on the wide columns, with a slab connecting them at their bottom to form a gutter. Assume a suitable thickness of the shell, and design the main shell, edge beams, and tie beams.

9.5. An unsymmetrical inverted umbrella hypar shell canopy is to be constructed according to the plan layout shown in Fig. P9.5. Points *A*, *B*, *C*, *D*, *E*, *F*, *G*, and *H* are all at the same elevation. Point *O* is a support column located 6 ft below the other points. The thickness of the portion *BCDEF* of the shell is 3 in., and the loading on this portion is 37.5 lb/ft^2 self-weight plus 7.5 lb/ft^2 finish load (which is over the entire shell). The thickness, *t*, of the remaining portion of the shell is to be designed such that there is no moment on the column under dead load (self-weight plus finish load). Find this required thickness. Also, obtain the forces in the edge beams, as well as the loads on the column. Assume that the dead weight is uniformly distributed over the plan area, and neglect the weight of the beams.

(For the Ciba laboratories in Mexico, Candela has designed a very impressive entrance canopy from an inverted, umbrella-type hypar shell. This shell dramatically cantilevers

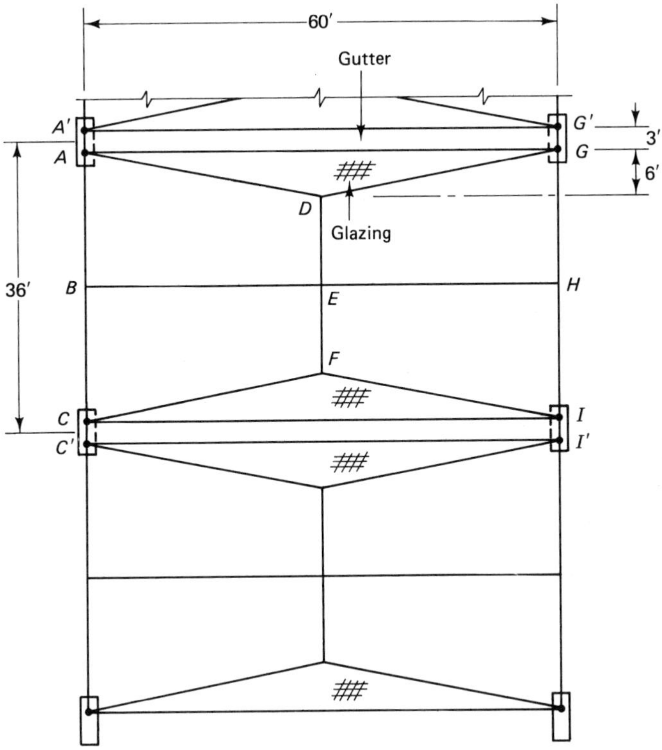

Figure P9.4 Roof structure, consisting of hipped hypar shells, used to cover an industrial factory.

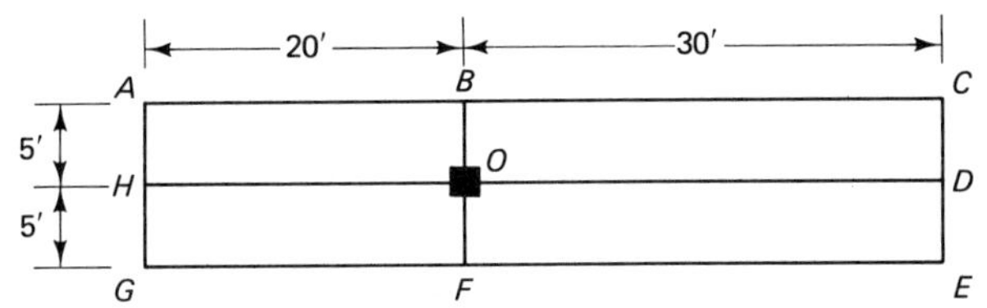

Figure P9.5 Plan layout for the inverted umbrella hypar shell of Problem 9.5.

35 ft unsymmetrically on one side of the column. Candela has used lightweight perlite concrete in the 35-ft arm and ordinary 8-in.-thick concrete in the shorter 9-ft arm.)

9.6. A 3-in.-thick hypar shell has points *A, B, C, D, E, F, G,* and *H* located at the same elevation, and point *O* raised up by 15 ft, as shown in Fig. P9.6. The loading can be assumed to be 70 lb/ft^2 over the *plan area*. Find the membrane stresses in the shell, the forces in the edge beams, and the forces on the columns at *B, D, F,* and *H*. Find the maximum shear stress in the shell if the load is considered to be 70 lb/ft^2 over the *surface area* of the shell.

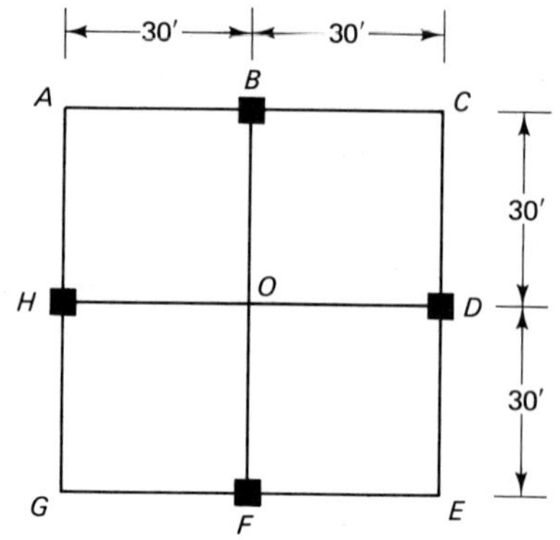

Figure P9.6 Plan layout for the hypar shell of Problem 9.6.

9.7. Occasionally, umbrella hypar shells have been used economically as footings below columns [13,14]. For the construction of such footings, the soil can easily be cut to the desired shell profile by using straight edges. Then reinforcement and concrete for the shell are constructed over an ordinary concrete smoothing base. Precast footings have also been used, where sand is poured in the center of the footing in order to fill the void between the shell and ground.

Design an umbrella footing for an 18 in. × 18 in. column which carries a vertical load of 250 kips. Assume that the safe bearing capacity of the soil is 5 kip/ft^2.

10

Fourier Series

10.1 INTRODUCTION

Most problems encountered in the theory of elasticity, as well as problems of shells or plates, involve the solution of a partial differential equation. The solution of such an equation has to satisfy prescribed boundary conditions. Fourier series gives us a powerful tool for solving many of these partial differential equations. By assuming the solution in the form of a Fourier series in one variable, such that this series satisfies the differential equation and some boundary conditions, we can separate the variables in the partial differential equation, and thus obtain an ordinary differential equation in the remaining variable. This ordinary differential equation can be solved directly, and then the constants determined to satisfy the remaining boundary conditions. We shall briefly review in this chapter the theory of Fourier series [1], so that we can readily employ this theory in solving further shell problems.

10.2 FOURIER SERIES FOR PERIODIC FUNCTIONS

A function $f(x)$ is said to be periodic, with period p, if

$$f(x + p) = f(x)$$

Hence the graph of the function can be obtained simply by duplicating or translating the portion of its curve in an interval $x = a$ to $x = a + p$.

It can be easily proved that a function, periodic with period p, is also periodic with period kp, where k is an integer. For example, if $k = 2$, we can see that

$$f(x + 2p) = f[(x + p) + p] = f(x + p) = f(x)$$

Hence

$$f(x + 2p) = f(x)$$

i.e., the function is periodic with period $2p$.

It is also easy to prove that the sums, products, and quotients of any two periodic functions having the same period p are also periodic with a period p. We also note that

$$\int_a^{a+p} f(x)\,dx = \int_a^0 f(x)\,dx + \int_0^p f(x)\,dx + \int_p^{a+p} f(x)\,dx \tag{a}$$

Changing the variable to $\eta = x - p$, in the third term on the right-hand side above, we obtain

$$\int_p^{a+p} f(x)\,dx = \int_0^a f(\eta + p)\,d\eta = \int_0^a f(\eta)\,d\eta = \int_0^a f(x)\,dx$$

This third term will cancel with the first term in Eq. (a), which then becomes

$$\int_a^{a+p} f(x)\,dx = \int_0^p f(x)\,dx$$

Hence we can also conclude that

$$\int_a^{a+p} f(x)\,dx = \int_b^{b+p} f(x)\,dx$$

If n is an integer, we know that $\cos nx$ and $\sin nx$ are periodic functions with period 2π. Therefore, we can consider expanding conveniently a given periodic function $f(x)$ with a period 2π, in terms of a series or summation of these well-known functions. Let us assume an expansion of the form

$$f(x) = \frac{a_o}{2} + \sum_{n=1}^{\infty} a_n \cos nx + \sum_{n=1}^{\infty} b_n \sin nx \tag{10.1}$$

The constants a_n and b_n can be evaluated as follows. First, we integrate with respect to x, both sides of Eq. (10.1), from $-\pi$ to π. Then since

$$\int_{-\pi}^{\pi} \cos nx\,dx = 0$$

and

$$\int_{-\pi}^{\pi} \sin nx\,dx = 0$$

only the first term on the right-hand side remains nonzero, and we have

$$a_o = \frac{1}{\pi} \int_{-\pi}^{\pi} f(x)\, dx \tag{10.2a}$$

Now, to find a_n, we multiply both sides of Eq. (10.1) by $\cos mx$, (m being an integer), and integrate from $-\pi$ to π. Since

$$\int_{-\pi}^{\pi} \sin nx \cos mx = 0 \qquad \text{for all } m$$

and

$$\int_{-\pi}^{\pi} \cos nx \cos mx = 0 \qquad \text{for all } m\text{, except } m = n$$

the right-hand side of Eq. (10.1) reduces to only a single nonzero term when $m = n$. We thus get

$$\int_{-\pi}^{\pi} f(x) \cos nx\, dx = a_n \int_{-\pi}^{\pi} \cos^2 nx\, dx = a_n\pi$$

or

$$a_n = \frac{1}{\pi} \int_{-\pi}^{\pi} f(x) \cos nx\, dx \tag{10.2b}$$

Similarly, we can obtain

$$b_n = \frac{1}{\pi} \int_{-\pi}^{\pi} f(x) \sin nx\, dx \tag{10.2c}$$

It may be noted that we can derive expression (10.2a) for a_o, by putting $n = 0$ into Eq. (10.2b), and this is the reason why the first term of the series Eq. (10.1) was taken as $a_o/2$ and not a_o.

Thus any function $f(x)$, periodic with a period 2π, can be expanded as an infinite series given by Eq. (10.1), and the constants in this series are given by Eqs. (10.2b,c). This series is called the Fourier series of the function $f(x)$.

If $f(x)$ is defined only in the interval $-\pi \leqslant x \leqslant \pi$, we can take its "periodic extension" over the remaining interval. This periodic extension is obtained simply by translating the graph of the function between $-\pi \leqslant x \leqslant \pi$, over the remaining interval. This is shown in Fig. 10.1 for a function $f(x) = x^2 + c$, which is defined over $-\pi \leqslant x \leqslant \pi$ only. We thus obtain a function defined over $-\infty \leqslant x \leqslant \infty$ which has a period 2π. This function can then be expanded into the Fourier series of Eq. (10.1). It should be noted, however, that at any point $x = x_o$ in the intervals outside $-\pi \leqslant x \leqslant \pi$, the series will converge to a point B on the periodic extension of the function and not to the value of the function itself [i.e., not to the point A obtained by substituting $x = x_o$ into $f(x) = x^2 + c$]. However, this does not matter to us, since we are interested in the Fourier expansion of the function only in the interval $-\pi \leqslant x \leqslant \pi$ over which the function has been defined.

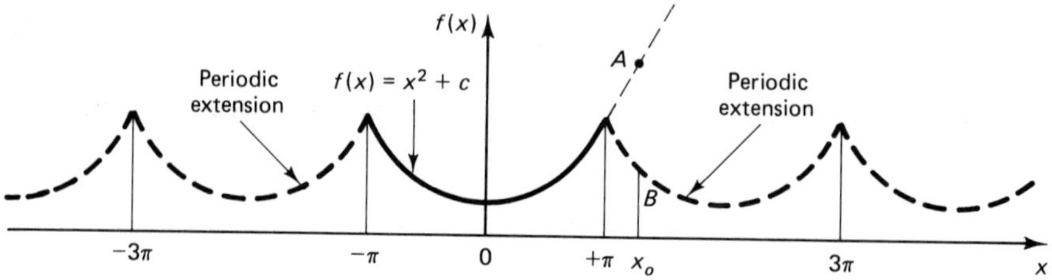

Figure 10.1 Periodic extension, obtained by translating the graph of the function in the interval $-\pi \leq x \leq \pi$, at periodic spacings, over the entire axis $-\infty \leq x \leq \infty$.

Example 10.1

Let us expand a function $f(x) = 1 + x$, defined on the interval $[-\pi, \pi]$, as a Fourier series. We define

$$f(x) = 1 + x \qquad (-\pi \leq x \leq \pi)$$

The periodic extension of $f(x)$ is shown in Fig. 10.2. From Eq. (10.2b), we can write

$$\begin{aligned} a_n &= \frac{1}{\pi} \int_{-\pi}^{\pi} (1 + x) \cos nx \, dx \\ &= \frac{1}{\pi} \left[\frac{(1 + x) \sin nx}{n} + \frac{\cos nx}{n^2} \right]_{-\pi}^{\pi} \qquad (n \neq 0) \\ &= 0 \qquad (n \neq 0) \end{aligned}$$

For $n = 0$ we obtain

$$a_o = \frac{1}{\pi} \int_{-\pi}^{\pi} (1 + x) \, dx = \frac{1}{\pi} \left[x + \frac{x^2}{2} \right]_{-\pi}^{\pi} = 2$$

Also, from Eq. (10.2c), we have

$$b_n = \frac{1}{\pi} \int_{-\pi}^{\pi} (1 + x) \sin nx \, dx$$

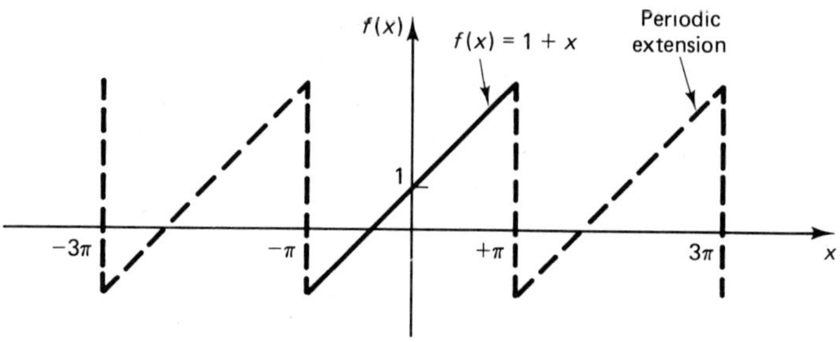

Figure 10.2 Periodic extension of the ramp function $f(x) = 1 + x$.

$$= \frac{1}{\pi}\left[-\frac{(1 + x)\cos nx}{n} + \frac{\sin nx}{n^2}\right]_{-\pi}^{\pi}$$

$$= -\frac{2\cos n\pi}{n}$$

$$= \frac{2(-1)^{n+1}}{n} \qquad n = 1, 2, \ldots, \infty.$$

Substituting for a_n, b_n in Eq. (10.1), we obtain the Fourier series as

$$(1 + x) = 1 + \sum_{n=1}^{\infty} \frac{2(-1)^{n+1}}{n} \sin nx \qquad (-\pi \leqslant x \leqslant \pi) \tag{b}$$

It may be noted that beyond $x = \pm\pi$, the series converges to a point on the periodic extension of the function, not to a value given by $1 + x$. Hence in the interval $-\infty \leqslant x \leqslant \infty$, the Fourier series represents the extended function shown in Fig. 10.2.

10.3 CONVERGENCE

The power of n in the denominator of the series expansion indicates how fast the series will converge. The higher the power of n, the faster is the convergence. In Example 10.1 we had n^1 in the denominator, indicating not a very rapid convergence. Hence we will be required to take the summation over a large number of the terms in the series to get good accuracy.

How fast the series will converge depends on the characteristics of the function to be expanded in the interval $-\infty \leqslant x \leqslant \infty$. Without proof, we can broadly summarize that if the function $f(x)$ is discontinuous itself, the convergence will be of the $1/n$ type. If the function is continuous in $-\infty \leqslant x \leqslant \infty$, but its first derivative (slope) $f'(x)$ is discontinuous, the convergence will be $1/n^2$ type. If $f(x)$ and $f'(x)$ are continuous but $f''(x)$ has discontinuities, the convergence is of the type $1/n^3$, and so on.

In Example 10.1 (Fig. 10.2), we see that the function itself (periodic extension) has discontinuities at $x = \pi, 3\pi$, etc., and hence we had a $1/n$ type of convergence in the series expansion. On the other hand, in Fig. 10.1 we see that the function is continuous throughout, but its first derivative has discontinuities at $x = \pi, 3\pi$, etc. Hence we expect that its Fourier series expansion will have a $1/n^2$ type of convergence.

It can be proved that at points of discontinuities of a function, its Fourier series will converge to an average of the two values of the function at the discontinuity [1]. Thus in the preceding example, at $x = \pi$, the average value of the function being expanded is

$$\frac{1}{2}[(1 + \pi) + (1 - \pi)] = 1$$

We thus verify from Eq. (b) that the series converges to (1) when we substitute $x = \pi$.

10.4 FOURIER SERIES OVER AN INTERVAL $-L$ TO L

If a function $f(x)$ is defined in the interval $(-L, L)$, then, as before, by taking its periodic extension over the remaining interval, we can generate a function which is periodic with period $2L$ over the interval $(-\infty, \infty)$. We can consider expanding this function in a Fourier series of the form

$$f(x) = \frac{a_o}{2} + \sum_{n=1}^{\infty} a_n \cos\frac{n\pi x}{L} + \sum_{n=1}^{\infty} b_n \sin\frac{n\pi x}{L} \tag{10.3a}$$

The trigonometric functions in the series also now have period $2L$. Constants a_n and b_n can be obtained directly by integration, according to our earlier procedure. Alternatively, we can change the variable from x to ξ such that

$$x = \frac{L}{\pi}\xi \tag{c}$$

The function

$$f(x) = f\left(\frac{L}{\pi}\xi\right)$$

then becomes a function of ξ defined over $(-\pi \leqslant \xi \leqslant \pi)$, and its extension is periodic with period 2π in ξ, since

$$f\left[\frac{L}{\pi}(\xi + 2\pi)\right] = f\left(\frac{L}{\pi}\xi + 2L\right) = f(x + 2L) = f(x) = f\left(\frac{L}{\pi}\xi\right)$$

Hence this function can be expanded into a Fourier series whose coefficients are given by Eqs. (10.2a–c). We shall obtain

$$a_n = \frac{1}{\pi}\int_{-\pi}^{\pi} f\left(\frac{L\xi}{\pi}\right) \cos n\xi \, d\xi$$

Changing the variable back to x, using Eq. (c), we have

$$a_n = \frac{1}{L}\int_{-L}^{L} f(x) \cos\frac{n\pi x}{L}\, dx \tag{10.3b}$$

Similarly,

$$b_n = \frac{1}{L}\int_{-L}^{L} f(x) \sin\frac{n\pi x}{L}\, dx \tag{10.3c}$$

Using Eqs. (10.3a–c), we can expand any function defined on the interval $(-L, L)$ as a Fourier series.

10.5 FOURIER SINE AND COSINE SERIES

A function $f(x)$ defined on the interval $(-L, L)$ is called even (or symmetric) if

$$f(-x) = f(x)$$

and is called odd (or antisymmetric) if

$$f(-x) = -f(x)$$

Figure 10.3 shows graphs of even and odd functions. It can be seen that the graph of the even function in the interval $(-L, 0)$ can be obtained by taking a mirror image about the vertical axis of its graph in the interval $(0, L)$. For the odd function, its graph in the interval $(-L, 0)$ is obtained by taking first a reflection about the vertical axis, and then again reflecting it about the horizontal axis.

It can be easily proved that the sums or products of even functions are themselves even, while the sums of odd functions are odd. Also, we can easily prove that the product of even and odd functions is odd, and the product of two odd functions is even.

It is obvious and can also be easily proved mathematically [1] that the integral from $-L$ to L of an even function will be two times its integral from 0 to L; for odd functions, the integral from $-L$ to L will be zero.

Now since the cosine is an even function and the sine is an odd function, the first two terms in Eq. (10.3a) are even, while the last term is odd. Hence if $f(x)$ is an even function, we expect that its Fourier series will not contain the last term; i.e., b_n will be zero. This can also be seen from Eq. (10.3c), wherein the integrand is an odd function (being a product of even and odd functions), and hence its integral over $(-L, L)$ will be zero. Also, from Eq. (10.3b), we can write

$$\begin{aligned} a_n &= \frac{1}{L} \int_{-L}^{L} f(x) \cos \frac{n\pi x}{L}\, dx \\ &= \frac{2}{L} \int_{0}^{L} f(x) \cos \frac{n\pi x}{L}\, dx \end{aligned}$$

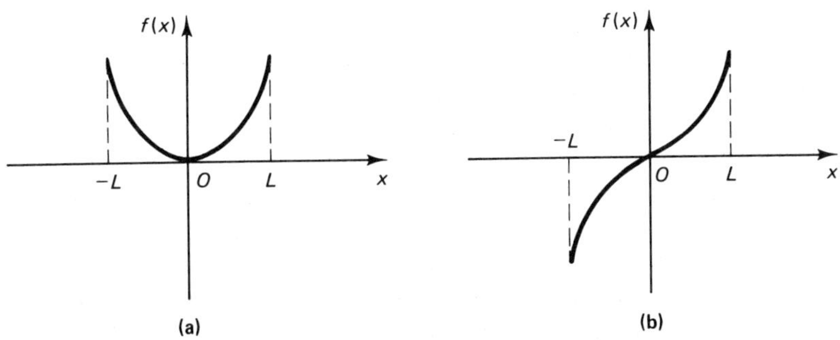

Figure 10.3 Even and odd functions. (a) Even function of x, symmetric about the vertical axis. (b) Odd function of x, antisymmetric about the vertical axis.

In a similar manner, we can check that if $f(x)$ is an odd function, its Fourier series will contain only the sine terms; i.e., $a_n = 0$. Hence

$$b_n = \frac{2}{L}\int_0^L f(x)\sin\frac{n\pi x}{L}\,dx$$

We can now summarize the results just discussed, as follows.

Fourier cosine series for even functions defined in the interval $(-L, L)$

$$f(x) = \frac{a_0}{2} + \sum_{n=1}^{\infty} a_n \cos\frac{n\pi x}{L} \tag{10.4a}$$

where

$$a_n = \frac{2}{L}\int_0^L f(x)\cos\frac{n\pi x}{L}\,dx \tag{10.4b}$$

Fourier sine series for odd functions defined in the interval $(-L, L)$

$$f(x) = \sum_{n=1}^{\infty} b_n \sin\frac{n\pi x}{L} \tag{10.5a}$$

where

$$b_n = \frac{2}{L}\int_0^L f(x)\sin\frac{n\pi x}{L}\,dx \tag{10.5b}$$

The advantage of these series is seen when we consider a function $f(x)$ which is defined only in the interval $(0, L)$. We can then generate an even function in the interval $(-L, L)$ by taking an "even extension" of $f(x)$ which is obtained from its mirror image about the vertical axis [Fig. 10.4(a)]. This new even function, which matches with the given $f(x)$ in the interval $(0, L)$, can be expanded as a Fourier cosine series given by Eqs. (10.4a,b).

Alternatively, we can generate an odd function in the interval $(-L, L)$ by taking an "odd extension" of $f(x)$ obtained by taking its mirror image about the vertical axis, and then about the horizontal axis [Fig. 10.4(b)]. This new odd function also matches the given $f(x)$ in the interval $(0, L)$ and can be expanded as a Fourier sine series given by Eqs. (10.5a,b).

Thus we can expand $f(x)$ defined in the interval $(0, L)$ in either a Fourier cosine or sine series, by suitably choosing either an even or an odd extension of the function in the interval $(-L, 0)$. Which extension we choose outside the interval $(0, L)$ is usually immaterial to us for problems defined over $(0, L)$ only, since the resulting series will always converge to the given function in $(0, L)$, irrespective of our choice of the extension.

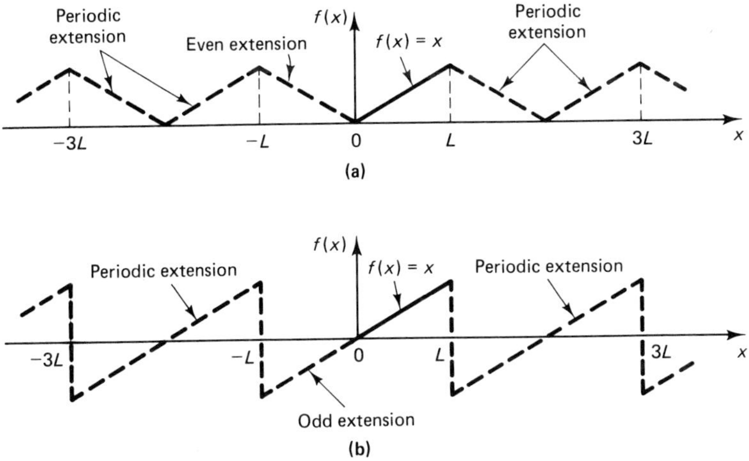

Figure 10.4 Even and odd extensions. (a) An even function formed from an even extension of $f(x) = x$. (b) An odd function formed from an odd extension of $f(x) = x$.

Example 10.2

We shall expand a function $f(x) = x$ defined in the interval $(0, L)$ into (a) a Fourier cosine series and (b) a Fourier sine series.

Case (a). Referring to Fig. 10.4(a), we can obtain an even function in $(-L, L)$ by taking an even extension as shown. Then, substituting for $f(x)$ in Eq. (10.4b), we have

$$a_n = \frac{2}{L}\int_0^L x \cos\frac{n\pi x}{L}\,dx = -\frac{2L}{n^2\pi^2}(1 - \cos n\pi)$$

or

$$a_n = -\frac{4L}{n^2\pi^2} \qquad n = 1, 3, 5, \ldots, \infty$$

and

$$a_n = 0 \qquad n = 2, 4, 6, \ldots, \infty$$

Also, for $n = 0$ we obtain by direct integration; $a_o = L$. Hence the Fourier cosine series [Eq. (10.4a)], for $f(x)$, becomes

$$f(x) = x = \frac{L}{2} - \sum_{n=1,3,5\ldots}^{\infty} \frac{4L}{n^2\pi^2}\cos\frac{n\pi x}{L} \qquad (0 \leqslant x \leqslant L) \tag{d}$$

Case (b). Referring to Fig. 10.4(b), we have an odd extension in the interval $(-L, L)$. Hence, substituting $f(x) = x$ into Eq. (10.5b), we obtain

$$b_n = \frac{2}{L}\int_0^L x \sin\frac{n\pi x}{L}\,dx$$

$$= -\frac{2L}{n\pi}\cos n\pi \qquad n = 1, 2, 3, \ldots, \infty$$

or

$$b_n = \frac{2L}{n\pi}(-1)^{n+1} \qquad n = 1, 2, 3, \ldots, \infty$$

Hence the Fourier sine series [Eq. 10.5(a)] for $f(x)$ becomes

$$f(x) = x = \sum_{n=1,2,3..}^{\infty} (-1)^{n+1}\frac{2L}{n\pi}\sin\frac{n\pi x}{L} \qquad (0 \leqslant x < L) \tag{e}$$

Thus we are able to expand $f(x) = x$, defined in the interval $(0, L)$, either as a Fourier cosine or a sine series. The actual functions that the series represent in the domain $(-\infty, \infty)$ are as shown in Fig. 10.4 for the two cases. In each case, however, they represent the given function only in $(0, L)$, which is the domain of our interest. The choice of the series will depend on the type of problem. It is interesting to note, however, that in the present example, the function $f(x)$ over the interval $(-\infty, \infty)$ in Fig. 10.4(a) is continuous, but $f'(x)$ is discontinuous. Hence its Fourier series [Eq. (d)], has a $1/n^2$ type of convergence. On the other hand, $f(x)$ over $(-\infty, \infty)$ in Fig. 10.4(b) is discontinuous at L, $3L$, etc. Hence its Fourier series [Eq. (e)], has a $1/n$ type of convergence, as expected. In this particular example, therefore, the Fourier cosine series will give a more rapid convergence than the sine series. We should also note for this example that the cosine series will converge to the correct value of $f(x)$ at $x = \pm L$, but the sine series will converge to the average value of zero at the points of discontinuity, $x = \pm L$.

Example 10.3

We shall expand into a Fourier sine series the function $f(x) = p_o$ defined in the interval $(0, L)$. This function could well represent a uniform load on a beam of span L. For expanding as a sine series, we take an odd extension, as shown in Fig. 10.5. From Eq. (10.5b) we have

$$b_n = \frac{2}{L}\int_0^L p_o \sin\frac{n\pi x}{L}\,dx$$

$$= \frac{2p_o}{n\pi}(1 - \cos n\pi) \qquad n = 1, 2, 3, \ldots, \infty$$

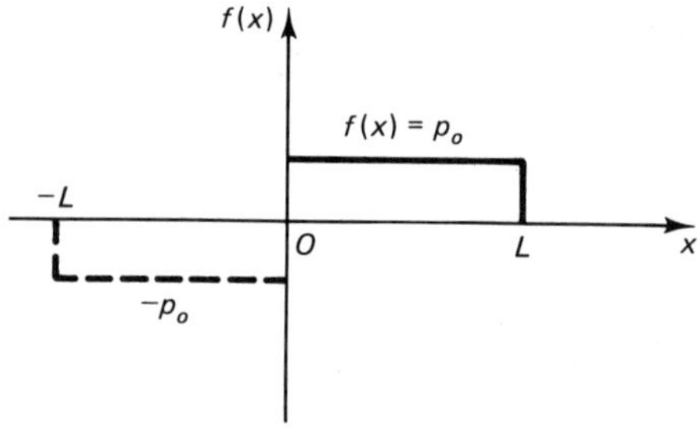

Figure 10.5 Odd extension of a function representing a uniform beam load.

or

$$b_n = \frac{4p_o}{n\pi} \qquad n = 1, 3, 5, \ldots, \infty$$

and

$$b_n = 0 \qquad n = 2, 4, 6, \ldots, \infty$$

Hence, from Eq. (10.5a) we have

$$p_o = \sum_{n=1,3,5,\ldots}^{\infty} \frac{4p_o}{n\pi} \sin \frac{n\pi x}{L} \tag{10.6}$$

or

$$p_o = \frac{4p_o}{\pi}\left(\sin \frac{\pi x}{L} + \frac{1}{3} \sin \frac{3\pi x}{L} + \frac{1}{5} \sin \frac{5\pi x}{L} + \cdots\right)$$

The convergence of this series is of the type $1/n$ since the function has discontinuities at $x = 0, L$. Figure 10.6(a) shows how the first term of the series alone will approximate $f(x)$ in $(0, L)$. Obviously, the first term does not give a good representation of the function, and we must consider more terms. The graph of the second term of the series ($n = 3$), is shown in Fig. 10.6(b). If this term is added to the first term, the two terms together will represent

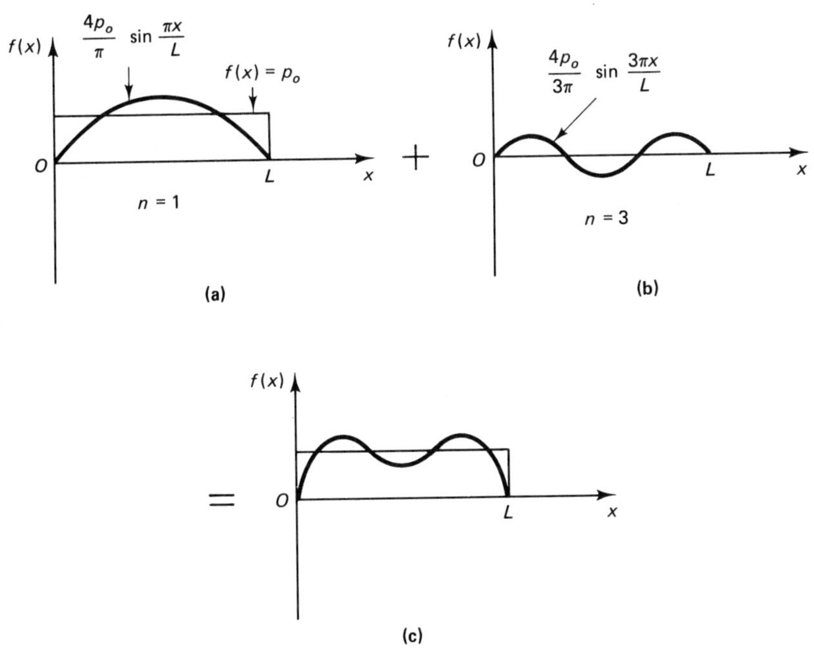

Figure 10.6 Approximation of a uniform load by a Fourier sine series. (a) First term in series. (b) Second term in series. (c) Summation of first and second terms.

the function a little better, as shown in Fig. 10.6(c). After adding a sufficient number of terms, the function will be represented by a wavy line, as shown in Fig. 10.7.

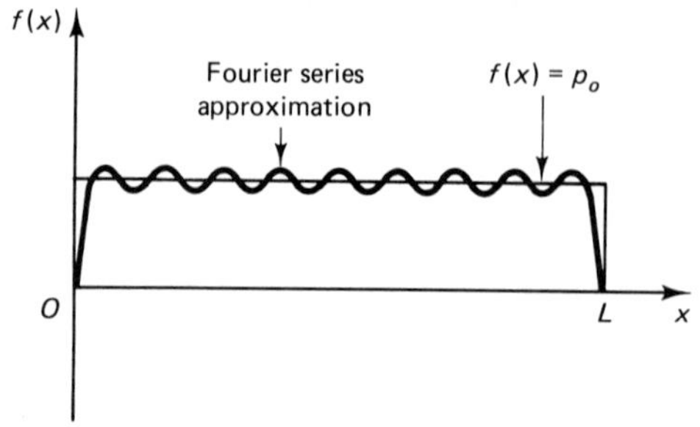

Figure 10.7 Fourier sine series representation of a uniform beam load, constructed from several terms in the series.

It may be noted that at the points of discontinuity of $f(x)$, i.e., at $x = 0$ and $x = L$, the series converges to the average value (zero) of the extended function, not to p_o. This discrepancy, however, is not significant for practical problems. For example, in the case of a beam carrying a uniform load p_o, the Fourier series graph of Fig. 10.7 will almost represent the full load on the beam, although the value of the intensity of the load at the ends is zero in the series representation.

We shall now consider a few applications of Fourier series to solving structural engineering problems.

10.6 SOME PRACTICAL APPLICATIONS

Fourier series can be used conveniently to solve problems of simple beams, buckling, beams on elastic foundations, plates, shells, and other problems in the theory of elasticity. We shall consider a few simple examples.

Example 10.4: Deflection of a Beam

Figure 10.8 shows a beam of span L, simply supported at its ends $x = 0$ and $x = L$, and subjected to a uniform loading p_o. The deflection w of the beam is governed by the equation (refer to Chapter 1)

$$Kw'''' = p_o$$

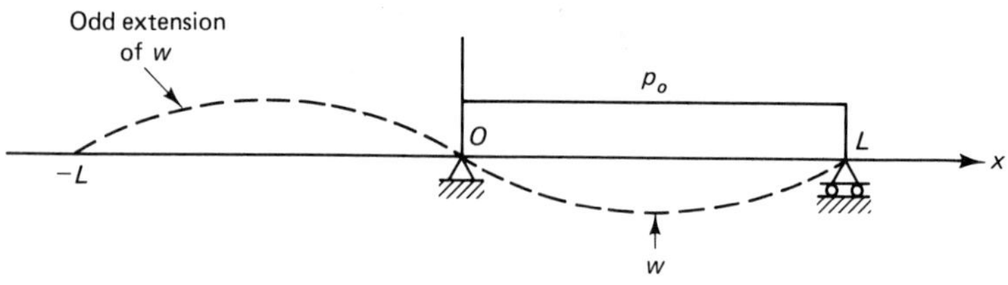

Figure 10.8 Simply supported beam of length L, subjected to a uniform load p_o.

where $K = Et^3/12$ for a beam of thickness t with unit width, and $(\cdot)' = d(\cdot)/dx$. We can expand p_o in a Fourier sine series according to Eq. (10.6). We shall also assume the solution of w in series form as*

$$w = \sum_{n=1,2,\ldots}^{\infty} A_n \sin \frac{n\pi x}{L} \tag{f}$$

such that it automatically satisfies the boundary conditions $w = w'' = 0$ at $x = 0$ and $x = L$. Substituting for w from above and for p_o from Eq. (10.6) into the beam differential equation, we obtain

$$K\left(\sum_{n=1,2,\ldots}^{\infty} \frac{n^4\pi^4}{L^4} A_n \sin \frac{n\pi x}{L} \right) = \sum_{n=1,3,\ldots}^{\infty} \frac{4p_o}{n\pi} \sin \frac{n\pi x}{L}$$

Equating term by term, we obtain

$$A_n = 0 \qquad \text{for } n = 2, 4, 6, \ldots$$

$$A_n = \frac{4p_o L^4}{n^5\pi^5 K} \qquad \text{for } n = 1, 3, 5, \ldots, \infty$$

and hence, we have

$$w = \sum_{n=1,3,\ldots}^{\infty} \frac{4p_o L^4}{n^5\pi^5 K} \sin \frac{n\pi x}{L} \tag{10.7}$$

This is the series solution of the problem. In expanded form, we have

$$w = \frac{4p_o L^4}{K\pi^5}\left(\sin \frac{\pi x}{L} + \frac{1}{3^5} \sin \frac{3\pi x}{L} + \frac{1}{5^5} \sin \frac{5\pi x}{L} + \cdots \right)$$

Due to the presence of the term n^5 in the denominator of Eq. (10.7), we expect a very rapid convergence of this series. Indeed, if we substitute $x = L/2$, the deflection at midspan, taking only the first term of the series, is

$$w\left(x = \frac{L}{2}\right) = \frac{4p_o L^4}{K\pi^5} = \frac{5.02}{384} \frac{p_o L^4}{K}$$

This is very close to the exact solution of

$$w = \frac{5}{384} \frac{p_o L^4}{K}$$

With an error of less than 1%, this is sufficiently accurate for all practical purposes. Now, if we take the first two terms of the series ($n = 1$ and $n = 3$), we get

$$w\left(x = \frac{L}{2}\right) = \frac{p_o L^4}{K}\left[\frac{4}{\pi^5} - \frac{4}{\pi^5}\left(\frac{1}{3^5}\right)\right] = \frac{4.999}{384} \frac{p_o L^4}{K}$$

*The differential equation has constant coefficients, and hence it can have e^{ix} type of solutions; i.e., cosine and sine solutions of the equation are possible.

which is nearly the same as the exact solution. We can see that even if we take only one term of the series, which represents the loading p_o rather crudely [Fig. 10.6(a)],* the deflection obtained is very accurate, due to the $1/n^5$ type of convergence of the series for w. This fact is very useful for solving cylindrical shell problems, where we can obtain good results even by taking only the first term of the Fourier series of the loading, since taking additional terms becomes very laborious when solving shell problems by hand calculations.

The $1/n^5$ type of convergence obtained for w can be explained as follows. Figure 10.8 shows the deflected shape of w in $(0, L)$ and its odd extension in the interval $(-L, 0)$. This function is continuous throughout. Now w', w'', and w''' represent, respectively, the slope, bending moment, and shear (after multiplication by some constants) in the beam (refer to Chapter 1). Since w is assumed as a sine series, w' and w''' will be cosine series and w'' will be a sine series. Knowing the slope, shear force diagram, and bending moment diagram of this simply supported beam with a uniformly distributed load, we can easily visualize the plots of the curves w' and w''', with their even extensions, and of the curve w'', with its odd extension. It can then be readily observed that all functions w', w'', and w''' are continuous over $(-L, L)$ and beyond. Similarly, w'''', like w and w'', will be a sine series. From the differential equation for the beam, we know that w'''' corresponds to the loading. Now the odd extension of the loading p_o over $(-L, L)$ in the present case is discontinuous at $x = 0$, $x = L$, etc. Hence we conclude that over the interval $(-L, L)$, functions w, w', w'', and w''' are continuous, whereas w'''' has discontinuities. Hence the Fourier series of w has a $1/n^5$ type of convergence.

Although, in this example we obtained excellent convergence for the deflection w with just the first term in the series, the situation is not as good for convergence of the series for the shear Q_x. For the bending moment M_x, however, we get good convergence. This can be explained as follows.

Using the expressions $Q_x = -Kw'''$ and $M_x = -Kw''$ with Eq. (10.7), we can easily obtain the following series for Q_x and M_x:

$$Q_x = \sum_{n=1,3,5,\ldots}^{\infty} \frac{4p_oL}{n^2\pi^2} \cos \frac{n\pi x}{L}$$

and

$$M_x = \sum_{n=1,3,5,\ldots}^{\infty} \frac{4p_oL^2}{n^3\pi^3} \sin \frac{n\pi x}{L}$$

Using only the first term of the series, at $x = 0$, we obtain

*The total load on the beam is $W = \int_0^L p_o \, dx$. Substituting the series of Eq. (10.6) for p_o, we obtain

$$W = \sum_{n=1,3,\ldots}^{\infty} \frac{8}{n^2\pi^2} (p_oL)$$

Thus the total load on the beam, using only the first term of this series, is $(8/\pi^2)(p_oL)$, i.e., 81% of the actual load (p_oL).

$$Q_x = \frac{8}{\pi^2}\left(\frac{p_oL}{2}\right) = 0.81\left(\frac{p_oL}{2}\right)$$

This value is only 81% of the exact value for Q_x at $x = 0$. The reduction is the same as that also obtained for the total load on the beam, as would be expected in satisfying static equilibrium. Similarly, the midspan bending moment, using only the first term of the above series, is

$$M_x = \frac{32}{\pi^3}\left(\frac{p_oL^2}{8}\right) = 1.03\left(\frac{p_oL^2}{8}\right)$$

Because of the $1/n^3$ type of convergence of the series for M_x, the moment above is only about 3% higher than the exact value. Thus if we use only the first harmonic of the Fourier series for solving a problem, the shear force obtained could be increased by multiplying by $\pi^2/8 = 1.234$, in order to arrive at a more correct value. Similarly, the moment (or longitudinal stress) could be multiplied by 1/1.03 to obtain a more accurate value. We shall use this approach when solving barrel shell problems using the first harmonic only.

We were able to obtain a Fourier series solution of the last problem only because we could assume w in a form [Eq. (f)] which automatically satisfied the boundary conditions; $w = w'' = 0$ at $x = 0$ and L, for the simply supported beam. If the beam were fixed at one or both of its ends, we could not have solved the problem using Fourier sine or cosine series, since we cannot satisfy the boundary conditions $w = w' = 0$, using sine or cosine functions only. In such cases we have to use series comprised of more complex orthogonal functions than the simple sine and cosine functions of Fourier series.

Example 10.5: Column Buckling

For a column hinged at its ends $x = 0$ and $x = L$, and subjected to an axial load P, the governing equation is [2] (see also Chapter 1, Problem 1.5)

$$Kw'''' + Pw'' = 0 \qquad \text{where} \qquad K = \frac{Et^3}{12}$$

We can again assume the deflection w in the series form of Eq. (f), so that is satisfies the boundary conditions $w = w'' = 0$ automatically. Substituting Eq. (f) in the differential equation, we have

$$\sum_{n=1,2,\ldots}^{\infty} \left(K\frac{n^4\pi^4}{L^4} - P\frac{n^2\pi^2}{L^2}\right)A_n \sin\frac{n\pi x}{L} = 0$$

For a nontrivial solution, we must have

$$P = \frac{n^2\pi^2}{L^2}K$$

Taking $n = 1$, we find $P = \pi^2K/L^2$, which is the well-known Euler buckling load. The corresponding buckled shape is given by $w = A_1[\sin(\pi x/L)]$, where A_1 is indeterminate.

By taking larger values of n, we can obtain higher buckling loads, corresponding to higher mode shapes.

Example 10.6: Plate Bending

The differential equation governing the deflection of a rectangular plate is [3]

$$\frac{\partial^4 w}{\partial x^4} + 2\frac{\partial^4 w}{\partial x^2\,\partial y^2} + \frac{\partial^4 w}{\partial y^4} = \frac{p(x,y)}{K}$$

where p is the distributed loading and K is the flexural rigidity of the plate. If the plate is simply supported on edges $x = 0$ and $x = L$ (see Fig. 10.9), we can separate the variables in the partial differential equation by assuming the solution in the form

$$w(x, y) = \sum_{n=1,2,\ldots}^{\infty} Y_n(y) \sin \frac{n\pi x}{L} \tag{g}$$

This solution automatically satisfies the boundary conditions $w = w'' = 0$ at $x = 0$ and L. We can also expand the loading $p(x, y)$ into a Fourier sine series in x, such that

$$p(x, y) = \sum_{n=1,2,\ldots}^{\infty} p_n(y) \sin \frac{n\pi x}{L}$$

Substituting for w and p into the partial differential equation, we obtain

$$Y_n^{\cdot\cdot\cdot\cdot} - \frac{2n^2\pi^2}{L^2} Y_n^{\cdot\cdot} + \frac{n^4\pi^4}{L^4} Y_n = \frac{p_n(y)}{K}$$

where $(\cdot)^{\cdot} = d(\cdot)/dy$. This is an ordinary differential equation for the nth harmonic $Y_n(y)$. The solution for each harmonic Y_n can now be obtained by solving this equation for a known $p_n(y)$, and by determining the constants in the solution which satisfy the boundary conditions at $y = 0$ and b. The complete solution is then given by Eq. (g).

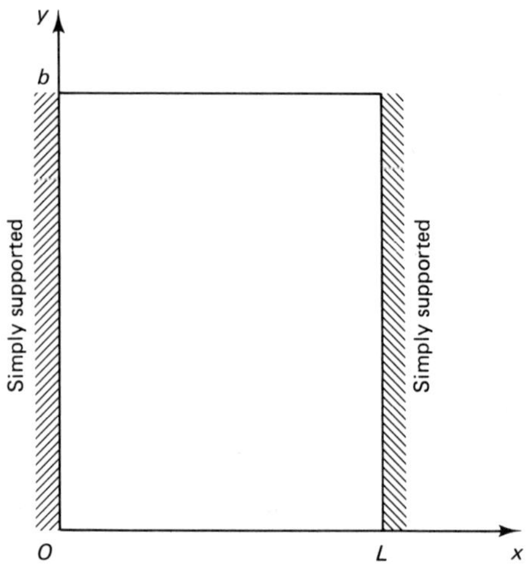

Figure 10.9 Rectangular plate, simply supported at edges $x = 0$ and $x = L$.

Solutions of this type, for rectangular plates, are called Levy solutions, and they allow us to solve problems with simply supported boundary conditions on $x = 0$ and L and any arbitrary boundary conditions on $y = 0$ and b, for any given loading $p(x, y)$. [We shall be using similar procedures for solving partial differential equations for shells of double curvature, such as elliptic paraboloids (Chapter 11) and cylindrical barrel shells (Chapter 13).]

If the plate is simply supported on all four edges $x = 0$, $x = L$, and $y = 0$, $y = b$, we can assume a Fourier sine series also for each $Y_n(y)$ in the y-direction. Such a series will automatically satisfy boundary conditions at $y = 0$ and b, and will also satisfy the differential equation which has constant coefficients in both x and y. Deflection w will then have the following form of a double Fourier series:

$$w(x, y) = \sum_{n=1,2...}^{\infty} \sum_{m=1,2...}^{\infty} A_{mn} \sin \frac{n\pi x}{L} \sin \frac{m\pi y}{b}$$

The constants A_{mn} can now be obtained directly by substituting this expression into the partial differential equation wherein $p(x, y)$ is also expanded into a double Fourier sine series. Such a solution is called the Navier solution for simply supported rectangular plates [3].

10.7 OTHER APPLICATIONS

In earlier chapters we studied the bending of circular plates under axisymmetric loading and also membrane stresses in shells of revolution under axisymmetric loads. In both cases, the governing equations were ordinary differential equations without the independent variable θ. When the loads are not axisymmetric, the governing equations will be partial differential equations containing derivatives with respect to r and θ for circular plates, and ϕ and θ for shells of revolution. It so happens, however, that these equations have constant coefficients in θ, and hence they admit solutions of the type $e^{in\theta}$, i.e., $\cos n\theta$ or $\sin n\theta$. Also, due to circularity of the structure, all deflections, stresses, etc. will satisfy periodicity relations of the type

$$w(r, \theta) = w(r, \theta + 2\pi) = w(r, \theta + 4\pi) = \cdots$$

or

$$N_\phi(\phi, \theta) = N_\phi(\phi, \theta + 2\pi) = N_\phi(\phi, \theta + 4\pi) = \cdots$$

i.e., all functions are periodic in θ with a period 2π.* Hence we can separate the variable θ in the governing partial differential equations, by expanding the given loads, the deflections, and the stresses as Fourier series in the θ-direction. This will

*This is the reason why the mathematical equations have constant coefficients in θ, such that they admit solutions of the exponential type $e^{in\theta}$, i.e., $\cos n\theta$ or $\sin n\theta$ in θ.

lead to an ordinary differential equation in the other variable for each nth harmonic, as in the Levy solutions for rectangular plates. These equations can readily be solved and adapted to the given boundary conditions at the edges $r = constant$ or $\phi = constant$. This is the general procedure for solving nonaxisymmetric load problems for shells of revolution or circular plates, and is similar to the Levy solutions for rectangular plates. Similar procedures are adopted for solving rectangular or circular plate problems in plane stress or plane strain solutions in the theory of elasticity. We shall be using this solution scheme for solving problems of shells of revolution under nonaxisymmetric loadings in the next chapter.

REFERENCES

[1] Berg, P. W., and McGregor, J. L., *Elementary Partial Differential Equations,* Holden-Day, Inc., San Francisco, 1964.

[2] Timoshenko, S. P., *Strength of Materials,* 3rd ed., Part I, D. Van Nostrand Company, Inc., Princeton, N.J., 1955.

[3] Timoshenko, S. P., and Woinowsky-Krieger, S., *Theory of Plates and Shells,* 2nd ed., McGraw-Hill Book Company, New York, 1959.

PROBLEMS

10.1. Expand the triangular loading shown in Figure P10.1 into Fourier cosine and sine series. Discuss the type of convergence obtained in each case and the reasons for such convergence.

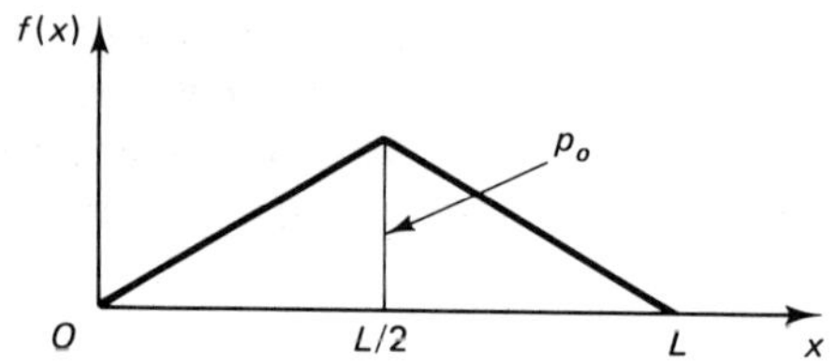

Figure P10.1 Triangular loading over $(0, L)$ to be expanded as Fourier cosine and sine series.

10.2. A beam of span L is hinged at both ends $x = \pm(L/2)$ (Fig. P10.2). Obtain the deflection of the beam as a Fourier cosine series, under a uniform load p_o. Compare the calculated midspan deflection with its exact value. [*Hint:* Assume a suitable

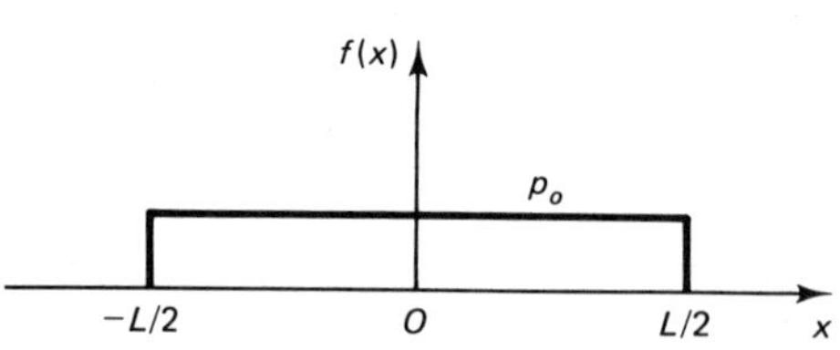

Figure P10.2 Uniform load p_o, acting on a beam of length L with hinges at $x = \pm L/2$.

extension of the loading function in the intervals $(L/2, L)$ and $(-L, -L/2)$, and expand it into a Fourier series, given by

$$p_o = \sum_{n=1,3,5,\ldots}^{\infty} (-1)^{(1/2)(n-1)} \frac{4p_o}{n\pi} \cos \frac{n\pi x}{L}$$

10.3. A prestress force loading on a beam of length L is represented by opposite uniform axial loads F/c over small lengths c at the beam's ends (Fig. P10.3). F is the total prestress force at each end. Expand this loading in a Fourier cosine series.

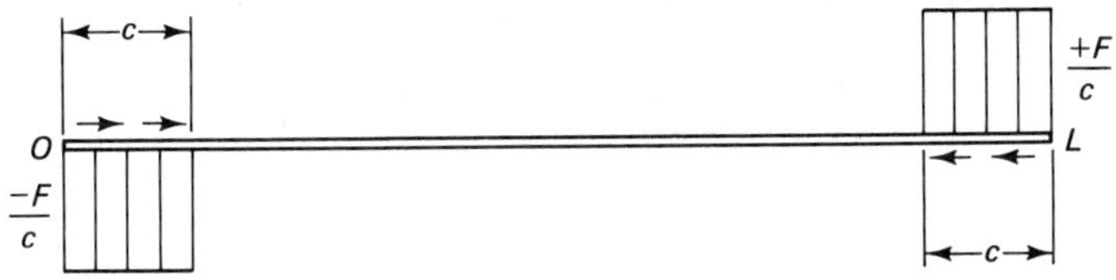

Figure P10.3 Representation of compressive loading at the ends of a prestressed beam.

10.4. A rectangular plate with sides of length a and b in the x- and y-directions, respectively, is simply supported on its edges $x = 0$, $x = a$ and $y = 0$, $y = b$. It is loaded by a distributed load whose intensity is constant along the y-direction and varies linearly from a value p_o at $x = a/2$ to zero at $x = 0$ and $x = a$. Obtain an expression for the deflection w of the plate using a double Fourier series (Navier solution).

10.5. If the plate of Problem 10.4 above is fixed at $y = 0$ and $y = b$, obtain a solution using the Levy method.

11

Membrane Analysis of Shells of Double Curvature Using Fourier Series

11.1 INTRODUCTION

In Chapter 8 we derived the basic differential equation governing the membrane stress resultants in a general shell of double curvature. Because of the special geometry of hypar shells with straight-line boundaries, the differential equation becomes simplified for such shells. Hence we were able to solve several problems of hypar shells with straight boundaries in Chapter 9 by direct integration of the simplified equation. Although for particular paraboloids of revolution a direct simple solution of the equation is possible [1], for other shells of double curvature (e.g., the elliptic paraboloids), the equation cannot be solved directly. In such cases we have to resort to the use of Fourier series for solving problems. Having reviewed the basic concepts of Fourier series in Chapter 10, we can now derive expressions for the membrane stress resultants in elliptic paraboloid shells, as well as for hypar shells with generator parabolas as boundaries.

In Chapter 6 we studied the membrane analysis of shells of revolution under axisymmetric loads. Under general nonaxisymmetric loadings, such as wind loading, the shear stress resultant $N_{\phi\theta}$ will be nonzero, in addition to resultants N_ϕ and N_θ. The resulting equilibrium equations give rise to a partial differential equation which we are also now in a position to solve using Fourier expansions, in the θ-direction, in order to separate the variables.

11.2 MEMBRANE ANALYSIS OF ELLIPTIC PARABOLOIDS

Figure 11.1 shows an elliptic paraboloid whose plan projection is a rectangular area with sides of length a and b. We can write the equation of this surface as

$$z = \frac{x^2}{2h_1} + \frac{y^2}{2h_2} \tag{11.1}$$

where h_1 and h_2 are constants. Sections $x = constant$ or $y = constant$ are parabolas, whereas sections $z = constant$ are ellipses. Taking $y = 0$ and $x = a/2$, we obtain

$$z = h_x = \frac{a^2}{8h_1} \tag{11.2a}$$

Similarly, for $x = 0$ and $y = b/2$, we have

$$z = h_y = \frac{b^2}{8h_2} \tag{11.2b}$$

The dimension h_x represents the "drop" of the parabola from O to C as we go along the x-direction, while h_y represents the drop of the parabola as we go along the y-direction from O to A or from C to B. Thus the total rise of the shell from B to O will be $(h_x + h_y)$.

The general differential equation for the stress function Φ was derived in Chapter 8 [Eq. (8.10)]. We shall assume that $\bar{p}_x = \bar{p}_y = 0$ and that $\bar{p}_z$ is a constant load p, uniformly distributed over the plan area. In practice, the dead load of the shell will not be uniformly distributed over the plan area, but we assume it as such for shells of low rise, since this assumption will lead to a simple yet accurate solution.

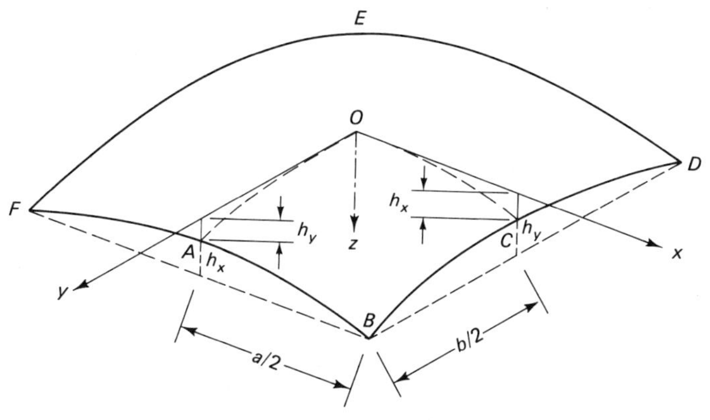

Figure 11.1 Elliptic paraboloid shell with a rectangular plan projection.

With the assumptions above, we now rewrite below the relevant equations from Chapter 8 for convenient reference:

$$\overline{N}_x \frac{\partial^2 z}{\partial x^2} + 2\overline{N}_{xy} \frac{\partial^2 z}{\partial x\, \partial y} + \overline{N}_y \frac{\partial^2 z}{\partial y^2} = -\overline{p}_z \tag{8.7c}$$

$$\left(\frac{\partial^2 z}{\partial y^2}\right)\left(\frac{\partial^2 \Phi}{\partial x^2}\right) - 2\left(\frac{\partial^2 z}{\partial x\, \partial y}\right)\left(\frac{\partial^2 \Phi}{\partial x\, \partial y}\right) + \left(\frac{\partial^2 z}{\partial x^2}\right)\left(\frac{\partial^2 \Phi}{\partial y^2}\right) = -\overline{p}_z \tag{8.10}$$

where $\overline{p}_x = \overline{p}_y = 0$ in Eq. (8.10). We also had

$$\overline{N}_{xy} = -\frac{\partial^2 \Phi}{\partial x\, \partial y} \tag{8.9a}$$

$$\overline{N}_x = \frac{\partial^2 \Phi}{\partial y^2} \tag{8.9b}$$

$$\overline{N}_y = \frac{\partial^2 \Phi}{\partial x^2} \tag{8.9c}$$

where $\overline{p}_x = \overline{p}_y = 0$ in Eqs. (8.9a–c).

From Eq. (11.1) we note that $\partial^2 z/\partial x \partial y$ equals zero, and hence Eq. (8.7c) above becomes

$$\overline{N}_x \frac{\partial^2 z}{\partial x^2} + \overline{N}_y \frac{\partial^2 z}{\partial y^2} = -\overline{p}_z$$

This equation indicates that the load $\overline{p}_z$ at any point is supported by the forces in the arches in the x- and y-directions, while the shears N_{xy} in the shell do not contribute to supporting the load because the twist of the surface $\partial^2 z/\partial x \partial y$ is zero. The exact opposite was the case for hypar shells with straight-line boundaries, as considered in Chapter 9, where the shell surfaces had zero curvatures $\partial^2 z/\partial x^2$ and $\partial^2 z/\partial y^2$ but a nonzero twist. For that case, the load was resisted by the shear forces alone.

Now, substituting $\overline{p}_z = p$ and the appropriate derivatives of z from Eq. (11.1) into Eq. (8.10) above, we obtain the following differential equation:

$$\frac{1}{h_2} \frac{\partial^2 \Phi}{\partial x^2} + \frac{1}{h_1} \frac{\partial^2 \Phi}{\partial y^2} = -p \tag{11.3}$$

By making a simple coordinate transformation $\overline{x} = \sqrt{h_2}\, x$ and $\overline{y} = \sqrt{h_1}\, y$, we can put this equation into the standard form of Laplace's equation. However, we can also solve the equation above directly.

A simple solution of this equation can be taken as $\Phi = -(ph_1/2)y^2$. This solution gives $\overline{N}_y$ and $\overline{N}_{xy}$ as zero, and only $\overline{N}_x$ will be nonzero, as seen from Eqs. (8.9a–c) reproduced above. Thus the vertical load is carried by a series of parabolic arches in the x-direction. At their edges, these arches must be supported by inclined supports which resist the vertical and horizontal components of N_x from the arches. A similar solution can also be considered in the y-direction.

For a general solution of Eq. (11.3) under vertical loading, we assume that at the boundaries $x = \pm(a/2)$ and $y = \pm(b/2)$, the shell is supported by thin diaphragms that can resist loads in their own planes but cannot resist loads that are out of their planes. This means that at edges $x = \pm(a/2)$ and $y = \pm(b/2)$, respectively, the stress resultants N_x and N_y will be zero, and the diaphragms will carry only the shears from the shell. In practice, it is likely that the shell will be supported only at its four corners. Along the four edges, then, we will have to provide edge beams that will span as parabolic arches between the corner supports and transfer shears from the shell to the supports. Using the relations of Eqs. (8.9b,c) with $\overline{p}_x = \overline{p}_y = 0$, we can write the boundary conditions just discussed as: at $x = \pm(a/2)$,

$$\frac{\partial^2 \Phi}{\partial y^2} = 0 \tag{11.4a}$$

and at $y = \pm(b/2)$,

$$\frac{\partial^2 \Phi}{\partial x^2} = 0 \tag{11.4b}$$

Equation (11.4a) implies that at $x = \pm(a/2)$, Φ must be a linear function of y, say $\Phi = c_1 + c_2 y$, where c_1 and c_2 are constants. We can choose any values for c_1 and c_2, and still the boundary condition, Eq. (11.4a), will be satisfied. Hence we choose $c_1 = c_2 = 0$, i.e., $\Phi = 0$, along the edge $x = +(a/2)$. Similarly, from Eq. (11.4b) we can obtain $\Phi = 0$ along the edge $y = +(b/2)$, and by symmetry, Φ will then be zero on all four edges of the shell. We can therefore write the boundary conditions, Eqs. (11.4a,b), as

$$\Phi = 0 \quad \text{at} \quad x = \pm(a/2) \quad \text{and} \quad y = \pm(b/2) \tag{11.5}$$

It may be noted that even if Φ was not zero at the boundaries, we could always add to Φ a function $\Phi_o = c_1 + c_2 x + c_3 y$, which satisfies the differential equation and the boundary conditions of Eqs. (11.4a,b). The three constants c_1, c_2, and c_3 could be so chosen that values of $(\Phi + \Phi_o)$ at three corners, say B, D, and F (Fig. 11.1), are zero. This would then give $(\Phi + \Phi_o)$ as zero along edges BD and BF [because $(\Phi + \Phi_o)$ varies linearly along the boundaries, as seen from Eqs. (11.4a,b)], and by symmetry, it would also be zero on the remaining two edges. The addition of the function Φ_o will not affect the stress resultants in the shell, since they are related to second derivatives of the stress function, which are all zero for Φ_o.

We can now solve the differential equation (11.3), with the boundary conditions of Eq. (11.5), by a Levy-type solution discussed briefly for rectangular plates in Chapter 10. These types of solutions are commonly used to solve boundary value problems in engineering mechanics. We shall expand the loading and stress function Φ as Fourier series in y, so that the boundary conditions are automatically satisfied on edges $y = \pm(b/2)$. The resulting ordinary differential equation in x can be solved to satisfy boundary conditions on edges $x = \pm(a/2)$. By assuming an extension of the constant load p, with negative values in $-b \leqslant y \leqslant -(b/2)$ and $(b/2) \leqslant y \leqslant b$, we can expand this load as a Fourier cosine series in y, which is given by (refer to Problem 10.2)

$$p = \sum_{n=1,3,5,\ldots}^{\infty} (-1)^{(n-1)/2} \frac{4p}{n\pi} \cos \frac{n\pi y}{b} \tag{11.6}$$

We also assume the solution of Eq. (11.3) as

$$\Phi = \sum_{n=1}^{\infty} \phi_n(x) \cos \frac{n\pi y}{b} \tag{11.7}$$

This solution automatically satisfies the boundary condition $\Phi = 0$ at $y = \pm(b/2)$. Substituting Eqs. (11.6) and (11.7) into Eq. (11.3), we get the following differential equation for the nth harmonic, $\phi_n(x)$,

$$\frac{1}{h_2}\phi_n'' - \frac{1}{h_1}\frac{n^2\pi^2}{b^2}\phi_n = -(-1)^{(n-1)/2}\frac{4p}{n\pi} \qquad n = 1, 3, 5, \ldots$$

where $(\cdot)'$ denotes $d(\cdot)/dx$. We can rewrite this equation as

$$\phi_n'' - \lambda_n^2\phi_n = -(-1)^{(n-1)/2}\frac{4p}{n\pi}h_2 \qquad n = 1, 3, 5, \ldots \tag{11.8a}$$

where

$$\lambda_n^2 = \frac{h_2}{h_1}\frac{n^2\pi^2}{b^2} \tag{11.8b}$$

To solve this ordinary differential equation, we first obtain a particular solution, ϕ_{np}, as

$$\phi_{np} = (-1)^{(n-1)/2}\frac{4p}{n\pi}\frac{h_2}{\lambda_n^2} = (-1)^{(n-1)/2}\frac{4pb^2}{n^3\pi^3}h_1$$

We can also write down the solution of the homogeneous equation (11.8a), by taking its right-hand side as zero. Since λ_n^2 is always positive, this solution will be of the form

$$\phi_{no} = C_n \cosh \lambda_n x + D_n \sinh \lambda_n x$$

Since our solution has to be symmetric in x, we can drop the second term in the above, and add the particular solution, in order to find the complete solution for ϕ_n, as

$$\phi_n(x) = C_n \cosh \lambda_n x + (-1)^{(n-1)/2}\frac{4pb^2}{n^3\pi^3}h_1$$

Substituting the boundary condition $\Phi = 0$, i.e., $\phi_n = 0$ at $x = +(a/2)$, we obtain

$$C_n = -(-1)^{(n-1)/2} \frac{4pb^2h_1}{n^3\pi^3 \cosh(\lambda_n a/2)}$$

Substituting this term into the expression for $\phi_n(x)$ above, and using Eq. (11.7), we obtain the complete solution for Φ as

$$\Phi = \sum_{n=1,3,5}^{\infty} (-1)^{(n-1)/2} \frac{4pb^2h_1}{n^3\pi^3}\left[1 - \frac{\cosh \lambda_n x}{\cosh(\lambda_n a/2)}\right] \cos\frac{n\pi y}{b} \tag{11.9}$$

Now, substituting the above into Eqs. (8.9a–c), we can write the following expressions for the stress resultants $\overline{N}_x$, $\overline{N}_y$, and $\overline{N}_{xy}$:

$$\overline{N}_x = \frac{4ph_1}{\pi} \sum_{n=1,3,\ldots}^{\infty} (-1)^{(n+1)/2} \frac{1}{n}\left[1 - \frac{\cosh \lambda_n x}{\cosh(\lambda_n a/2)}\right] \cos\frac{n\pi y}{b} \tag{11.10a}$$

$$\overline{N}_y = \frac{4ph_2}{\pi} \sum_{n=1,3,\ldots}^{\infty} (-1)^{(n+1)/2} \frac{1}{n} \frac{\cosh \lambda_n x}{\cosh(\lambda_n a/2)} \cos\frac{n\pi y}{b} \tag{11.10b}$$

and

$$\overline{N}_{xy} = \frac{4p}{\pi}\sqrt{h_1h_2} \sum_{n=1,3,\ldots}^{\infty} (-1)^{(n+1)/2} \frac{1}{n} \frac{\sinh \lambda_n x}{\cosh(\lambda_n a/2)} \sin\frac{n\pi y}{b} \tag{11.10c}$$

It may be noted that in deriving the equations above, we have substituted for λ_n from Eq. (11.8b) and have also used the relation

$$-(-1)^{(n-1)/2} = (-1)^{(n+1)/2} \qquad n = 1, 3, 5, \ldots$$

Expressions for N_x, N_y, and N_{xy} can now be obtained from the above by using Eqs. (8.1a–c). We have here

$$\tan\phi = \frac{\partial z}{\partial x} = \frac{x}{h_1}$$

Hence

$$\cos\phi = \frac{1}{\sqrt{1 + \tan^2\phi}} = \frac{h_1}{\sqrt{h_1^2 + x^2}}$$

Similarly,

$$\cos\theta = \frac{h_2}{\sqrt{h_2^2 + y^2}}$$

and hence

$$N_x = \frac{\cos \theta}{\cos \phi} \overline{N}_x = \frac{h_2}{h_1} \sqrt{\frac{h_1^2 + x^2}{h_2^2 + y^2}} \overline{N}_x \tag{11.11a}$$

$$N_y = \frac{\cos \phi}{\cos \theta} \overline{N}_y = \frac{h_1}{h_2} \sqrt{\frac{h_2^2 + y^2}{h_1^2 + x^2}} \overline{N}_y \tag{11.11b}$$

and

$$N_{xy} = \overline{N}_{xy} \tag{11.11c}$$

Knowing the geometry of the shell surface from Eq. (11.1), and the loading p, we can now estimate the stress resultants by using Eqs. (11.10a–c) and (11.11a–c). A simple computer program can be written to generate values of N_x, N_y, and N_{xy} for known parameters and different values of x/a and y/b. Parme [2] has given ready tables for calculating these membrane stresses. Let us denote

$$\alpha = \frac{h_2}{h_1} \sqrt{\frac{h_1^2 + x^2}{h_2^2 + y^2}}$$

and substitute for h_1 and h_2, in terms of h_x and h_y from Eqs. (11.2a,b), in order to obtain

$$\alpha = \sqrt{\frac{1 + 4(2x/a)^2(2h_x/a)^2}{1 + 4(2y/b)^2(2h_y/b)^2}}$$

Then, also substituting for h_1 and h_2 in terms of h_x and h_y in Eqs. (11.10a–c), we can write the stress resultants in the following form [using Eqs. (11.11a–c)]:

$$N_x = -\frac{pa^2}{4h_x} \alpha R_1 \tag{11.12a}$$

$$N_y = -\frac{pb^2}{4h_y} \frac{1}{\alpha} R_2 \tag{11.12b}$$

and

$$N_{xy} = -\frac{pab}{4\sqrt{h_x h_y}} R_3 \tag{11.12c}$$

where R_1, R_2, and R_3 are coefficients which depend only on h_x/h_y, x/a, and y/b. Parme [2] has given tables containing values of R_1, R_2, and R_3 for h_x/h_y varying from 0.2 to 1.0 and for x/a and y/b values at every eighth-interval point on the shell. These values are reproduced here in Table 11.1. Using these tables, one can readily obtain the variation of membrane stress resultants in an elliptic paraboloid shell. Except near the corners, the stresses in the shell are compressive, and are usually so small under normal loads that the factor α in Eqs. (11.12a,b) may be taken as equal to 1 for simplicity. One could also easily write a simple computer program to obtain the membrane stress resultants at every eighth-interval point for any given shell, using Eqs. (11.10a–c) and (11.11a–c).

From Table 11.1 it is noticed that the shear stress goes to infinity at the corners $x = \pm(a/2)$ and $y = \pm(b/2)$. We can see this from Eq. (11.10c) also. For large values of n (i.e., large values of λ_n), sinh $\lambda_n x$ and cosh $(\lambda_n a/2)$ will be equal to $e^{(\lambda_n x)}$ and $e^{(\lambda_n a/2)}$, respectively. Hence their ratio will be equal to $e^{\lambda_n[x-(a/2)]}$, which will be unity at $x = a/2$ [the same result can also be proved for $x = -(a/2)$]. Also, the product of the terms $(-1)^{(n+1)/2}$ and sin $(n\pi y/b)$ at $y = b/2$ gives the same sign for successive values of n in the series. The series thus corresponds to a summation of $1/n$ for $n = 1, 3, 5, \ldots$, which is divergent. Hence N_{xy} does not converge at the corners.

The reason for infinite shear stresses at the corners can also be explained physically. We had seen for the hypar shell with straight-line boundaries that the shell element has a "twist"; i.e., opposite sides of the element do not have the same slope (refer to Chapter 9). Hence vertical components of the shear stresses on opposite faces of the element do not cancel each other, and thus can support a vertical load. For an elliptic paraboloid, however, opposite sides of its element have the same slope, and hence the shear stresses do not support a vertical load. At any internal point on the shell, the vertical load is supported by the compressive forces N_x and N_y in the two parabolic arches at right angles to each other. At a diaphragm support, say $x = a/2$, N_x is zero and the vertical load is still carried by the arch in the y-direction. At the corners, however, both N_x and N_y are zero, and hence the vertical load has to be resisted by N_{xy}, which becomes infinite due to the lack of any twist of the shell element. In reality, however, the vertical load at the corners will be resisted by the development of transverse shear forces (which will also give rise to bending moments), and thus the membrane stress N_{xy} will not be infinite.

For design of such elliptic paraboloid shells, the membrane stresses at interior points are small, and hence only nominal reinforcement for temperature and shrinkage will usually be necessary in the x- and y-directions. As an example, for a shell with $a = b = 80$ ft and $h_x = h_y = 10$ ft (i.e., a total rise of 20 ft), we see from Table 11.1 that the maximum coefficient for N_x or N_y is 0.5. Hence taking $\alpha \simeq 1$, and for a loading of 65 lb/ft^2, we have $N_x^{\max} = N_y^{\max} = -0.5(65)(80)^2/(4)(10) = -5200$ lb/ft. This will give a maximum compressive stress of only 144.4 psi in a 3-in. thick shell. In the zones near the corners, however, due to larger values of N_{xy}, principal stresses can be calculated, and diagonal steel, as shown in Fig. 11.2, should be provided.

TABLE 11.1 Coefficients for Computing Force Components of Elliptical Paraboloid Shells

		Value of $y/(b/2)$									
	Force Compo-	(a) $h_x/h_y = 1.0$					(d) $h_x/h_y = 0.8$				
$x/(a/2)$	nent	0	0.25	0.50	0.75	1.0	0	0.25	0.50	0.75	1.0
	R_2	0.250	0.233	0.182	0.101	0	0.289	0.270	0.213	0.119	0
0.00	R_1	0.250	0.267	0.318	0.399	0.500	0.211	0.230	0.287	0.381	0.500
	R_3	0	0	0	0	0	0	0	0	0	0
	R_2	0.267	0.250	0.199	0.111	0	0.304	0.285	0.228	0.130	0
0.25	R_1	0.233	0.250	0.301	0.389	0.500	0.196	0.215	0.272	0.370	0.500
	R_3	0	0.029	0.068	0.096	0.108	0	0.034	0.069	0.100	0.114
	R_2	0.318	0.301	0.250	0.150	0	0.347	0.331	0.277	0.169	0
0.50	R_1	0.182	0.199	0.250	0.350	0.500	0.153	0.169	0.223	0.331	0.500
	R_3	0	0.068	0.140	0.210	0.244	0	0.065	0.139	0.215	0.255
	R_2	0.399	0.389	0.350	0.250	0	0.416	0.406	0.369	0.270	0
0.75	R_1	0.101	0.111	0.150	0.250	0.500	0.084	0.094	0.131	0.230	0.500
	R_3	0	0.096	0.210	0.356	0.465	0	0.091	0.201	0.353	0.480
	R_2	0.500	0.500	0.500	0.500	0	0.500	0.500	0.500	0.500	0
1.00	R_1	0	0	0	0	0	0	0	0	0	0
	R_3	0	0.108	0.243	0.465	∞	0	0.101	0.229	0.443	∞
		(b) $h_x/h_y = 0.6$					(e) $h_x/h_y = 0.4$				
	R_2	0.336	0.316	0.252	0.143	0	0.395	0.374	0.307	0.180	0
0.00	R_1	0.164	0.184	0.248	0.357	0.500	0.105	0.126	0.193	0.320	0.500
	R_3	0	0	0	0	0	0	0	0	0	0
	R_2	0.348	0.329	0.267	0.155	0	0.403	0.383	0.319	0.192	0
0.25	R_1	0.152	0.171	0.233	0.345	0.500	0.097	0.117	0.181	0.308	0.500
	R_3	0	0.031	0.067	0.103	0.120	0	0.026	0.060	0.101	0.125
	R_2	0.383	0.367	0.312	0.197	0	0.425	0.410	0.357	0.235	0
0.50	R_1	0.117	0.133	0.188	0.304	0.500	0.075	0.090	0.143	0.265	0.500
	R_3	0	0.060	0.132	0.216	0.265	0	0.049	0.115	0.208	0.274
	R_2	0.436	0.426	0.392	0.296	0	0.459	0.451	0.419	0.331	0
0.75	R_1	0.064	0.074	0.108	0.204	0.500	0.041	0.049	0.081	0.169	0.500
	R_3	0	0.081	0.185	0.342	0.494	0	0.065	0.156	0.316	0.506
	R_2	0.500	0.500	0.500	0.500	0	0.500	0.500	0.500	0.500	0
1.00	R_1	0	0	0	0	0	0	0	0	0	0
	R_3	0	0.089	0.208	0.413	∞	0	0.070	0.173	0.363	∞
		(c) $h_x/h_y = 0.2$									
	R_2	0.462	0.446	0.388	0.248	0					
0.00	R_1	0.038	0.054	0.112	0.252	0.500					
	R_3	0	0	0	0	0					
	R_2	0.465	0.451	0.396	0.261	0					
0.25	R_1	0.035	0.049	0.104	0.239	0.500					
	R_3	0	0.014	0.040	0.088	0.128					
	R_2	0.473	0.462	0.414	0.303	0					
0.50	R_1	0.027	0.038	0.086	0.197	0.500					
	R_3	0	0.027	0.074	0.174	0.280					
	R_2	0.485	0.480	0.456	0.383	0					
0.75	R_1	0.015	0.020	0.044	0.117	0.500					
	R_3	0	0.034	0.098	0.246	0.510					
	R_2	0.500	0.500	0.500	0.500	0					
1.00	R_1	0	0	0	0	0					
	R_3	0	0.038	0.108	0.262	∞					

Source: Ref. [2].

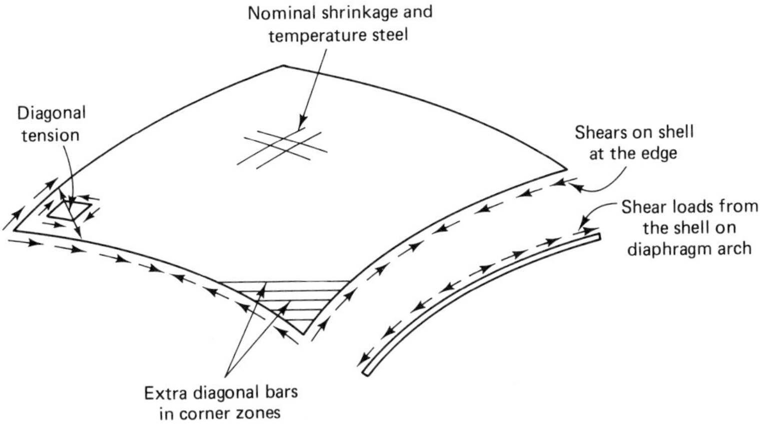

Figure 11.2 Reinforcement design considerations for an example elliptic paraboloid structure.

The diaphragm supports for the shell will usually be parabolic arches spanning between the corner supports. The loading on these arches will be the shear stresses from the shell plus their own weight. Shears from the shell at the edges can be obtained from Table 11.1. For additional values at tenth-interval points along the edges, Parme [2] has given a table of coefficients which may also be used, or values may be obtained directly from a simple computer program.

11.3 MEMBRANE ANALYSIS OF SHELLS WITH NEGATIVE κ

In Chapter 9 we were easily able to obtain membrane stress resultants in hypar shells with straight-line boundaries, without the use of Fourier series. This was possible because the equation of the surface of such shells simplified the differential equation [Eq. (8.10)] to such an extent that it could be directly integrated. When, however, we consider hypar shells with generating parabolas as the boundaries, we must use Fourier series, as we did for elliptic paraboloids, in order to solve the partial differential equations by separation of variables.

The equation of the surface of the hyperbolic paraboloid shown in Fig. 11.3 can be written as

$$z = \frac{x^2}{2h_1} - \frac{y^2}{2h_2} \tag{11.13}$$

where, again, h_1 and h_2 are constants. Sections x or y = *constant* are parabolas, whereas sections z = *constant* give hyperbolas on the surface.

As before, h_x and h_y, the drop or rise of the shell as we move along the x- and y-directions, can be obtained in terms of h_1 and h_2 from Eqs. (11.2a,b) [where we now have a negative sign in Eq. (11.2b)]. Substituting the appropriate derivatives

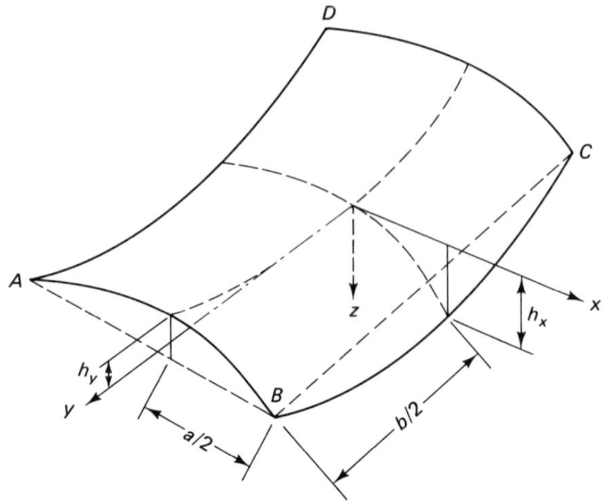

Figure 11.3 Hyperbolic paraboloid with generating parabolas as boundaries.

of z from Eq. (11.13) into Eq. (8.10), and also taking $\bar{p}_x = \bar{p}_y = 0$ and $\bar{p}_z$ as a constant p, we obtain the differential equation for the shell as

$$\frac{1}{h_2}\frac{\partial^2 \Phi}{\partial x^2} - \frac{1}{h_1}\frac{\partial^2 \Phi}{\partial y^2} = p \tag{11.14}$$

Because of the negative sign in the second term on the left-hand side, this becomes a partial differential equation of the hyperbolic type, similar to that for axial vibrations of a bar or lateral vibrations of a beam. Again, a simple solution of this equation can be taken as $\Phi = -(ph_1/2)y^2$. From Eqs. (8.9a–c) we see that this solution will give only N_x as nonzero, and the shell will thus behave only as a series of arches in the x-direction. These arches must be supported at the edges *AD* and *BC* (Fig. 11.3) to resist the N_x forces there. Similarly, if we keep the edges *AD* and *BC* free, the shell will act as a series of parabolic cables and transfer the load to the supports along *AB* and *CD*.

Let us now consider the solution when edges *AD* and *BC* are free, while *AB* and *CD* are supported by diaphragms that resist loads only in the vertical plane. Thus we have the boundary conditions:

$$N_y = \bar{N}_y = 0 \quad \text{or} \quad \frac{\partial^2 \Phi}{\partial x^2} = 0 \quad \text{at} \quad y = \pm\frac{b}{2} \tag{11.15a}$$

$$N_x = \bar{N}_x = 0 \quad \text{or} \quad \frac{\partial^2 \Phi}{\partial y^2} = 0 \quad \text{at} \quad x = \pm\frac{a}{2} \tag{11.15b}$$

$$N_{xy} = 0 \quad \text{or} \quad \frac{\partial^2 \Phi}{\partial x\,\partial y} = 0 \quad \text{at} \quad x = \pm\frac{a}{2} \tag{11.15c}$$

As before, we can assume a solution of Eq. (11.14) in the form

$$\Phi = \sum_{n=1,3,5,\ldots}^{\infty} \phi_n(x) \cos \frac{n\pi y}{b} \tag{11.7}$$

This solution automatically satisfies the boundary conditions of Eq. (11.15a) at $y = \pm(b/2)$, as ensured by considering only the odd values of n in the summation. Expanding the load p in the Fourier series given by Eq. (11.6), and substituting Eqs. (11.6) and (11.7) into Eq. (11.14), we obtain for ϕ_n the following equation:

$$\frac{1}{h_2}\phi_n'' + \frac{1}{h_1}\frac{n^2\pi^2}{b^2}\phi_n = (-1)^{(n-1)/2}\frac{4p}{n\pi} \qquad n = 1, 3, 5, \ldots$$

where $(\cdot)' = d(\cdot)/dx$. We can rewrite this equation as

$$\phi_n'' + \lambda_n^2\phi_n = (-1)^{(n-1)/2}\frac{4p}{n\pi}h_2 \qquad n = 1, 3, 5, \ldots \tag{11.16a}$$

where

$$\lambda_n^2 = \frac{h_2}{h_1}\frac{n^2\pi^2}{b^2} \tag{11.16b}$$

The particular solution of this equation is

$$\phi_{np} = (-1)^{(n-1)/2}\frac{4p}{n\pi}\frac{h_2}{\lambda_n^2} = (-1)^{(n-1)/2}\frac{4pb^2}{n^3\pi^3}h_1 \tag{11.16c}$$

Also, noting the positive sign of the second term on the left-hand side in Eq. (11.16a), we can now write the homogeneous solution as

$$\phi_{no} = C_n \cos \lambda_n x + D_n \sin \lambda_n x$$

Due to symmetry, the second term above will be zero, and hence the complete solution for ϕ_n becomes

$$\phi_n = C_n \cos \lambda_n x + (-1)^{(n-1)/2}\frac{4pb^2}{n^3\pi^3}h_1$$

and thus

$$\Phi = \sum_{n=1,3,5}^{\infty}\left[C_n \cos \lambda_n x + (-1)^{(n-1)/2}\frac{4pb^2}{n^3\pi^3}h_1\right]\cos\frac{n\pi y}{b} \tag{11.17}$$

Substituting this solution into the boundary condition of Eq. (11.15b), we obtain

$$C_n = -(-1)^{(n-1)/2}\frac{4pb^2}{n^3\pi^3}h_1\frac{1}{\cos(\lambda_n a/2)}$$

and hence

$$\Phi = \frac{4pb^2h_1}{\pi^3} \sum_{n=1,3,5,\ldots}^{\infty} \frac{(-1)^{(n-1)/2}}{n^3} \left[1 - \frac{\cos \lambda_n x}{\cos (\lambda_n a/2)}\right] \cos \frac{n\pi y}{b} \tag{11.18}$$

The boundary condition for the shears [Eq. (11.15c)], when substituted into Eq. (11.17), gives

$$\lambda_n C_n \sin \frac{\lambda_n a}{2} \frac{n\pi}{b} \sin \frac{n\pi y}{b} = 0$$

which means

$$\sin \frac{\lambda_n a}{2} = 0$$

Substituting for λ_n from Eq. (11.16b), we rewrite this equation as

$$\sin \left(\sqrt{\frac{h_2}{h_1}} \frac{a}{2b} n\pi\right) = 0 \tag{11.19}$$

This relation will always be satisfied for any value of $n = 1, 3, 5, \ldots$ if we take $(\sqrt{h_2/h_1})(a/2b)$ equal to 1 or a higher integer, say m. This condition can be written as

$$a = m \sqrt{\frac{h_1}{h_2}}\, 2b \tag{11.20}$$

where m is an integer. Thus we can obtain a membrane solution under a uniformly distributed load $\overline{p}_z = p$, for the shell with sides AB and CD supported by shear diaphragms and sides AD and BC free, provided that the geometric relation above is satisfied by the shell surface. Having obtained an expression for Φ [Eq. (11.18)], one can readily write expressions for N_x, N_y, and N_{xy} using Eqs. (8.9a–c) and (8.1a–c). We thus obtain

$$\overline{N}_x = \frac{4ph_1}{\pi} \sum_{n=1,3,5,\ldots}^{\infty} (-1)^{(n+1)/2} \frac{1}{n} \left[1 - \frac{\cos \lambda_n x}{\cos (\lambda_n a/2)}\right] \cos \frac{n\pi y}{b} \tag{11.21a}$$

$$\overline{N}_y = \frac{4ph_2}{\pi} \sum_{n=1,3,5,\ldots}^{\infty} (-1)^{(n-1)/2} \frac{1}{n} \frac{\cos \lambda_n x}{\cos (\lambda_n a/2)} \cos \frac{n\pi y}{b} \tag{11.21b}$$

$$\overline{N}_{xy} = \frac{4p\sqrt{h_1h_2}}{\pi} \sum_{n=1,3,5,\ldots}^{\infty} (-1)^{(n-1)/2} \frac{1}{n} \frac{\sin \lambda_n x}{\cos (\lambda_n a/2)} \sin \frac{n\pi y}{b} \tag{11.21c}$$

Expressions for N_x, N_y, and N_{xy} can now be obtained from Eqs. (11.11a–c), which remain the same as for elliptic paraboloids.

The additional condition imposed by Eq. (11.20) for a membrane solution to be possible can also be interpreted physically as follows. We have seen in Chapter 9 that asymptotes to the various hyperbolas, obtained by substituting

$z = constant$ into Eq. (11.13), lie in the planes $\bar{x} = 0$ and $\bar{y} = 0$, which make angles ψ with the x-axis, such that $\tan \psi \doteq \sqrt{h_2/h_1}$ [Fig. 11.4(a)]. The equation of the surface with $\bar{x}$, $\bar{y}$, and z as coordinate axes was derived as $z = 2\bar{x}\,\bar{y}/(h_1 + h_2)$. Planes $\bar{x} = constant$ thus give a series of straight lines on the surface. These lines are the straight-line generators of the surface that lie in planes parallel to the $\bar{y}$-z plane. These planes make an angle ψ with the x-axis, and hence the plan projections of the straight-line generators will also be inclined at an angle ψ with the x-axis. Similarly, planes $\bar{y} = constant$ will intersect the surface in a series of straight-line generators which lie in planes parallel to the $\bar{x}$-z plane, and thus have their plan projections inclined at an angle ψ with the x-axis. We have also seen in Chapter 9 that forces $\bar{N}_x$ or $\bar{N}_y$, applied at a boundary of the shell, remain constant along the respective generator lines, for a uniform load $\bar{p}_z = p$.

Now consider the plan projection of the shell, as shown in Fig. 11.4(b). Under uniform loading, we want to make $\bar{N}_x$ and $\bar{N}_{xy}$ zero along *AD* and *BC*, while *AB* and *DC* will have $\bar{N}_y = 0$ and nonzero resultant $\bar{N}_{xy}$, due to diaphragm supports. A particular solution of the equation, given by Eqs. (11.16c) and (11.17), will, in general, give nonzero resultants $\bar{N}_x$ and $\bar{N}_{xy}$ along *AD* and *BC*. At any point *P* along *BC*, these nonzero forces will have a nonzero resultant. There will also be a corresponding nonzero force at *P*′, situated symmetrically opposite to *P*. To nullify this force at *P* on edge *BC*, we first apply an opposite force, say *R*, which can be resolved into two components R_1 and R_2 along the generators PP_1 and PP_2 emanating from *P*. Components R_1 and R_2 are carried by the generators to the edges *AB* and *CD* at points P_1 and P_2. Since these edges can only carry shears, we must apply $-R_1$ and $-R_2$ at P_1 and P_2, respectively, along generators P_1P_3 and P_2P_4. Thus we will have only resultant shear forces at P_1 and P_2, as required by the boundary conditions. These forces are again carried to points P_3 and P_4 by the generators. Again, we must apply forces R_1 and R_2 at P_3 and P_4, respectively, along generators P_3P' and P_4P', so that there are no normal forces at P_3 or P_4. Finally, at *P*, we obtain forces R_1 and R_2 which have a combined resultant *R*, and which will nullify the symmetrical nonzero force at *P*′.

Thus, using this procedure, it seems feasible that we can make edges *BC* and *AD* free of all forces, and also have *AB* and *CD* carry only shear forces. This is possible, however, only if by going along generators starting from any point *P* along *BC*, we are able to reach the corresponding point *P*′ on *AD* after touching side *AB* or *CD*. This condition will obviously require certain restrictions on the ratio of the lengths of the sides *AB* and *BC* in relation to the slope ($\tan \psi$) of the generator lines. From Fig. 11.4(c), we can see that starting along the generator at point *C*, we must reach the corresponding point *D* on the opposite edge *AD*. Since the generators make an angle ψ with the x-axis, the distance *CD* must therefore be equal to ($2b \cot \psi$). Hence we have

$$a = 2b \cot \psi = 2b \sqrt{\frac{h_1}{h_2}}$$

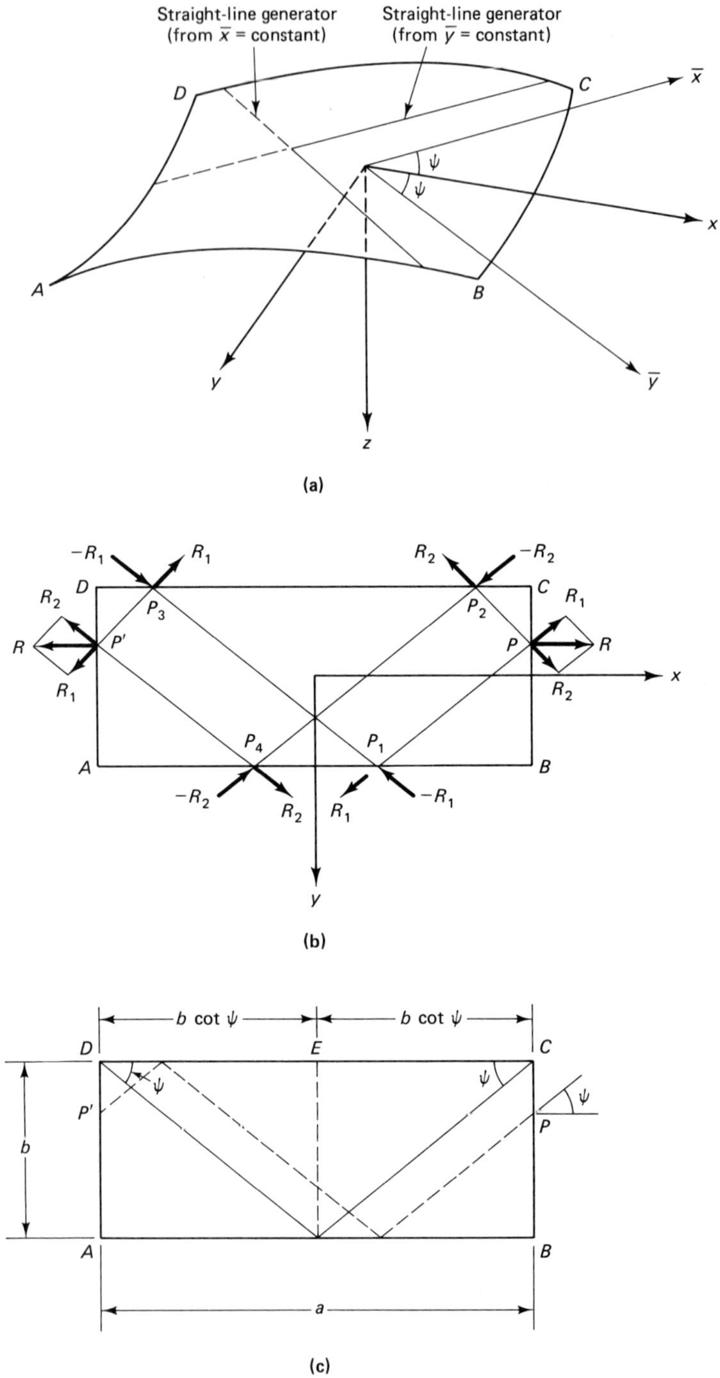

Figure 11.4 Illustration of the response of a hyperbolic paraboloid. (a) Coordinates along straight-line generators. (b) Transference of edge forces along straight-line generators. (c) Geometrical demonstration of the need for Eq. (11.20) in obtaining a membrane solution.

which is the same as Eq. (11.20) with $m = 1$. The same conclusions can be drawn if side *CD* is an integer multiple of ($2b \cot \psi$), i.e., for any integer value of m in Eq. (11.20). Thus we can find a membrane solution, for a shell under uniform load, with two opposite edges free and two edges supported by diaphragms, only when the plan dimensions of the shell satisfy Eq. (11.20), i.e., when a is an even multiple of ($b \cot \psi$).* Whenever this condition is not satisfied, we shall still be able to eliminate forces on the edge *BC* and have only shears on *AB* and *CD;* however, we must provide for whatever normal and shear forces are transferred to edge *AD*.

Another membrane solution of practical interest is possible when we have shear diaphragm supports on all four edges. In this case, Eqs. (11.15a,b) are the boundary conditions. Expressions for Φ and the stress resultants remain the same as before, given by Eqs. (11.18) and (11.21a–c). However, this solution is possible only if the denominator, $\cos (\lambda_n a/2)$, in Eq. (11.18) is not zero. Substituting for λ_n, we get

$$\cos \left(n\pi \frac{a}{2b} \sqrt{\frac{h_2}{h_1}} \right) \neq 0 \qquad \text{or} \qquad \cos \left(\frac{n\pi}{2} \frac{a}{b \cot \psi} \right) \neq 0$$

Since n varies as 1, 3, 5, . . . , the term above will be zero if $[a/(b \cot \psi)]$ is any odd number [i.e., if length a is an odd multiple of ($b \cot \psi$)]. Also, in the previous case, the boundary condition $N_{xy} = 0$ on edges $x = \pm(a/2)$ gave us the condition that a should be an even multiple of $b \cot \psi$. In the present case, when we have shear diaphragm supports on the edges $x = \pm(a/2)$, N_{xy} will be nonzero on these edges. Hence a should not be an even multiple of ($b \cot \psi$). We must therefore have length a as neither an even nor an odd multiple of ($b \cot \psi$); i.e., the plan dimensions of the shell should be such that a generator emanating from one corner should not end up in any other corner. It is left to the reader to visualize, by drawing figures similar to Fig. 11.4(b) and 11.4(c), how the normal stresses can be nullified on the four edges when the shell is supported by diaphragms on all edges [1].

The membrane analysis of hypar shells with their generator parabolas as boundaries can be used to design roofs of civic halls or industrial sheds, where long strips of such shells, with $a \gg b$ (Fig. 11.5), are simply supported on edges *BC* and *AD,* while edges *AB* and *CD* are kept free. A number of such strips can be placed next to each other to cover a large area. Intermediate skylight openings can also be provided between adjacent strips. These strips can be precast, prestressed, and stacked one above another. They can then be lifted and placed into position. The strips have very light curvatures in both directions, and hence the dead load can be considered as uniformly distributed over the plan area. To obtain membrane stress resultants in a typical unit *ABCD,* we take axes x, y, and z as

*It may be noted that the term $\cos (\lambda_n a/2) = \cos [(n\pi a/2b) \sqrt{h_2/h_1}] = \cos [n\pi a/(2b \cot \psi)]$ appears in the denominator of Eq. (11.18). Hence this term should not be zero. Because $\cos [(n\pi/2)(a/b \cot \psi)]$ is not zero if $[a/(b \cot \psi)]$ is equal to 2, 4, 6, etc. (i.e., if a is an even multiple of $b \cot \psi$), only these even multiple terms are reflected in Eq. (11.20).

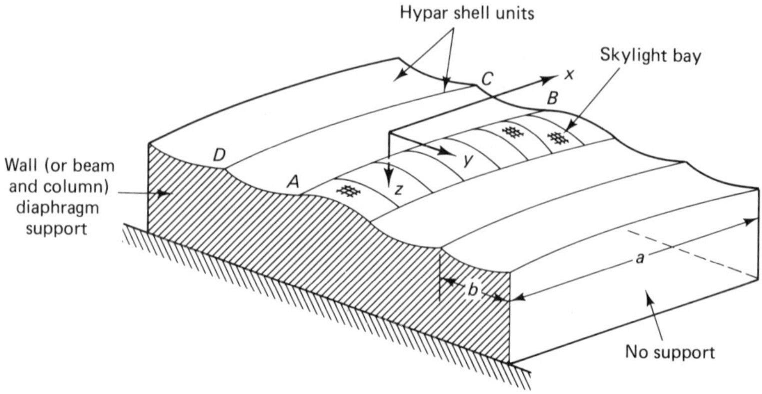

Figure 11.5 Roof structure, consisting of long hypar shell units with generator parabolas as boundaries.

shown in Fig. 11.5. Then the equation of the surface remains the same as before, Eq. (11.13). Also, the boundary conditions are as given by Eqs. (11.15a–c). Hence we can obtain the membrane stress resultants from the previous solution.

Such shells can also be analyzed approximately by treating them as beams of curved cross section, spanning along the x-direction. A "beam and transverse arch analysis," similar to that discussed later in Chapter 14 for barrel shells and folded plates, can easily be performed in order to obtain the internal stresses. The only difference in such an analysis is that one must here consider the effects of longitudinal curvature when calculating transverse stresses. The beam method has an advantage over the membrane analysis in that it also gives the transverse moments in the shell [3–6]. Shells of this type can be economically constructed, without prestressing, for spans up to 10 m [6]. For longer spans, however, prestressing is desirable.

Another form of double curvature shells with negative κ, which is popular in construction, is the conoid. It is generated by moving a straight line with one of its ends on a transverse straight line and the other end on a transverse plane curve, usually a parabola (see Chapter 3). The advantage of the conoid in civil engineering construction is that it allows the entry of light in an industrial unit. The equation of such a surface will be of the type $z = f_1(y^2)f_2(x)$, so that sections $x = constant$ are parabolas, while sections $y = constant$ are the straight-line generators. It can be seen that the derivatives $\partial^2 z/\partial y^2$ and $\partial^2 z/\partial x \partial y$ will be linear functions in x and y, respectively, for this surface. Hence the partial differential equation (8.10) will now have variable coefficients in both x and y, which will thus not permit any exponential solutions in x and y (i.e., a Fourier series solution of such an equation is not possible). However, other forms of solutions are possible. Wilby and Naqui [7] have given solutions for membrane as well as bending stress resultants in conoids, using orthogonal beam functions obtained from the solutions of lateral vibration problems for beams. This reference also gives exhaustive tables, from which internal stress resultants in conoids of different dimensions can readily be calculated for practical design.

11.4 SHELLS OF REVOLUTION UNDER NONAXISYMMETRIC LOADING: GENERAL EQUATIONS

In Chapters 4 and 6 we have studied the membrane analysis of circular cylinders, cones, spheres, and other general shells of revolution under axisymmetric loads. These loads may adequately describe the majority of problems encountered in design practice. However, there will still be cases when the shell is subjected to wind loads, or when it is not supported in an axisymmetric fashion, and we thus must be able to obtain the nonaxisymmetric membrane stress resultants.

We again consider an element of the shell bounded by sections ϕ, $\phi + d\phi$, and θ, $\theta + d\theta$, as shown in Fig. 11.6(a). All three stress resultants N_ϕ, N_θ, and $N_{\phi\theta} = N_{\theta\phi}$ will now be nonzero and will vary with both ϕ and θ, as shown. Also, the loads per unit area of the shell surface in the ϕ-, θ-, and radial directions are taken as p_ϕ, p_θ, and p_r, respectively.

We can now derive the equations of equilibrium in the ϕ, r, and θ directions. The forces $(N_\theta r_1\, d\phi)$ on opposite faces θ and $\theta + d\theta$ = *constant* have a resultant in the horizontal plane of $(N_\theta r_1\, d\phi)\, d\theta$. This resultant has components $(N_\theta r_1\, d\phi)$ $d\theta \cos\phi$ and $(N_\theta r_1\, d\phi)\, d\theta \sin\phi$ in the meridional and radial directions, respectively, as shown in Fig. 11.6(b). Here we have neglected the components of the term $(\partial N_\theta/\partial\theta)\, d\theta$, which are of smaller order. We also see from Fig. 11.6(c) that forces $(N_\phi r\, d\theta)$, on opposite faces of the element, have a resultant $(N_\phi r\, d\theta)\, d\phi$ in the radial direction. Also, the forces $N_{\theta\phi} r_1\, d\phi$, on opposite faces, have components $(N_{\theta\phi} r_1\, d\phi) \cos\phi$ in the horizontal plane, directed along the radii θ and $\theta + d\theta$ = *constant* of the parallel circle, as seen from Fig. 11.6(d). These components have a resultant $(N_{\theta\phi} r_1\, d\phi) \cos\phi\, d\theta$ directed in the θ-direction.

The equilibrium equations in the ϕ-, θ-, and r-directions can now be readily written. We have in the ϕ-direction

$$\frac{\partial(N_\phi r\, d\theta)}{\partial\phi} d\phi - (N_\theta r_1\, d\phi)\, d\theta \cos\phi + \frac{\partial(N_{\theta\phi} r_1\, d\phi)}{\partial\theta} d\theta + p_\phi r_1\, d\phi\, r\, d\theta = 0$$

In the θ-direction, we can write

$$\frac{\partial(N_\theta r_1\, d\phi)}{\partial\theta} d\theta + \frac{\partial(N_{\phi\theta} r\, d\theta)}{\partial\phi} d\phi + (N_{\theta\phi} r_1\, d\phi) \cos\phi\, d\theta + p_\theta r_1\, d\phi\, r\, d\theta = 0$$

Also, in the radial direction, we have

$$-(N_\phi r\, d\theta)\, d\phi - (N_\theta r_1\, d\phi)\, d\theta \sin\phi + p_r r_1\, d\phi\, r\, d\theta = 0$$

Dropping $(d\phi)(d\theta)$ throughout, and taking $N_{\theta\phi} = N_{\phi\theta}$, we have the first two equations as

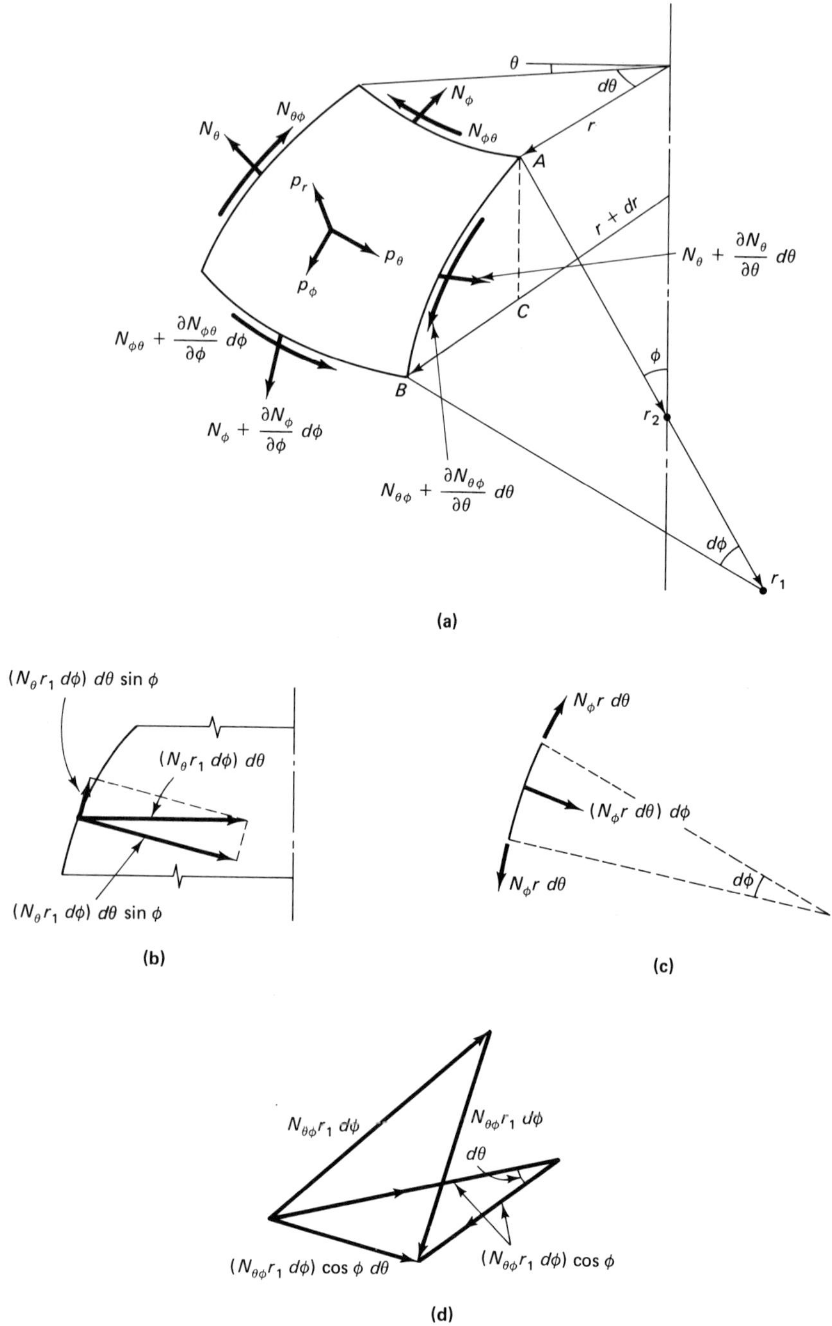

Figure 11.6 Shell of revolution under nonaxisymmetric loading. (a) Stress resultants acting on the shell element. (b) Meridional and radial components of force due to N_θ. (c) Radial component of force due to N_ϕ. (d) Resultant force in the θ-direction due to shear $N_{\theta\phi}$.

$$\frac{\partial(N_\phi r)}{\partial\phi} + \frac{\partial(N_{\phi\theta} r_1)}{\partial\theta} - N_\theta r_1 \cos\phi = -p_\phi r_1 r \tag{11.22a}$$

$$\frac{\partial(N_\theta r_1)}{\partial\theta} + \frac{\partial(N_{\phi\theta} r)}{\partial\phi} + N_{\phi\theta} r_1 \cos\phi = -p_\theta r_1 r \tag{11.22b}$$

Also, the third equation simplifies to the same form as we had for axisymmetric problems, after we divide it throughout by $r_1 r$ and substitute $r_2 = r/\sin\phi$. Hence we have

$$\frac{N_\phi}{r_1} + \frac{N_\theta}{r_2} = p_r \tag{11.22c}$$

Thus the normal load on the shell, at any point, is still carried by "hoop" forces in the rings in the directions of the two principal sections at that point. Equations (11.22a,b) are now partial differential equations, and to solve them, we shall first eliminate N_θ by substituting $N_\theta r_1 = p_r r_1 r_2 - N_\phi r_2$ from Eq. (11.22c). We also note from Fig. 11.6(a) that in triangle ABC, side AB has length $(r_1\, d\phi)$, while BC has length dr since the radius of the parallel circle at B will be $r + dr$. We therefore have the relation $dr = (r_1\, d\phi)\cos\phi$. Hence we can write

$$\frac{\partial(N_\phi r)}{\partial\phi} = \frac{\partial N_\phi}{\partial\phi} r + N_\phi \frac{dr}{d\phi} = \frac{\partial N_\phi}{\partial\phi} r + N_\phi r_1 \cos\phi$$

Similarly,

$$\frac{\partial(N_{\phi\theta} r)}{\partial\phi} = \frac{\partial N_{\phi\theta}}{\partial\phi} r + N_{\phi\theta} r_1 \cos\phi$$

Substituting in these relations, eliminating N_θ, and dividing throughout by $r = r_2 \sin\phi$, we obtain Eqs. (11.22a,b) in the following form:

$$\frac{\partial N_\phi}{\partial\phi} + N_\phi\left(\frac{r_1}{r_2} + 1\right)\cot\phi + \frac{\partial N_{\phi\theta}}{\partial\theta}\frac{r_1}{r_2\sin\phi} = (p_r\cot\phi - p_\phi)r_1 \tag{11.23a}$$

$$\frac{\partial N_{\phi\theta}}{\partial\phi} + 2N_{\phi\theta}\frac{r_1}{r_2}\cot\phi - \frac{\partial N_\phi}{\partial\theta}\frac{1}{\sin\phi} = \left(-p_\theta - \frac{\partial p_r}{\partial\theta}\frac{1}{\sin\phi}\right)r_1 \tag{11.23b}$$

These are two partial differential equations for N_ϕ and $N_{\phi\theta}$. They have variable coefficients in ϕ, but not in θ. Hence it is possible for us to separate the variables by expanding N_ϕ, $N_{\phi\theta}$, and the loads as Fourier series in θ. This is also expected physically, since any of these functions will have the same values at θ, $\theta + 2\pi$, . . . , etc., which corresponds to the same point on the shell (i.e., these functions are periodic in θ, with period 2π). After separating the variables, we shall obtain two ordinary differential equations for the nth harmonics, $N_{\phi n}$ and $N_{\phi\theta n}$, with ϕ as the independent variable.

We can assume general Fourier expansions of the loads in the following form:

$$p_\phi = \sum_{n=0}^{\infty} p_{\phi n} \cos n\theta + \sum_{n=1}^{\infty} \bar{p}_{\phi n} \sin n\theta \tag{11.24a}$$

$$p_\theta = \sum_{n=1}^{\infty} p_{\theta n} \sin n\theta + \sum_{n=0}^{\infty} \bar{p}_{\theta n} \cos n\theta \tag{11.24b}$$

$$p_r = \sum_{n=0}^{\infty} p_{rn} \cos n\theta + \sum_{n=1}^{\infty} \bar{p}_{rn} \sin n\theta \tag{11.24c}$$

It can be easily seen that the loads represented by the first set of terms above are symmetric in θ, while those represented by the second set are antisymmetric in θ. We shall obtain the solution for only the symmetric terms of the above, since the solution for the other terms can be obtained in a similar manner. We shall also assume the stress resultants to be expanded as Fourier series in θ. For the loading that is symmetric about $\theta = 0$, the stress resultants will also be symmetric about $\theta = 0$, and hence they can be assumed in the form

$$N_\phi = \sum_{n=0}^{\infty} N_{\phi n} \cos n\theta \tag{11.25a}$$

$$N_{\phi\theta} = \sum_{n=1}^{\infty} N_{\phi\theta n} \sin n\theta \tag{11.25b}$$

$$N_\theta = \sum_{n=0}^{\infty} N_{\theta n} \cos n\theta \tag{11.25c}$$

Substituting the first set of terms of Eqs. (11.24a–c) and Eqs. (11.25a–c) into Eqs. (11.23a,b), dropping $\cos n\theta$ or $\sin n\theta$ terms, and equating term by term on the left- and right-hand sides, we get the following ordinary differential equations for the nth harmonic of the stress resultants:

$$\frac{dN_{\phi n}}{d\phi} + N_{\phi n}\left(\frac{r_1}{r_2} + 1\right) \cot \phi + N_{\phi\theta n} \frac{r_1 n}{r_2 \sin \phi} = (p_{rn} \cot \phi - p_{\phi n}) r_1 \tag{11.26a}$$

and

$$\frac{dN_{\phi\theta n}}{d\phi} + 2N_{\phi\theta n} \frac{r_1}{r_2} \cot \phi + N_{\phi n} \frac{n}{\sin \phi} = \left(-p_{\theta n} + p_{rn} \frac{n}{\sin \phi}\right) r_1 \tag{11.26b}$$

These equations have variable coefficients, and hence a solution in an explicit form for a general shell of revolution is not easily obtainable. Results can usually be obtained by numerical integration [1]. However, for special cases of circular cylinders, cones, or spheres, the solutions can be obtained more easily. Here we shall consider the solution of two problems for a spherical shell subjected to nonaxisymmetric loading. We shall separately derive detailed solutions for circular cylindrical shells in the following chapters.

11.5 SPHERICAL SHELLS: WIND LOADING

In the design of roof domes, stresses due to wind loading will generally be small and will usually not govern the design. In any case, we must be able to obtain the membrane stress resultants under such a loading. As for roof trusses of industrial sheds, wind loading can be taken as a normal pressure to the shell at all points. Hence we can describe it by taking $p_\phi = p_\theta = 0$, and by taking p_r as nonzero. Also, when wind is assumed to blow along the direction $\theta = 0$, the pressure will be symmetric about the $r-\phi$ plane at $\theta = 0$ and will be a direct pressure on the windward side ($\theta = 0$ to $\pi/2$) and a suction pressure on the leeward side ($\theta = \pi/2$ to π). Hence the wind loading can be described by a function that varies as $\cos\theta$, and changes sign for $\theta > \pi/2$. Thus in the simplest form, we can assume the wind load to be given by

$$p_r = -p \sin\phi \cos\theta \qquad \text{and} \qquad p_\phi = p_\theta = 0$$

This expression for p_r corresponds to the first term of the Fourier expansion in the θ-direction of the symmetric wind load. The $\sin\phi$ term reduces the wind pressure to zero at the top of the dome, as would be expected when the wind blows horizontally.

From Fig. 11.7(a) we can see that all loads pass through point O at the center of the dome. We take any section $\phi = constant$, and consider the freebody of the shell above [Fig. 11.7(b)]. On this section we have two unknown stress resultants, N_ϕ and $N_{\phi\theta}$. Since the loads vary as $\cos\theta$, N_ϕ will also vary as $\cos\theta$. Therefore, the vertical components of all loads will add up to zero, as do the vertical components of the N_ϕ forces, $\cos\theta$ being antisymmetric about $\theta = \pi/2$. Hence writing the equation of equilibrium of vertical forces will not yield N_ϕ, as it did for the axisymmetric loading case. We note, however, that the loads will have a resultant moment about a diameter perpendicular to the plane of the paper through point O'. This moment due to the wind pressure can be equated to the corresponding moment due to the N_ϕ forces (since $N_{\phi\theta}$ forces on the section will not give any such moment), in order to obtain N_ϕ. We can write this moment balance equation easily as

$$2\int_0^\pi (N_\phi \sin\phi)(r\,d\theta) r\cos\theta = 2\int_0^\phi \int_0^\pi [p_{\bar{r}} \cos\bar{\phi}\,\bar{r}\cos\theta - p_{\bar{r}}\sin\bar{\phi} \times \cos\theta\,(a\cos\bar{\phi} - a\cos\phi)]\,\bar{r}\,d\theta\,a\,d\bar{\phi}$$

where $\bar{\phi}$ is a dummy variable and $\bar{r}$ is r evaluated at $\phi = \bar{\phi}$. Substituting $r = a \sin\phi$ and $\bar{r} = a\sin\bar{\phi}$, we obtain

$$2a^2 \sin^3\phi \int_0^\pi N_\phi \cos\theta\,d\theta = 2a^3 \int_0^\phi \int_0^\pi p_{\bar{r}} \sin^2\bar{\phi}\cos\phi\cos\theta\,d\theta\,d\bar{\phi}$$

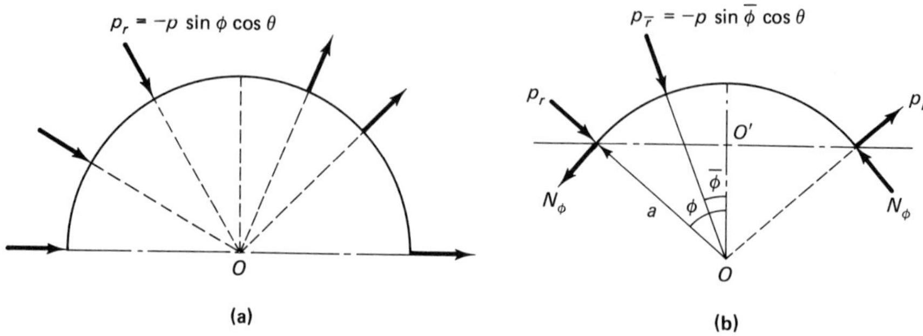

Figure 11.7 Wind pressure acting on a spherical dome. (a) Elevation view. (b) Freebody of section above $\phi = constant$.

Taking $\cos \phi$ out of the integral on the right-hand side above, since it is independent of $\bar{\phi}$, and substituting $p_{\bar{r}} = -p \sin \bar{\phi} \cos \theta$ and $N_\phi = N_{\phi 1} \cos \theta$ into the above, we have

$$\sin^3 \phi \, N_{\phi 1} \int_0^\pi \cos^2 \theta \, d\theta = -pa \cos \phi \int_0^\phi \int_0^\pi \sin^3 \bar{\phi} \cos^2 \theta \, d\theta \, d\bar{\phi}$$

or

$$N_{\phi 1} \sin^3 \phi \, \frac{\pi}{2} = -pa \frac{\pi}{2} \cos \phi \int_0^\phi \sin^3 \bar{\phi} \, d\bar{\phi}$$

This gives

$$N_{\phi 1} = \frac{pa}{\sin^3 \phi} \left(-\frac{2}{3} \cos \phi + \cos^2 \phi - \frac{\cos^4 \phi}{3} \right)$$

and hence

$$N_\phi = \frac{-pa \cos \theta}{3 \sin^3 \phi} (\cos^4 \phi - 3 \cos^2 \phi + 2 \cos \phi) \qquad (11.27)$$

Now, considering equilibrium of forces in the horizontal direction, that is, in the direction of the wind, we can obtain $N_{\phi\theta}$, since N_ϕ is known. Also, knowing N_ϕ, N_θ can then be obtained from Eq. (11.22c). Thus, by simply considering statics of the shell, we can determine all internal stress resultants. This is expected, since membrane theory problems are "statically determinate." We can also obtain the stress resultants by directly solving Eqs. (11.26a,b). Taking $r_1 = r_2 = a$, these equations become (for the sphere)

$$\frac{dN_{\phi n}}{d\phi} + 2 \cot \phi \, N_{\phi n} + \frac{n}{\sin \phi} N_{\phi\theta n} = a(-p_{\phi n} + p_{rn} \cot \phi) \qquad (11.28a)$$

$$\frac{dN_{\phi\theta n}}{d\phi} + 2 \cot \phi \, N_{\phi\theta n} + \frac{n}{\sin \phi} N_{\phi n} = a\left(-p_{\theta n} + p_{rn} \frac{n}{\sin \phi} \right) \qquad (11.28b)$$

We shall solve these equations following a procedure described by Flügge [1]. Noting the similarity of the equations, we can assume two new functions, U_n and V_n, in the following form:

$$U_n(\phi) = N_{\phi n} + N_{\phi\theta n} \tag{11.29a}$$

and

$$V_n(\phi) = N_{\phi n} - N_{\phi\theta n} \tag{11.29b}$$

Adding and subtracting Eqs. (11.28a,b) to and from each other, we can obtain two equations: one containing U_n only, and the other containing V_n only. These equations are

$$\frac{dU_n}{d\phi} + \left(2 \cot \phi + \frac{n}{\sin \phi}\right) U_n = a\left(-p_{\phi n} - p_{\theta n} + \frac{n + \cos \phi}{\sin \phi} p_{rn}\right) \tag{11.30a}$$

and

$$\frac{dV_n}{d\phi} + \left(2 \cot \phi - \frac{n}{\sin \phi}\right) V_n = a\left[-p_{\phi n} + p_{\theta n} - \frac{(n - \cos \phi)}{\sin \phi} p_{rn}\right] \tag{11.30b}$$

Both of these equations are first-order ordinary differential equations of the type

$$\frac{dy}{dx} + P(x)y = Q(x)$$

We know that if we multiply such equations by an "integrating factor" exp $[\int P(x)\, dx]$, the left-hand side becomes a complete derivative. We can easily check that the equation thus becomes

$$\frac{d}{dx}\left(e^{\int P(x)dx} y\right) = e^{\int P(x)dx} Q$$

and hence y can be obtained by integrating this equation. Proceeding in this manner, we see that for the first equation [Eq. (11.30a)], the integrating factor will be

$$\begin{aligned} \exp\left[\int \left(2 \cot \phi + \frac{n}{\sin \phi}\right) d\phi \right] &= \exp\left[2 \ln(\sin \phi) + n \ln\left(\tan \frac{\phi}{2}\right)\right] \\ &= e^{\ln[\sin^2 \phi \tan^n(\phi/2)]} \\ &= \sin^2 \phi \tan^n \frac{\phi}{2} \end{aligned}$$

Similarly, for the second equation [Eq. (11.30b)], the integrating factor is $\sin^2 \phi \cot^n (\phi/2)$. When we multiply the two equations [Eqs. (11.30a,b)] by their respective integrating factors, the left-hand side of each equation becomes a complete derivative. The equations then take on the following forms:

$$\frac{d}{d\phi}\left(\sin^2\phi \tan^n\frac{\phi}{2}U_n\right) = -a\left(p_{\phi n} + p_{\theta n} - \frac{n+\cos\phi}{\sin\phi}p_{rn}\right)\sin^2\phi\tan^n\frac{\phi}{2} \tag{11.31a}$$

$$\frac{d}{d\phi}\left(\sin^2\phi \cot^n\frac{\phi}{2}V_n\right) = -a\left(p_{\phi n} - p_{\theta n} + \frac{n-\cos\phi}{\sin\phi}p_{rn}\right)\sin^2\phi\cot^n\frac{\phi}{2} \tag{11.31b}$$

We shall integrate these equations, from $\phi = \phi_o$ to $\phi = \phi$ (following the definite integral approach of Section 6.4), in order to obtain the following solutions for U_n and V_n:

$$\left[\sin^2\phi\tan^n\frac{\phi}{2}U_n\right]_{\phi_o}^{\phi} = -a\int_{\phi_o}^{\phi}\left(p_{\phi n} + p_{\theta n} - \frac{n+\cos\phi}{\sin\phi}p_{rn}\right)\sin^2\phi\tan^n\frac{\phi}{2}\,d\phi \tag{11.32a}$$

and

$$\left[\sin^2\phi\cot^n\frac{\phi}{2}V_n\right]_{\phi_o}^{\phi} = -a\int_{\phi_o}^{\phi}\left(p_{\phi n} - p_{\theta n} + \frac{n-\cos\phi}{\sin\phi}p_{rn}\right)\sin^2\phi\cot^n\frac{\phi}{2}\,d\phi \tag{11.32b}$$

For simplicity, we have not used any dummy variable within the integral signs on the right-hand sides of the equations above. Using these equations, we can obtain the solutions for U_n and V_n, and hence the solutions for $N_{\phi n}$ and $N_{\phi\theta n}$ for any given loading.

For the special case of wind loading which we are considering, we substitute $p_{\phi n} = p_{\theta n} = 0$ and $p_r = -p\sin\phi\cos\theta$ into the equations above. We note that this load corresponds to $n = 1$ in the symmetric terms of Eqs. (11.24a–c) with $p_{r1} = -p\sin\phi$. We then integrate Eqs. (11.32a,b) from $\phi = 0$ to $\phi = \phi$. Since $\sin\phi = 0$ at $\phi = \phi_o = 0$, and also the limit (as $\phi\to 0$) of $[\sin^2\phi\cot^n(\phi/2)]$ is zero for $n = 1$, the terms at $\phi = \phi_o$ on the left-hand sides of both equations vanish, and we obtain

$$\begin{aligned}\sin^2\phi\tan\frac{\phi}{2}U_1 &= -pa\int_0^{\phi}(1+\cos\phi)\sin^2\phi\tan\frac{\phi}{2}\,d\phi\\ &= -pa\int_0^{\phi}\sin^3\phi\,d\phi\end{aligned}$$

or

$$U_1 = \frac{-pa}{3\sin^2\phi\tan(\phi/2)}(\cos^3\phi - 3\cos\phi + 2)$$

Similarly, from the second equation, we have

$$V_1 = \frac{pa}{3\sin^2\phi\cot(\phi/2)}(\cos^3\phi - 3\cos\phi + 2)$$

Hence

$$N_{\phi 1} = \frac{U_1 + V_1}{2} = \frac{-pa}{3 \sin^3 \phi} (\cos^3 \phi - 3 \cos \phi + 2) \cos \phi$$

which is the same as the coefficient of $\cos \theta$ in Eq. (11.27). Now we also have

$$N_{\phi\theta 1} = \frac{U_1 - V_1}{2} = \frac{-pa}{3 \sin^3 \phi} (\cos^3 \phi - 3 \cos \phi + 2)$$

Multiplying these two equations by $\cos \theta$ and $\sin \theta$, respectively, we obtain expressions for N_ϕ and $N_{\phi\theta}$. When N_ϕ and p_r are substituted into Eq. (11.22c), the expression for N_θ is also obtained. We can summarize the expressions for the stress resultants, as follows:

$$N_\phi = \frac{-pa}{3 \sin^3 \phi} (\cos^3 \phi - 3 \cos \phi + 2) \cos \phi \cos \theta \tag{11.33a}$$

$$N_{\phi\theta} = \frac{-pa}{3 \sin^3 \phi} (\cos^3 \phi - 3 \cos \phi + 2) \sin \theta \tag{11.33b}$$

$$N_\theta = \frac{-pa}{3 \sin^3 \phi} (3 \sin^4 \phi - \cos^4 \phi + 3 \cos^2 \phi - 2 \cos \phi) \cos \theta \tag{11.33c}$$

It is noticed that all the stress resultants are zero at the top of the dome. The value of N_θ will be $\pm(pa)$ at $\phi = \pi/2$ and $\theta = 0, \pi$. For the roof dome considered in Example 6.1 and Section 7.7, we had $a = 60$ ft. If the wind-pressure intensity is 20 lb/ft^2, we would take $p = 10$ lb/ft^2, since our loading equation considers pressure on the windward side as well as suction on the leeward side. From Eq. (11.33c) we can easily check that the maximum numerical value of N_θ in this shell would be obtained at its base, $\phi = 30°$, and for $\theta = 0$ or π. This value for N_θ is 228.72 lb/ft, which will give a stress of only 6.35 lb/in.^{2} in a 3-in.-thick shell. The maximum value of N_ϕ will be even smaller than that of N_θ. Being very small, wind stresses would thus not govern the design of the shell.

Wind pressures on circular structures, such as cylindrical shafts, hyperbolic cooling towers, etc., vary along θ, from $\theta = 0$ to $\theta = \pi$, where the wind blows in the direction defined by the diameter $\theta = 0$. Some codes specify a given function for the variation of this wind pressure with θ. At $\theta = 0$, the pressure p_r is negative (radially inward), and p_r becomes positive (i.e., suction) as we move away from $\theta = 0$. The curve has very small values for pressure on the leeward side. Since the pressure is symmetric about $\theta = 0$, we can represent it by a Fourier cosine series. The terms of this series for the pressure will be of the form $p_{rn} \cos n\theta$, for which a solution can be easily obtained using the procedure described above.

11.6 SPHERICAL SHELL SUPPORTED NONAXISYMMETRICALLY

A dome may not be completely supported around its entire lower periphery. The supporting structure may have openings to allow for doors or windows. In such cases, the solution obtained for the axisymmetric problem will not be strictly valid, since the support condition is not axisymmetric. The nonzero N_ϕ forces over the openings, obtained from the membrane solution, will therefore have to be applied in the opposite direction, as edge loads on the shell (along with the support reactions caused by them). When this edge-load solution is superimposed on the axisymmetric solution, we shall get the final solution, where $N_\phi = 0$ over the openings. Before considering a specific problem, we shall first need to obtain the solution of the spherical shell under edge loads.

For the edge-load solution, we substitute $p_{\phi n} = p_{\theta n} = p_{rn} = 0$ into Eqs. (11.31a,b). Hence the right-hand sides of these equations are zero. Integrating their left-hand sides, we obtain

$$U_n = A_n \frac{\cot^n (\phi/2)}{\sin^2 \phi}$$

and

$$V_n = B_n \frac{\tan^n (\phi/2)}{\sin^2 \phi}$$

where A_n and B_n are constants of integration to be determined from the boundary conditions at the top and bottom edges of the shell. We also obtain

$$N_{\phi n} = \frac{U_n + V_n}{2} = \frac{1}{2 \sin^2 \phi} \left(A_n \cot^n \frac{\phi}{2} + B_n \tan^n \frac{\phi}{2} \right) \tag{11.34a}$$

and

$$N_{\phi\theta n} = \frac{U_n - V_n}{2} = \frac{1}{2 \sin^2 \phi} \left(A_n \cot^n \frac{\phi}{2} - B_n \tan^n \frac{\phi}{2} \right) \tag{11.34b}$$

Also, with $p_{rn} = 0$, we have

$$N_{\theta n} = -N_{\phi n} \tag{11.34c}$$

It may be seen that the term U_n goes to infinity at $\phi = 0$ for all values of n, while the term V_n is finite at $\phi = 0$, except when $n = 0$ or 1. Hence for a dome without a top opening, we take $A_n = 0$, since we do not expect infinite stresses at its top, due to edge loads at its lower edge. Constant B_n is then determined from the edge loads applied at the lower edge. Here we have the freedom of prescribing either N_ϕ or $N_{\phi\theta}$, and we must accept whatever values we get for the other stresses, by providing appropriate supports. When $n = 0$ or 1, the term V_n is also singular at $\phi = 0$, where we get concentrated vertical and horizontal forces and a couple, corresponding to these U_n and V_n terms. Physically, we see that for a shell closed at the top, the edge loads we apply at the lower edge will have to be self-equilibrating with zero resultant vertical and horizontal forces or moments, since there is no other support or load on the shell to resist any unbalanced force or moment due to loads at the lower edge. We shall see that the Fourier expansion of such self-equilibrating loads will give the value of harmonics, corresponding to $n = 0$ and 1, as zero. Hence we shall not need to consider the terms corresponding to $n = 0$ or 1 in Eqs. (11.34a,b). When, however, a non-self-equilibrating load is applied at the lower edge, we will need a support to balance the resultant force or moment. This support can only exist in the form of concentrated forces and moment at $\phi = 0$. In such cases, the first two harmonics will be nonzero in the Fourier expansion of the loading, and the singular terms corresponding to $n = 0$ and $n = 1$ in Eqs. (11.34a,b) will give the required concentrated support reactions at $\phi = 0$.

Let us consider a domed roof, similar to that in Example 6.1 and Section 7.7, under its own weight. The dome is supported on a cylindrical wall which will resist the vertical component of N_ϕ. A ring beam is provided at the junction to resist the horizontal component of N_ϕ. To allow for entrance to this structure, an opening has to be made in the cylindrical wall. We assume that this opening subtends an angle of 30°, as shown in Fig. 11.8. The ring beam is, of course, assumed to be continuous all around. Thus the dome is assumed not to have a vertical support at its edge $\phi = 30°$, for $-15° \leqslant \theta \leqslant 15°$.

To solve this problem, we first assume that there is a support all around. Then, under an axisymmetric dead load p, resultants N_ϕ and N_θ will be given by [see Eqs. (6.6a,b)]

$$N_\phi = \frac{-pa}{1 + \cos\phi} \qquad N_\theta = pa\left(\frac{1}{1 + \cos\phi} - \cos\phi\right)$$

Substituting $p = 65$ lb/ft^2, $\phi = 30°$, and $a = 60$ ft, we get at the lower edge $N_\phi = -2090$ lb/ft. To nullify the vertical component of these forces over the opening AB, we must superimpose a second solution, wherein we apply a downward force of 2090 sin 30° = 1045 lb/ft on AB. But this load by itself will not be in equilibrium, and we must calculate its support reactions over BC and AC, and apply them also as loads. To find these vertical support reactions R (lb/ft) corresponding to a downward load of 1045 lb/ft over AB, we assume approximately that they vary linearly with y from B to C, and that they are symmetrical about OC. We can thus write

$$\begin{aligned} R &= \alpha + \beta y \\ &= \alpha + \beta(r\cos\theta) \end{aligned}$$

where α and β are constants that we can determine from equilibrium of vertical forces and moments about the diameter $\theta = 90°$. Equating the sum of vertical forces to zero, we have

$$2\int_{\theta=\pi/12}^{\pi} Rr\,d\theta + 2\int_0^{\pi/12} (1045)\,r\,d\theta = 0$$

Substituting for R and $r = 30$ ft and integrating, we obtain

$$172.79\,\alpha - 465.87\beta = -16{,}414.82 \tag{a}$$

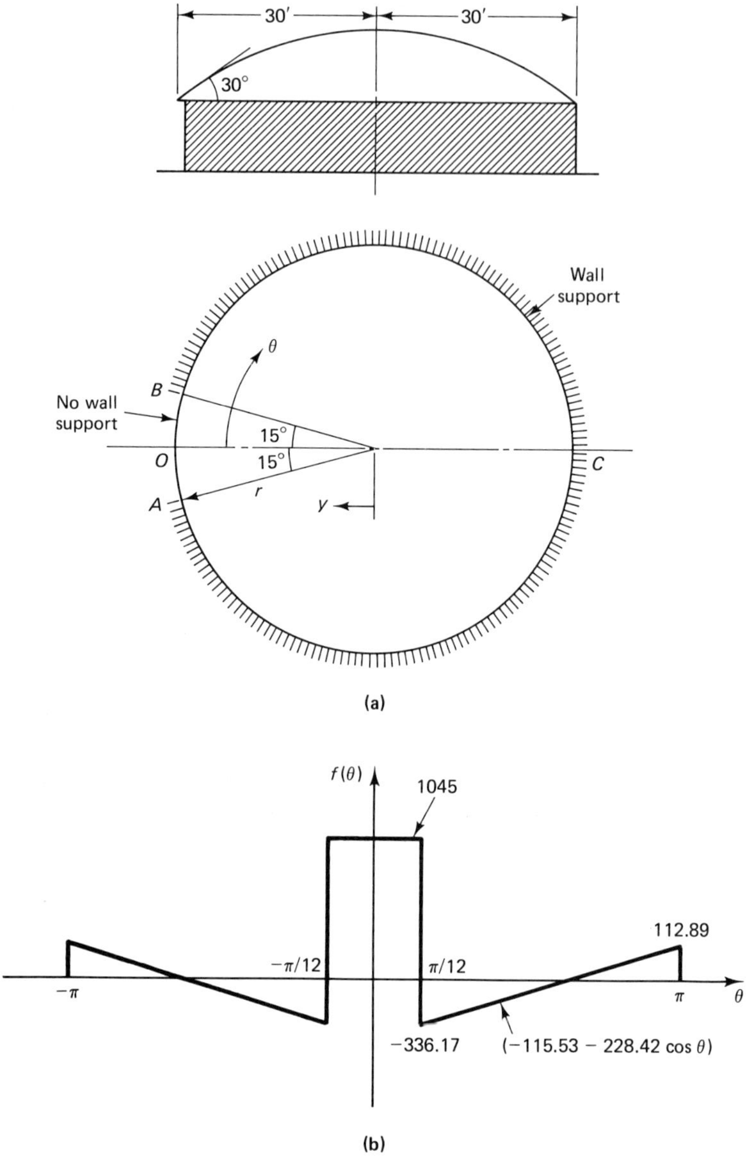

Figure 11.8 Dome roof supported on a cylindrical wall with an opening. (a) Elevation and plan configurations. (b) Edge loads to be applied on the shell.

Now equating to zero the moments about diameter $\theta = 90°$, we have

$$2\int_{\pi/12}^{\pi} R(r\,d\theta)r\cos\theta + 2\int_{0}^{\pi/12} 1045\,(r\,d\theta)(r\cos\theta) = 0$$

Substituting for R and r and integrating, we have

$$-0.5176\alpha + 78.894\beta = -540.93 \tag{b}$$

Solving Eqs. (a) and (b) simultaneously, we obtain $\alpha = -115.53$ and $\beta = -7.614$. Hence we have $R = -115.53 - 228.42\cos\theta$. As expected, the support reaction is tensile (downward) near $\theta = \pi$, and compressive near A and B. The vertical edge load to be applied on the shell, $f(\theta)$, is as shown in Fig. 11.8(b) and is given by

$$f(\theta) = 1045 \qquad \left(-\frac{\pi}{12} \leqslant \theta \leqslant +\frac{\pi}{12}\right)$$

$$f(\theta) = -115.53 - 228.42\cos\theta \qquad \left(\frac{\pi}{12} \leqslant \theta \leqslant \pi \text{ and } -\frac{\pi}{12} \geqslant \theta \geqslant -\pi\right)$$

This is a self-equilibrating loading which is symmetrical about $\theta = 0$, and hence it can be expanded in a Fourier cosine series over $-\pi \leqslant \theta \leqslant \pi$. This expansion can be written as

$$f(\theta) = \frac{f_o}{2} + \sum_{n=1}^{\infty} f_n \cos n\theta$$

where

$$f_n = \frac{2}{\pi}\int_0^{\pi} f(\theta)\cos n\theta\,d\theta$$

Since the terms corresponding to $n = 0$ and $n = 1$ (i.e., f_o and f_1) give a resultant vertical force and a moment about $\theta = 90°$, it is obvious that we shall have $f_o = f_1 = 0$ for our self-equilibrating edge load. This fact can also be easily verified by calculating f_o and f_1 from the equations above. Now, for $n \geqslant 2$, we have

$$\begin{aligned} f_n &= \frac{2}{\pi}\left[\int_0^{\pi/12} 1045\cos n\theta\,d\theta + \int_{\pi/12}^{\pi}(-115.53 - 228.42\cos\theta)\cos n\theta\,d\theta\right] \\ &= 738.82\,\frac{\sin(n\pi/12)}{n} + 72.71\left[\frac{\sin[(n+1)\pi/12]}{n+1} + \frac{\sin[(n-1)\pi/12]}{n-1}\right] \end{aligned} \tag{11.35a}$$

As expected, since the loading has discontinuities at $\theta = \pm(\pi/12)$, we get a $1/n$ type of convergence in its Fourier series above. To obtain the stress resultants, we now substitute into Eq. (11.34a), $\phi = 30°$, and equate $N_{\phi n}\sin 30°$ to f_n above, after taking U_n as zero. This gives us constants B_n as

$$B_n = \frac{f_n}{\tan^n(\pi/12)} \tag{11.35b}$$

Knowing the constants B_n, then $N_{\phi n}$, $N_{\phi\theta n}$, and $N_{\theta n}$ are given by Eqs. (11.34a–c), with $A_n = 0$. The stress resultants due to edge loads are then obtained when we substitute these quantities into Eqs. (11.25a–c). Final stress resultants in the shell are obtained when we add the axisymmetric solutions for N_ϕ and N_θ to the edge-load solutions. Finally, then, we have

$$N_\phi = \sum_{n=2,3,\ldots}^{\infty} \frac{B_n}{2\sin^2\phi} \tan^n\frac{\phi}{2}\cos n\theta - \frac{pa}{1+\cos\phi}$$

$$N_{\phi\theta} = \sum_{n=2,3,\ldots}^{\infty} \frac{-B_n}{2\sin^2\phi} \tan^n\frac{\phi}{2}\sin n\theta$$

$$N_\theta = \sum_{n=2,3,\ldots}^{\infty} \frac{-B_n}{2\sin^2\phi} \tan^n\frac{\phi}{2}\cos n\theta + pa\left(\frac{1}{1+\cos\phi} - \cos\phi\right)$$

where B_n is given by Eq. (11.35b). For numerical calculation, the term $\sin^2\phi$ in the denominator should be expanded in terms of $\sin(\phi/2)$ and $\cos(\phi/2)$, so that $\sin^2(\phi/2)$ can be canceled with that in the numerator (n being greater than or equal to 2).

In practice, one may assume that the shell is also supported over the opening. A beam is then provided to span over the opening AB. This beam is designed to resist force $N_{\phi\theta}$ and the vertical component of N_ϕ. The beam may also be combined with the circular ring beam, mentioned earlier, which is required to resist the horizontal component of N_ϕ.

REFERENCES

[1] Flügge, W., *Stresses in Shells,* 2nd ed., Springer-Verlag, New York, 1973.

[2] Parme, A. L., "Shells of Double Curvature," *Trans. Am. Soc. Civ. Eng.*, Vol. 123, 1958, Paper No. 2951.

[3] Rahman, A. P. M., Deshmukh, R. S., and Jain, S. S., "Precast Prestressed Hyperboloid Shells for Industrial Roofs," *Indian Concr. J.*, Vol. 42, Nov. 1968, pp. 452–460.

[4] Nori, V. V., "Simplified Analysis of Bending Effects in Doubly Curved Saddle Shells," *ASCE J. Struct. Div.*, Vol. 4, July 1976, pp. 71–75.

[5] Nori, V. V., "Precast, Pretensioned Hypar Shell Units for Factory Roof," *Indian Concr. J.*, Vol. 52, Jan. 1978, pp. 18–20.

[6] Deshmukh, R. S., and Sreenath, H. G., "Precast Reinforced Concrete Hyperboloid Shells for Roofing Industrial Buildings," *Indian Concr. J.*, Vol. 47, July 1973, pp. 255–258.

[7] Wilby, C. B., and Naqui, M. M., "Reinforced Concrete Conoidal Shell Roofs—Flexural Theory, Design Tables," Cement and Concrete Association, Slough, England, 1973.

PROBLEMS

11.1. Hypar shells, similar to those shown in Fig. 11.5, are to be designed as a cover for an industrial shed. The cross-sectional width of each unit is 6 ft, and the span length is 35 ft. The drop of the concave parabolas in the transverse direction is 2 ft, while the rise of the longitudinal convex parabolas is 6 in. The thickness of the shell unit is 2.5 in. Obtain the membrane stresses in the shell unit under a total load of 60 lb/ft^2 over the plan area. Assume that the longitudinal edges of the shell are free and that the transverse edges are simply supported.

11.2. Derive expressions for the longitudinal stresses, longitudinal shears, and transverse bending moments and shears in hypar shell units of the type shown in Fig. 11.5. Use "beam theory" similar to that described in Chapter 14 for barrel shells. Use these formulas to obtain the transverse moments, shears, etc., in the shell roof of Problem 11.1. Design the reinforcement for this shell. (*Note:* The reader should consider Sections 14.5 to 14.7 in order to solve this problem.)

11.3. Derive differential equations for the membrane stress resultants N_θ, $N_{x\theta}$, and N_x (or N_θ, $N_{s\theta}$, and N_s) in (a) a circular cylinder, and (b) a conical shell, both under non-axisymmetric loading. These equations may be derived directly or they may be obtained from Eqs. (11.23a,b) already derived.

11.4. For a spherical roof dome of radius a, obtain expressions for the membrane stress resultants under a radial wind load described by a general term in a Fourier expansion: $p_r = -p_{rn} \sin \phi \cos n\theta$. Assume that $p_\phi = p_\theta = 0$.

11.5. A circular cylindrical chimney shell of height L is subjected to a wind pressure p_r, the Fourier expansion of which is described by the terms $p_r = -p_{rn} \cos n\theta$. Obtain the membrane stress resultants in the shell. Assume the chimney to be fixed at its base and free at the top.

11.6. The spherical shell of Section 11.6 or Example 6.1 is supported on a wall which has four symmetrically located openings at $\theta = 0°, 90°, 180°$, and $270°$. The width of each opening subtends an angle of 30° in plan. Obtain the membrane stress resultants in the shell under a load of 65 lb/ft^2 over the surface area.

12

Membrane Theory of General Cylindrical Shells

12.1 INTRODUCTION

In Chapter 4 we studied the membrane theory for circular cylinders only. We shall now consider cylindrical shells with any arbitrary shape of cross section: e.g., circular, elliptical, parabolic and cylinders of open or closed type. Of all these shells, only the circular cylinder is a shell of revolution, whereas the others can be called shells of translation, since they can be generated by the translation of a straight line. All such shells will have zero Gaussian curvatures and hence have developable surfaces.

12.2 DERIVATION OF EQUATIONS

Let us consider an element of the general cylindrical shell, bounded by sections x, $x + dx$, and ϕ, $\phi + d\phi$ [Fig. 12.1(a)]. The internal stress resultants N_x, $N_{x\phi} = N_{\phi x}$, N_ϕ and the loads defined per unit area of the shell midsurface, acting on the element, are shown in Fig. 12.1(b). The radius of curvature r_1 is infinite, while r_2 is the second principal radius of curvature. We can write the equations for equilibrium of forces in the r-, ϕ-, and x-directions, as follows.

Equilibrium in the radial direction. The forces $(N_\phi \cdot dx)$, acting on opposite faces of the element, make an angle $d\phi$ with each other, and hence their resultant along the radial direction will be $(N_\phi \cdot dx)\, d\phi$. Thus summing forces in this direction and neglecting terms of higher order, we can write

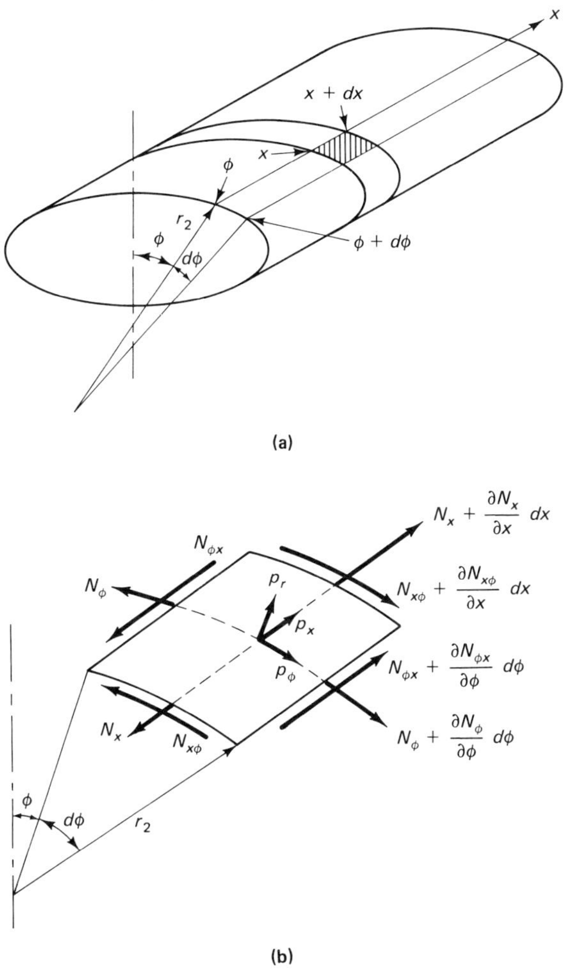

Figure 12.1 General cylindrical shell. (a) Coordinate system. (b) Forces acting on an element.

$$-N_\phi \, dx \, d\phi + p_r \, dx(r_2 \, d\phi) = 0$$

or

$$N_\phi = p_r r_2 \tag{12.1a}$$

Equilibrium in the ϕ-direction. Considering equilibrium of forces in the ϕ-direction, we can derive

$$\frac{\partial N_\phi}{\partial \phi} \, d\phi \, dx + \frac{\partial N_{x\phi}}{\partial x} \, dx(r_2 \, d\phi) + p_\phi \, dx(r_2 \, d\phi) = 0$$

or

$$\frac{\partial N_\phi}{\partial \phi} + \frac{\partial N_{x\phi}}{\partial x} r_2 = -p_\phi r_2 \tag{12.1b}$$

Equilibrium in the *x*-direction. For equilibrium in the x-direction, we can also obtain

$$\frac{\partial N_x}{\partial x} dx(r_2\, d\phi) + \frac{\partial N_{x\phi}}{\partial \phi} d\phi\, dx + p_x\, dx(r_2\, d\phi) = 0$$

or

$$\frac{\partial N_x}{\partial x} + \frac{1}{r_2}\frac{\partial N_{x\phi}}{\partial \phi} = -p_x \tag{12.1c}$$

Knowing p_r, N_ϕ can be determined directly from Eq. (12.1a), and then $N_{x\phi}$ and N_x can be determined by successively integrating Eqs. (12.1b) and (12.1c), respectively. We can therefore summarize the equilibrium equations and their solutions, as follows.

General Equilibrium Equations: Summary

$$N_\phi = p_r r_2 \tag{12.1a}$$

$$\frac{\partial N_{x\phi}}{\partial x} + \frac{1}{r_2}\frac{\partial N_\phi}{\partial \phi} = -p_\phi \tag{12.1b}$$

$$\frac{\partial N_x}{\partial x} + \frac{1}{r_2}\frac{\partial N_{x\phi}}{\partial \phi} = -p_x \tag{12.1c}$$

General Solution of Equilibrium Equations

$$N_\phi = p_r\, r_2 \tag{12.2a}$$

$$N_{x\phi} = -\int \left(p_\phi + \frac{1}{r_2}\frac{\partial N_\phi}{\partial \phi}\right) dx + f_1(\phi) \tag{12.2b}$$

$$N_x = -\int \left(p_x + \frac{1}{r_2}\frac{\partial N_{x\phi}}{\partial \phi}\right) dx + f_2(\phi) \tag{12.2c}$$

where f_1 and f_2 are unknown functions of ϕ, to be determined from the boundary conditions. For most practical problems, we shall have $p_x = 0$, and p_ϕ and p_r will be independent of x. In such cases, since N_ϕ will then also be independent of x, we shall find from Eq. (12.2b)

$$N_{x\phi} = -\left(p_\phi + \frac{1}{r_2}\frac{\partial N_\phi}{\partial \phi}\right)x + f_1(\phi)$$

or

$$N_{x\phi} = -x\,F(\phi) + f_1(\phi)$$

where

$$F(\phi) = p_\phi + \frac{1}{r_2}\frac{\partial N_\phi}{\partial \phi}$$

Hence taking the partial derivative of $N_{x\phi}$ with respect to ϕ, we obtain

$$\frac{\partial N_{x\phi}}{\partial \phi} = -x\frac{dF}{d\phi} + \frac{df_1(\phi)}{d\phi}$$

Substituting this expression, and $p_x = 0$, into Eq. (12.2c), we find

$$N_x = \left(\frac{x^2}{2r_2}\right)\frac{dF}{d\phi} - \left(\frac{x}{r_2}\right)\frac{df_1}{d\phi} + f_2(\phi)$$

We shall henceforth consider only this special case of loading, and we therefore summarize the corresponding solutions below.

Solution for $p_x = 0$, and p_r, p_ϕ independent of x

$$N_\phi = p_r r_2 \tag{12.3a}$$

$$N_{x\phi} = -xF(\phi) + f_1(\phi) \tag{12.3b}$$

$$N_x = \frac{x^2}{2r_2}\frac{dF(\phi)}{d\phi} - \frac{x}{r_2}\frac{df_1(\phi)}{d\phi} + f_2(\phi) \tag{12.3c}$$

where

$$F(\phi) = p_\phi + \frac{1}{r_2}\frac{\partial N_\phi}{\partial \phi} \tag{12.3d}$$

Let us now consider the application of these equations to a cantilever shell which is free at edge $x = 0$ and "fixed" at $x = L$, as shown in Fig. 12.2. We have the boundary conditions $N_x = N_{x\phi} = 0$ at $x = 0$. Substituting these conditions into Eqs. (12.3b,c), we obtain $f_1(\phi) = f_2(\phi) = 0$. Hence we obtain

$$N_{x\phi} = -xF(\phi)$$

and

$$N_x = \frac{x^2}{2r_2}\frac{dF(\phi)}{d\phi}$$

Figure 12.2(b) and (c) show the variations of $N_{x\phi}$ and N_x with x. It is seen that these curves are very similar to the variations of shear force and axial force (bending

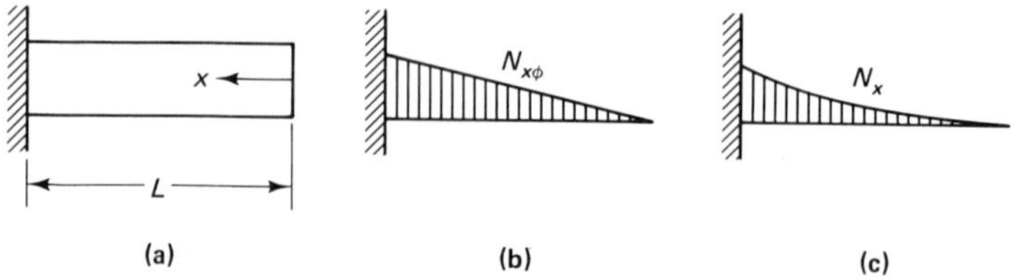

Figure 12.2 Cantilevered cylindrical shell. (a) Shell configuration. (b) Variation of force $N_{x\phi}$ with x. (c) Variation of force N_x with x.

moment) in a cantilever beam. The variation of stresses around the cross section can be determined for given loadings p_r and p_ϕ.

12.3 SIMPLY SUPPORTED CYLINDRICAL SHELLS

General Equations

Shells spanning a distance L may be supported at their ends by ring beams or solid walls—supports that are stiff against deformations in their own planes, but are nearly perfectly flexible for deformations perpendicular to their planes. Such supports are called diaphragms. They can thus resist $N_{x\phi}$ forces from the shell, but they cannot resist N_x forces which are perpendicular to their own planes. Thus the boundary condition at such a support will be $N_x = 0$. We shall now consider the application of Eqs. (12.3a–d) to such "diaphragm supported" (or "simply supported") shells.

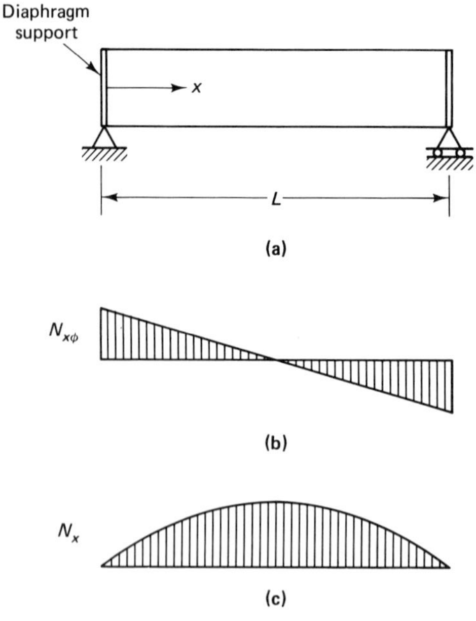

Figure 12.3 Cylindrical shell supported by diaphragms. (a) Shell configuration. (b) Variation of $N_{x\phi}$ with x. (c) Variation of N_x with x.

Figure 12.3(a) shows a cylindrical shell simply supported at $x = 0$ and L. The boundary conditions are $N_x = 0$ at $x = 0$ and $x = L$. Substituting these into Eq. (12.3c), we have at $x = 0$,

$$N_x(0) = f_2(\phi) = 0$$

and at $x = L$,

$$N_x(L) = \frac{L^2}{2r_2}\frac{dF}{d\phi} - \frac{L}{r_2}\frac{df_1}{d\phi} = 0$$

Therefore,

$$f_1(\phi) = \frac{L}{2}F(\phi) + C$$

where C is a constant. Substituting $f_1(\phi)$ into Eq. (12.3b), we have

$$N_{x\phi} = \left(\frac{L}{2} - x\right)F(\phi) + C$$

For loadings p_ϕ and p_r which are constant along x, we expect that the shears on cross sections $x = 0$ and $x = L$ will be mirror images of one another, since the shell and the loading are symmetrical about $x = L/2$. However, we see that the constant term in the expression above will give shears which are in opposite directions on the faccs $x = 0$ and $x = L$. Hence C must be taken as zero. It can also be seen that $N_{x\phi} = C$ will give a torsional moment about the x-axis on the shell, which will not be in equilibrium with the loadings p_ϕ, p_r, which are constant in x and symmetric about the r-x plane. Hence we must have

$$N_{x\phi} = \left(\frac{L}{2} - x\right)F(\phi) \qquad \text{and} \qquad f_1(\phi) = \frac{L}{2}F(\phi)$$

Then, from Eq. (12.3c), we have

$$N_x = -\frac{x(L - x)}{2r_2}\frac{dF}{d\phi}$$

We can plot the variations of $N_{x\phi}$ and N_x along x, as shown in Fig. 12.3(b) and (c). It is seen again that the variations of $N_{x\phi}$ and N_x along the span are similar, respectively, to those of the shear force and bending moment in a simply supported beam.

We will find similar beam-like behavior if we assume that the cylinder in Fig. 12.3(a) has an overhang L_2, beyond the support at $x = L$. In this case we can find $N_{x\phi}$ and N_x in the overhang portion of the shell, as for the cantilever shell. At the support $x = L$, whatever value of $N_{x\phi}$ is obtained from the overhang shell will be transmitted to the diaphragm support which must resist it. N_x forces obtained from the overhang, however, will be transmitted as a boundary condition at $x = L$ for the main span of the shell. This condition, along with the condition

$N_x = 0$ at $x = 0$, will be sufficient to determine N_x and $N_{x\phi}$ in the main span of the shell. It will be seen again, that in this case also, the variations of $N_{x\phi}$ and N_x are respectively similar to those of the shear and moment in a simply supported beam with an overhang [1]. This analogy will also apply when the shell has overhangs on both sides. However, for "statically indeterminate" problems, such as when the shell is supported on three supports, for determining the stress resultants, the deformations of the shell have to be considered. These deformations are quite different from those in an ordinary beam. Hence, in such cases, the variations of N_x and $N_{x\phi}$ in the shell will be different from those in a corresponding beam.

We summarize below the expressions for the membrane stress resultants in simply supported shells, and then consider some specific loading examples.

Membrane stress resultants in shells simply supported at x = 0 and L
(With $p_x = 0$ and p_r, p_ϕ independent of x)

$$N_\phi = p_r r_2 \tag{12.4a}$$

$$N_{x\phi} = \left(\frac{L}{2} - x\right)F(\phi) \tag{12.4b}$$

$$N_x = -\frac{x(L - x)}{2r_2}\frac{dF}{d\phi} \tag{12.4c}$$

where

$$F(\phi) = p_\phi + \frac{1}{r_2}\frac{\partial N_\phi}{\partial \phi} \tag{12.4d}$$

Simply Supported Shells of Circular Cross Section

Example 12.1: Circular Pipe Filled with Water

Figure 12.4 shows a circular pipe, filled with water, and simply supported at its ends $x = 0$ and $x = L$. The loading is

$$p_r = \gamma a(1 - \cos\phi) \qquad \text{and} \qquad p_\phi = 0$$

where γ is the density of water, and $r_2 = a$ is the radius of the circular cylinder. Substituting this loading into Eqs. (12.4a–d), we obtain

$$N_\phi = \gamma a^2(1 - \cos\phi)$$

$$F(\phi) = \frac{1}{a}\frac{\partial N_\phi}{\partial \phi} = \gamma a \sin\phi$$

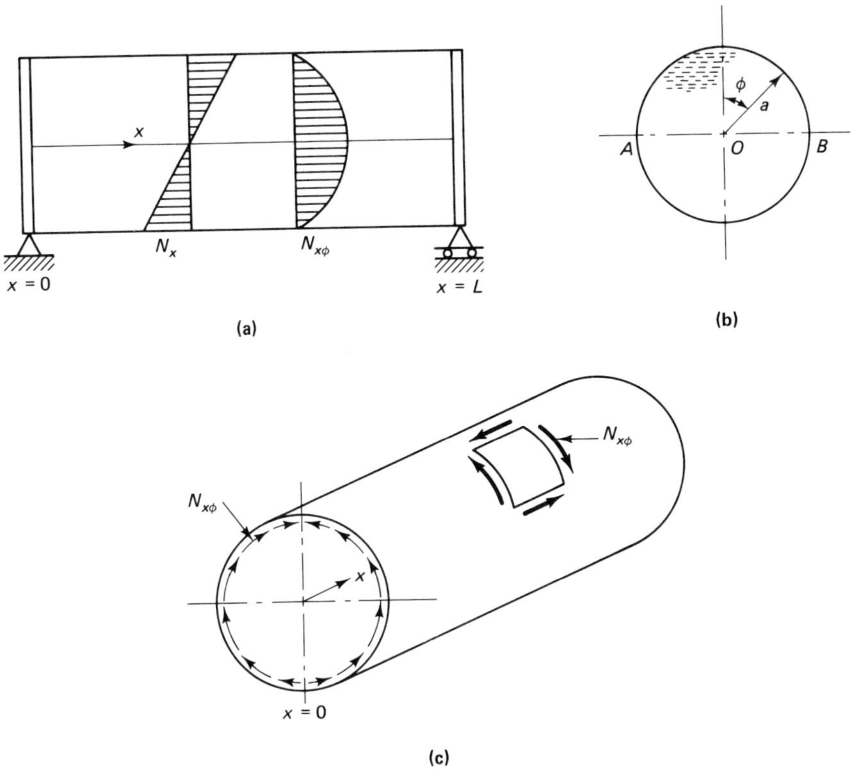

Figure 12.4 Circular cylindrical shell filled with water. (a) Shell configuration showing typical variations in $N_{x\phi}$ and N_x with depth. (b) Cross section. (c) Variation of $N_{x\phi}$ over the cross section at $x = 0$.

$$N_{x\phi} = \gamma a\left(\frac{L}{2} - x\right) \sin \phi$$

$$N_x = -\frac{\gamma}{2} x(L - x) \cos \phi$$

The variations of N_x and $N_{x\phi}$ over the depth of any cross section $x = constant$ is shown in Fig. 12.4(a). It is seen that, just as is the case in the direction of the span, the variations of N_x and $N_{x\phi}$ over the depth of the cross section are also similar, respectively, to those of axial stress and shear stress in a beam.* Thus the pipe carries the water load by bending, as a beam between the two supports. The load is transmitted to the supports by the resultant $N_{x\phi}$, whose variation over the cross section at $x = 0$ is shown in Fig. 12.4(c). In the right half of the cross section, $N_{x\phi}$ is positive (since $\sin \phi$ is positive), while in the left half, $N_{x\phi}$ is

*If z is measured upward from the center O, then $\cos \phi = z/a$ and thus $\sin \phi = \sqrt{1 - (z/a)^2}$. Hence N_x varies linearly with z. But $N_{x\phi}$ does not vary parabolically over the depth, as does the shear stress in a beam of rectangular cross section (although the variation is very similar).

negative. The directions of $N_{x\phi}$ acting on the shell are, therefore, as shown. The horizontal components from both halves will cancel each other, while the total vertical component is given by

$$V = 2\int_0^{\pi} N_{x\phi}\,(a\,d\phi)\sin\phi$$

$$= \gamma a^2 L\int_0^{\pi}\sin^2\phi\,d\phi$$

$$= \frac{\pi}{2}a^2\gamma L$$

which is, as expected, half the weight of water in the pipe.

The assumed simply supported boundary conditions will be valid only for a pipe between two ring supports, having flexible expansion joints beyond the supports. Otherwise, the pipe will carry N_x forces due to the pressure on the circular plates at the closed ends. Values of these N_x forces can be obtained by solving the problem of the end circular plate subjected to water pressure and then finding the reactions on the pipe. These reactions can be applied as boundary conditions on the shell to obtain the new stress distribution of N_x and $N_{x\phi}$ [1]. It can be verified that the variation of $N_{x\phi}$ along the span remains the same as for the simply supported pipe, while the variation of N_x is similar to that of the axial stress in a beam subjected to vertical load plus an eccentric axial force. This is true because the water pressure on the end plate gives a resultant eccentric force on the pipe cross section.

If the cross section of the shell is elliptic, instead of circular, membrane stresses can be obtained by substituting p_r, p_ϕ, and r_2 as functions of ϕ (or z) into Eqs. (12.4a–d). It is found that in this case, the variation of N_x and $N_{x\phi}$ over the depth is generally much different than that in a beam [1].

Example 12.2: Circular Cylindrical Barrel Shell under Self-Weight

Open circular cylindrical shells, or "barrel" shells, are commonly used for roof structures, and hence are of great interest in practice. If p is the dead load per unit area of the shell surface, we have

$$p_r = -p\cos\phi \qquad \text{and} \qquad p_\phi = p\sin\phi$$

Substituting this loading and $r_2 = a$ into Eqs. (12.4a–d), we get

$$N_\phi = -pa\cos\phi \tag{12.5a}$$

$$F(\phi) = 2p\sin\phi \tag{12.5b}$$

$$N_{x\phi} = 2p\left(\frac{L}{2} - x\right)\sin\phi \tag{12.5c}$$

$$N_x = -\frac{px(L - x)}{a}\cos\phi \tag{12.5d}$$

The variations of N_ϕ, $N_{x\phi}$, and N_x over the depth of the cross section, for a semicircular barrel, are shown in Fig. 12.5. The variations of N_x and $N_{x\phi}$ are, respectively, linear and

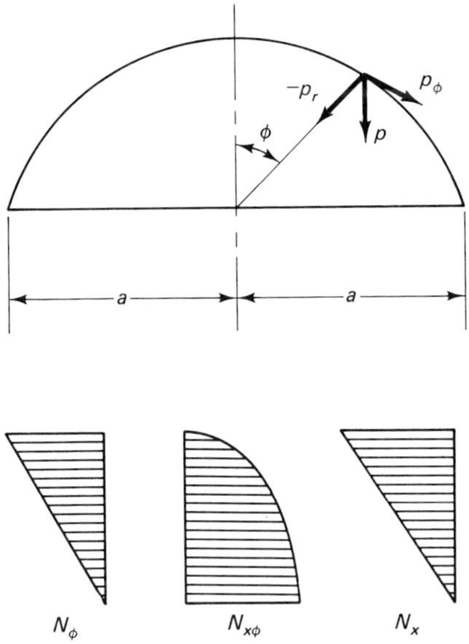

Figure 12.5 Open circular cylinder or "barrel shell" under self-weight, showing typical variations of N_ϕ, $N_{x\phi}$, and N_x over the depth.

nearly parabolic (sine curve), similar to those for a simple beam. It is interesting to note that at the edges $\phi = \pm\pi/2$, N_ϕ is zero. At these edges, $N_{x\phi} = 2p[(L/2) - x]$ is, however, nonzero. The variation of $N_{x\phi}$ from $x = 0$ to $x = L$ at the edge $\phi = \pi/2$ is shown in Fig. 12.6. These edge forces are self-equilibrating and can be resisted by providing an edge member. Forces equal and opposite to the edge forces on the shell will act as loads on the edge members. The edge members have free ends at diaphragm supports, and the loads will be resisted by development of tensile forces which will be zero at $x = 0$ and L, but at other sections will have a value $T(x)$, given by

$$T(x) = \int_0^x 2p\left(\frac{L}{2} - x\right) dx = 2p\left(\frac{L}{2}x - \frac{x^2}{2}\right)$$

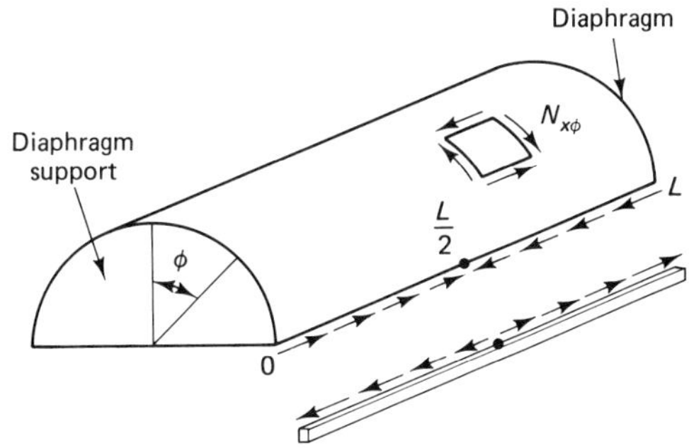

Figure 12.6 Variation of $N_{x\phi}$ along the edge $\phi = \pi/2$ of a barrel shell.

This force has a maximum value at $x = L/2$, which is given by

$$T_{max} = \frac{pL^2}{4}$$

At $x = L/2$, the total compressive force on the cross section of the shell is obtained by substituting $x = L/2$ into Eq. (12.5d), and integrating over the cross section. We have, at $x = L/2$,

$$C = 2\int_0^{\pi/2} N_x(a\,d\phi) = -\frac{pL^2}{2}\int_0^{\pi/2} \cos\phi\,d\phi = -\frac{pL^2}{2}$$

It is seen that this unbalanced compressive force on the shell cross section is balanced by the tensile forces T_{max} in the two edge beams which must be provided. In a similar manner, equilibrium will be satisfied at any other section along the span of the shell. It can be readily checked that these compressive and tensile forces together form a couple which is equal to the external bending moment at the section. This action is thus similar to the action of a reinforced concrete beam where the external moment is resisted by the couple formed from compression in concrete above the neutral axis and the tensile force concentrated in the steel reinforcement at the bottom. It can also be easily verified that the resultant of $N_{x\phi}$, on any section of the shell, will be a vertical force equal to the external shear force at that section.

Thus a semicircular barrel shell can support dead load under membrane conditions if tension edge members are provided at its bottom edges. If no such member is actually provided, a small portion of the shell at the edge will itself carry the tensile forces (i.e., forces opposite to the unbalanced $N_{x\phi}$ forces at the edges will have to be applied as edge forces on the shell to make the edges free).

In reality, however, there are complications in the ideal membrane behavior of the semicircular barrel shell, due to the incompatibility of deformations at the edges. Equation (12.5d) shows that the shell has $N_x = 0$ at $\phi = \pm\pi/2$. On the other hand, the edge member has large tensile forces which will cause it to extend in the axial direction. In addition, due to its own weight, the edge member will tend to deflect vertically. These deflections will be different than the vertical deformations of the shell at the edges. These incompatibilities of vertical and longitudinal deformations of the shell and edge members will require application of redundant forces between the shell and edge members. The values of these redundant edge forces can be found such that compatibility is satisfied at the edges. Since edge loads cannot be considered using the membrane theory, we must account for such effects by means of bending theory (see Chapter 13).

When the cylindrical shell cross section has edges with angle ϕ less than 90°, there are further complications, since N_ϕ [Eq. 12.5(a)] will now also be nonzero at the longitudinal edges. Here the edge member has to resist N_ϕ, in addition to $N_{x\phi}$ forces from the shell edges. This gives rise to further incompatibilities of deformations.

It should be clear, though, that such barrel shells, combined with edge beams, can carry a load over a span L by bending and shear action similar to that in a simple beam.

Example 12.3: Circular Cylindrical Barrel Shell under Live Load or Snow Load

If a barrel shell is loaded by a pressure p_o, which is uniformly distributed over its plan area, we have (see Fig. 12.7)

$$p(a\,d\phi) = p_o(a\,d\phi)\cos\phi \qquad \text{or} \qquad p = p_o\cos\phi$$

where p is the vertical load defined per unit area of shell surface. Hence, we have

$$p_r = -p_o\cos^2\phi$$

and

$$p_\phi = p_o\cos\phi\sin\phi$$

Substituting into Eqs. (12.4a–d), we obtain

$$N_\phi = -p_o a\cos^2\phi \tag{12.6a}$$

$$F(\phi) = 3p_o\sin\phi\cos\phi \tag{12.6b}$$

$$N_{x\phi} = 3p_o\left(\frac{L}{2} - x\right)\sin\phi\cos\phi \tag{12.6c}$$

$$N_x = -\frac{3p_o}{2a}x(L - x)(\cos^2\phi - \sin^2\phi) \tag{12.6d}$$

In this case we note that both N_ϕ and $N_{x\phi}$ are zero at edges $\phi = \pm\pi/2$. Hence no edge member will be required to carry such loads under membrane conditions. When the shell is not semicircular in section (that is, when ϕ at the edges is less than 90°), however, edge members will be required as discussed earlier.

12.4 SIMPLY SUPPORTED BARREL SHELLS WITH NONCIRCULAR CROSS SECTIONS

Equations (12.4a–d) derived by us apply to simply supported barrel shells of any cross section. By substituting values of r_2 and the loadings p_r, p_ϕ into these equa-

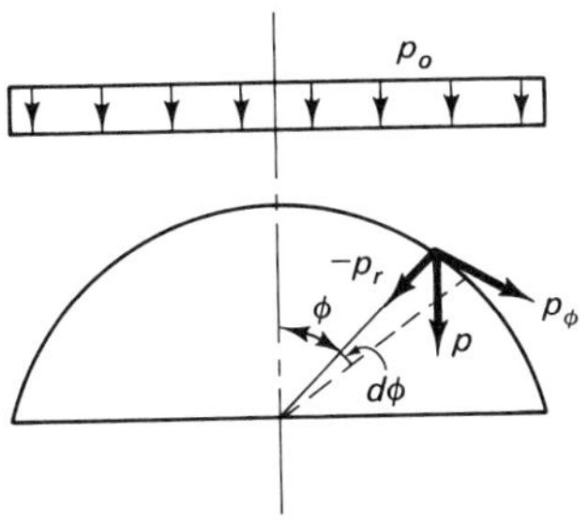

Figure 12.7 Barrel shell subjected to a load p_o uniformly distributed over the plan area.

tions, we can obtain expressions for membrane stress resultants. A variety of open cross-sectional shapes, such as elliptic, parabolic, cycloidal, catenary, etc., may be used. It can be shown that these various profiles of the cross section can be described by a single equation, $r_2 = r_o \cos^n \phi$, where r_o is the radius of curvature of the cross section at the crown $\phi = 0$ and n is a constant [2]. Different profiles, such as a circle, parabola, etc., can be defined by giving different values to r_o and n. For example, for a circular cross section, we must take $n = 0$ and $r_o = a$, whereas for a parabolic cross section defined by $y^2 = 4az$, we must take $r_o = 2a$ and $n = -3$.

Substituting $r_2 = r_o \cos^n \phi$ into Eqs. (12.4a–d), we can obtain general expressions for membrane stresses under dead or snow load [2]. Using these general expressions we can derive stress resultants in a shell of given cross-sectional profile. Since barrel shells other than those of circular cross section are not commonly used in practice, except for structures such as roofs of large storage silos, we shall not go into the details of determining membrane stresses in such shells. It can be verified, however, that for catenary-shaped cross sections, we shall find under dead load, $N_x = N_{x\phi} = 0$ throughout the shell, and only N_ϕ is nonzero. This means that such a shell does not span as a beam in the longitudinal direction, but transfers the load to the lower edges by acting merely as a series of catenary arches in the transverse direction. Supports must necessarily be provided at the lower edges to take up the load from these arches. The catenary arches are well known to be the funicular curves for dead loads; i.e., the arches carry dead loads by in-plane stresses only, without development of any bending moments or shears.

Similarly, for a cylindrical shell with a parabolic cross section, we shall find $N_{x\phi} = N_x = 0$ under snow or live load and the shell acts in this case merely as a series of parabolic arches. These parabolic arches are known to be funicular curves under loadings which are distributed uniformly over the plan area (see also Chapter 8).

As an example of a shell with a noncircular cross section, let us consider the

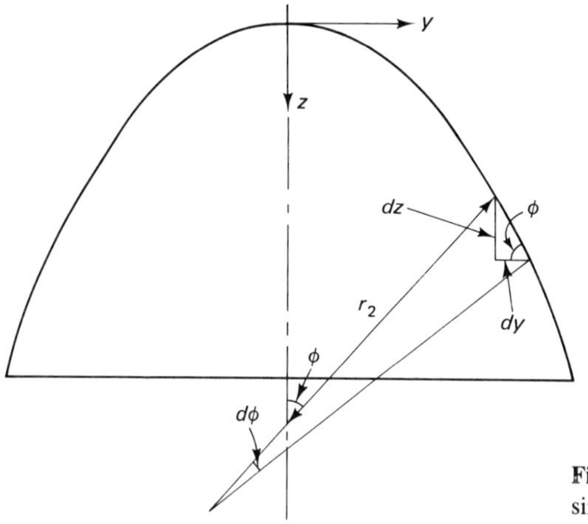

Figure 12.8 Parabolic cross section of a simply supported shell.

shell of parabolic cross section shown in Fig. 12.8. Such shells are commonly used for storage silos or aircraft hangars, which require a relatively large height. Parabolic or catenary cross sections provide a greater height than that of a circular cross section with the same bay width (see also Fig. 15.9). Such shells usually have short spans of about 20 to 30 ft and are supported on a series of ribbed arches at regularly spaced intervals.

Let us assume that the shell of Fig. 12.8 is simply supported over a span L and is subjected to a uniform load p over its surface area. The equation describing the middle surface of the parabola can be written as

$$y^2 = 4az$$

where the parameter a is known for a given cross section. We have noted earlier that the equation of this surface, in terms of r_2 and ϕ, is given by

$$r_2 = 2a \cos^{-3}\phi$$

This expression can be derived easily, as follows. From the expression above for y^2 in rectangular Cartesian coordinates, we can obtain derivatives as

$$\frac{dy}{dz} = \frac{2a}{y}$$

and

$$\frac{d^2y}{dz^2} = -\frac{2a}{y^2}\left(\frac{dy}{dz}\right) = -\frac{4a^2}{y^3}$$

From Fig. 12.8 we also see that tan ϕ is equal to dz/dy, and hence we can write the following relations using the derivatives above:

$$\frac{dy}{dz} = \cot\phi = \frac{2a}{y}$$

and

$$\frac{d^2y}{dz^2} = -\frac{4a^2}{y^3} = -\frac{1}{2a}\left(\frac{2a}{y}\right)^3 = -\frac{\cot^3\phi}{2a}$$

Now if the curvature, κ, of the middle surface is taken to be positive, as shown in Fig. 12.8, then using calculus we can write the following:

$$\kappa = \frac{1}{r_2} = -\frac{d^2y/dz^2}{[1 + (dy/dz)^2]^{3/2}}$$

The negative sign above is required because for the curve shown in Fig. 12.8 (for which κ is assumed positive), the derivative d^2y/dz^2 is negative. From the above, we have

$$r_2 = -\frac{[1 + (dy/dz)^2]^{3/2}}{d^2y/dz^2}$$

Substituting expressions for the derivatives dy/dz and d^2y/dz^2 in terms of ϕ into the expression above for r_2, we obtain

$$r_2 = 2a \frac{\csc^3\phi}{\cot^3\phi}$$

or

$$r_2 = 2a \cos^{-3}\phi$$

as noted earlier.

Now, under a self-weight loading of p, we have

$$p_r = -p \cos \phi$$

and

$$p_\phi = p \sin \phi$$

Substituting these values and the expression for r_2 above into Eqs. (12.4a–d), we can readily obtain the membrane stress resultants in this simply supported parabolic shell. The expressions for these stress resultants are

$$\begin{aligned} N_\phi &= -2ap \cos^{-2}\phi \\ F(\phi) &= -p \sin \phi \\ N_{x\phi} &= -\left(\frac{L}{2} - x\right)p \sin \phi \\ N_x &= \frac{p}{4a}(x)(L - x)\cos^4\phi \end{aligned}$$

For a silo shell, such as that shown in Fig. 15.9, where the equation of the middle surface is $y^2 = 48z$ and the self-weight loading is 70 lb/ft^2 over the surface area, we can substitute $a = 12$ and $p = 70$ into the expressions above in order to obtain the membrane stress resultants.

As an exercise, the reader should similarly obtain stress resultants under snow/live load which is uniformly distributed over the plan area, and verify that N_x and $N_{x\phi}$ are indeed zero for this case, as stated previously.

12.5 MEMBRANE STRESSES USING FOURIER SERIES

The partial differential equations for the membrane stress resultants in cylindrical shells are given by Eqs. (12.1a–c). We note that these equations have constant coefficients in x, and hence they will admit solutions in the form of cosine and sine functions in x. For simply supported barrel shells, which are commonly constructed as roof structures, we can therefore consider a Fourier series solution for the stress resultants.

We can assume the solutions of Eqs. (12.1a–c) expanded as Fourier series in x. If such solutions also satisfy the boundary conditions at $x = 0$ and $x = L$ automatically, they will then form the complete solution. Let us assume that

$$N_\phi = \sum_{n=1,2\ldots}^{\infty} N_{\phi n} \sin \frac{n\pi x}{L} \tag{12.7a}$$

$$N_{x\phi} = \sum_{n=0,1,2\ldots}^{\infty} N_{x\phi n} \cos \frac{n\pi x}{L} \tag{12.7b}$$

$$N_x = \sum_{n=1,2\ldots}^{\infty} N_{xn} \sin \frac{n\pi x}{L} \tag{12.7c}$$

It is seen from the above that $N_x = 0$ at $x = 0$ and L, while $N_{x\phi}$ is not zero at the ends. Hence the boundary conditions for the simply supported shells are satisfied by these expansions. For solving Eqs. (12.1a–c), we must expand the loads also into Fourier series in x. This can be done by assuming odd extensions of the loads in the interval $(-L, 0)$ (refer to Chapter 10). We shall now solve the equations for specific loading conditions of dead load and live load, for which closed-form solutions have already been obtained by us in Examples 12.2 and 12.3, respectively.

Example 12.4: Simply Supported Circular Cylindrical Barrel Shell under Self-Weight—Fourier Series Solution

We take $r_2 = a$, $p_x = 0$, and expand the dead load p, which is constant in x and ϕ, as a Fourier sine series over the interval $(0, L)$. Referring to Example 10.3, we have

$$p = \sum_{n=1,3,5,\ldots}^{\infty} \frac{4p}{n\pi} \sin \frac{n\pi x}{L}$$

We also have the relations (Fig. 12.5) $p_r = -p \cos \phi$ and $p_\phi = p \sin \phi$. Hence

$$p_x = 0 \tag{12.8a}$$

$$p_\phi = \frac{4p \sin \phi}{\pi} \sum_{n=1,3,\ldots}^{\infty} \frac{1}{n} \sin \frac{n\pi x}{L} \tag{12.8b}$$

$$p_r = -\frac{4p \cos \phi}{\pi} \sum_{n=1,3,\ldots}^{\infty} \frac{1}{n} \sin \frac{n\pi x}{L} \tag{12.8c}$$

Introducing Eqs. (12.7) and (12.8) into Eq. (12.1a), and equating the left and right-hand sides term by term, we obtain

$$N_{\phi n} = -\frac{4pa}{n\pi} \cos \phi \qquad n = 1, 3, \ldots, \infty \tag{12.9a}$$

Hence, from Eqs. (12.1b,c), we can also easily obtain

$$N_{x\phi n} = \frac{8pL}{n^2\pi^2}\sin\phi \qquad n = 1, 3, \ldots, \infty \tag{12.9b}$$

and

$$N_{xn} = -\frac{8pL^2}{an^3\pi^3}\cos\phi \qquad n = 1, 3, \ldots, \infty \tag{12.9c}$$

The complete solution is then given by substitution of these expressions into Eqs. (12.7a–c). By taking a sufficient number of terms in the series, one can see that this solution agrees closely with the closed-form solution of Eqs. (12.5a–d) obtained earlier. For many practical applications, even the first term alone will give acceptable results.

The real advantage of Fourier series solution will be seen when we consider the bending theory of such shells in Chapter 13. There we shall obtain complex partial differential equations for stress resultants or displacements, with x and ϕ as the independent variables. It is not possible to obtain a closed-form solution of these equations by direct integration, as in membrane analysis problems. Hence to solve these equations for simply supported barrel shells, we assume Fourier sine series solutions in the x-direction, which automatically satisfy the boundary conditions at $x = 0$ and L, and give us ordinary differential equations in ϕ. The particular solution of the partial differential equation, corresponding to a given loading, has also to be expressed as a Fourier sine series. This should be the solution of the partial differential equation for the bending theory, but commonly, it is taken as the solution [Eqs. (12.7a–c)] of the membrane theory equations (12.1a–c) without much loss of accuracy, as we shall see in Chapter 13.

It is interesting to note from Eqs. (12.7a–c) and (12.9a–c) that the series for N_ϕ, $N_{x\phi}$, and N_x have, respectively, $1/n$, $1/n^2$, and $1/n^3$ types of convergence. These series actually represent expansions of the functions given by Eqs. (12.5a,c,d). It can be easily seen that after taking an odd extension required for expansion into a sine series, the function of Eq. (12.5a) will have a discontinuity at $x = 0$ in the interval $(-L, L)$, and hence its Fourier series [Eqs. (12.7a) and (12.9a)] has the $1/n$ type of convergence. It can also be seen that the function of Eq. (12.5c) and its even extension (since $N_{x\phi}$ is expanded in a cosine series) will be continuous over $(-L, L)$, but its first derivative will be discontinuous at $x = -L$, O, L, etc. Hence the series for $N_{x\phi}$ has a $1/n^2$ type of convergence. The reader can also see that the function of Eq. (12.5d) and its first derivative, with their odd extensions, will be continuous over $(-L, L)$, but will have a discontinuous second derivative. Hence N_x has $1/n^3$-type convergence. Thus even considering only the first term of the series in our membrane solutions, the accuracy obtained for the $N_{x\phi}$ and N_x solutions will be good, while that for N_ϕ will be as good as that obtained for the loading from the first term of its series.

Example 12.5: Live or Snow Load on Simply Supported Barrel Shells—Fourier Series Solution

In Example 12.3 we obtained a closed-form solution of this problem of the barrel shell under live or snow load. Now, expanding the load p_o in a Fourier sine series, we can write

$$p_r = -p_o\cos^2\phi = -\frac{4p_o}{\pi}\cos^2\phi\sum_{n=1,3,\ldots}^{\infty}\frac{1}{n}\sin\frac{n\pi x}{L}$$

and

$$p_\phi = p_o \cos\phi \sin\phi = \frac{4p_o}{\pi} \cos\phi \sin\phi \sum_{n=1,3,\ldots}^{\infty} \frac{1}{n} \sin\frac{n\pi x}{L}$$

Introducing these series and Eqs. (12.7a–c) into Eqs. (12.1a–c), with $r_2 = a$, and equating term by term, we first obtain

$$N_{\phi n} = -\frac{4p_o a}{n\pi} \cos^2\phi \qquad n = 1, 3, \ldots, \infty \tag{12.10a}$$

Then, substituting this into Eq. (12.1b), we have

$$N_{x\phi n} \frac{-n\pi}{L} + \frac{1}{a} \frac{-4p_o a}{n\pi} (-2\cos\phi \sin\phi) = \frac{-4p_o}{n\pi} \cos\phi \sin\phi$$

Hence

$$N_{x\phi n} = \frac{12p_o L}{n^2\pi^2} \sin\phi \cos\phi = \frac{6p_o L}{n^2\pi^2} \sin 2\phi \qquad n = 1, 3, \ldots, \infty \tag{12.10b}$$

Also, from Eq. (12.1c), we obtain

$$\frac{n\pi}{L} N_{xn} + \frac{1}{a} \frac{6p_o L}{n^2\pi^2} 2\cos 2\phi = 0$$

Therefore,

$$N_{xn} = -\frac{12p_o L^2}{n^3\pi^3 a} \cos 2\phi \qquad n = 1, 3, \ldots, \infty \tag{12.10c}$$

Substitution of Eqs. (12.10a–c) into Eqs. (12.7a–c) gives the complete solution. The reader can easily verify that the observations regarding convergence of the series noted in the dead load solutions will be applicable here as well.

12.6 FOURIER SERIES EXPANSION IN THE ϕ-DIRECTION

Barrel shell roofs constructed in practice are commonly simply supported at their ends in the x-direction. Hence expansion of the stress resultants as Fourier series in x is ideal for solving such problems. In the ϕ-direction, however, a variety of boundary conditions can exist at the edges ϕ = *constant,* depending on whether these edges are free, continuous with other barrels (as in the case of the inner barrel of a multibarrel shell), or whether they are continuous with edge beams, etc.

In addition, since r_2 is a function of ϕ, Eqs. (12.1b,c) do not have constant coefficients in ϕ, and hence may not admit solutions consisting of simple cosine and sine functions in ϕ (i.e., of the type $e^{\lambda\phi}$) for a general shell cross section. For circular cross sections, however, r_2 is constant, and hence a Fourier series solution is possible in the ϕ-direction. However, such an expansion for the circular shell may still not prove to be very useful, because it will not permit us to satisfy the different possible boundary conditions at the edges $\phi = constant$ that we are likely to encounter in practical problems.

For closed circular cylinders, however, all stress resultants will be periodic with a period 2π, since, for example, $N_\phi(\phi) = N_\phi(\phi + 2\pi)$, etc. We can therefore obtain Fourier series solutions in ϕ for this case and satisfy different boundary conditions on the edges $x = constant$. Assuming the loads to be symmetrical about $\phi = 0$, the solutions for the stress resultants can be assumed in the form

$$N_\phi = \sum_{m=0}^{\infty} N_{\phi m} \cos m\phi \tag{12.11a}$$

$$N_{x\phi} = \sum_{m=1}^{\infty} N_{x\phi m} \sin m\phi \tag{12.11b}$$

$$N_x = \sum_{m=0}^{\infty} N_{xm} \cos m\phi \tag{12.11c}$$

$N_{x\phi}$ is chosen with a sine variation in ϕ, because for loads symmetrical about $\phi = 0$, $N_{x\phi}$ must also be symmetrical about the axis $\phi = 0$ [see Fig. 12.4(c)]. The same can be said for the choice of cosine functions for N_ϕ and N_x. The loads, which are symmetrical about $\phi = 0$ can also be expanded in Fourier series as*

$$p_r = \sum_{m=0}^{\infty} p_{rm} \cos m\phi \tag{12.12a}$$

$$p_\phi = \sum_{m=1}^{\infty} p_{\phi m} \sin m\phi \tag{12.12b}$$

$$p_x = \sum_{m=0}^{\infty} p_{xm} \cos m\phi \tag{12.12c}$$

Introducing Eqs. (12.11a–c) and (12.12a–c) into Eqs. (12.1a–c), solutions for membrane stress resultants can be obtained (after solving the resulting ordinary differential equations), and these can be adapted to the known boundary conditions at $x = 0$ and L. If these edges are simply supported, a solution in the form of a double Fourier series in x and ϕ can be obtained [1].

REFERENCES

[1] Flügge, W., *Stresses in Shells,* 2nd ed., Springer-Verlag, New York, 1973.

[2] Ramaswamy, G. S., *Design and Construction of Concrete Shell Roofs,* Revised ed., R. E. Krieger, Malabar, Florida, 1984.

*For loads that are antisymmetric about $\phi = 0$, the sine and cosine terms will have to be interchanged in Eqs. (12.11a–c) and (12.12a–c). For other general loads, the series for each load and stress resultant will contain the summation of both sine and cosine terms.

PROBLEMS

12.1. A circular cylindrical barrel shell of span L_1 is supported by diaphragms at its ends A and B. The shell cantilevers out beyond B for a length L. Find the membrane stress resultants, as closed-form solutions, in the shell under (a) dead load p uniformly distributed over its surface area, and (b) snow load p_o uniformly distributed over its plan area.

12.2. A circular cylindrical barrel shell is simply supported at $x = 0$ and L. It carries a snow load p_o uniformly distributed over its plan area between $x = L/4$ and $x = 3L/4$. Find expressions for the membrane stress resultants, using only the first term of the Fourier series.

12.3. Obtain the solution of Problem 12.2 in closed form.

12.4. For a cantilever circular cylindrical barrel shell, of semicircular cross section with radius a, the thickness increases linearly from zero at the tip of the cantilever ($x = 0$) to a value t at the fixed end ($x = L$). Find the membrane stress resultants in this shell, under its own weight.

12.5. For Problem 12.4, obtain the solution by expanding the loading as a Fourier series in the ϕ-direction.

12.6. For the water pipe in Example 12.1, obtain the solution for membrane stress resultants using a double Fourier series in ϕ and x.

12.7. Solve Example 12.4 again, assuming that the shell spans from $x = -L/2$ to $x = +L/2$ with the origin of coordinates $x = 0$ at its midspan.

12.8. A vertical chimney of circular cross section with radius a is subjected to a wind load, defined by $p_x = p_\phi = 0$ and $p_r = A_n \cos n\theta$. Find the membrane stress resultants in the chimney shell. Check that the axial stresses at the base of a chimney of height H match with those obtained by treating the chimney as a cantilever beam.

12.9. For the simply supported shell with parabolic cross section shown in Fig. 12.8, obtain the membrane stress resultants in the shell under a load p_o which is uniformly distributed over the plan area.

12.10. For a simply supported shell with a catenary cross section, the equation of the middle surface is given by

$$z = -a\left(1 - 85h\frac{y}{a}\right)$$

Obtain the membrane stress resultants in this shell under a load p which is uniformly distributed over (a) its surface area, and (b) its plan area.

13

Bending Theory of Circular Cylindrical Barrel Shells

13.1 INTRODUCTION

In Chapter 12 we have seen that with the membrane theory alone, we could satisfy boundary conditions at edges $x = 0$ and $x = L$ of simply supported open cylindrical shells, but we were unable to satisfy the boundary conditions at the edges $\phi = constant$. To satisfy compatibility and equilibrium at these edges, we must consider the bending stresses in the shell. In a general bending theory derivation, we will have 10 unknown stress resultants: N_x, N_ϕ, $N_{x\phi}$, $N_{\phi x}$, M_x, M_ϕ, $M_{x\phi}$, $M_{\phi x}$, Q_x, and Q_ϕ. The six possible equations of equilibrium will not be sufficient to solve for these 10 unknowns. By considering equations of kinematics and Hooke's law, however, we can derive a sufficient number of equations to solve for the unknown functions, which will then include deflections, as well.

Due to the variety of applications of barrel shells, for the construction of long-span column-free roof structures, an early interest in the solutions of problems involving these shells was generated (mainly in Germany in the period 1920–1935). Flügge [1] derived detailed equations of the bending theory, which consisted of three partial differential equations in terms of the displacements u, v, and w. His derivation considered the "trapezoidal" shape of the cross section of the shell element in which the (z/r) terms were not assumed to be negligible with respect to unity (refer to Section 1.6 and Fig. 1.15).

Considering the complications involved in using these "exact" equations for solving practical problems, efforts were made to simplify these equations. For solving problems of buckling of cylinders, Donnel neglected the trapezoidal shape of the cross section (as we did in Chapter 5 for the bending theory of circular cylinders), and he arrived at three simplified partial differential equations in u, v and w. These equations could be reduced to a single eighth-order equation in w.

Finsterwalder neglected some of the less important stress resultants, such as M_x, Q_x and $M_{x\phi} = M_{\phi x}$, and presented a somewhat simplified theory. Schorer further assumed Poisson's ratio as zero, neglected transverse strains in comparison with the longitudinal strains, and thus arrived at a simplified theory called "Schorer theory."

All of these theories are, however, based on the following common assumptions:

1. The material of the shell is homogeneous, isotropic, and linearly elastic, following Hooke's law.
2. Stresses normal to the shell surface can be neglected.
3. Deformations and slopes are small (so that approximate expressions for curvatures can be used).
4. Plane sections remain plane after bending (i.e., shear deformations due to transverse shear are neglected).

The assumptions above are similar to those we made in deriving the bending theory equations for circular plates, circular cylinders, and shells of revolution in previous chapters.

We shall henceforth restrict our attention to only simply supported shells—barrel shells supported at $x = 0$ and $x = L$ by "diaphragms" which are rigid in their own planes but are perfectly flexible against forces out of their planes. Such shell roofs are commonly constructed in civil engineering practice. The "simply supported" end conditions enable us to derive a simple mathematical solution of the complex equations. Thus for equations derived by any of the theories mentioned above, the procedure for the solution of simply supported shells is to assume each pertaining function as a Fourier series expansion in the x-direction. This solution will satisfy the differential equations and also the boundary conditions at $x = 0$ and L. The resulting ordinary differential equations in coordinate ϕ can be solved, and the solution can be adapted to given boundary conditions at the edges $\phi = constant$. We shall follow this procedure in this chapter, using the equations derived by Schorer theory.

13.2 DERIVATION OF EQUATIONS: SCHORER THEORY

If the longitudinal "span" L of the shell is large compared to the radius of its cross section, we expect that the shell will behave as a beam with curved cross section, which is supported by end diaphragms at $x = 0$ and L. Indeed, we can design circular tubes by simply using beam formulas derived in strength of materials. For such "long" shells, whose ratio of span to radius of cross section, L/a, is greater than at least three, simplified equations of Schorer theory can be applied.

When such a "long" shell spans as a beam of curved cross section, the external bending moment at any section $x = constant$ is resisted by the couple formed from compressive and tensile forces, respectively, at the top and bottom of the shell cross section, with a "lever arm" on the order of the depth of the cross section. Thus, due to its curvature and depth, the shell will resist the loads more efficiently than a plate, which resists loads by developing moments, say M_x, formed by compressive and tensile forces having a lever arm less than the thickness t. Such plate action for resisting loads would restrict the possible span length to an impractical small dimension. Thus we can reasonably assume that such plate action in the longitudinal direction is negligible in the shell, and hence the moments M_x over the shell thickness will be very small or zero. Correspondingly, the shears Q_x will also be zero. We can further assume that the twisting moments $M_{x\phi}$ and $M_{\phi x}$ are also zero.

As in beams, we can neglect the transverse and shear strains in the shell as compared to the longitudinal strain ε_x. Hence Schorer assumed ε_ϕ and $\gamma_{x\phi}$ as zero. He further simplified the equations by assuming Poisson's ratio ν to be zero. Thus Schorer's simplifying assumptions, which are in addition to the basic assumptions listed in the preceding section, can be summarized as follows.

Schorer's Simplifications

1. The actual "trapezoidal" shape of the cross section of the shell element is assumed to be rectangular.
2. M_x, Q_x, $M_{x\phi}$, and $M_{\phi x}$ are all equal to zero.
3. ε_ϕ and $\gamma_{x\phi}$ are equal to zero.*
4. $\nu = 0$.

Using these assumptions, we shall now derive the differential equations governing the bending behavior of circular cylindrical barrel shells. Figure 13.1(a) shows the stress resultants and distributed loads acting on an element of the shell bounded by sections x, $x + dx$, and ϕ, $\phi + d\phi$. We define the displacements along the x, ϕ, and radial directions as u, v, and w, respectively. Displacements u and v are positive along the positive x- and ϕ-directions, and w is positive when radially outward.

Equilibrium Equations

Considering equilibrium of forces in the x-, ϕ-, and r-directions, we can readily write down the following equations:

$$\frac{\partial N_x}{\partial x}\,dx(a\,d\phi) + \frac{\partial N_{\phi x}}{\partial \phi}\,d\phi\,dx + p_x(a\,d\phi)\,dx = 0$$

*It may be noted that although we assume ε_ϕ and $\gamma_{x\phi}$ as zero, we do not consider N_ϕ and $N_{x\phi}$ to be zero. In the theory of beams, we also neglected the shear deformations but not the shear forces.

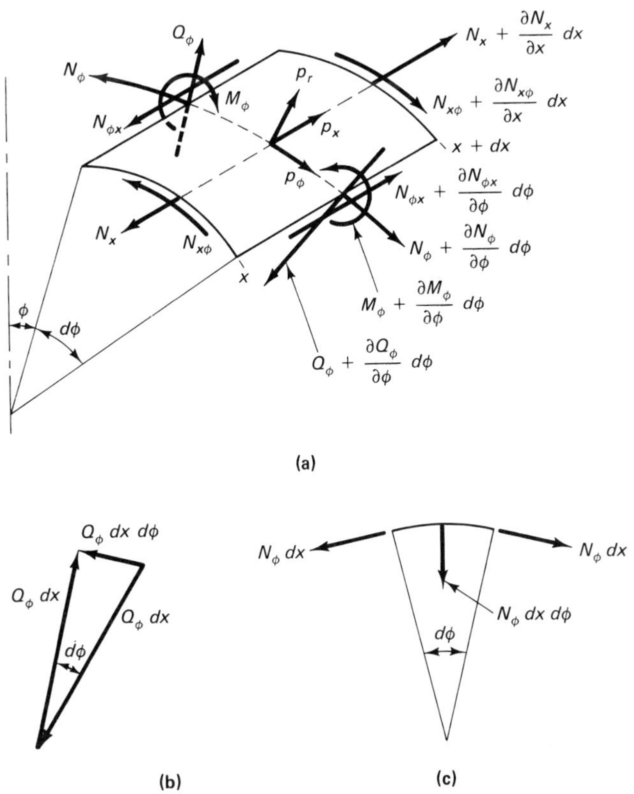

Figure 13.1 Element of a circular cylindrical barrel shell. (a) Stress resultants and distributed loading. (b) Tangential resultant of shears ($Q_\phi \cdot dx$). (c) Radial resultant of forces ($N_\phi \cdot dx$).

$$\frac{\partial N_\phi}{\partial \phi} d\phi\, dx + \frac{\partial N_{x\phi}}{\partial x} dx(a\, d\phi) - Q_\phi\, dx\, d\phi + p_\phi(a\, d\phi)\, dx = 0$$

$$-\frac{\partial Q_\phi}{\partial \phi} d\phi\, dx - N_\phi\, dx\, d\phi + p_r(a\, d\phi)\, dx = 0$$

The shears ($Q_\phi \cdot dx$) on the opposite edges are inclined at an angle $d\phi$ with one another. Taking the triangle of forces, as shown in Fig. 13.1(b), their resultant is $Q_\phi\, dx\, d\phi$, along the negative ϕ-direction.* This resultant is taken as the third term in the second equation above. Similarly, the second term in the third equation is obtained by taking the resultant of forces ($N_\phi \cdot dx$) acting on opposite faces of

*The contribution due to the increment in Q_ϕ, i.e., $(\partial Q_\phi/\partial\phi)\, d\phi\, dx$, is neglected, as it will be equal to $(\partial Q_\phi/\partial\phi) d\phi\, dx \sin(d\phi/2) \approx (\partial Q_\phi/\partial\phi)\, d\phi\, dx\, d\phi/2$, which is much smaller than the other terms.

the element, as shown in Fig. 13.1(c). Now, by taking moments about the edge $\phi + d\phi = constant$, we can write one more equation of equilibrium, as

$$\frac{\partial M_\phi}{\partial \phi} d\phi\, dx - Q_\phi\, dx(a\, d\phi) = 0$$

Since we have assumed M_x, Q_x, $M_{\phi x}$, and $M_{x\phi}$ as zero, the remaining two moment equilibrium equations will not give us any new information, other than $N_{x\phi} = N_{\phi x}$. Denoting $\partial(\cdot)/\partial x$ by $(\cdot)'$ and $\partial(\cdot)/\partial\phi$ by $(\cdot)^{\bullet}$, and simplifying, we can summarize the four equations above as

$$N_x' + \frac{1}{a} N_{x\phi}^{\bullet} = -p_x \tag{13.1a}$$

$$N_{x\phi}' + \frac{1}{a} N_\phi^{\bullet} = -p_\phi \tag{13.1b}$$

$$\frac{1}{a} N_\phi + \frac{1}{a} Q_\phi^{\bullet} = p_r \tag{13.1c}$$

$$\frac{1}{a} M_\phi^{\bullet} - Q_\phi = 0 \tag{13.1d}$$

It may be noted that in writing the second equation above, we have neglected the term containing Q_ϕ. This is a simplification, and it can be verified that this term gives only a negligible additional contribution to the final differential equation for w. The preceding four equations contain five unknowns, and hence we must consider the deformations of the shell in order to obtain additional equations for a solution.

Kinematics

Let ε_x and ε_ϕ be the two elemental axial strains and $\gamma_{x\phi}$ be the in-plane elemental shear strain, all constant over the thickness of the element. Referring to Fig. 13.2, we can write

$$\varepsilon_x = \frac{u + \dfrac{\partial u}{\partial x} dx - u}{dx} = \frac{\partial u}{\partial x} = u' \tag{13.2a}$$

Strain ε_ϕ will be caused by displacement v, as well as by the radial deflection w. From Fig. 13.2(b) we can write

$$\varepsilon_\phi = \frac{\partial v}{a\, \partial \phi} + \frac{w}{a} = \frac{1}{a} v^{\bullet} + \frac{1}{a} w \tag{13.2b}$$

Also, from Fig. 13.2(c), we can obtain

$$\gamma_{x\phi} = \frac{\partial v}{\partial x} + \frac{1}{a}\frac{\partial u}{\partial \phi} = v' + \frac{1}{a} u^{\bullet} \tag{13.2c}$$

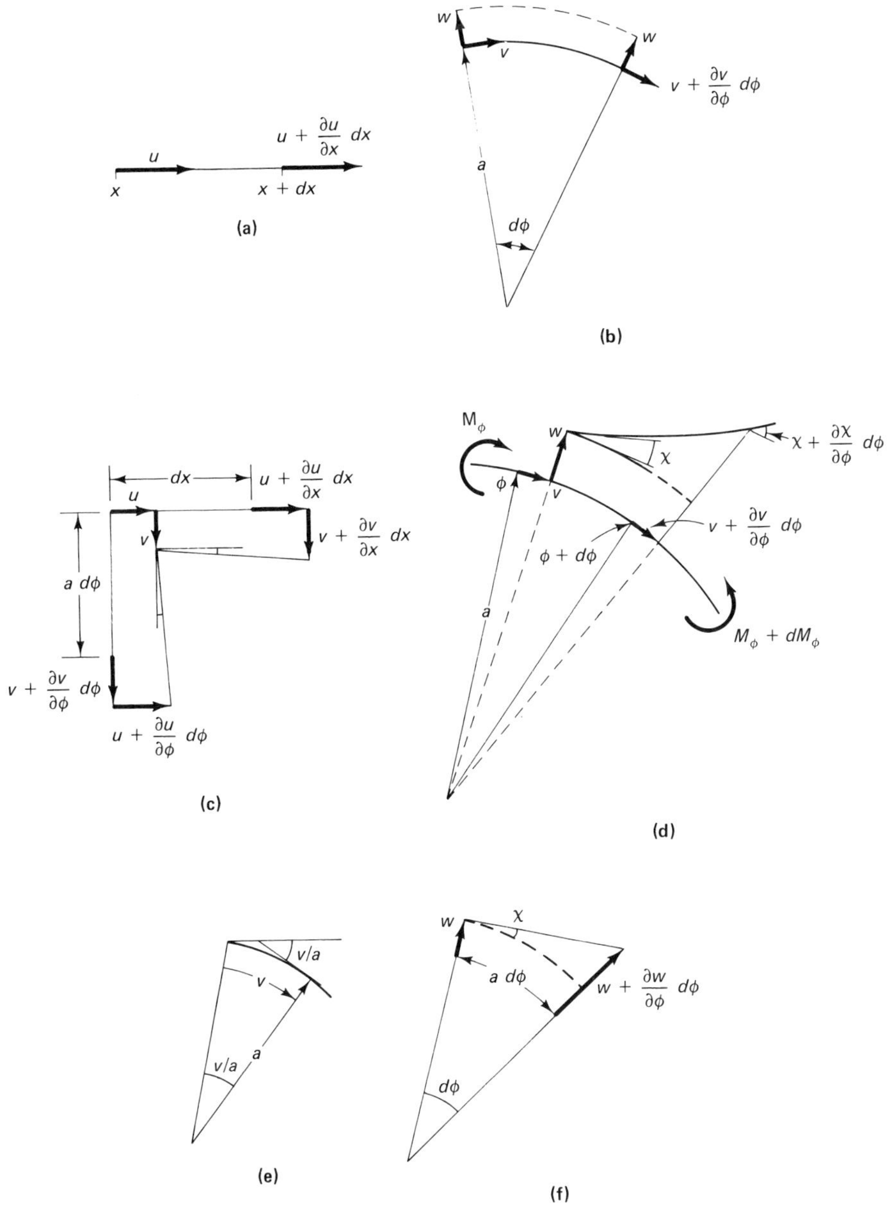

Figure 13.2 Deformations of the barrel shell. (a) Axial deformations. (b) Tangential and radial deformations. (c) Deformations leading to shear strain. (d) Deformations leading to a change in curvature κ_ϕ. (e) Rotation of tangent due to displacement v. (f) Rotation of tangent due to variation in displacement w.

By our earlier assumption, we take $\varepsilon_\phi = \gamma_{x\phi} = 0$. Hence from Eqs. (13.2b,c), we obtain

$$v^\bullet = -w \tag{13.3a}$$

and

$$u^\bullet = -av'$$

which gives

$$u^{\bullet\bullet} = -a(v')^\bullet = -a(v^\bullet)' = +aw' \tag{13.3b}$$

Also, from Eq. (13.2a), we have

$$\varepsilon_x^{\bullet\bullet} = (u')^{\bullet\bullet} = aw'' \tag{13.3c}$$

If χ denotes the rotation of the tangent (anticlockwise positive), as shown in Fig. 13.2(d), the change of curvature, κ_ϕ, is given by

$$\kappa_\phi = \frac{\chi + \dfrac{\partial\chi}{\partial\phi}\,d\phi - \chi}{a\,d\phi} = \frac{1}{a}\frac{\partial\chi}{\partial\phi} \tag{13.3d}$$

where κ_ϕ is taken positive as caused by positive M_ϕ; i.e., κ_ϕ is positive when the shell curvature decreases. This change in curvature corresponds to a positive $\partial\chi/\partial\phi$, and hence the positive sign in the expression above is correct. Now, due to a displacement v, the tangent will rotate clockwise by an amount v/a [Fig. 13.2(e)], and due to displacements w, it will rotate anticlockwise by an amount $(1/a)(\partial w/\partial\phi)$ [Fig. 13.2(f)]. Hence we find

$$\chi = \frac{w^\bullet - v}{a}$$

In Eq. (13.3a) we had $w = -v^\bullet$. Hence we can expect that $w^\bullet$ (or $-v^{\bullet\bullet}$) $\gg v^\bullet \gg v$, since our solutions will be of the type $e^{\lambda\phi}$, with $\lambda \gg 1$. We can therefore neglect the second term in the expression above, and write

$$\chi \approx \frac{w^\bullet}{a} \tag{13.3e}$$

Hooke's Law Equations

From the elasticity relations

$$\varepsilon_x = \frac{1}{Et}(N_x - \nu N_\phi) \qquad \text{and} \qquad \varepsilon_\phi = \frac{1}{Et}(N_\phi - \nu N_x)$$

we can obtain

$$N_x = \frac{Et}{1 - \nu^2} (\varepsilon_x + \nu\varepsilon_\phi)$$

With $\nu = 0$, this becomes*

$$N_x = Et\varepsilon_x \tag{13.4a}$$

Now, proceeding as in Chapters 5 and 7, we can derive the relation

$$M_\phi = K(\kappa_\phi + \nu\kappa_x)$$

or, taking $\nu = 0$,

$$M_\phi = K\kappa_\phi \tag{13.4b}$$

where $K = Et^3/12$ is the flexural rigidity of the shell. We shall now summarize below the equilibrium equations (13.1a–d), kinematic relations (13.3a–e), and Hooke's law relations (13.4a,b).

Equilibrium Equations

$$N_x' + \frac{1}{a} N_{x\phi}^{\bullet} = -p_x \tag{13.1a}$$

$$N_{x\phi}' + \frac{1}{a} N_\phi^{\bullet} = -p_\phi \tag{13.1b}$$

$$\frac{1}{a} N_\phi + \frac{1}{a} Q_\phi^{\bullet} = p_r \tag{13.1c}$$

$$\frac{1}{a} M_\phi^{\bullet} - Q_\phi = 0 \tag{13.1d}$$

Kinematic Relations

$$v^{\bullet} = -w \tag{13.3a}$$

$$u^{\bullet\bullet} = aw' \tag{13.3b}$$

$$\varepsilon_x^{\bullet\bullet} = aw'' \tag{13.3c}$$

*We shall also have $N_\phi = [Et/(1 - \nu^2)](\varepsilon_\phi + \nu\varepsilon_x)$, which appears to be zero when $\nu = 0$ and ε_ϕ is neglected. However, N_ϕ is not zero, and its value will be obtained from the equilibrium equations (similar to obtaining the shear force in a beam wherein shear strains are neglected). Similarly, $N_{x\phi}$ and Q_ϕ will also be obtained from the equations of equilibrium.

$$\kappa_\phi = \frac{\chi^{\bullet}}{a} \tag{13.3d}$$

$$\chi = \frac{w^{\bullet}}{a} \tag{13.3e}$$

(χ is taken as positive counterclockwise when considering the right-half of the shell cross section).

Hooke's Law

$$N_x = Et\varepsilon_x \tag{13.4a}$$

$$M_\phi = K\kappa_\phi \tag{13.4b}$$

where

$$K = \frac{Et^3}{12}$$

The 11 equations above contain 11 unknowns. We can eliminate all the unknowns except w, and thus obtain a single differential equation for w, as follows.

Differentiating Eqs. (13.1a) and (13.1b) with respect to x and ϕ, respectively, we can eliminate $N_{x\phi}$ from these two equations. We thus get

$$N_x'' - \frac{1}{a^2} N_\phi^{\bullet\bullet} = -p_x' + \frac{1}{a} p_\phi^{\bullet} \tag{13.5a}$$

Also, differentiating Eq. (13.1d) with respect to ϕ, we can eliminate Q_ϕ from Eqs. (13.1c) and (13.1d), in order to obtain

$$\frac{1}{a^2} M_\phi^{\bullet\bullet} + \frac{1}{a} N_\phi = p_r \tag{13.5b}$$

Differentiating Eq. (13.5b) twice with respect to ϕ, dividing by a, and adding to Eq. (13.5a), we obtain

$$\frac{1}{a^3} M_\phi^{\bullet\bullet\bullet\bullet} + N_x'' = -p_x' + \frac{1}{a} p_\phi^{\bullet} + \frac{1}{a} p_r^{\bullet\bullet} \tag{13.5c}$$

Combining Eqs. (13.4a) and (13.3c), we have

$$N_x^{\bullet\bullet} = Et\varepsilon_x^{\bullet\bullet} = Etaw'' \tag{13.5d}$$

Also, from Eqs. (13.4b) and (13.3d,e), we can obtain

$$M_\phi = K\kappa_\phi = K\frac{\chi^{\bullet}}{a} = K\frac{w^{\bullet\bullet}}{a^2} \tag{13.5e}$$

We now differentiate Eq. (13.5c) twice, with respect to ϕ, to obtain

$$\frac{1}{a^3} M_\phi^{\cdot\cdot\cdot\cdot\cdot\cdot} + N_x^{\prime\prime\cdot\cdot} = -p_x^{\prime\cdot\cdot} + \frac{1}{a} p_\phi^{\cdot\cdot\cdot} + \frac{1}{a} p_r^{\cdot\cdot\cdot\cdot}$$

Substituting for $N_x^{\cdot\cdot}$ and M_ϕ in terms of w, from Eqs. (13.5d,e), and simplifying, we obtain the following final equation for w:

$$w^{\cdot\cdot\cdot\cdot\cdot\cdot\cdot\cdot} + \Psi^8 w'''' = -\frac{a^5}{K} p_x^{\prime\cdot\cdot} + \frac{a^4}{K} (p_\phi^{\cdot\cdot\cdot} + p_r^{\cdot\cdot\cdot\cdot}) \tag{13.6}$$

where

$$\Psi^8 = \frac{Eta^6}{K} = \frac{12a^6}{t^2}$$

This equation is an eighth-order partial differential equation for w. For a given loading, we desire to obtain its solution, which must also satisfy the boundary conditions at the four edges of the shell. In a general case, this would be a formidable task. However, for simply supported shells, by using Fourier series expansions in the x-direction, we not only satisfy the boundary conditions at $x = 0$ and L, but we also separate the variables in Eq. (13.6) to obtain an eighth-order ordinary differential equation in the ϕ-direction. The eight constants of integration obtained in the solution of this equation can be determined from the boundary conditions on edges $\phi = constant$.

The general procedure for solution of this equation would thus be to obtain a particular solution of Eq. (13.6) which would satisfy the boundary conditions on edges $x = 0$ and L, but not necessarily on edges $\phi = constant$. Then, to this solution, we add the solution of the homogeneous equation (with right-hand side being zero), so that the two solutions together satisfy the boundary conditions on edges $\phi = constant$.

13.3 SOLUTION OF THE DIFFERENTIAL EQUATION

Edge-Load Solution

With the right-hand side taken as zero, i.e., with no distributed loads on the shell, Eq. (13.6) becomes

$$w^{\cdot\cdot\cdot\cdot\cdot\cdot\cdot\cdot} + \Psi^8 w'''' = 0 \tag{13.7}$$

where

$$\Psi^8 = \frac{12a^6}{t^2}$$

For a simply supported shell, w must be zero at the diaphragm supports, i.e., at $x = 0$ and $x = L$. Hence we assume a Fourier sine series expansion of w, such as

$$w(x, \phi) = \sum_{n=1}^{\infty} W_n(\phi) \sin \frac{n\pi x}{L} \tag{13.8}$$

This representation automatically satisfies the boundary conditions at $x = 0$ and L for w, as well as for other functions, as we shall see later. Substituting Eq. (13.8) into Eq. (13.7), we obtain an eighth-order ordinary differential equation for the nth harmonic $W_n(\phi)$. We thus have

$$\frac{d^8W_n}{d\phi^8} + \frac{n^4\pi^4}{L^4} \Psi^8 W_n = 0 \tag{13.9}$$

Since this equation has constant coefficients, we can assume its solutions in the form

$$W_n = Ae^{m\phi}$$

Substituting this into Eq. (13.9), we obtain the characteristic equation to determine m, as

$$m^8 + \lambda^8 = 0$$

where

$$\lambda^8 = \frac{n^4\pi^4}{L^4} \Psi^8 = 12 \frac{n^4\pi^4a^6}{L^4t^2} \tag{13.10}$$

It can be seen that λ^8 is a nondimensional quantity and will have a large value for thin shells. Now, solving the characteristic equation for m, we have

$$m = \lambda(-1)^{1/8}$$

The eight roots of (-1) are obtained using deMoivre's theorem of complex variables, after writing (-1) as a complex number:

$$\begin{aligned}(-1)^{1/8} &= (\cos \pi + i \sin \pi)^{1/8} \\ &= \cos \frac{\pi + 2\pi k}{8} + i \sin \frac{\pi + 2\pi k}{8}\end{aligned}$$

where

$$k = 0, 1, \ldots, 7$$

Alternatively, these roots can be written as

$$\pm \cos \frac{\pi}{8} \pm i \sin \frac{\pi}{8} \qquad \text{and} \qquad \pm \sin \frac{\pi}{8} \pm i \cos \frac{\pi}{8}$$

Hence the eight roots of m can be summarized as

$$m = \pm(\alpha \pm i\beta) \qquad \text{and} \qquad m = \pm(\beta \pm i\alpha)$$

where

$$\alpha = \lambda \cos\frac{\pi}{8} = 0.9238795\lambda \tag{13.11a}$$

$$\beta = \lambda \sin\frac{\pi}{8} = 0.3826834\lambda \tag{13.11b}$$

Then we can express the solution for the nth harmonic of Eq. (13.9) as

$$\begin{aligned} W_n(\phi) &= A_1 e^{(\alpha+i\beta)\phi} + A_2 e^{(\alpha-i\beta)\phi} + A_3 e^{-(\alpha+i\beta)\phi} + A_4 e^{-(\alpha-i\beta)\phi} \\ &\quad + A_5 e^{(\beta+i\alpha)\phi} + A_6 e^{(\beta-i\alpha)\phi} + A_7 e^{-(\beta+i\alpha)\phi} + A_8 e^{-(\beta-i\alpha)\phi} \end{aligned}$$

where $A_1, \ldots, A_8$ are constants to be determined from the boundary conditions at the edges $\phi = constant$.* By expanding the exponential terms, and combining them suitably, we can also write the solution as

$$\begin{aligned} W_n(\phi) = &\ [A \cos\beta\phi \cosh\alpha\phi - B \sin\beta\phi \sinh\alpha\phi \\ &+ C \cos\alpha\phi \cosh\beta\phi - D \sin\alpha\phi \sinh\beta\phi] \\ &+ [E \cos\beta\phi \sinh\alpha\phi - F \sin\beta\phi \cosh\alpha\phi \\ &+ G \cos\alpha\phi \sinh\beta\phi - H \sin\alpha\phi \cosh\beta\phi] \end{aligned} \tag{13.12}$$

where $A, B, \ldots, H$ are new unknown constants. It is seen that the first four terms of this solution are even functions of ϕ, symmetrical about $\phi = 0$, whereas the last four terms are odd functions of ϕ. In most practical problems, the shell, as well as its loading (such as dead load and snow load), are symmetrical about the crown $\phi = 0$. For such problems, constants E, F, G, H will be zero, and we are left with only four constants A, B, C, D to be determined from the boundary conditions at edge $\phi = \phi_k$. The solution thus becomes much more simplified, and henceforth we shall restrict our discussion to only symmetrical loading.

Symmetrical Loading

For loads symmetric about $\phi = 0$, the solution for w becomes

$$w(x, \phi) = \sum_{n=1}^{\infty} W_n(\phi) \sin\frac{n\pi x}{L} \tag{13.8}$$

*Strictly speaking, we should write λ, α, and β as λ_n, α_n, and β_n, and the constants as $A_{1n}, \ldots, A_{8n}$, to represent the solution of the nth harmonic. However, this is not done for the sake of simplicity.

where

$$W_n(\phi) = A\cos\beta\phi\cosh\alpha\phi - B\sin\beta\phi\sinh\alpha\phi + C\cos\alpha\phi\cosh\beta\phi - D\sin\alpha\phi\sinh\beta\phi \tag{13.13}$$

For each harmonic n, we must also express the stress resultants and displacements in terms of the constants A, B, C, and D. To do this, we proceed as follows.

From Eq. (13.5e), for any harmonic n, we have

$$M_\phi = \frac{K}{a^2} w^{\bullet\bullet} = \frac{K}{a^2}\frac{d^2W_n}{d\phi^2}\sin\frac{n\pi x}{L} \tag{13.14a}$$

Then, from Eqs. (13.1a–d), with loads taken as zero for the homogeneous solution, we obtain

$$Q_\phi = \frac{1}{a}M^\bullet_\phi = \frac{K}{a^3}\frac{d^3W_n}{d\phi^3}\sin\frac{n\pi x}{L} \tag{13.14b}$$

$$N_\phi = -Q^\bullet_\phi = -\frac{K}{a^3}\frac{d^4W_n}{d\phi^4}\sin\frac{n\pi x}{L} \tag{13.14c}$$

$$N'_{x\phi} = -\frac{1}{a}N^\bullet_\phi = \frac{K}{a^4}\frac{d^5W_n}{d\phi^5}\sin\frac{n\pi x}{L}$$

Integrating this last equation gives*

$$N_{x\phi} = -\frac{K}{a^4}\frac{L}{n\pi}\frac{d^5W_n}{d\phi^5}\cos\frac{n\pi x}{L} \tag{13.14d}$$

and we also find

$$N'_x = -\frac{1}{a}N^\bullet_{x\phi} = \frac{K}{a^5}\frac{L}{n\pi}\frac{d^6W_n}{d\phi^6}\cos\frac{n\pi x}{L}$$

or

$$N_x = \frac{K}{a^5}\frac{L^2}{n^2\pi^2}\frac{d^6W_n}{d\phi^6}\sin\frac{n\pi x}{L} \tag{13.14e}$$

We can also obtain expressions for the displacements. From Eqs. (13.2a) and (13.4a), we have

$$u' = \varepsilon_x = \frac{N_x}{Et} = \frac{K}{Eta^5}\frac{L^2}{n^2\pi^2}\frac{d^6W_n}{d\phi^6}\sin\frac{n\pi x}{L}$$

*Constants of integration with respect to x will here be functions of ϕ and will add terms $f_1(\phi)$ in $N_{x\phi}$ and $(x/a)(df_1/d\phi) + f_2(\phi)$ in N_x similar to the terms in Eqs. (12.3b,c) of Chapter 12. However, expression (13.14e) automatically satisfies the boundary conditions $N_x = 0$ at $x = 0$ and L. Hence we must have $f_2 = 0$ and $df_1/d\phi = 0$ or $f_1 = constant$. This constant will appear in the expression for $N_{x\phi}$, but will also be zero for the same reasons as stated in Section 12.3.

or

$$u = -\frac{K}{Eta^5}\frac{L^3}{n^3\pi^3}\frac{d^6W_n}{d\phi^6}\cos\frac{n\pi x}{L} \tag{13.14f}$$

where the constant of integration, which gives only a rigid-body displacement, is taken equal to zero. Now, from Eq. (13.2c), equating $\gamma_{x\phi}$ to zero, we have

$$v' = -\frac{1}{a}u^{\bullet} = +\frac{K}{Eta^6}\frac{L^3}{n^3\pi^3}\frac{d^7W_n}{d\phi^7}\cos\frac{n\pi x}{L}$$

Hence

$$v = \frac{K}{Eta^6}\frac{L^4}{n^4\pi^4}\frac{d^7W_n}{d\phi^7}\sin\frac{n\pi x}{L} \tag{13.14g}$$

Again, the integration constant is taken as zero, since Eq. (13.14g) satisfies the boundary conditions $v = 0$ at $x = 0$ and L. The rotation of the tangent, χ, is obtained by using Eq. (13.3e) as

$$\chi \approx \frac{w^{\bullet}}{a} = \frac{1}{a}\frac{dW_n}{d\phi}\sin\frac{n\pi x}{L} \tag{13.14h}$$

This completes the evaluation of the various quantities of interest to us, required for solving most practical problems. It may be noted that M_ϕ, N_x, w, and v are zero, and $N_{x\phi}$ and u are nonzero at $x = 0$ and L, as required by the simply supported conditions.

Substituting Eq. (13.13) into Eqs. (13.14a–h), we can express all desired quantities in terms of the unknown constants A, B, C, D. Using the recurrence relations for the derivatives of trigonometric and hyperbolic functions and their products, we can check that all odd derivatives; $dW_n/d\phi$, $d^3W_n/d\phi^3$, . . . , etc. will have the same form, and that all even derivatives W_n, $d^2W_n/d\phi^2$, . . . , etc. will also have a similar form. Hence the solutions can be conveniently summarized into two groups, in a tabular format similar to the one used by Gibson [2]. The form of solution for a function $F_1(x, \phi)$, with even derivatives in ϕ, will be

$$\begin{aligned} F_1(x, \phi) &= f_1(x)g_1(\phi) \\ &= f_1(x)[(c_1A - c_2B)\cos\beta\phi\cosh\alpha\phi - (c_2A + c_1B)\sin\beta\phi\sinh\alpha\phi \\ &\quad + (c_3C - c_4D)\cos\alpha\phi\cosh\beta\phi - (c_4C + c_3D)\sin\alpha\phi\sinh\beta\phi] \end{aligned} \tag{13.15a}$$

where c_1, c_2, c_3, c_4 are constants having different numerical values for the different functions. Also, $f_1(x)$ will be different for different functions. Equation (13.15a) can also be written as

$$\begin{aligned} F_1(x, \phi) = f_1(x)[&A(c_1R_1 - c_2R_2) - B(c_2R_1 + c_1R_2) \\ &+ C(c_3R_3 - c_4R_4) - D(c_4R_3 + c_3R_4)] \end{aligned} \tag{13.15b}$$

where

$$R_1 = \cos\beta\phi \cosh\alpha\phi \qquad R_2 = \sin\beta\phi \sinh\alpha\phi$$

$$R_3 = \cos\alpha\phi \cosh\beta\phi \qquad R_4 = \sin\alpha\phi \sinh\beta\phi$$

Similarly, the form of the solution for any function $F_2(x, \phi)$, with odd derivatives in ϕ, will be

$$\begin{aligned} F_2(x, \phi) &= f_2(x) \cdot g_2(\phi) \\ &= f_2(x)[(c_5A - c_6B) \cos\beta\phi \sinh\alpha\phi - (c_6A + c_5B) \sin\beta\phi \cosh\alpha\phi \\ &\quad + (c_7C - c_8D) \cos\alpha\phi \sinh\beta\phi - (c_8C + c_7D) \sin\alpha\phi \cosh\beta\phi] \end{aligned}$$

or

$$F_2(x, \phi) = f_2(x)[A(c_5R_5 - c_6R_6) - B(c_6R_5 + c_5R_6) \tag{13.15c}$$

$$+ C(c_7R_7 - c_8R_8) - D(c_8R_7 + c_7R_8)] \tag{13.15d}$$

where

$$R_5 = \cos\beta\phi \sinh\alpha\phi \qquad R_6 = \sin\beta\phi \cosh\alpha\phi$$

$$R_7 = \cos\alpha\phi \sinh\beta\phi \qquad R_8 = \sin\alpha\phi \cosh\beta\phi$$

Expressions for F_1 and F_2, for various stress resultants and displacements, obtained by substitution of Eq. (13.13) into Eqs. (13.14a–h), are summarized in Tables 13.1 and 13.2.

We shall also need expressions for the vertical and horizontal stress resultants, and for the corresponding displacements at the edge $\phi = \phi_k$, in order that the boundary conditions can be applied more conveniently in some problems. We denote

TABLE 13.1 $F_1(x, \phi)$: Functions with Even Derivatives in ϕ[a]

$F_1(x, \phi) = f_1(x)g_1(\phi)$ [Refer to Eq. (13.15a or b)]

$F_1(x, \phi)$	$f_1(x)$	c_1	c_2	c_3	c_4
M_ϕ	$\frac{K}{a^2}\lambda^2 \sin\frac{n\pi x}{L}$	$1/\sqrt{2}$	$1/\sqrt{2}$	$-1/\sqrt{2}$	$1/\sqrt{2}$
N_ϕ	$-\frac{K}{a^3}\lambda^4 \sin\frac{n\pi x}{L}$	0	1	0	-1
N_x	$\frac{K}{a^5}\frac{L^2}{n^2\pi^2}\lambda^6 \sin\frac{n\pi x}{L}$	$-1/\sqrt{2}$	$1/\sqrt{2}$	$1/\sqrt{2}$	$1/\sqrt{2}$
u	$-\frac{K}{Eta^5}\frac{L^3}{n^3\pi^3}\lambda^6 \cos\frac{n\pi x}{L}$	$-1/\sqrt{2}$	$1/\sqrt{2}$	$1/\sqrt{2}$	$1/\sqrt{2}$
w	$\sin\frac{n\pi x}{L}$	1	0	1	0

[a] $K = Et^3/12$; $\lambda^8 = 12n^4\pi^4a^6/L^4t^2$.

TABLE 13.2 $F_2(x, \phi)$: Functions with Odd Derivatives in ϕ[a]

$F_2(x, \phi) = f_2(x)g_2(\phi)$ [Refer to Eq. (13.15c or d)]
$[\rho_1 = \cos(\pi/8) = 0.9238795$ and $\rho_2 = \sin(\pi/8) = 0.3826834]$

$F_2(x, \phi)$	$f_2(x)$	c_5	c_6	c_7	c_8
Q_ϕ	$\frac{K}{a^3}\lambda^3 \sin\frac{n\pi x}{L}$	ρ_2	ρ_1	$-\rho_1$	$-\rho_2$
$N_{x\phi}$	$\frac{-K}{a^4}\frac{L}{n\pi}\lambda^5 \cos\frac{n\pi x}{L}$	$-\rho_2$	ρ_1	ρ_1	$-\rho_2$
χ	$\frac{\lambda}{a}\sin\frac{n\pi x}{L}$	ρ_1	ρ_2	ρ_2	ρ_1
v	$\frac{K}{Eta^6}\frac{L^4}{n^4\pi^4}\lambda^7 \sin\frac{n\pi x}{L}$	$-\rho_1$	ρ_2	$-\rho_2$	ρ_1

[a] $K = Et^3/12$; $\lambda^8 = 12n^4\pi^4a^6/L^4t^2$; from Eqs. (13.11a,b), $\alpha = \rho_1\lambda$ and $\beta = \rho_2\lambda$.

the vertical and horizontal stress resultants per unit length, at the edge $\phi = \phi_k$, by V and H, respectively, with their positive directions as shown in Fig. 13.3. The corresponding displacements Δ_V and Δ_H are also as shown. So, from Fig. 13.3, we can write the relations

$$V = Q_\phi \cos\phi_k + N_\phi \sin\phi_k \tag{13.16a}$$

$$H = -Q_\phi \sin\phi_k + N_\phi \cos\phi_k \tag{13.16b}$$

$$\Delta_V = -w \cos\phi_k + v \sin\phi_k \tag{13.16c}$$

$$\Delta_H = w \sin\phi_k + v \cos\phi_k \tag{13.16d}$$

Using the expressions from Tables 13.1 and 13.2, where $\phi = \phi_k$, we can also express the quantities above in terms of the unknown constants A, B, C, and D. This completes the solution of the homogeneous equation. To this solution we must add a particular solution of Eq. (13.6) for any given loading, in order to obtain the complete solution.

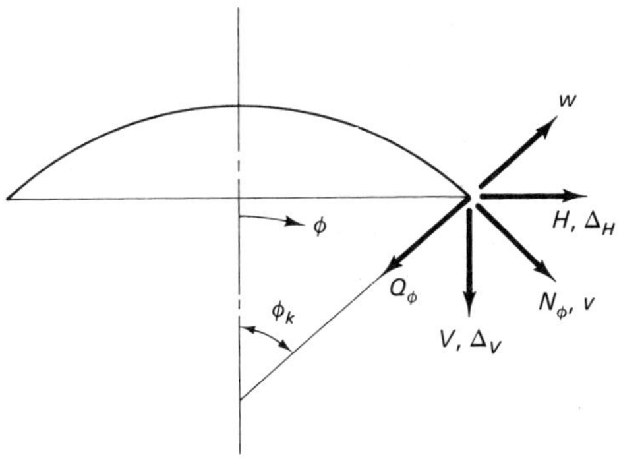

Figure 13.3 Components of shell deformations and forces at the edge $\phi = \phi_k$.

Particular Solution

Let us consider a constant dead load, p, per unit area of the shell surface. As in the membrane analysis of Chapter 12, we shall have

$$p_x = 0 \qquad p_\phi = p \sin \phi \qquad p_r = -p \cos \phi$$

Expanding p as a Fourier sine series in x, we can write the loads above as [refer to Eqs. (12.8b,c)]

$$p_\phi = \frac{4p \sin \phi}{\pi} \sum_{n=1,3,\ldots}^{\infty} \frac{1}{n} \sin \frac{n\pi x}{L}$$

$$p_r = -\frac{4p \cos \phi}{\pi} \sum_{n=1,3,\ldots}^{\infty} \frac{1}{n} \sin \frac{n\pi x}{L}$$

We shall also assume the particular solution, w_p, as a Fourier sine series in the interval $(0, L)$ so that it satisfies the boundary conditions $w = 0$ at $x = 0$ and L. Thus we have

$$w_p = \sum_{n=1}^{\infty} W_{pn} \sin \frac{n\pi x}{L}$$

Substituting this expression, and the series for the loadings, into Eq. (13.6), and dropping $\sin (n\pi x/L)$, we get the following ordinary differential equation for the nth harmonic W_{pn}:

$$\frac{d^8 W_{pn}}{d\phi^8} + \lambda^8 W_{pn} = -\frac{8pa^4}{n\pi K} \cos \phi$$

The particular solution of this equation can be taken as $W_{pn} = A \cos \phi$, where constant A can be determined by substitution into the equation above. We shall then get

$$W_{pn} = -\frac{8pa^4}{n\pi K} \frac{1}{1 + \lambda^8} \cos \phi \tag{13.17}$$

where $\lambda^8 = 12n^4\pi^4a^6/L^4t^2$. For thin shells, with $a/t \gg 1$, and hence $\lambda^8 \gg 1$, we can take $\lambda^8 + 1 \approx \lambda^8$ in the denominator above. Indeed, we shall see that by doing so, this solution matches that obtained in the membrane theory. This leads us to consider adopting the membrane solution itself as a particular solution. Such a solution will not exactly satisfy Eq. (13.6), since it will be a solution of an approximate differential equation obtained by neglecting bending moments and transverse shears. However, for practical problems, this solution will be sufficiently close to the correct particular solution. It is therefore a common practice to consider, for simplicity, that the membrane solution for a given loading is the particular solution, and then add to it the homogeneous solution obtained earlier. Henceforth, therefore, we shall adopt only the membrane solution as the particular solution.

Membrane solution as the particular solution. In Example 12.4 we found the membrane stress resultants under uniform dead load p. Referring to Eqs. (12.9a–c), we can rewrite these stress resultants for the nth harmonic, dropping the subscript n for simplicity, as

$$N_\phi = -\frac{4pa}{n\pi}\cos\phi\sin\frac{n\pi x}{L}$$

$$N_{x\phi} = \frac{8pL}{n^2\pi^2}\sin\phi\cos\frac{n\pi x}{L}$$

$$N_x = -\frac{8pL^2}{an^3\pi^3}\cos\phi\sin\frac{n\pi x}{L}$$

where $n = 1, 3, \ldots, \infty$. We shall also have

$$M_\phi = Q_\phi = 0$$

To obtain the displacements, we proceed as follows:

$$u' = \varepsilon_x = \frac{N_x}{Et} = -\frac{8pL^2}{Et\,an^3\pi^3}\cos\phi\sin\frac{n\pi x}{L}$$

Hence

$$u = \frac{8pL^3}{Et\,an^4\pi^4}\cos\phi\cos\frac{n\pi x}{L}$$

where the constant of integration, which gives a simple rigid-body displacement, is taken as zero. Also, from Eq. (13.3b), we write

$$u^{\cdot\cdot} = aw'$$

Hence

$$w' = -\frac{8pL^3}{Eta^2n^4\pi^4}\cos\phi\cos\frac{n\pi x}{L}$$

or

$$w = -\frac{8pL^4}{Eta^2n^5\pi^5}\cos\phi\sin\frac{n\pi x}{L}$$

where the constant of integration is again zero from the boundary conditions at $x = 0$. From Eq. (13.3a), we have $v^{\cdot} = -w$. Hence

$$v = \frac{8pL^4}{Eta^2n^5\pi^5}\sin\phi\sin\frac{n\pi x}{L}$$

Also, from Eq. (13.3e), we recall that $\chi = w^{\cdot}/a$; so

$$\chi = \frac{8pL^4}{Eta^3n^5\pi^5}\sin\phi\sin\frac{n\pi x}{L}$$

We shall summarize the membrane solutions obtained above as follows.

Particular solution for uniform dead load: summary for *n*th harmonic

$$M_\phi = 0 \tag{13.18a}$$

$$N_\phi = -\frac{4pa}{n\pi}\cos\phi\sin\frac{n\pi x}{L} \tag{13.18b}$$

$$N_x = -\frac{8pL^2}{an^3\pi^3}\cos\phi\sin\frac{n\pi x}{L} \tag{13.18c}$$

$$u = \frac{8pL^3}{Et\,an^4\pi^4}\cos\phi\cos\frac{n\pi x}{L} \tag{13.18d}$$

$$w = -\frac{8pL^4}{Eta^2n^5\pi^5}\cos\phi\sin\frac{n\pi x}{L} \tag{13.18e}$$

$$Q_\phi = 0 \tag{13.18f}$$

$$N_{x\phi} = \frac{8pL}{n^2\pi^2}\sin\phi\cos\frac{n\pi x}{L} \tag{13.18g}$$

$$\chi = \frac{8pL^4}{Eta^3n^5\pi^5}\sin\phi\sin\frac{n\pi x}{L} \tag{13.18h}$$

$$v = \frac{8pL^4}{Eta^2n^5\pi^5}\sin\phi\sin\frac{n\pi x}{L} \tag{13.18i}$$

The vertical and horizontal stress resultants and displacements at the edge ϕ = *constant* can be obtained from Eqs. (13.16a–d). We can now compare Eq. (13.18e) with the solution of the bending theory differential equation [Eq. (13.17)]. Neglecting (1) in comparison with λ^8, and substituting the value of λ^8, Eq. (13.17) becomes

$$W_{pn} = -\frac{8pa^4}{n\pi(Et^3/12)}\,\frac{\cos\phi}{12n^4\pi^4a^6/(L^4t^2)} = -\frac{8pL^4}{Eta^2n^5\pi^5}\cos\phi$$

This matches exactly with the coefficient of sin $(n\pi x/L)$ in the membrane solution for w found in Eq. (13.18e). Hence using the membrane solution as the particular solution appears to be sufficiently accurate for solving practical problems.

In a similar manner, the particular (membrane) solution for live/snow loads, or any other loadings, can be derived. Combining the particular solution with the homogeneous solutions of Tables 13.1 and 13.2, we obtain the complete solution, wherein constants A, B, C, and D are determined from the boundary conditions at $\phi = \phi_k$ for each harmonic. The final solution is then obtained as the summation of such solutions over a sufficient number of harmonics. Considering the lengthy calculations involved, when solving practical problems by hand computations, one would consider only the first harmonic $n = 1$, or at most, summation over $n = 1$ and $n = 3$. We shall now consider a simple example in the next section.

13.4 SIMPLY SUPPORTED SINGLE BARREL WITH FREE EDGES

Figure 13.4 shows a single-barrel shell, simply supported at $x = 0$ and $x = L$. The L/a ratio for this shell is 3.36. Since this ratio is greater than 3, we can apply, with reasonable accuracy, the Schorer theory for solving this problem. The rise of the shell is $h = a(1 - \cos \phi_k) = 2.924$ ft, which is about $\frac{1}{14}$ of its span. We take the loading as

3-in.-thick shell	=	37.5 lb/ft²
Finish	=	10.0
Live load	=	15.0
Total		62.5 lb/ft²

For practical applications, we can consider the total load as uniformly distributed over the surface area of the shell, although strictly speaking, this is not true for the live load. We shall analyze the shell only for the first harmonic ($n = 1$) of the Fourier series expansion for the loading. The shell parameters are (refer to Tables 13.1 and 13.2)

$$\lambda^8 = \frac{12\pi^4(12.5)^6}{42^4(0.25)^2} = 22{,}927.87 \qquad \text{or} \qquad \lambda = 3.508$$

$$\alpha = \rho_1\lambda = 3.241 \qquad \text{and} \qquad \beta = \rho_2\lambda = 1.342$$

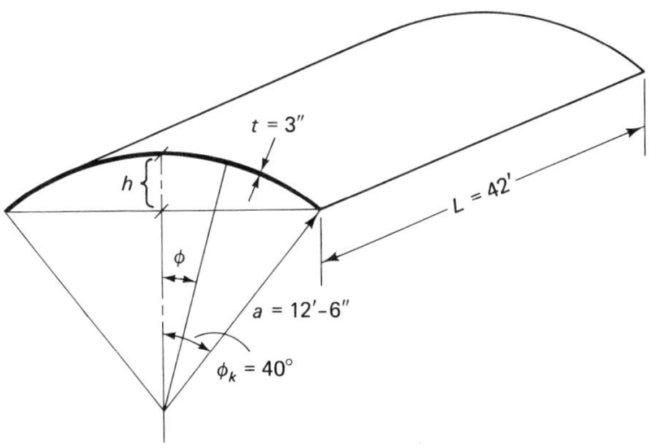

Figure 13.4 Simply supported single-barrel shell with free edges.

TABLE 13.3 Particular (Membrane) Solution for Example Barrel Shell

ϕ (deg)	$N_\phi/\sin\frac{\pi x}{L}$	$N_{x\phi}/\cos\frac{\pi x}{L}$	$N_x/\sin\frac{\pi x}{L}$	$M_\phi/\sin\frac{\pi x}{L}$	$Q_\phi/\sin\frac{\pi x}{L}$
0	−994.72	0.0	−2275.67	0.0	0.0
10	−979.61	369.48	−2241.10	0.0	0.0
20	−934.73	727.73	−2138.43	0.0	0.0
30	−861.45	1063.87	−1970.79	0.0	0.0
40	−762.00	1367.69	−1743.26	0.0	0.0

Particular Solution

We take the membrane solution as the particular solution, which can be obtained by substituting $p = 62.5$ into Eqs. (13.18a–i). We can summarize the values of the stress resultants thus obtained, for various values of ϕ, as given in Table 13.3. To these values we must add the homogenous solutions from Tables 13.1 and 13.2 which contain the four unknown constants A, B, C, and D that must be determined from the boundary conditions. These boundary conditions are $N_\phi = N_{x\phi} = M_\phi = Q_\phi = 0$ at $\phi = \phi_k = 40°$. To write the values of the stress resultants, in terms of A, B, C, and D, we first evaluate the functions $R_1, R_2, \ldots, R_8$ in tabular form, as shown in Tables 13.4(a,b).

Using the values at $\phi = 40°$, we can write expressions for the stress resultants at the edge, from Tables 13.1 and 13.2 and Eqs. (13.15a–d). Adding the membrane values at $\phi = 40°$ from Table 13.3 to these expressions, and equating the total stress resultants to zero at the free edge, we obtain the following equations for the constants:

$$\frac{M_\phi}{\sin(\pi x/L)} = (-8.301A - 58.371B + 0.940C + 15.419D)\frac{K}{a^2} + 0 = 0$$

$$\frac{N_\phi}{\sin(\pi x/L)} = (46.413A + 34.855B - 10.080C + 11.388D)\frac{K}{a^2} - 762.0 = 0$$

$$\frac{Q_\phi}{\sin(\pi x/L)} = (-8.767A - 14.153B + 3.695C + 2.704D)\frac{K}{a^2} + 0 = 0$$

$$\frac{N_{x\phi}}{\cos(\pi x/L)} = (213.298A + 50.124B + 9.225C + 59.555D)\frac{K}{a^2} + 1367.69 = 0$$

The last term in each of these equations represents the membrane solution. The common factor K/a^2 is taken out from the homogeneous solution, so that the value of E need not be specified. Solving these equations gives

$$A = \frac{-36.1241}{K/a^2} \qquad B = \frac{28.0868}{K/a^2}$$

$$C = \frac{-43.6514}{K/a^2} \qquad D = \frac{89.5366}{K/a^2}$$

TABLE 13.4(a) ***R*-Functions Even in ϕ, for Example Barrel Shell**

ϕ (deg)	$\alpha\phi$	$\beta\phi$	$R_1 = \cos \beta\phi \cosh \alpha\phi$	$R_2 = \sin \beta\phi \sinh \alpha\phi$	$R_3 = \cos \alpha\phi \cosh \beta\phi$	$R_4 = \sin \alpha\phi \sinh \beta\phi$
0	0	0	1.0	0.0	1.0	0.0
10	0.566	0.234	1.133	0.138	0.867	0.127
20	1.131	0.468	1.527	0.626	0.473	0.439
30	1.697	0.703	2.152	1.705	−0.158	0.756
40	2.263	0.937	2.877	3.831	−0.940	0.832

TABLE 13.4(b) ***R*-Functions Odd in ϕ, for Example Barrel Shell**

ϕ (deg)	$\alpha\phi$	$\beta\phi$	$R_5 = \cos \beta\phi \sinh \alpha\phi$	$R_6 = \sin \beta\phi \cosh \alpha\phi$	$R_7 = \cos \alpha\phi \sinh \beta\phi$	$R_8 = \sin \alpha\phi \cosh \beta\phi$
0	0	0	0.0	0.0	0.0	0.0
10	0.566	0.234	0.580	0.270	0.199	0.551
20	1.131	0.468	1.239	0.772	0.207	1.006
30	1.697	0.703	2.012	1.823	−0.096	1.247
40	2.263	0.937	2.815	3.914	−0.689	1.133

TABLE 13.5 Procedure to Evaluate N_ϕ for Example Barrel Shell

1	2	3	4	5	6	7	$N_\phi/\sin(\pi x/L)$
ϕ (deg)	$A\left(\frac{K}{a^2}\right)(c_1R_1 - c_2R_2)$	$-B\left(\frac{K}{a^2}\right)(c_2R_1 + c_1R_2)$	$C\left(\frac{K}{a^2}\right)(c_3R_3 - c_4R_4)$	$-D\left(\frac{K}{a^2}\right)(c_4R_3 + c_3R_4)$	2 + 3 + 4 + 5	$(6)\left(-\frac{\lambda^4}{a}\right)$	7 + membrane
0	0.0	−28.087	0.0	89.537	61.450	−744.47	−1,739.19
10	4.985	−31.822	−5.544	77.628	45.247	−548.18	−1,527.79
20	22.614	−42.889	−19.163	42.351	2.913	−35.29	−970.02
30	61.592	−60.443	−33.000	−14.147	−45.998	557.28	−304.17
40	138.391	−80.806	−36.318	−84.164	−62.897	762.00	~0.0

We can substitute these values into Eqs. (13.15a–d) and into Tables 13.1 and 13.2, in order to evaluate the values of stress resultants and displacements for various values of ϕ. These calculations can be done in tabular form for each quantity, as shown for N_ϕ in Table 13.5. Proceeding in a similar manner we can obtain the values for other stress resultants for various values of ϕ. These are summarized in Table 13.6.

As expected, the values of M_ϕ, N_ϕ, Q_ϕ, and $N_{x\phi}$ at $\phi = 40°$ are zero. The functions vary along x as per the sine or cosine function carried with them. Since $\sin(\pi x/L) = 1$ at $x = L/2$, the numerical values given in Table 13.6 are the actual values at midspan for M_ϕ, N_ϕ, Q_ϕ, and N_x. For $N_{x\phi}$ they represent the values at the supports. The results above can be readily used to design reinforcement in the barrel shell. We shall see the details of this in Chapter 15.

For hand calculations, the procedure just followed becomes very laborious, even with the simplest case of free-edge boundary conditions. However, a simple computer program can be written incorporating the expressions of the homogeneous solution, as well as a general expression for membrane stress resultants and displacements (particular solutions) under Fourier expansion, for any loading. Solutions can then be easily obtained for any loading, with summation over one or more terms of the series.

13.5 OTHER BOUNDARY CONDITIONS

In the preceding example involving a single-barrel shell with free edges, the boundary conditions led to a simple formulation of equations for constants A, B, C, and D. In practice, we may also have multibarrel shells, shells with edge beams, etc., for which the boundary conditions are more complex. We shall consider such cases in this section.

Multibarrel Shells

Consider a typical internal barrel of a multibarrel shell structure (Fig. 13.5). We can assume symmetry about the vertical axes through the valleys. Hence the horizontal displacement, the rotation, and the shear stress $N_{x\phi}$, at the valleys, will

TABLE 13.6 Final Stress Resultants for Example Barrel Shell

ϕ (deg)	$M_\phi/\sin\frac{\pi x}{L}$ (lb-ft/ft)	$N_\phi/\sin\frac{\pi x}{L}$ (lb/ft)	$Q_\phi/\sin\frac{\pi x}{L}$ (lb/ft)	$N_{x\phi}/\cos\frac{\pi x}{L}$ (lb/ft)	$N_x/\sin\frac{\pi x}{L}$ (lb/ft)
0	−958.02	−1,739.19	0.0	0.0	−17,367.59
10	−822.39	−1,527.79	118.32	2,692.64	−14,702.15
20	−489.50	−970.02	172.82	4,465.80	−5,837.23
30	−143.91	−304.17	125.66	4,125.73	11,671.60
40	0.0	0.0	0.0	~0.0	41,203.12

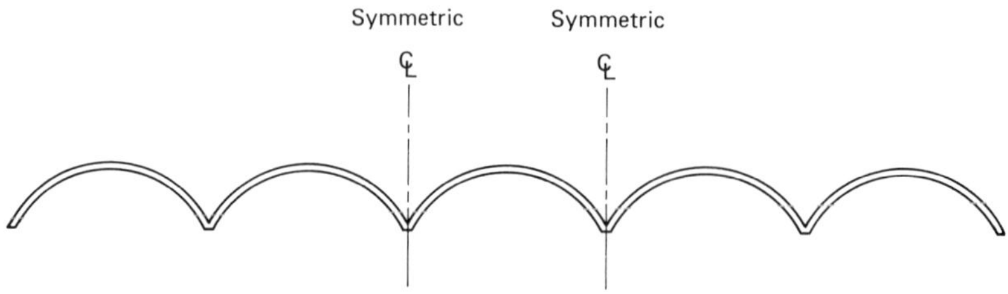

Figure 13.5 Shell roof consisting of identical multiple barrels.

be zero. Also, since there is no edge beam or any other support along the valleys which can resist vertical forces, we must have the resultant vertical force V at the valleys equal to zero. Hence we can summarize the four boundary conditions as

$$\text{At } \phi = \phi_k: \quad \Delta_H = 0 \quad \chi = 0 \quad N_{x\phi} = 0 \quad V = 0$$

With the help of Eqs. (13.16a–d), we can again formulate four equations for A, B, C, and D and solve the problem directly.

The boundary conditions above will not be valid at the valleys of the external shell of this structure. Hence the solution obtained can be used only for internal shell units and not for "end" shells. A practical approach adopted to solve such end-shell problems is to assume that the inner half of the external shell will have a stress distribution similar to that in an internal barrel unit, and that its outer half will have stresses similar to those in a single shell with free edges. This means that for designing such structures completely, we must solve two problems: one of an internal shell unit, and the other of a single shell with free edges.

Single Shell with Edge Beams

Since a "long" shell behaves similar to a beam with a curved cross section, its depth should generally be between $\frac{1}{8}$ to $\frac{1}{16}$ of the span, as for most reinforced concrete beams. When the depth of the shell cross section itself does not satisfy this requirement, the extra depth required can be obtained by providing edge beams, as shown in Fig. 13.6(a).

Typical edge beams have narrow widths, generally 8 to 12 in., and they can thus be considered as having zero stiffness in resisting horizontal loads or torsional moments. Hence, at the valleys, we will have $H = 0$ and $M_\phi = 0$. Also, to satisfy compatibility, the vertical and longitudinal deflections of the shell and the beam must match at their junction. We can thus write the boundary conditions as:

$$\text{At } \phi = \phi_k: \quad H = M_\phi = 0 \quad \Delta_V \text{ (shell)} = \Delta_V \text{ (beam)} \quad u_{\text{shell}} = u_{\text{beam}}$$

To obtain Δ_V(beam) and u_{beam} in terms of A, B, C, and D, we must apply on the beam forces opposite to V and $N_{x\phi}$ acting at the edge of the shell. Considering only the first harmonic, the loading on the beam will be as shown in Fig. 13.7.

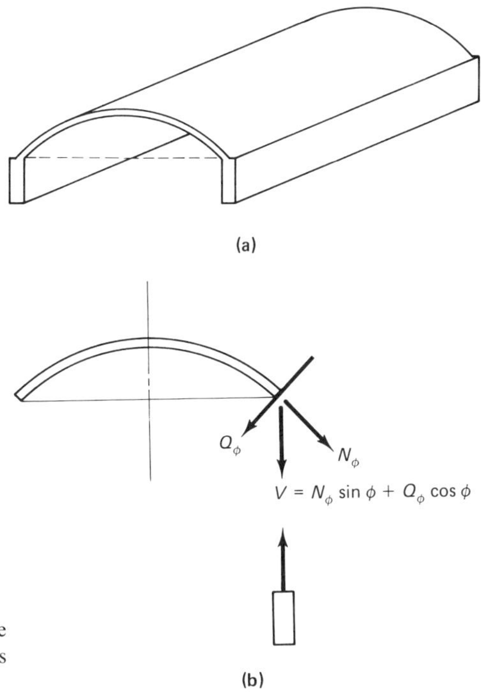

Figure 13.6 Single-barrel shell with edge beams. (a) Shell configuration. (b) Forces acting at the shell/edge beam junction.

Since the beam is also supported by diaphragms at the ends $x = 0$ and L, expressions for vertical and horizontal deflections at the top face of this simply supported beam can be obtained under the two loads V and $N_{x\phi}$ which come from the shell. Since V and $N_{x\phi}$ will be in terms of the four unknown constants A, B, C, and D, so also will be Δ_V(beam) and u_{beam}. Finally, from the boundary conditions, we can obtain the four equations for A, B, C, and D.

Multibarrel Shell with Edge Beams

For a typical internal barrel of a multibarrel shell with edge beams (Fig. 13.8), we shall have the boundary conditions:

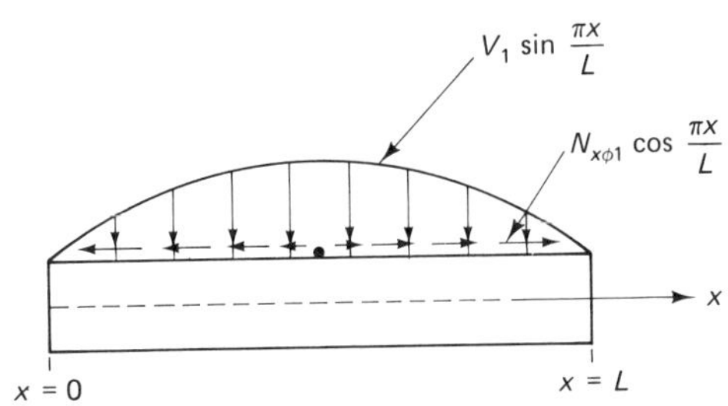

Figure 13.7 First harmonic approximation to the loading on the edge beam of a barrel shell.

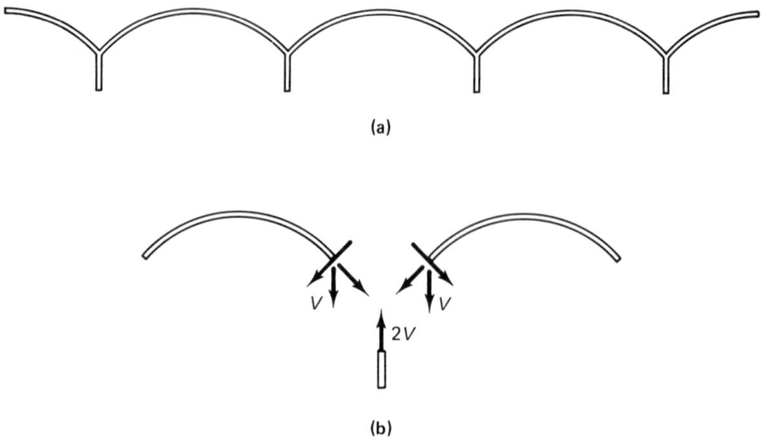

Figure 13.8 Multibarrel shell with edge beams. (a) Configuration of multiple barrels. (b) Forces acting at an internal junction.

$$\text{At } \phi = \phi_k: \quad \Delta_H = \chi = 0$$
$$\Delta_V(\text{shell}) = \Delta_V(\text{beam})$$
$$u_{\text{shell}} = u_{\text{beam}}$$

In applying V and $N_{x\phi}$ on the edge beam, we must remember that the beam receives these forces from two adjoining shells, and hence we must apply $2V$ and $2N_{x\phi}$ on the edge beam, in order to obtain Δ_V(beam) and u_{beam}. Here again, the outer half of the external shell must be designed by using the results of a single shell with edge beams.

13.6 SHORT SHELLS

The equations derived in this chapter, assuming Schorer's simplifications, can only be applied for solving problems of "long" shells—shells that have a ratio L/a greater than at least 3. For multibarrel shells without edge beams, where horizontal displacements and rotations at the valleys are restrained, this ratio could be less than 3. There will exist a wide range of shells constructed in practice which do not satisfy this aspect ratio requirement, and hence cannot be solved by the Schorer theory. For such short shells, the beam action in the span direction is less predominant. The shell does not act as a beam of curved cross section, with longitudinal stresses varying from compressive to tensile from the crown to the valleys. Only the lower portion of the shell will act as a beam in the longitudinal direction in very short shells. Hence, for solving such problems, we must use more refined formulations, such as those of Flügge or Donnel [1].

In Donnel's formulation, M_x, Q_x, $M_{x\phi}$, ε_ϕ, $\gamma_{x\phi}$, and ν, which were taken as zero by Schorer, are now taken into consideration. However, the "trapezoidal" cross-sectional shape of the shell element, as considered by Flügge, is taken as rectangular by Donnel. Proceeding as in the Schorer theory, we can derive a partial differential equation for w, which can then be solved in a similar fashion as done for the Schorer equation. The solution of a shell problem using these equations will, however, involve considerable complexities unless the method is programmed on a computer.

Even before analyzing any shell by a refined computer program, one would like to have a quick solution by approximate methods. Hence a number of tables and approximate methods are available which can be used for solving cylindrical shell problems when a computer program is not readily available, or when a quick solution is desired, or a check on computer results is to be done. In the next chapter, we study the analysis of cylindrical shells by using such methods.

REFERENCES

[1] Flügge, W., *Stresses in Shells,* 2nd ed., Springer-Verlag, New York, 1973.

[2] Gibson, J. E., *Linear Elastic Theory of Thin Shells,* Pergamon Press Ltd., Oxford, 1965.

PROBLEMS

13.1. Obtain the particular solution for a live load of p_o, uniformly distributed over the plan area of a barrel shell, by using the differential equation of bending theory. Compare your solution with that obtained in the membrane analysis in Chapter 12.

13.2. Solve the shell problem of Section 13.4 for a live load of 15 lb/ft^2, uniformly distributed over the plan area, using Schorer theory.

13.3. Solve the shell problem of Section 13.4 as if it were an internal shell of a multibarrel shell structure without edge beams. Compare the variations of N_x and M_ϕ with those obtained in Section 13.4.

13.4. Write a computer program for finding the internal stress resultants in a cylindrical barrel shell using Schorer theory. The program should be able to analyze a single shell or an internal shell of a multibarrel shell (both without edge beams). It should also allow for summation over one or more harmonics, as specified. Consider only loads uniformly distributed over the surface area.

14

Analysis of Cylindrical Barrel Shells by Tables, Approximate Methods, and Computer Programs

14.1 INTRODUCTION

In Chapter 13 we saw the complexities involved in solving a simply supported cylindrical shell problem, even by the simpler Schorer theory. Hence, to aid the design engineer, a number of tables, charts, and approximate methods have been developed. These tables are derived from computer analyses of shells with varying parameters. Early tables published by ASCE [1] cover a wide range of shell parameters, yet the designer must still perform lengthy calculations for analyzing a shell, even with the use of these tables. Other tables [2–4] or charts [5] may cover specific types of shells with certain restrictions on the parameters, such as span-to-chord-width ratios, presence of edge beams, depth of edge beams, etc.; however, these are much simpler to use and one can directly obtain the stress resultants in the shell, provided that the shell parameters fit the requirements of a particular set of tables. For a more detailed analysis, or for shells not considered in existing tables, one can develop a computer program for solving such problems. In this chapter we study the analysis of shells by such methods, i.e., by use of tables and computer programs.

14.2 ANALYSIS USING THE ASCE MANUAL

To aid in the design of shell roof structures, the American Society of Civil Engineers (ASCE) has published an extensive set of tables for calculating stress resultants in barrel shells, since as early as 1952 [1]. Later, the Portland Cement Association

[6] published additional tables as an extension of the ASCE tables, in order to cover some shell parameters not previously considered.

The ASCE Manual simplifies the exact shell equations (Flügge's equations) in a manner similar to Donnel's simplifications, for solving problems of shells with $L/a > 1.67$. Further simplifications are made in preparing tables for short shells, with $L/a < 1.67$. The procedure adopted for solving these shell problems is somewhat different from the one we used in Chapter 13. In our use of the Schorer equations, we always solved for the unknown integration constants A, B, C, and D. In the ASCE manual, unknown stress resultants at the boundary $\phi = constant$ are assumed as the redundant forces. These forces are then obtained by solving compatibility (or equilibrium) equations at the boundary. Assuming a Fourier series expansion in the x-direction, and considering only the first term, the redundant stress resultants and corresponding displacements are taken as

$$\overline{X}_1 = X_1 \sin \frac{\pi x}{L} \qquad \overline{\delta}_1 = \delta_1 \sin \frac{\pi x}{L}$$

$$\overline{X}_2 = X_2 \sin \frac{\pi x}{L} \qquad \overline{\delta}_2 = \delta_2 \sin \frac{\pi x}{L}$$

$$\overline{X}_3 = X_3 \cos \frac{\pi x}{L} \qquad \overline{\delta}_3 = \delta_3 \cos \frac{\pi x}{L}$$

$$\overline{X}_4 = X_4 \sin \frac{\pi x}{L} \qquad \overline{\delta}_4 = \delta_4 \sin \frac{\pi x}{L}$$

These quantities are shown in their assumed positive directions in Fig. 14.1. Redundants $\overline{X}_1$ and $\overline{X}_2$ are conveniently taken in the vertical and horizontal directions, respectively, rather than taking them in the directions of N_ϕ and Q_ϕ. Redundant $\overline{X}_3$ is the longitudinal shear, whereas $\overline{X}_4$ is the edge moment. The ASCE Manual measures coordinate ϕ, starting from $\phi = 0$ at the edge, to $\phi = \phi_k$ at the crown. Hence the direction of positive $N_{x\phi}$ is as shown on the shell element in Fig. 14.1(c). Since $\cos(\pi x/L)$ is negative for $x > L/2$, $X_3 \cos(\pi x/L)$ is also negative for $x > L/2$, when X_3 is positive. Hence the direction of this edge shear is as shown in Fig. 14.1(c).

For solving the problem of a shell with edge beams, as shown in Fig. 14.2, we first separate the shell and the beam into two primary structures: (1) the shell with its loading, and (2) the edge beam with its self-weight. At the junction of the shell and edge beam, in general, redundants $\overline{X}_1$, $\overline{X}_2$, $\overline{X}_3$, and $\overline{X}_4$, as shown in Fig. 14.1, will act on the shell. Equal and opposite forces will act on the beam, as shown in Fig. 14.3. Four similar unknown stress resultants, $\overline{X}_5$, $\overline{X}_6$, $\overline{X}_7$, and $\overline{X}_8$, will also develop at the other edge of the shell. For long shells, which predominantly act as beams of curved cross section in the longitudinal direction, the effect of edge loads applied at one edge will also be felt at the other edge. Hence, in general,

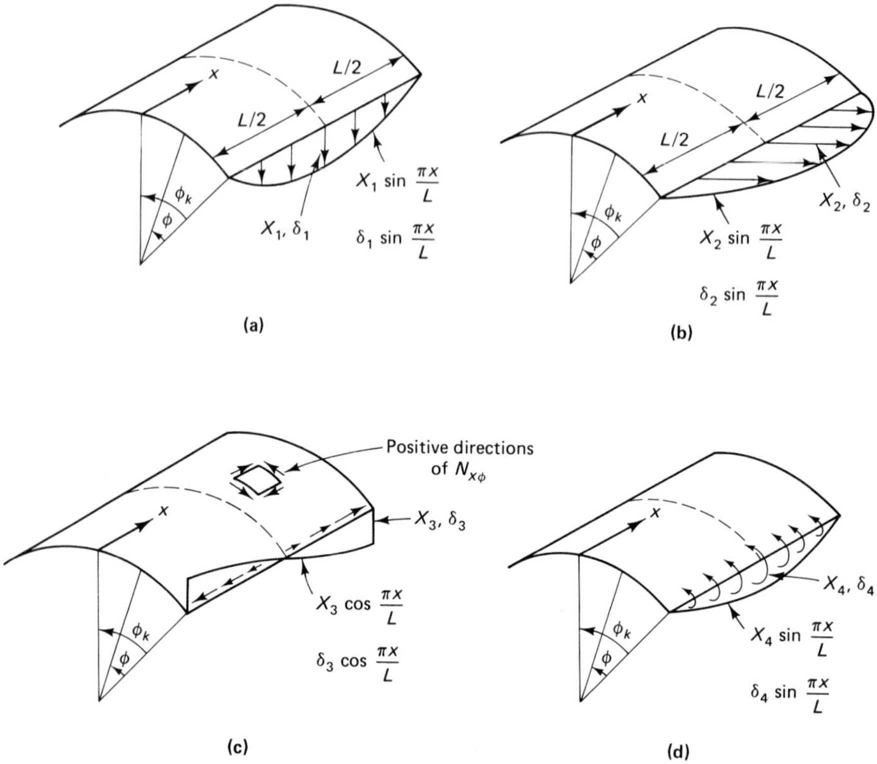

Figure 14.1 Redundant forces and corresponding displacements at $\phi = 0$, used in the ASCE Manual. (a) Redundants $\overline{X}_1$ and displacements $\overline{\delta}_1$. (b) Redundants $\overline{X}_2$ and displacements $\overline{\delta}_2$. (c) Redundants $\overline{X}_3$ and displacements $\overline{\delta}_3$. (d) Redundants $\overline{X}_4$ and displacements $\overline{\delta}_4$.

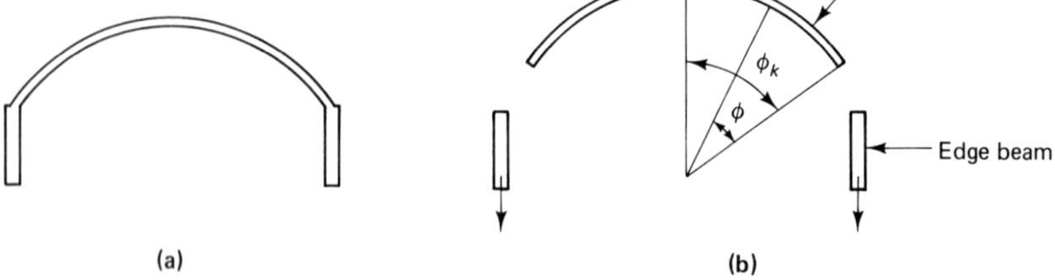

Figure 14.2 Barrel shell with edge beams. (a) Shell section configuration. (b) Primary structures of shell and edge beams.

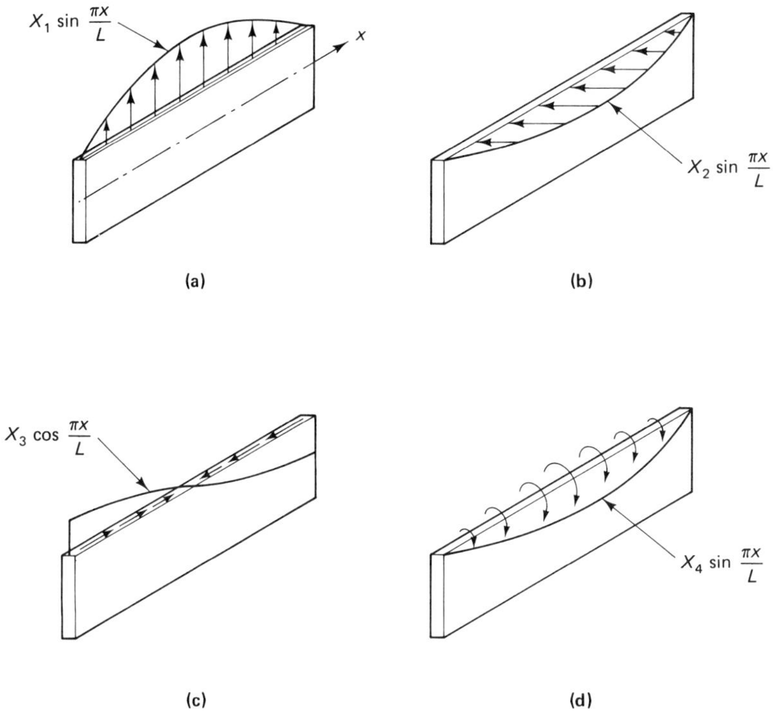

Figure 14.3 Redundant forces acting on the edge beam at $\phi = 0$ (see Fig. 14.2). (a) Vertical redundant forces. (b) Horizontal redundants. (c) Axial redundants. (d) Redundant moments.

eight compatibility equations will have to be written to solve for $\overline{X}_1, \ldots, \overline{X}_8$. However, in most practical problems, the shells, as well as their loadings, will be symmetrical about the crown. Hence we need only keep the symmetrical terms in the solutions and solve four equations for $\overline{X}_1, \ldots, \overline{X}_4$ acting at one edge, as we did in Chapter 13 using Schorer theory.

For short shells (e.g., aircraft hangars, grain silos, etc.), the effects of loads $\overline{X}_1, \ldots, \overline{X}_4$ at one edge generally die out rapidly and are not felt at the other edge. Hence for such short shells, even with unsymmetrical loadings, we need solve for only four redundants at one edge.*

To solve for redundants $\overline{X}_1, \overline{X}_2, \overline{X}_3, \overline{X}_4$, and obtain the stress resultants in the shell, we proceed as follows. Let $\overline{\delta}_{ij}$ ($i = 1, \ldots, 4, j = 1, \ldots, 4$) represent the displacement in the direction of $\overline{X}_i$ due to a unit value of $\overline{X}_j$, and let $\overline{\delta}_{i0}$ ($i = 1, \ldots, 4$) be the displacements of the primary structures in the direction of

*It may be recalled from Eq. (13.10) of Schorer theory that λ^8 was proportional to $(a^4/L^4)\,(a^2/t^2)$. Our solutions were of the type $\exp\,(\pm C\lambda\phi)$, where C is a constant number. At the edge $\phi = 0$, the dominant solutions will be of the type $e^{-C\lambda\phi}$, which die out slower for long shells which have smaller a/L values in comparison to short shells.

$\overline{X}_i$. The four compatibility equations at the junction of the shell and the edge beam can then be written as

$$\overline{\delta}_{i0} + \overline{\delta}_{i1}\overline{X}_1 + \overline{\delta}_{i2}\overline{X}_2 + \overline{\delta}_{i3}\overline{X}_3 + \overline{\delta}_{i4}\overline{X}_4 = 0 \qquad (i = 1, \ldots, 4) \qquad (14.1\text{a})$$

$\overline{\delta}_{i0}$ and $\overline{\delta}_{ij}$ are relative displacements between the shell and the edge beam, so that

$$\overline{\delta}_{i0} = \overline{\delta}^S_{i0} + \overline{\delta}^B_{i0} \qquad \text{and} \qquad \overline{\delta}_{ij} = \overline{\delta}^S_{ij} + \overline{\delta}^B_{ij} \qquad (14.1\text{b,c})$$

where superscripts S and B represent, respectively, quantities of the shell and the edge beam, taken positive in the assumed positive directions of their respective redundant forces (i.e., in opposite directions for the shell and edge beam). After calculating $\overline{X}_1, \ldots, \overline{X}_4$ by solving Eqs. (14.1a–c), we can obtain the final stress resultants in the shell and the beam by superimposing the effects of edge loads $\overline{X}_1, \ldots, \overline{X}_4$ onto the stress resultants in the primary structures.

Stress resultants and displacements in the primary shell structure are usually obtained from membrane analysis. Table 1A of the ASCE Manual gives the membrane solution (stress resultants and displacements) for various values of ϕ, under uniform dead and live loads, in closed form. Since edge-load solutions will be in the form of Fourier series, the membrane stress resultants and displacements are also given for the first ($n = 1$) and third ($n = 3$) terms of the Fourier series in Tables 1B and 1C,* respectively. For solving practical problems by hand calculations, only the first term of the series will be used, and generally, the total load (including live load) will be treated as dead load (i.e., uniformly distributed over the surface area). Hence $\overline{\delta}_{i0}$ values can be obtained from Table 1B.† This table is reproduced here as Table 14.1.

Tables 2A and 2B of the ASCE Manual, respectively, give stress resultants at various values of ϕ, and edge displacements at $\phi = 0$, for the shell under the action of unit edge loads: $\overline{X}_1 = (1)\sin(\pi x/L)$, $\overline{X}_2 = (1)\sin(\pi x/L)$, $\overline{X}_3 = (1)\cos(\pi x/L)$, and $\overline{X}_4 = (1)\sin(\pi x/L)$, which are symmetrically applied at both edges. The shell parameters considered are $a/t = 100, 200$, and $a/L = 0.1, 0.2, \ldots, 0.6$, where ϕ_k ranges from 30 to 50°. (The PCA extended tables give additional values for $a/t = 50$ and 150.) These tables are thus useful in obtaining edge load solutions for shells with symmetrical loads and with $L/a > 1.67$. Tables 2A and 2B of the ASCE Manual are here reproduced in Tables 14.2 and 14.3.

Tables 3A and 3B of the ASCE Manual give stress resultants and edge displacements for edge loads applied at only one edge, and for various values of at/L^2 from 0.002 to 0.1. They can thus be used for short shells with symmetrical or unsymmetrical loadings, since edge loads at one edge do not affect the other edge in such shells. Hence only four equations at any edge need to be solved

*In this chapter, Tables 1A, 1B, 1C, 2A, 2B, 3A, and 3B refer to tables given in the ASCE Manual. The reader may note that the ASCE tables use the symbols T_x, T_ϕ, and S in place of N_x, N_ϕ, and $N_{x\phi}$, respectively.

†The ASCE tables do not give values of $\overline{\delta}_{30}$ and $\overline{\delta}_{40}$, which have to be calculated from stress resultants.

TABLE 14.1 (ASCE Table 1B) Membrane Forces and Displacements in Simply Supported Cylindrical Shells; Loads Varying Longitudinally from Zero at the Ends to Maximum Positive at the Middle

(a) Uniform Transverse Load

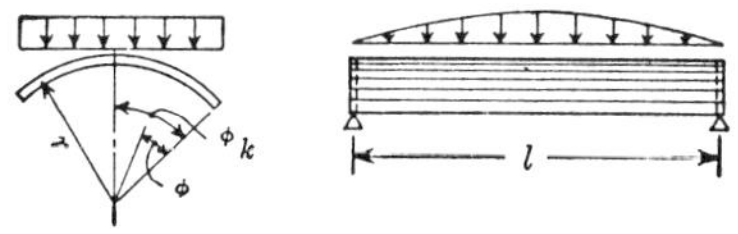

Longitudinal Force T_x—

$$p_u r \left[\left(\frac{l}{r}\right)^2 \times \text{Col. (1)} \right] \sin \frac{\pi x}{l}$$

Shearing Force S—

$$p_u r \left[\left(\frac{l}{r}\right) \times \text{Col. (2)} \right] \cos \frac{\pi x}{l}$$

Transverse Force T_ϕ—

$$p_u r \times \text{Col. (3)} \times \sin \frac{\pi x}{l}$$

Vertical Displacement Δ_V—

$$p_u r \frac{l^4}{r^3 t E} \left[\left(1 + \frac{1}{2} (\pi r/l)^2 + \frac{1}{12} (\pi r/l)^4\right) \times \text{Col. (4)} \right] \sin \frac{\pi x}{l}$$

Horizontal Displacement Δ_H—

$$+ p_u r \frac{l^4}{r^3 t E} \left\{ \left(\frac{r}{l}\right)^4 \times \text{Col. (5)} + \left[1 + \frac{1}{2} \left(\frac{\pi r}{l}\right)^2 + \frac{1}{12} \left(\frac{\pi r}{l}\right)^4\right] \times \text{Col. (6)} \right\} \times \sin \frac{\pi x}{l}$$

(b) Dead Weight Load

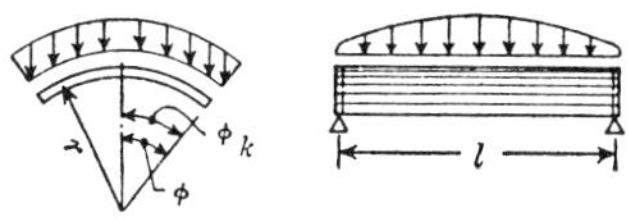

Longitudinal Force T_x—

$$p_d r \left[\left(\frac{l}{r}\right)^2 \times \text{Col. (7)} \right] \sin \frac{\pi x}{l}$$

Shearing Force S—

$$p_d r \left[\left(\frac{l}{r}\right) \times \text{Col. (8)} \right] \cos \frac{\pi x}{l}$$

Transverse Force T_ϕ—

$$p_d r \times \text{Col. (9)} \times \sin \frac{\pi x}{l}$$

Vertical Displacement Δ_V—

$$p_d r \frac{l^4}{r^3 t E} \left[\left(\frac{2 r}{\pi l}\right)^2 + \frac{2}{\pi^4} + \left(\frac{r}{l}\right)^4 \times \text{Col. (10)} \right] \sin \frac{\pi x}{l}$$

Horizontal Displacement Δ_H—

$$+ p_d r \frac{l^4}{r^3 t E} \left[\left(\frac{r}{l}\right)^4 \times \text{Col. (11)} \right] \times \sin \frac{\pi x}{l}$$

$\phi_k - \phi$	T_x (1)	S (2)	T_ϕ (3)	Δ_V (4)	Δ_H (5)	Δ_H (6)	T_x (7)	S (8)	T_ϕ (9)	Δ_V (10)	Δ_H (11)
0	−0.3040	0	−1.0000	0.12319	0	0	−0.2026	0	−1.000	1.0000	0
5	−0.2993	−0.0829	−0.9924	0.12180	0.0872	−0.00009	−0.2019	−0.0555	−0.9962	0.9924	0.0868
10	−0.2856	−0.1633	−0.9698	0.11766	0.1736	−0.00064	−0.1996	−0.1105	−0.9848	0.9698	0.1710
15	−0.2623	−0.2387	−0.9330	0.11102	0.2588	−0.00213	−0.1957	−0.1648	−0.9659	0.9330	0.2500
20	−0.2329	−0.3069	−0.8830	0.10222	0.3420	−0.00490	−0.1904	−0.2178	−0.9397	0.8830	0.3214
25	−0.1954	−0.3658	−0.8214	0.09170	0.4226	−0.00930	−0.1837	−0.2690	−0.9063	0.8214	0.3830
30	−0.1520	−0.4140	−0.7500	0.08001	0.5000	−0.01539	−0.1754	−0.3183	−0.8660	0.7500	0.4330
35	−0.1040	−0.4487	−0.6710	0.06771	0.5736	−0.02325	−0.1660	−0.3652	−0.8191	0.6710	0.4698
40	−0.0528	−0.4702	−0.5868	0.05537	0.6428	−0.03272	−0.1552	−0.4092	−0.7660	0.5868	0.4924
45	0	−0.4775	−0.5000	0.04355	0.7071	−0.04355	−0.1433	−0.4502	−0.7071	0.5000	0.5000
50	0.0528	−0.4702	−0.4132	0.03272	0.7660	−0.05537	−0.1302	−0.4877	−0.6428	0.4132	0.4924
55	0.1040	−0.4487	−0.3290	0.02325	0.8191	−0.06771	−0.1162	−0.5215	−0.5736	0.3290	0.4698
60	0.1520	−0.4140	−0.2500	0.01539	0.8660	−0.08001	−0.1013	−0.5513	−0.5000	0.2500	0.4330
65	0.1954	−0.3658	−0.1786	0.00930	0.9063	−0.09170	−0.0856	−0.5769	−0.4226	0.1786	0.3830
70	0.2329	−0.3069	−0.1170	0.00490	0.9397	−0.10222	−0.0693	−0.5982	−0.3420	0.1170	0.3214
75	0.2623	−0.2387	−0.0669	0.00213	0.9659	−0.11102	−0.0524	−0.6149	−0.2588	0.0669	0.2500
80	0.2856	−0.1633	−0.0301	0.00064	0.9848	−0.11766	−0.0351	−0.6269	−0.1736	0.0301	0.1710
85	0.2993	−0.0829	−0.0076	0.00009	0.9962	−0.12180	−0.0177	−0.6342	−0.0872	0.0076	0.0868
90	0.3040	0	0	0	1.0000	−0.12319	0	−0.6366	0	0	0

Source: Ref. [1].

TABLE 14.2 (ASCE Table 2A) Symmetrical Edge Loads on Simply Supported Cylindrical Shells; Stress Resultants at Various Values of ϕ

(*a*) Basic Formulas and Loading Diagrams

	Vertical Edge Load	Horizontal Edge Load	Shear Edge Load	Edge Moment Load
	Longitudinal Force T_x— $V_L\left[\left(\frac{l}{r}\right)^2 \times \text{Col. (1)}\right]\sin\frac{\pi x}{l}$	Longitudinal Force T_x— $H_L\left[\left(\frac{l}{r}\right)^2 \times \text{Col. (5)}\right]\sin\frac{\pi x}{l}$	Longitudinal Force T_x— $S_L\left[\left(\frac{l}{r}\right)^2 \times \text{Col. (9)}\right]\sin\frac{\pi x}{l}$	Longitudinal Force T_x— $\frac{M_L}{r}\left[\left(\frac{l}{r}\right)^2 \times \text{Col. (13)}\right]\sin\frac{\pi x}{l}$
	Shearing Force S— $V_L\left[\frac{l}{r} \times \text{Col. (2)}\right]\cos\frac{\pi x}{l}$	Shearing Force S— $H_L\left[\frac{l}{r} \times \text{Col. (6)}\right]\cos\frac{\pi x}{l}$	Shearing Force S— $S_L\left[\frac{l}{r} \times \text{Col. (10)}\right]\cos\frac{\pi x}{l}$	Shearing Force S— $\frac{M_L}{r}\left[\frac{l}{r} \times \text{Col. (14)}\right]\cos\frac{\pi x}{l}$
	Transverse Force T_ϕ— $V_L \times \text{Col. (3)} \times \sin\frac{\pi x}{l}$	Transverse Force T_ϕ— $H_L \times \text{Col. (7)} \times \sin\frac{\pi x}{l}$	Transverse Force T_ϕ— $S_L \times \text{Col. (11)} \times \sin\frac{\pi x}{l}$	Transverse Force T_ϕ— $\frac{M_L}{r} \times \text{Col. (15)} \times \sin\frac{\pi x}{l}$
	Transverse Moment M_ϕ— $V_L\,[r \times \text{Col. (4)}]\sin\frac{\pi x}{l}$	Transverse Moment M_ϕ— $H_L\,[r \times \text{Col. (8)}]\sin\frac{\pi x}{l}$	Transverse Moment M_ϕ— $S_L\,[r \times \text{Col. (12)}]\sin\frac{\pi x}{l}$	Transverse Moment M_ϕ— $M_L \times \text{Col. (16)} \times \sin\frac{\pi x}{l}$
	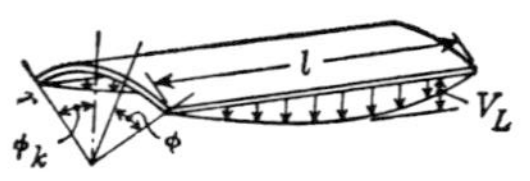	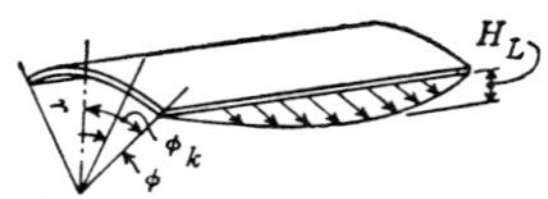	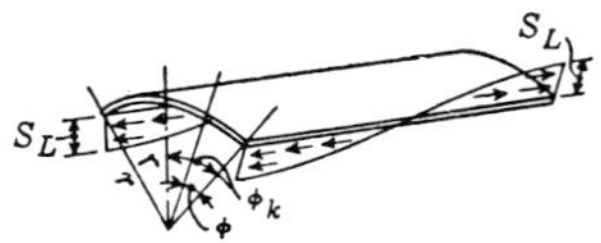	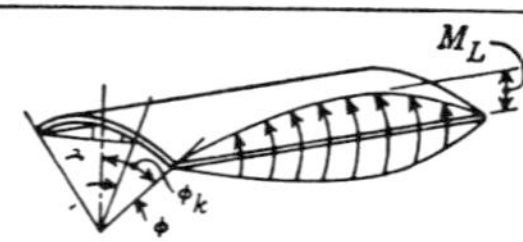

ϕ	T_x (1)	S (2)	T_ϕ (3)	M_ϕ (4)	T_x (5)	S (6)	T_ϕ (7)	M_ϕ (8)	T_x (9)	S (10)	T_ϕ (11)	M_ϕ (12)	T_x (13)	S (14)	T_ϕ (15)	M_ϕ (16)
(*b*) $r/t = 100$ and $r/l = 0.1$																
$\phi_k = 30$:																
30	− 5.382	0	−3.471	−0.3278	− 0.0013	0	+0.9997	+0.1310	−0.0904	0	−0.0207	−0.0010	− 0.0038	0	+ 0.0008	+ 0.9738
20	− 3.575	−2.619	−2.656	−0.2779	+ 0.0017	−0.0001	+0.9848	+0.1162	−0.0396	−0.0403	−0.0081	−0.0007	+ 0.0172	+ 0.0021	+ 0.0022	+ 0.9767
10	+ 1.809	−3.265	−0.775	−0.1512	+ 0.0034	+0.0017	+0.9399	+0.0722	+0.1118	−0.0250	+0.0142	+0.0002	+ 0.0264	+ 0.0173	+ 0.0014	+ 0.9854
0	+10.69	0	+0.500	0	− 0.0151	0	+0.8660	0	+0.3600	+0.1000	0	0	− 0.1370	0	0	+ 1.000

TABLE 14.2 (ASCE Table 2A) (*continued*)

ϕ	Vertical Edge Load				Horizontal Edge Load				Shear Edge Load				Edge Moment Load			
	T_x	S	T_ϕ	M_ϕ	T_x	S	T_ϕ	M_ϕ	T_x	S	T_ϕ	M_ϕ	T_x	S	T_ϕ	M_ϕ
	(1)	(2)	(3)	(4)	(5)	(6)	(7)	(8)	(9)	(10)	(11)	(12)	(13)	(14)	(15)	(16)
								(b) (Continued)								
$\phi_k=35$:																
35	− 3.430	0	−2.959	−0.3703	− 0.0061	0	+0.9986	+0.1754	−0.0774	0	−0.0243	−0.0015	− 0.0310	0	− 0.0051	+0.9642
30	− 3.220	−0.921	−2.820	−0.3594	− 0.0043	−0.0068	+0.9951	+0.1717	−0.0694	−0.0205	−0.0213	−0.0014	− 0.0211	− 0.0076	− 0.0037	+0.9650
20	− 1.538	−2.302	−1.812	−0.2781	+ 0.0060	−0.0136	+0.9663	+0.1425	−0.0054	−0.0439	−0.0015	−0.0008	+ 0.0373	− 0.0043	+ 0.0040	+0.9710
10	+ 1.822	−2.301	−0.330	−0.1444	+ 0.0080	+0.0041	+0.9071	+0.0848	+0.1212	−0.0149	+0.0185	−0.0002	+ 0.0469	+ 0.0251	+ 0.0045	+0.9825
0	+ 6.845	0	+0.574	0	− 0.0316	0	+0.8192	0	+0.3076	+0.1000	0	0	− 0.2051	0	0	+1.000
$\phi_k=40$:	+															
40	− 2.210	0	−2.534	−0.4048	− 0.0165	0	+0.9954	+0.2247	−0.0675	0	−0.0279	−0.0022	− 0.0712	0	− 0.0184	+0.9530
30	− 1.806	−1.138	−2.168	−0.3683	− 0.0049	−0.0068	+0.9836	+0.2041	−0.0461	−0.0331	−0.0179	−0.0018	− 0.0196	− 0.0293	− 0.0074	+0.9570
20	− 0.578	−1.830	−1.205	−0.2691	+ 0.0176	−0.0032	+0.9420	+0.1672	+0.0181	−0.0423	+0.0054	−0.0009	+ 0.0815	− 0.0118	+ 0 0112	+0.9653
10	+ 1.516	−1.614	−0.045	−0.1352	+ 0.0156	+0.0081	+0.8685	+0.1047	+0.1236	−0.0057	+0.0221	−0.0002	+ 0.0724	+ 0.0998	+ 0.0106	+0.9798
0	+ 4.534	0	+0.643	0	− 0.0631	0	+0.7660	0	+0.2682	+0.1000	0	0	− 0.3099	0	0	+1.000
$\phi_k=45$:																
45	− 1.602	0	−2.231	−0.4355	− 0.0359	0	+0.9873	+0.2779	−0.0596	0	−0.0315	−0.0031	− 0.1280	0	− 0.0437	0.9392
40	− 1.546	−0.434	−2.163	−0.4274	− 0.0309	−0.0094	+0.9849	+0.2743	−0.0559	−0.0160	−0.0292	−0.0030	− 0.1102	− 0.0325	− 0.0385	+0.9402
30	− 1.098	−1.180	−1.646	−0.3654	+ 0.0033	−0.0181	+0.9642	+0.2461	−0.0258	−0.0398	−0.0123	−0.0022	+ 0.0128	− 0.0642	− 0.0047	−0.9469
20	− 0.166	−1.550	−0.763	−0.2562	+ 0.0412	−0.0050	+0.9138	+0.1899	+0.0338	−0.0389	+0.0120	−0.0010	+ 0.1523	− 0.0158	+ 0.0277	+0.9595
10	+ 1.310	−1.263	+0.179	−0.1255	+ 0.0260	+0.1175	+0.8251	+0.1072	+0.1222	+0.0025	+0.0250	−0.0002	+ 0.0981	+ 0.0680	+ 0.0212	+0.9771
0	+ 3.407	0	+0.707	0	− 0.1185	0	+0.7071	0	+0.2378	+0.1000	0	0	− 0.4613	0	0	+1.000
$\phi_k=50$:																
50	− 1.051	0	−1.931	−0.4541	− 0.0693	0	+0.9696	+0.3336	−0.0531	0	−0.0350	−0.0042	− 0.2048	0	− 0.0885	+0.9218
40	− 0.948	−0.558	−1.741	−0.4265	− 0.0213	−0.0326	+0.9650	+0.3198	−0.0423	−0.0271	−0.0268	−0.0037	− 0.1192	− 0.0964	− 0.0573	+0.9258
30	− 0.619	−0.999	−1.208	−0.3492	+ 0.0284	−0.0366	+0.9430	+0.2786	−0.0098	−0.0424	−0.0057	−0.0025	+ 0.0849	− 0.1086	+ 0.0111	+0.9371
20	− 0.001	−1.184	−0.451	−0.2379	+ 0.0820	−0.0039	+0.8851	+0.2104	+0.0440	−0.0340	+0.0181	−0.0010	+ 0.2503	− 0.0099	+ 0.0576	+0.9537
10	+ 0.998	−0.930	+0.317	−0.1141	+ 0.0374	+0.0357	+0.7738	+0.1167	+0.1188	+0.0077	+0.0275	−0.0002	+ 0.1176	+ 0.1119	+ 0.0373	+0.9745
0	+ 2.488	0	+0.766	0	− 0.2047	0	+0.6428	0	+0.2138	+0.1000	0	0	− 0.6695	0	0	+1.000
								(c) $r/t = 100$ and $r/l = 0.2$								
$\phi_k=30$:																
30	− 5.160	0	−3.420	−0.3028	− 0.0742	0	+0.9815	+0.1219	−0.1946	0	−0.0414	−0.0005	− 0.549	0	− 0.0813	+0.9009
20	− 3.440	−2.531	−2.632	−0.2565	− 0.0123	−0.0292	+0.9757	+0.1084	−0.0755	−0.0776	−0.0145	0	− 0.073	− 0.2142	− 0.0153	+0.9124
10	− 1.763	−3.228	−0.787	−0.1395	+ 0.0700	−0.0087	+0.9427	+0.0678	+0.2208	−0.0469	+0.0287	+0.0009	+ 0.557	− 0.0429	+ 0.0568	+0.9456
0	+10.92	0	+0.500	0	− 0.1140	0	+0.8660	0	+0.7119	+0.2000	0	0	− 1.067	0	0	+1.000

TABLE 14.2 (ASCE Table 2A) *(continued)*

ϕ	Vertical Edge Load				Horizontal Edge Load				Shear Edge Load				Edge Moment Load			
	T_x (1)	S (2)	T_ϕ (3)	M_ϕ (4)	T_x (5)	S (6)	T_ϕ (7)	M_ϕ (8)	T_x (9)	S (10)	T_ϕ (11)	M_ϕ (12)	T_x (13)	S (14)	T_ϕ (15)	M_ϕ (16)
(c) (Continued)																
$\phi_k = 35$:																
35	− 3.087	0	−2.871	−0.3347	− 0.1552	0	+0.9603	+0.1591	−0.1487	0	−0.0458	−0.0024	− 0.912	0	− 0.1914	+0.8637
30	− 2.939	−0.833	−2.744	−0.3248	− 0.1260	−0.0399	+0.9622	+0.1559	−0.1334	−0.0394	−0.0401	−0.0022	− 0.742	− 0.2343	− 0.1585	+0.8669
20	− 1.639	−2.156	−1.807	−0.2514	+ 0.0561	−0.0634	+0.9649	+0.1301	−0.0106	−0.0845	−0.0018	−0.0012	+ 0.327	− 0.3747	+ 0.0315	+0.8914
10	+ 1.501	−2.296	−0.365	−0.1302	+ 0.1526	+0.0074	+0.9224	+0.0781	+0.2378	−0.0280	+0.0369	−0.0001	+ 0.928	+ 0.0486	+ 0.1212	+0.9362
0	+ 7.434	0	+0.574	0	− 0.3244	0	+0.8192	0	+0.6159	+0.2000	0	0	− 2.045	0	0	+1.000
$\phi_k = 40$:																
40	− 1.764	0	−2.398	−0.3556	− 0.3096	0	+0.9087	+0.1962	−0.1273	0	−0.0525	−0.0035	− 1.421	0	− 0.3885	+0.8156
30	− 1.640	−0.947	−2.105	−0.3239	− 0.1336	−0.1389	+0.9351	+0.1845	−0.0880	−0.0626	−0.0335	−0.0029	− 0.631	− 0.6461	− 0.2007	+0.8306
20	− 0.965	−1.697	−1.251	−0.2374	+ 0.2243	−0.1133	+0.9680	+0.1487	+0.0323	−0.0817	+0.0109	−0.0013	+ 1.015	− 0.5396	+ 0.1574	+0.8707
10	+ 1.111	−1.753	−0.109	−0.1188	+ 0.2753	+0.0557	+0.9101	+0.0870	+0.2408	−0.0111	+0.0434	−0.0007	+ 1.316	+ 0.2523	+ 0.2259	+0.9271
0	+ 5.836	0	+0.643	0	− 0.7210	0	+0.7660	0	+0.5460	+0.2000	0	0	− 3.452	0	0	+1.000
$\phi_k = 45$:																
45	− 0.775	0	−1,934	−0.3619	− 0.5679	0	+0.7941	+0.2293	−0.1067	0	−0.0577	−0.0050	− 2.074	0	− 0.7365	+0.7509
40	− 0.820	−0.217	−1.897	−0.3556	− 0.4987	−0.1493	+0.8118	+0.2269	−0.0988	−0.0287	−0.0535	−0.0048	− 1.828	− 0.5460	− 0.6582	+0.7558
30	− 1.062	−0.731	−1.584	−0.3068	− 0.0210	−0.3078	+0.9229	+0.2071	−0.0506	−0.0726	−0.0228	−0.0034	− 0.116	− 1.138	− 0.1410	+0.7920
20	− 0.934	−1.317	−0.910	−0.2175	+ 0.5381	−0.1552	+1.008	+0.1641	+0.0570	−0.0737	+0.0224	−0.0014	+ 1.958	− 0.5987	+ 0.3956	+0.8510
10	+ 0.700	−1.478	+0.039	−0.1073	+ 0.4245	+0.1611	+0.9154	+0.0946	+0.2353	+0.0028	+0.0482	0	+ 1.641	+ 0.5811	+ 0.3778	+0.9196
0	+ 5.344	0	+0.707	0	− 1.392	0	+0.7071	0	+0.4995	+0.2000	0	0	− 5.299	0	0	+1.000
$\phi_k = 50$:																
50	+ 0.052	0	−1.437	−0.3493	− 0.9469	0	+0.5813	+0.2514	−0.0848	0	−0.0601	−0.0065	− 2.814	0	− 1.245	+0.6613
40	− 0.291	−0.028	−1.413	−0.3315	− 0.5827	−0.4585	+0.7040	+0.2452	−0.0714	−0.0439	−0.0468	−0.0056	− 1.762	− 1.381	− 0.8336	+0.6869
30	− 1.016	−0.383	−1.246	−0.2786	+ 0.2885	−0.5526	+0.9656	+0.2228	−0.0258	−0.0723	−0.0113	−0.0035	+ 0.797	− 1.690	+ 0.0859	+0.7523
20	− 1.244	−1.059	−0.745	−0.1953	+ 0.1015	−0.1620	+1.112	+0.1764	+0.0664	−0.0639	+0.0312	−0.0013	+ 3.038	− 0.536	+ 0.7607	+0.8338
10	+ 0.381	−1.427	+0.110	−0.0943	+ 0.5624	+0.3610	+0.9471	+0.1015	+0.2257	+0.0125	+0.0513	0	+ 1.811	+ 1.083	+ 0.5718	+0.9154
0	+ 5.670	0	+0.766	0	− 2.398	0	+0.6428	0	+0.4748	+0.2000	0	0	− 7.473	0	0	+1.000
(d) $r/t = 100$ and $r/l = 0.3$																
$\phi_k = 30$:																
30	− 4.709	0	−3.362	−0.2769	− 0.2662	0	+0.9571	+0.1099	−0.2618	0	−0.0596	−0.0024	− 2.101	0	− 0.3444	+0.7859
20	− 3.568	−2.381	−2.632	−0.2347	− 0.0094	−0.0961	+0.9752	+0.0982	−0.1182	−0.1174	−0.0231	−0.0016	− 0.108	− 0.7669	− 0.0802	+0.8119
10	+ 1.053	−3.290	−0.837	−0.1275	+ 0.2942	+0.0035	+0.9639	+0.0622	+0.3251	−0.0750	+0.0421	−0.0003	+ 2.372	+ 0.0141	+ 0.1945	+0.8845
0	+12.43	0	+0.500	0	− 0.6606	0	+0.8660	0	+1.102	+0.3000	0	0	− 5.456	0	0	+1.000

TABLE 14.2 (ASCE Table 2A) (*continued*)

ϕ	Vertical Edge Load				Horizontal Edge Load				Shear Edge Load				Edge Moment Load			
	T_x	S	T_ϕ	M_ϕ	T_x	S	T_ϕ	M_ϕ	T_x	S	T_ϕ	M_ϕ	T_x	S	T_ϕ	M_ϕ
	(1)	(2)	(3)	(4)	(5)	(6)	(7)	(8)	(9)	(10)	(11)	(12)	(13)	(14)	(15)	(16)
								(*d*) (Continued)								
$\phi_k=35$:																
35	− 2.156	0	−2.682	−0.2903	− 0.6039	0	+0.8706	+0.1362	−0.2140	0	−0.0685	−0.0036	− 3.493	0	− 0.7613	+0.7084
30	− 2.203	−0.595	−2.591	−0.2821	− 0.4823	−0.1544	+0.8892	+0.1338	−0.1937	−0.0568	−0.0603	−0.0033	− 2.808	− 0.8945	− 0.6319	+0.7161
20	− 2.106	−1.817	−1.850	−0.2201	+ 0.2698	−0.2300	+0.9859	+0.1137	−0.0244	−0.1248	−0.0046	−0.0018	+ 1.506	− 1.358	+ 0.1098	+0.7726
10	+ 0.513	−2.456	−0.482	−0.1148	+ 0.6242	+0.0755	+0.9800	+0.0698	+0.3456	−0.0471	+0.0535	−0.0002	+ 3.764	+ 0.4236	+ 0.4372	+0.8689
0	+10.03	0	+0.574	0	− 1.543	0	+0.8192	0	+0.9676	+0.3000	0	0	− 9.403	0	0	+1.000
$\phi_k=40$:																
40	− 0.216	0	−1.977	−0.2853	− 1.176	0	+0.6733	+0.1553	−0.1668	0	−0.0736	−0.0048	− 5.193	0	− 1.469	+0.6071
30	− 1.020	−0.275	−1.892	−0.2631	− 0.4865	−0.5149	+0.8208	+0.1488	−0.1238	−0.0837	−0.0485	−0.0039	− 2.222	− 2.288	− 0.7475	+0.6444
20	− 2.210	−1.203	−1.430	−0.1982	+ 0.9141	−0.3958	+1.069	+0.1252	+0.0263	−0.1164	+0.0125	−0.0018	+ 3.993	− 1.809	+ 0.5716	+0.7382
10	− 0.295	−2.141	−0.338	−0.1018	+ 1.073	+0.2676	+1.040	+0.0762	+0.3397	−0.0252	+0.0611	−0.0001	+ 5.004	+ 1.180	+ 0.7966	+0.8572
0	+10.01	0	+0.643	0	− 3.068	0	+0.7660	0	+0.8934	+0.3000	0	0	− 14.38	0	0	+1.000
$\phi_k=45$:																
45	+ 1.400	0	−1.178	−0.2565	− 1.970	0	+0.3055	+0.1591	−0.1142	0	+0.0719	−0.0059	− 6.844	0	− 2.463	+0.4746
40	+ 1.083	+0.354	−1.224	−0.2537	− 1.728	−0.5178	+0.3766	+0.1589	−0.1112	−0.0310	−0.0674	−0.0056	− 6.032	− 1.802	− 2.202	+0.4871
30	− 1.040	+0.432	−1.454	−0.2296	− 0.0472	−1.061	+0.8385	+0.1550	−0.0776	−0.0851	−0.0328	−0.0040	− 0.316	− 3.743	− 0.4779	+0.5768
20	− 3.110	−0.788	−1.326	−0.1743	+ 1.941	−0.5051	+1.279	+0.1338	+0.0382	−0.1013	+0.0241	−0.0016	+ 6.770	− 1.867	+ 1.312	+0.7124
10	− 0.946	−2.214	−0.330	−0.0899	+ 1.513	+0.6328	+1.158	+0.0821	+0.3209	−0.0128	+0.0645	−0.0001	+ 5.696	+ 2.230	+ 1.229	+0.8510
0	+11.24	0	+0.707	0	− 5.257	0	+0.7071	0	+0.8743	+0.3000	0	0	− 19.61	0	0	+1.000
$\phi_k=50$:																
50	+ 2.605	0	−0.326	−0.2300	− 2.821	0	−0.2407	+0.1409	−0.0565	0	−0.0611	−0.0063	− 7.871	0	− 3.553	+0.3153
40	+ 1.276	+1.180	−0.676	−0.2032	− 1.744	−1.346	+0.1577	+0.1474	−0.0651	−0.0327	−0.0514	−0.0056	− 4.994	− 3.782	− 2.394	+0.3754
30	− 1.837	+1.062	−1.342	−0.1926	+ 0.8732	−1.624	+1.033	+0.1561	−0.0643	−0.0700	−0.0208	−0.0036	+ 2.207	− 4.666	+ 0.2204	+0.5257
20	− 4.099	−0.690	−1.432	−0.1527	+ 3.129	−0.4486	+1.623	+0.1422	+0.0214	−0.0879	+0.0276	−0.0013	+ 8.917	− 1.422	+ 2.171	+0.6995
10	− 1.187	−2.497	−0.379	−0.0801	+ 1.766	+1.155	+1.315	+0.0885	+0.3005	−0.0113	+0.0639	+0.0002	+ 5.588	+ 3.317	+ 1.625	+0.8508
0	+12.66	0	+0.766	0	− 7.773	0	+0.6428	0	+0.8992	+0.3000	0	0	− 23.76	0	0	+1.000
								(*e*) $r/t = 100$ and $r/l = 0.4$								
$\phi_k=30$:																
30	− 3.508	0	−3.175	−0.2404	− 0.736	0	+0.881	+0.0947	−0.3318	0	−0.0772	−0.0028	− 5.33	0	− 0.879	+0.6558
20	− 3.500	−1.942	−2.592	−0.2048	− 0.038	−0.2682	+0.958	+0.0856	−0.1568	−0.1503	−0.0304	−0.0019	− 0.14	− 1.920	− 0.196	+0.6995
10	− 0.244	−3.281	−0.947	−0.1121	+ 0.804	−0.0008	+1.008	+0.0544	+0.4153	−0.0998	+0.0545	−0.0003	+ 6.27	+ 0.187	+ 0.502	+0.8169
0	+15.23	0	+0.500	0	− 1.772	0	+0.866	0	+1.508	+0.4000	0	0	− 15.27	0	0	+1.000

TABLE 14.2 (ASCE Table 2A) *(continued)*

ϕ	Vertical Edge Load				Horizontal Edge Load				Shear Edge Load				Edge Moment Load			
	T_x (1)	S (2)	T_ϕ (3)	M_ϕ (4)	T_x (5)	S (6)	T_ϕ (7)	M_ϕ (8)	T_x (9)	S (10)	T_ϕ (11)	M_ϕ (12)	T_x (13)	S (14)	T_ϕ (15)	M_ϕ (16)
(e) (Continued)																
$\phi_k=35$:																
35	-0.217	0	-2.261	-0.2351	-1.518	0	+0.670	+0.1091	-0.2518	0	-0.0842	-0.0040	-8.48	0	-1.890	+0.5364
30	-0.656	-0.102	-2.241	-0.2293	-1.214	-0.3878	+0.723	+0.1078	-0.2316	-0.0672	-0.0746	-0.0037	-6.82	-2.173	-1.574	+0.5500
20	-2.982	-1.086	-1.911	-0.1833	+0.680	-0.5777	+1.014	+0.0952	-0.0480	-0.1540	-0.0073	-0.0019	+3.69	-3.300	+0.248	+0.6463
10	-1.512	-2.715	-0.712	-0.0982	+1.596	+0.1982	+1.092	+0.0612	+0.4268	-0.0671	+0.0673	-0.0001	+9.37	+1.106	+1.064	+0.8000
0	+15.11	0	+0.574	0	-3.990	0	+0.819	0	+1.373	+0.4000	0	0	-23.99	0	0	+1.000
$\phi_k=40$:																
40	+2.485	0	-1.221	-0.2044	-2.664	0	+0.250	+0.1088	-0.1542	0	-0.0818	-0.0049	-11.15	0	-3.211	+0.3896
30	+0.117	+0.913	-1.512	-0.1948	-1.123	-1.171	+0.606	+0.1093	-0.1315	-0.0851	-0.0567	-0.0039	-4.83	-4.928	-0.164	+0.4541
20	-4.296	-0.278	-1.730	-0.1580	+2.058	-0.9139	+1.236	+0.1013	+0.0023	-0.1351	+0.0099	-0.0017	+8.71	-3.904	+1.250	+0.6098
10	-2.814	-2.749	-0.749	-0.0862	+2.495	+0.6036	+1.272	+0.0665	+0.4228	-0.0471	+0.0728	0	+11.30	+2.733	+1.764	+0.7918
0	+17.20	0	+0.643	0	-7.122	0	+0.766	0	+1.368	+0.4000	0	0	-33.39	0	0	+1.000
$\phi_k=45$:																
45	+4.329	0	-0.119	-0.1513	-3.817	0	-0.372	+0.0899	-0.0653	0	-0.0657	-0.0051	-12.50	0	-4.647	+0.2210
40	+3.663	+1.126	-0.277	-0.1526	-3.361	-1.005	-0.230	+0.0923	-0.0723	-0.0186	-0.0630	-0.0049	-11.09	-3.299	-4.168	+0.2412
30	-0.918	+2.029	-1.249	-0.1564	-0.150	-2.079	+0.703	+0.1060	-0.1030	-0.0673	-0.0384	-0.0035	-0.92	-6.970	-0.971	+0.3846
20	-6.034	-0.008	-1.889	-0.1377	+3.796	-1.015	+1.638	+0.1083	-0.0344	-0.1142	+0.0158	-0.0013	+12.56	-3.637	+2.441	+0.5941
10	-3.355	-3.194	-0.854	-0.0780	+3.104	+1.260	+1.497	+0.0731	+0.3651	-0.0449	+0.0716	+0.0003	+11.51	+4.262	+2.363	+0.7936
0	+19.64	0	+0.707	0	-10.78	0	+0.707	0	+1.388	+0.4000	0	0	-39.59	0	0	+1.000
$\phi_k=50$:																
50	+4.973	0	+0.784	-0.0881	-4.477	0	-1.043	+0.0545	+0.0216	0	-0.0395	-0.0045	-11.45	0	-5.493	+0.0676
40	+2.829	+2.328	+0.079	-0.1028	-2.858	-2.155	-0.394	+0.0716	-0.0276	+0.0025	-0.0408	-0.0041	-7.65	-4.182	-3.773	+0.1539
30	-2.433	+2.516	-1.399	-0.1279	+1.235	-2.673	+1.065	+0.1057	-0.1165	-0.0382	-0.0313	-0.0029	+2.61	-3.745	+0.216	+0.3594
20	-6.881	-0.213	-2.118	-0.1244	+5.125	-0.8202	+2.120	+0.1184	-0.0779	-0.1025	+0.0109	-0.0010	+13.82	-1.157	+3.391	+0.5981
10	-3.049	-3.540	-0.904	-0.0724	+3.212	+1.909	+1.707	+0.0810	+0.3428	-0.0541	+0.0667	+0.0003	+10.19	+5.359	+2.668	+0.7998
0	+20.35	0	+0.766	0	-13.59	0	+0.643	0	+1.459	+0.4000	0	0	-41.80	0	0	+1.000
(f) $- r/t = 100$ and $r/l = 0.5$																
$\phi_k=30$:																
30	-1.757	0	-2.879	-0.2019	-1.420	0	+0.767	+0.0789	-0.3849	0	-0.0912	-0.0030	-10.64	0	-1.779	+0.5185
20	-3.508	-1.323	-2.524	-0.1738	-0.037	-0.5108	+0.932	+0.0727	-0.1963	-0.1773	-0.0368	-0.0020	-0.48	-3.877	-0.421	+0.5825
10	-2.292	-3.378	-1.108	-0.0969	+1.616	+0.0381	+1.071	+0.0489	+0.4854	-0.1263	+0.0654	-0.0002	+12.53	+0.236	+0.998	+0.7480
0	+20.22	0	+0.500	0	-3.767	0	+0.866	0	+1.964	+0.5000	0	0	-30.35	0	0	+1.000

TABLE 14.2 (ASCE Table 2A) (*continued*)

ϕ	Vertical Edge Load				Horizontal Edge Load				Shear Edge Load				Edge Moment Load			
	T_x (1)	S (2)	T_ϕ (3)	M_ϕ (4)	T_x (5)	S (6)	T_ϕ (7)	M_ϕ (8)	T_x (9)	S (10)	T_ϕ (11)	M_ϕ (12)	T_x (13)	S (14)	T_ϕ (15)	M_ϕ (16)
(*f*) (Continued)																
$\phi_k=35$:																
35	+ 2.471	0	−1.677	−0.1788	− 2.776	0	+0.395	+0.0816	−0.2558	0	−0.0962	−0.0039	− 14.99	0	− 3.345	+0.3692
30	+ 1.493	+0.587	−1.758	−0.1758	− 2.223	−0.7101	+0.495	+0.0816	−0.2428	−0.0837	−0.0875	−0.0037	− 12.12	− 4.668	− 2.786	+0.3888
20	− 4.209	−0.060	−2.002	−0.1480	+ 1.246	−1.062	+1.056	+0.0775	−0.0879	−0.1706	−0.0227	−0.0018	+ 6.40	− 5.907	+ 0.453	+0.5266
10	− 4.426	−3.090	−1.055	−0.0837	+ 2.986	+0.3737	+1.252	+0.0535	+0.4733	−0.0921	+0.0629	+0.0000	+ 17.12	+ 1.978	+ 1.931	+0.7345
0	+22.54	0	+0.574	0	− 7.555	0	+0.819	0	+1.867	+0.5000	0	0	− 44.56	0	0	+1.000
$\phi_k=40$:																
40	+ 5.472	0	−0.358	−0.1324	− 4.282	0	−0.220	+0.0682	−0.1143	0	−0.0775	−0.0034	− 17.17	0	− 5.060	+0.2033
30	+ 1.404	+2.232	−1.069	−0.1354	− 1.839	−1.889	+0.363	+0.0754	−0.1408	−0.0683	−0.0585	−0.0034	− 7.86	− 7.678	− 2.645	+0.2912
20	− 6.661	+0.757	−2.062	−0.1254	+ 3.336	−1.488	+1.415	+0.0819	−0.0853	−0.1408	+0.0027	−0.0014	+ 13.14	− 6.372	+ 1.904	+0.5001
10	− 5.826	−3.503	−1.214	−0.0751	+ 4.196	+1.017	+1.528	+0.0593	+0.4218	−0.0802	+0.0780	+0.0003	+ 18.72	+ 4.129	+ 2.821	+0.7347
0	+26.12	0	+0.643	0	−12.17	0	+0.766	0	+1.908	+0.5000	0	0	− 54.72	0	0	+1.000
$\phi_k=45$:																
45	+ 6.655	0	+0.771	−0.0750	− 4.823	0	−0.923	+0.0405	+0.0153	0	−0.0477	−0.0038	− 15.73	0	− 6.128	+0.0523
40	+ 5.749	+1.741	+0.524	−0.0793	− 4.625	−1.375	−0.727	+0.0448	−0.0054	+0.0023	−0.0480	−0.0037	− 14.17	− 4.167	− 5.525	+0.0765
30	− 0.652	+3.364	−1.053	−0.1039	− 0.365	−2.902	+0.573	+0.0709	−0.1266	−0.0318	−0.0417	−0.0028	− 2.11	− 9.122	− 1.418	+0.2487
20	− 8.443	+0.730	−2.362	−0.1125	+ 5.266	−1.497	+1.929	+0.0901	−0.1395	−0.1180	+0.0012	−0.0011	+ 16.12	− 5.253	+ 3.200	+0.5015
10	− 5.758	−4.033	−1.326	−0.0708	+ 4.705	+1.781	+1.798	+0.0670	+0.3807	−0.0882	+0.0726	+0.0003	+ 17.30	+ 5.658	+ 3.294	+0.7426
0	+27.76	0	+0.707	0	−16.12	0	+0.707	0	+2.002	+0.5000	0	0	− 58.57	0	0	+1.000
$\phi_k=50$:																
50	+ 6.074	0	+1.416	−0.0253	− 5.122	0	−1.462	+0.0088	+0.1028	0	−0.0139	−0.0027	− 11.38	0	− 6.094	−0.0520
40	+ 3.738	+2.900	+0.542	−0.0469	− 3.478	−2.507	−0.713	+0.0302	+0.0147	+0.0398	−0.0269	−0.0027	− 8.59	− 5.742	− 4.363	+0.0339
30	− 2.427	+3.394	−1.371	−0.0900	+ 1.103	−3.275	+1.025	+0.0759	−0.1664	−0.0015	−0.0410	−0.0022	+ 0.92	− 8.235	− 0.093	+0.2533
20	− 8.536	+0.212	−2.512	−0.1065	+ 6.245	−1.164	+2.385	+0.1024	−0.1868	−0.1144	−0.0079	−0.0010	+ 15.24	− 3.756	+ 3.776	+0.5154
10	− 4.760	−4.221	−1.261	−0.0672	+ 4.596	+2.387	+1.966	+0.0752	+0.3719	−0.1018	+0.0671	+0.0003	+ 14.96	+ 6.079	+ 3.305	+0.7497
0	+26.68	0	+0.766	0	−18.51	0	+0.643	0	+2.065	+0.5000	0	0	− 56.51	0	0	+1.000
(*g*) $r/t = 100$ and $r/l = 0.6$																
$\phi_k=30$:																
30	+ 0.690	0	−2.477	−0.1646	− 2.375	0	+0.609	+0.0636	−0.4130	0	−0.1015	−0.0031	− 17.24	0	− 2.909	+0.3912
20	− 3.149	−0.438	−2.432	−0.1447	− 0.079	−0.8580	+0.895	+0.0605	−0.2349	−0.1448	−0.0424	−0.0010	− 0.99	− 6.327	− 0.707	+0.4746
10	− 5.088	−3.429	−1.334	−0.0837	+ 2.720	+0.0560	+1.160	+0.0431	+0.5284	−0.1530	+0.0739	−0.0001	+ 16.89	+ 0.297	+ 1.631	+0.6844
0	+26.76	0	+0.500	0	− 6.379	0	+0.866	0	+2.481	+0.6000	0	0	− 50.48	0	0	+1.000

TABLE 14.2 (ASCE Table 2A) (*continued*)

ϕ	Vertical Edge Load				Horizontal Edge Load				Shear Edge Load				Edge Moment Load			
	T_x (1)	S (2)	T_ϕ (3)	M_ϕ (4)	T_x (5)	S (6)	T_ϕ (7)	M_ϕ (8)	T_x (9)	S (10)	T_ϕ (11)	M_ϕ (12)	T_x (13)	S (14)	T_ϕ (15)	M_ϕ (16)
(*g*) (Continued)																
$\phi_k=35$:																
35	+ 5.530	0	+0.999	−0.1294	− 4.187	0	+0.085	+0.0577	−0.2245	0	−0.0937	−0.0036	− 21.54	0	− 4.900	+0.2286
30	+ 3.964	+1.371	−1.194	−0.1291	− 3.368	−1.072	+0.231	+0.0589	−0.2258	−0.0617	−0.0849	−0.0034	− 17.56	− 5.539	− 4.097	+0.2534
20	− 5.521	+1.142	−2.096	−0.1187	+ 1.835	−1.625	+1.098	+0.0628	−0.1434	−0.1738	−0.0163	−0.0014	+ 8.78	− 8.709	+ 0.609	+0.4257
10	− 7.862	−3.492	−1.452	−0.0731	+ 4.617	+0.5579	+1.438	+0.0477	+0.4825	−0.1215	+0.0821	+0.0005	+ 25.70	+ 2.778	+ 2.864	+0.6789
0	+31.29	0	+0.574	0	−11.74	0	+0.891	0	+2.453	+0.6000	0	0	− 67.87	0	0	+1.000
$\phi_k=40$:																
40	+ 8.029	0	+0.412	−0.0781	− 5.611	0	+0.632	+0.0378	−0.0378	0	−0.0653	−0.0034	− 20.91	0	− 6.413	+0.0741
30	+ 2.640	+3.389	−0.660	−0.0907	− 2.518	−2.498	+0.140	+0.0500	−0.1275	−0.0385	−0.0561	−0.0020	− 10.34	− 9.508	− 3.449	+0.1748
20	− 8.598	+1.754	−2.347	−0.1015	+ 4.328	−2.036	+1.563	+0.0673	−0.1799	−0.1354	−0.0078	−0.0012	+ 16.02	− 8.383	+ 2.316	+0.4147
10	− 8.783	−4.118	−1.651	−0.0674	+ 5.859	+1.341	+1.421	+0.0538	+0.4136	−0.1192	+0.0787	+0.0004	+ 25.53	+ 5.071	+ 3.686	+0.6842
0	+34.71	0	+0.643	0	−17.02	0	+0.766	0	+2.558	+0.6000	0	0	− 75.81	0	0	+1.000
$\phi_k=45$:																
45	+ 7.904	0	+1.346	−0.0293	− 5.855	0	−1.258	+0.0115	+0.1009	0	−0.0267	−0.0026	− 15.56	0	− 6.655	−0.0364
40	+ 6.935	+2.078	+1.051	−0.0352	− 5.259	−1.550	−1.038	+0.0166	+0.0669	+0.0245	−0.0302	−0.0025	− 14.36	− 4.156	− 6.057	−0.0122
30	− 0.196	+4.195	−0.884	−0.0712	− 0.727	−3.373	+0.459	+0.0490	−0.1460	+0.0082	−0.0439	−0.0022	− 3.96	− 9.685	− 1.817	+0.1633
20	− 9.841	+1.352	−2.657	−0.0959	+ 6.025	−1.905	+2.094	+0.0775	−0.2510	−0.1174	−0.0148	−0.0010	+ 16.43	− 6.537	+ 3.383	+0.4285
10	− 7.893	−4.563	−1.682	−0.0659	+ 6.150	+2.089	+2.020	+0.0623	+0.3813	−0.1339	+0.0714	+0.0004	+ 22.53	+ 6.087	+ 3.873	+0.6947
0	+34.56	0	+0.707	0	−20.64	0	+0.707	0	+2.659	+0.6000	0	0	− 74.68	0	0	+1.000
$\phi_k=50$:																
50	+ 6.029	0	+1.647	+0.0038	− 4.847	0	−1.571	−0.0114	+0.1678	0	+0.0073	−0.0015	− 8.26	0	− 5.666	−0.0950
40	+ 4.096	+2.955	+0.283	−0.0192	− 3.656	−2.446	−0.851	+0.0100	+0.0523	+0.0699	−0.0151	−0.0017	− 8.00	− 4.483	− 4.344	−0.0164
30	− 1.803	+3.754	−1.264	−0.0627	+ 0.458	−3.469	+0.919	+0.0574	−0.2030	+0.0305	−0.0491	−0.0018	− 1.93	− 7.812	− 0.623	+0.1835
20	− 9.145	+0.659	−2.736	−0.0932	+ 6.530	−1.539	+2.469	+0.0897	−0.2862	−0.1225	−0.0245	−0.0010	+ 13.86	− 5.044	+ 3.549	+0.4425
10	− 6.445	−4.528	−1.501	−0.0624	+ 6.010	+2.570	+2.141	+0.0696	+0.3899	−0.1463	+0.0675	+0.0003	+ 20.05	+ 5.942	+ 3.658	+0.6986
0	+31.83	0	+0.766	0	−22.64	0	+0.643	0	+2.688	+0.6000	0	0	− 70.37	0	0	+1.000
(*h*) $r/t = 200$ and $r/l = 0.1$																
$\phi_k=30$:																
30	− 5.388	0	−3.482	−0.3278	− 0.0129	0	+0.9979	+0.1310	−0.0905	0	−0.0207	−0.0009	− 0.1005	0	− 0.0164	+0.9730
20	− 3.600	−2.628	−2.666	−0.2779	+ 0.0014	−0.0043	+0.9845	+0.1162	−0.0397	−0.0403	−0.0080	−0.0006	+ 0.0123	− 0.0332	− 0.0024	+0.9761
10	+ 1.787	−3.290	−0.781	−0.1511	+ 0.0164	+0.0020	+0.9410	+0.0722	+0.1117	−0.0251	+0.0142	−0.0002	+ 0.1340	+ 0.0177	+ 0.0099	+0.9852
0	+10.83	0	+0.500	0	− 0.0447	0	+0.8660	0	+0.3604	+0.1000	0	0	− 0.3807	0	0	+1.000

TABLE 14.2 (ASCE Table 2A) (*continued*)

ϕ	Vertical Edge Load				Horizontal Edge Load				Shear Edge Load				Edge Moment Load			
	T_x (1)	S (2)	T_ϕ (3)	M_ϕ (4)	T_x (5)	S (6)	T_ϕ (7)	M_ϕ (8)	T_x (9)	S (10)	T_ϕ (11)	M_ϕ (12)	T_x (13)	S (14)	T_ϕ (15)	M_ϕ (16)
(h) (Continued)																
$\phi_k=35$:																
35	−3.373	0	−2.947	−0.3692	−0.0349	0	+0.9926	+0.1751	−0.0773	0	−0.0243	−0.0015	−0.2044	0	−0.0435	+0.9623
30	−3.175	−0.907	−2.810	−0.3584	−0.0272	−0.0089	+0.9901	+0.1715	−0.0693	−0.0205	−0.0212	−0.0014	−0.1601	−0.0520	−0.0356	+0.9632
20	−1.566	−2.282	−1.814	−0.2774	+0.0192	−0.0121	+0.9675	+0.1424	−0.0055	−0.0439	−0.0015	−0.0008	+0.1134	−0.0710	+0.0090	+0.9700
10	+1.763	−2.310	−0.337	−0.1441	+0.0378	+0.0075	+0.9107	+0.0832	+0.1211	−0.0150	+0.0185	−0.0002	+0.2305	+0.0464	+0.0256	+0.9823
0	+6.999	0	+0.574	0	−0.1053	0	+0.8192	0	+0.3082	+0.1000	0	0	−0.6605	0	0	+1.000
$\phi_k=40$:																
40	−2.193	0	−2.526	−0.4041	−0.0791	0	+0.9782	+0.2237	−0.0670	0	−0.0278	−0.0022	−0.3607	0	−0.1000	+0.9479
30	−1.834	−1.137	−2.168	−0.3677	−0.0303	−0.0341	+0.9742	+0.2095	−0.0458	−0.0329	−0.0178	−0.0018	−0.1407	−0.1563	−0.0487	+0.9521
20	−0.678	−1.866	−1.220	−0.2690	+0.0667	−0.0239	+0.9491	+0.1669	+0.0177	−0.0425	+0.0054	−0.0009	+0.3060	−0.1102	+0.0428	+0.9635
10	+1.483	−1.699	−0.048	−0.1354	+0.0718	+0.0226	+0.8777	+0.0965	+0.1231	−0.0058	+0.0220	−0.0002	+0.3419	+0.1068	+0.0540	+0.9795
0	+4.960	0	+0.643	0	−0.2243	0	+0.7660	0	+0.2696	+0.1000	0	0	−1.083	0	0	+1.000
$\phi_k=45$:																
45	−1.415	0	−2.167	−0.4305	−0.1591	0	+0.9446	+0.2747	−0.0585	0	−0.0311	−0.0031	−0.5815	0	−0.2039	+0.9270
40	−1.384	−0.385	−2.106	−0.4226	−0.1386	−0.0418	+0.9469	+0.2713	−0.0549	−0.0127	−0.0288	−0.0030	−0.5076	−0.1526	−0.1817	+0.9284
30	−1.097	−1.081	−1.635	−0.3622	+0.0019	−0.0838	+0.9561	+0.2441	−0.0259	−0.0392	−0.0122	−0.0021	+0.0014	−0.3085	−0.0344	+0.9392
20	−0.350	−1.505	−0.798	−0.2550	+0.1623	−0.0353	+0.9371	+0.1892	+0.0326	−0.0387	+0.0118	−0.0010	+0.5987	−0.1323	+0.1135	+0.9566
10	+1.175	−1.323	+0.148	−0.1254	+0.1167	+0.0570	+0.8460	+0.1071	+0.1213	+0.0021	+0.0248	−0.0002	+0.4491	+0.2141	+0.0996	+0.9768
0	+3.887	0	+0.707	0	−0.4371	0	+0.7071	0	+0.2407	+0.1000	0	0	−1.684	0	0	+1.000
$\phi_k=50$:																
50	−0.838	0	−1.831	−0.4457	−0.2911	0	+0.8743	+0.3250	−0.0506	0	−0.0340	−0.0041	−0.8725	0	−0.3803	+0.8948
40	−0.842	−0.461	−1.678	−0.4195	−0.1746	−0.1378	+0.9021	+0.3125	−0.0408	−0.0260	−0.0262	−0.0036	−0.5294	−0.4145	−0.2512	+0.9030
30	−0.759	−0.908	−1.225	−0.3455	+0.1012	−0.1615	+0.9516	+0.2745	−0.0107	−0.0411	−0.0176	−0.0024	+0.2947	−0.4901	+0.0344	+0.9238
20	−0.314	−1.227	−0.517	−0.2369	+0.3240	−0.0358	+0.9439	+0.2092	+0.0369	−0.0337	+0.0174	−0.0010	+0.9856	−0.1131	+0.2362	+0.9497
10	+0.907	−1.111	+0.281	−0.1142	+0.1645	+0.1245	+0.8204	+0.1167	+0.1175	+0.0087	+0.0271	−0.0001	+0.5268	+0.3830	+0.1664	+0.9741
0	+3.402	0	+0.766	0	−0.7874	0	+0.6428	0	+0.2199	+0.1000	0	0	−2.487	0	0	+1.000
(i) $r/t = 200$ and $r/l = 0.2$																
$\phi_k=30$:																
30	−4.804	0	−3.394	−0.3054	−0.2444	0	+0.9630	+0.1216	−0.1770	0	−0.0407	−0.0017	−1.972	0	−0.3043	+0.8904
20	−3.563	−2.414	−2.649	−0.2592	−0.0144	−0.0894	+0.9776	+0.1084	−0.0792	−0.0792	−0.0160	−0.0012	−0.154	−0.7298	−0.0618	+0.9051
10	+1.170	−3.280	−0.836	−0.1412	+0.2619	−0.0028	+0.9629	+0.0679	+0.2192	−0.0502	+0.0279	−0.0003	+2.142	+0.0427	+0.1925	+0.9435
0	+12.12	0	+0.500	0	−0.5594	0	+0.8660	0	+0.7297	+0.2000	0	0	−4.639	0	0	+1.000

TABLE 14.2 (ASCE Table 2A) (*continued*)

ϕ	Vertical Edge Load				Horizontal Edge Load				Shear Edge Load				Edge Moment Load			
	T_x	S	T_ϕ	M_ϕ	T_x	S	T_ϕ	M_ϕ	T_x	S	T_ϕ	M_ϕ	T_x	S	T_ϕ	M_ϕ
	(1)	(2)	(3)	(4)	(5)	(6)	(7)	(8)	(9)	(10)	(11)	(12)	(13)	(14)	(15)	(16)
								(*i*) (Continued)								
$\phi_k=35$:																
35	− 2.269	0	−2.720	−0.3304	− 0.5645	0	+0.8832	+0.1560	−0.1455	0	−0.0466	−0.0027	− 3.344	0	− 0.7035	+0.8373
30	− 2.290	−0.624	−2.624	−0.3211	− 0.4523	−0.1446	+0.9000	+0.1531	−0.1313	−0.0386	−0.0410	−0.0025	− 2.696	− 0.8582	− 0.5812	+0.8425
20	− 2.035	−1.854	−1.855	−0.2507	+ 0.2416	−0.2183	+0.9868	+0.1291	−0.0150	−0.0842	−0.0033	−0.0014	+ 1.373	− 1.319	+ 0.1200	+0.8787
10	+ 0.648	−2.423	−0.474	−0.1313	+ 0.5770	+0.0604	+0.9767	+0.0783	+0.2331	−0.0309	+0.0358	−0.0002	+ 3.519	+0.3350	+ 0.4303	+0.9335
0	+ 9.620	0	+0.574	0	− 1.373	0	+0.8191	0	+0.6376	+0.2000	0	0	− 8.444	0	0	+1.000
$\phi_k=40$:																
40	− 0.324	0	−2.021	−0.3363	− 1.133	0	+0.6921	+0.1847	−0.1144	0	−0.0503	−0.0037	− 5.147	0	− 1.422	+0.7549
30	− 1.057	−0.319	−1.919	−0.3098	− 0.4724	−0.4965	+0.8324	+0.1760	−0.0838	−0.0572	−0.0331	−0.0030	− 2.213	− 2.270	− 0.7127	+0.7838
20	− 2.100	−1.229	−1.425	−0.2331	+ 0.8653	−0.3869	+1.067	+0.1458	+0.0202	−0.0785	+0.0084	−0.0015	+ 3.880	− 1.816	+ 0.5816	+0.8519
10	− 0.178	−2.090	−0.327	−0.1202	+ 1.021	+0.2424	+1.036	+0.0873	+0.2295	−0.0158	+0.0411	−0.0002	+ 4.849	+ 1.085	+ 0.8003	+0.9255
0	+ 9.537	0	+0.643	0	− 2.851	0	+0.7660	0	+0.5858	+0.2000	0	0	− 13.55	0	0	+1.000
$\phi_k=45$:																
45	+ 1.358	0	−1.210	−0.3140	− 1.967	0	+0.3169	+0.1973	−0.0793	0	−0.0494	−0.0047	− 7.076	0	− 2.500	+0.6266
40	+ 1.046	+0.344	−1.254	−0.3103	− 1.725	−0.5171	+0.3877	+0.1967	−0.0769	−0.0215	−0.0462	−0.0045	− 6.231	− 1.862	− 2.231	+0.6378
30	− 1.033	−0.270	−1.469	−0.2789	− 0.050	−0.6113	+0.8459	+0.1884	−0.0516	−0.0642	−0.0223	−0.0032	− 0.316	− 2.536	− 0.5827	+0.7167
20	− 3.035	−0.788	−1.324	−0.2104	+ 1.916	−0.5110	+1.281	+0.1588	+0.0289	−0.0681	+0.0165	−0.0014	+ 6.897	− 1.941	+ 1.367	+0.8267
10	− 0.885	−2.169	−0.327	−0.1087	+ 1.486	+0.6094	+1.159	+0.0955	+0.2164	−0.0071	+0.0436	−0.0001	+ 5.708	+ 2.201	+ 1.273	+0.9200
0	+10.91	0	+0.707	0	− 5.085	0	+0.7071	0	+0.5724	+0.2000	0	0	− 19.32	0	0	+1.000
$\phi_k=50$:																
50	+ 2.686	0	−0.312	−0.2588	− 2.918	0	−0.2665	+0.1807	−0.0395	0	−0.0419	−0.0052	− 8.502	0	− 3.770	+0.4461
40	+ 1.320	+1.218	−0.671	−0.2562	− 1.798	−1.391	+0.1445	+0.1856	−0.0445	−0.0226	−0.0351	−0.0046	− 5.350	− 4.060	− 2.523	+0.5106
30	− 1.860	+1.105	−1.355	−0.2385	+ 0.904	−1.675	+1.046	+0.1889	−0.0420	−0.0476	−0.0140	−0.0031	+ 2.433	− 4.983	+ 0.2750	+0.6671
20	− 4.145	−0.669	−1.455	−0.1867	+ 3.198	−0.4671	+1.649	+0.1655	+0.0168	−0.0585	+0.0187	−0.0013	+ 9.459	− 1.527	+ 2.336	+0.8314
10	− 1.190	−2.490	−0.395	−0.0979	− 1.779	+1.160	+1.332	+0.0992	+0.2016	−0.0062	+0.0428	−0.0001	+ 5.734	+ 3.392	+ 1.737	+0.9393
0	+12.58	0	+0.766	0	− 7.777	0	+0.6428	0	+0.5914	+0.2000	0	0	− 24.30	0	0	+1.000
								(*j*) $r/t = 200$ and $r/l = 0.3$								
$\phi_k=30$:																
30	− 2.875	0	−3.097	−0.2688	− 1.018	0	+0.8433	+0.1065	−0.2481	0	−0.0585	−0.0023	− 8.01	0	− 1.251	+0.7583
20	− 3.544	−1.727	−2.595	−0.2300	− 0.025	−0.3675	+0.9558	+0.0961	−0.1187	−0.0759	−0.0236	−0.0016	− 0.32	− 2.906	− 0.242	+0.7953
10	− 1.006	−3.351	−1.021	−0.1270	+ 1.142	+0.0251	+1.038	+0.0619	+0.3098	−0.1126	+0.0402	−0.0003	+ 9.20	+ 0.158	+ 0.787	+0.8828
0	+17.18	0	+0.500	0	− 2.618	0	+0.8660	0	+1.138	+0.3000	0	0	− 21.66	0	0	+1.000

TABLE 14.2 (ASCE Table 2A) (*continued*)

ϕ	Vertical Edge Load				Horizontal Edge Load				Shear Edge Load				Edge Moment Load			
	T_x (1)	S (2)	T_ϕ (3)	M_ϕ (4)	T_x (5)	S (6)	T_ϕ (7)	M_ϕ (8)	T_x (9)	S (10)	T_ϕ (11)	M_ϕ (12)	T_x (13)	S (14)	T_ϕ (15)	M_ϕ (16)
(*j*) (Continued)																
$\phi_k=35$:																
35	+ 1.157	0	−2.003	−0.2639	− 3.540	0	+0.5377	+0.1233	−0.1777	0	−0.0616	−0.0033	− 12.73	0	− 2.708	+0.6320
30	+ 0.434	+0.250	−2.034	−0.2580	− 1.763	−0.5643	+0.6157	+0.1220	−0.1650	−0.0475	−0.0548	−0.0031	− 10.23	− 3.260	− 2.236	+0.6464
20	− 3.630	−0.571	−1.989	−0.2098	+ 1.001	−0.8379	+1.05	+0.1087	−0.0418	−0.1115	−0.0066	−0.0017	+ 5.57	− 4.934	+ 0.474	+0.7427
10	− 2.937	−2.907	−0.903	−0.1149	+ 2.318	+0.2927	+1.184	+0.0698	+0.3091	−0.0522	+0.0487	−0.0002	+ 13.92	+ 1.650	+ 1.650	+0.8689
0	+18.66	0	+0.574	0	− 5.780	0	+0.8192	0	+1.058	+0.3000	0	0	− 35.32	0	0	+1.000
$\phi_k=40$:																
40	+ 4.491	0	−0.695	−0.2230	− 4.689	0	−0.0553	+0.1198	−0.0958	0	−0.0550	−0.0039	− 16.80	0	− 4.723	+0.4470
30	+ 0.927	+1.787	−1.263	−0.2170	− 1.599	−1.680	+0.4612	+0.1225	−0.0958	−0.0529	−0.0397	−0.0032	− 7.30	− 7.430	− 2.380	+0.5253
20	− 5.904	+0.374	−1.979	−0.1836	+ 2.992	−1.296	+1.378	+0.1168	−0.0298	−0.0936	+0.0040	−0.0016	+ 12.97	− 5.929	+ 1.932	+0.6997
10	− 4.643	−3.248	−1.057	−0.1038	+ 3.576	+0.8937	+1.451	+0.0774	+0.2786	−0.0416	+0.0506	−0.0001	+ 16.75	+ 3.932	+ 2.673	+0.8621
0	+22.55	0	+0.643	0	−10.29	0	+0.7660	0	+1.069	+0.3000	0	0	− 48.57	0	0	+1.000
$\phi_k=45$:																
45	+ 6.463	0	+0.608	−0.1511	− 5.443	0	−0.8614	+0.0899	−0.0114	0	−0.0370	−0.0038	− 17.92	0	− 6.551	+0.2280
40	+ 5.541	+1.688	+0.369	−0.1543	− 4.617	−1.379	−0.6652	+0.0937	−0.0214	−0.0039	−0.0364	−0.0037	− 15.90	− 4.718	− 5.866	+0.2542
30	− 0.843	+3.190	−1.138	−0.1700	− 0.203	−2.859	+0.6257	+0.1158	−0.0772	−0.0304	−0.0273	−0.0028	− 1.34	− 9.994	− 1.292	+0.4354
20	− 8.160	+0.555	−2.324	−0.1610	+ 5.220	−1.401	+1.929	+0.1250	−0.0627	−0.0765	+0.0042	−0.0013	+ 17.97	− 5.233	+ 3.575	+0.6795
10	− 5.055	−3.891	−1.245	−0.0955	+ 4.276	+1.728	+1.762	+0.0858	+0.2474	−0.0456	+0.0471	−0.0000	+ 16.41	+ 6.050	+ 3.437	+0.8640
0	+25.47	0	+0.707	0	−14.77	0	+0.7071	0	+1.134	+0.3000	0	0	− 56.01	0	0	+1.000
$\phi_k=50$:																
50	+ 6.626	0	+1.510	−0.0742	− 5.740	0	−1.603	+0.0441	+0.0544	0	−0.0135	−0.0031	− 15.39	0	− 7.343	+0.0368
40	+ 3.900	+3.127	+0.564	−0.0969	− 3.696	−2.769	−0.7707	+0.0676	+0.0026	+0.0201	−0.0201	−0.0029	− 10.44	− 7.529	− 5.048	+0.1442
30	− 2.878	+3.516	−1.463	−0.1393	+ 1.552	−3.453	+1.111	+0.1158	−0.1000	−0.0070	−0.0254	−0.0023	+ 3.25	− 9.817	+ 0.304	+0.4060
20	− 8.823	+0.094	−2.598	+0.1467	+ 6.637	−1.072	+2.491	+0.1375	−0.1008	−0.0715	−0.0025	−0.0011	+ 18.60	− 3.576	+ 4.612	+0.6832
10	− 4.273	−4.274	−1.262	−0.0891	+ 4.214	+2.483	+1.987	+0.0953	+0.2354	−0.0563	+0.0424	−0.0000	+ 14.04	+ 7.094	+ 3.666	+0.8700
0	+25.53	0	+0.766	0	−17.75	0	+0.6428	0	+1.193	+0.3000	0	0	− 56.30	0	0	+1.000
(*k*) $r/t = 200$ and $r/l = 0.4$																
$\phi_k=30$:																
30	+ 1.124	0	−2.479	−0.2211	− 2.625	0	+0.5943	+0.0868	−0.2847	0	−0.0710	−0.0026	− 20.19	0	− 3.173	+0.5913
20	− 3.463	−0.295	−2.481	−0.1937	− 0.066	−0.9430	+0.9092	+0.0810	−0.1583	−0.1339	−0.0302	−0.0018	− 0.87	− 7.343	− 0.619	+0.6619
10	− 5.528	−3.468	−1.407	−0.1117	+ 2.974	+0.0700	+1.194	+0.0553	+0.3622	−0.1029	+0.0493	−0.0003	+ 23.57	+ 0.435	+ 1.999	+0.8155
0	+27.68	0	+0.500	0	− 6.899	0	+0.8660	0	+1.634	+0.4000	0	0	− 56.52	0	0	+1.000

TABLE 14.2 (ASCE Table 2A) (*continued*)

ϕ	Vertical Edge Load				Horizontal Edge Load				Shear Edge Load				Edge Moment Load			
	T_x	S	T_ϕ	M_ϕ	T_x	S	T_ϕ	M_ϕ	T_x	S	T_ϕ	M_ϕ	T_x	S	T_ϕ	M_ϕ
	(1)	(2)	(3)	(4)	(5)	(6)	(7)	(8)	(9)	(10)	(11)	(12)	(13)	(14)	(15)	(16)
								(k) (Continued)								
$\phi_k=35$:																
35	+ 6.873	0	−0.795	−0.1830	− 4.932	0	−0.0417	+0.0837	−0.1462	0	−0.0635	−0.0032	− 27.50	0	− 5.935	+0.3911
30	+ 5.012	+1.712	−1.039	−0.1820	− 3.951	−1.261	+0.1374	+0.0848	−0.1482	−0.0403	−0.0577	−0.0030	− 22.24	− 7.055	− 4.913	+0.4179
20	− 6.203	+1.614	−2.206	−0.1647	+ 2.208	−1.889	+1.153	+0.0862	−0.0986	−0.1152	−0.0124	−0.0017	+ 11.73	−10.85	+ 1.001	+0.5937
10	− 9.087	−3.682	−1.633	−0.1001	+ 5.303	+0.6597	+1.536	+0.0621	+0.3204	−0.0822	+0.0539	−0.0001	+ 31.27	+ 3.583	+ 3.651	+0.8049
0	+34.19	0	+0.574	0	−13.36	0	+0.8192	0	+1.639	+0.4000	0	0	− 80.78	0	0	+1.000
$\phi_k=40$:																
40	+11.34	0	+0.768	−0.1216	− 7.637	0	−0.8602	+0.0620	+0.0319	0	−0.0448	−0.0032	− 32.54	0	− 8.301	+0.2027
30	+ 3.392	+4.247	−0.465	−0.1293	− 2.992	−3.037	+0.0122	+0.0723	−0.0781	−0.0171	−0.0365	−0.0026	− 13.32	−12.80	− 4.292	+0.2888
20	−10.20	+2.361	−2.620	−0.1409	+ 5.333	−2.419	+1.731	+0.0915	−0.1367	−0.0848	−0.0086	−0.0014	+ 21.73	−10.76	+ 3.315	+0.5642
10	−10.34	−4.580	−1.955	−0.0931	+ 6.841	+1.629	+1.957	+0.0704	+0.2620	−0.0845	+0.0492	−0.0000	+ 31.74	+ 6.881	+ 4.858	+0.8083
0	+39.09	0	+0.643	0	−19.78	0	+0.7660	0	+1.744	+0.4000	0	0	− 92.98	0	0	+1.000
$\phi_k=45$:																
45	+ 9.787	0	+1.934	−0.0436	− 7.229	0	−1.692	+0.0199	+0.0957	0	−0.0099	−0.0065	− 22.45	0	− 8.938	−0.0169
40	+ 8.573	+2.572	+1.567	−0.0513	− 6.456	−1.911	−1.419	+0.0264	+0.0691	+0.0238	−0.0133	−0.0043	− 20.42	− 5.970	− 8.074	+0.0165
30	− 0.227	+5.188	−0.825	−0.0982	− 0.709	−4.101	+0.4128	+0.0678	−0.0987	+0.0199	−0.0292	−0.0020	− 4.08	−13.43	− 2.116	+0.2506
20	−11.70	+1.769	−3.045	−0.1296	+ 7.382	−2.215	+2.378	+0.1027	−0.1953	−0.0728	−0.0161	−0.0012	+ 23.35	− 8.260	+ 4.819	+0.5714
10	− 9.060	−5.153	−2.000	−0.0887	+ 7.017	+2.520	+2.255	+0.0800	+0.2396	−0.0981	+0.0429	−0.0000	+ 27.40	+ 8.293	+ 5.115	+0.8168
0	+38.68	0	+0.707	0	−23.69	0	+0.7071	0	+1.826	+0.4000	0	0	− 91.16	0	0	+1.000
$\phi_k=50$:																
50	+ 7.414	0	+2.207	+0.0042	− 6.022	0	−2.055	−0.0134	+0.1395	0	+0.0147	−0.0013	− 13.14	0	− 7.852	−0.1157
40	+ 4.947	+3.616	+1.124	−0.0267	− 4.389	−3.008	−1.161	+0.0151	+0.0505	+0.0598	−0.0042	−0.0015	− 11.30	− 6.932	− 5.801	−0.0061
30	− 2.266	+4.540	−1.356	−0.0910	+ 0.8355	−4.146	+0.9927	+0.0776	−0.1446	+0.0347	−0.0352	−0.0017	− 1.18	−10.95	− 0.390	+0.2725
20	−10.69	+0.849	−3.073	−0.1245	+ 7.857	−1.711	+2.811	+0.1179	−0.2195	−0.0793	−0.0240	−0.0011	+ 19.84	− 6.154	+ 5.079	+0.5892
10	− 7.142	−5.081	−1.774	−0.0835	+ 6.637	+3.045	+2.378	+0.0892	+0.2513	−0.1076	+0.0402	−0.0001	+ 23.56	+ 7.944	+ 4.801	+0.8219
0	+35.00	0	+0.766	0	−25.44	0	+0.6428	0	+1.840	+0.4000	0	0	− 84.42	0	0	+1.000
								(l) $r/t = 200$ and $r/l = 0.5$								
$\phi_k=30$:																
30	+ 6.908	0	−1.578	−0.1695	− 4.932	0	+0.2325	+0.0656	−0.2715	0	−0.0759	−0.0026	− 36.86	0	− 5.858	+0.4164
20	− 3.091	+1.791	−2.323	−0.1564	− 2.320	−1.780	+0.8452	+0.0654	−0.1974	−0.1375	−0.0355	−0.0018	− 2.00	−13.49	− 1.174	+0.5258
10	−12.18	−3.626	−1.972	−0.0957	+ 5.663	+0.128	+1.422	+0.0492	+0.3600	−0.1314	+0.05366	−0.0002	+ 43.95	+ 0.71	+ 3.711	+0.7505
0	+43.35	0	+0.500	0	−13.29	0	+0.8660	0	+2.253	+0.5000	0	0	−107.5	0	0	+1.000

TABLE 14.2 (ASCE Table 2A) (*continued*)

ϕ	Vertical Edge Load				Horizontal Edge Load				Shear Edge Load				Edge Moment Load			
	T_x (1)	S (2)	T_ϕ (3)	M_ϕ (4)	T_x (5)	S (6)	T_ϕ (7)	M_ϕ (8)	T_x (9)	S (10)	T_ϕ (11)	M_ϕ (12)	T_x (13)	S (14)	T_ϕ (15)	M_ϕ (16)
(*l*) Continued																
$\phi_k=35$:																
35	+12.92	0	+0.520	−0.1106	− 7.760	0	−0.6687	+0.0486	+0.0566	0	−0.0528	−0.0027	− 41.17	0	− 9.181	+0.1851
30	+10.14	+3.193	+0.037	−0.1151	− 6.371	−1.952	−0.3803	+0.0518	−0.0847	−0.0181	−0.0505	−0.0025	− 33.79	−10.62	− 7.611	+0.2222
20	− 8.763	+4.022	−2.425	−0.1266	+ 3.367	−3.036	+1.253	+0.0670	−0.1818	−0.0966	−0.0204	−0.0015	+ 16.53	−16.90	+ 1.429	+0.4668
10	−16.09	−4.484	−2.447	−0.0897	+ 8.692	+1.030	+1.926	+0.0564	+0.2700	−0.1200	+0.0516	−0.0001	+ 50.25	+ 5.26	+ 5.747	+0.7504
0	+52.21	0	+0.574	0	−22.16	0	+0.8192	0	+2.374	+0.5000	0	0	−132.8	0	0	+1.000
$\phi_k=40$:																
40	+13.93	0	+1.963	−0.0443	− 8.898	0	−1.529	+0.0185	+0.1093	0	−0.0175	−0.0020	− 32.96	0	−10.30	−0.0072
30	+ 5.361	+5.888	+0.146	−0.0720	− 4.097	−3.897	−0.3259	+0.0398	−0.0475	+0.0298	−0.0294	−0.0019	− 17.70	−15.29	− 5.586	+0.1389
20	−12.79	+3.938	−3.042	−0.1129	+ 6.622	−3.303	+1.951	+0.0737	−0.2613	−0.0652	−0.0242	−0.0012	+ 24.15	−14.33	+ 3.905	+0.4658
10	−15.31	−5.476	−2.659	−0.0866	+ 9.721	+2.099	+2.350	+0.0645	+0.2136	−0.1319	+0.0433	−0.0000	+ 44.95	+ 8.13	+ 6.422	+0.7611
0	+53.53	0	+0.643	0	−28.17	0	+0.7660	0	+2.509	+0.5000	0	0	−133.2	0	0	+1.000
$\phi_k=45$:																
45	+10.22	0	+2.435	+0.0045	− 7.196	0	−1.959	−0.0106	+0.1830	0	+0.0132	−0.0011	− 18.31	0	− 8.981	−0.1089
40	+ 9.385	+2.771	+2.035	−0.0052	− 6.757	−1.967	−1.675	−0.0034	+0.1446	+0.0466	+0.0066	−0.0012	− 17.78	− 4.98	− 8.263	−0.0771
30	+ 0.957	+5.868	−0.565	−0.0616	− 1.640	−4.438	+0.2286	+0.0431	−0.1100	+0.0637	−0.0303	−0.0014	− 8.90	−13.02	− 2.879	+0.1523
20	−12.61	+2.686	−3.298	−0.1102	+ 7.723	−2.850	+2.505	+0.0881	−0.3155	−0.0676	−0.0341	−0.0011	+ 20.73	−10.69	+ 4.696	+0.4876
10	−12.52	−5.653	−2.457	−0.0832	+ 9.498	+2.791	+2.554	+0.0750	+0.2211	−0.1467	+0.0389	−0.0001	+ 38.31	+ 8.33	+ 6.068	+0.7681
0	+48.96	0	+0.707	0	−30.94	0	+0.7071	0	+2.539	+0.5000	0	0	−123.1	0	0	+1.000
$\phi_k=50$:																
50	+ 5.720	0	+2.151	+0.0279	− 4.333	0	−1.931	−0.0295	+0.1739	0	+0.0295	+0.0002	− 4.66	0	− 6.575	−0.1404
40	+ 5.105	+3.070	+1.265	−0.0015	− 4.431	−2.438	−1.238	−0.0037	+0.0786	+0.0777	+0.0054	−0.0007	− 9.77	− 3.60	− 5.629	−0.0518
30	− 0.025	+4.652	−1.014	−0.0656	− 1.252	−4.161	+0.6846	+0.0569	−0.1578	+0.0593	−0.0387	−0.0013	− 10.31	− 9.96	− 1.682	+0.1945
20	−10.40	+1.897	−3.112	−0.1093	+ 7.442	−2.639	+2.813	+0.1043	−0.2949	−0.0794	−0.0364	−0.0011	+ 16.18	− 9.65	+ 4.675	+0.5157
10	−12.65	−5.039	−2.643	−0.0852	+11.50	+3.115	+3.155	+0.0913	+0.2141	−0.1221	+0.0366	−0.0000	+ 41.78	+ 9.30	+ 7.575	+0.8127
0	+42.10	0	+0.766	0	−31.89	0	+0.6428	0	+2.441	+0.5000	0	0	−116.1	0	0	+1.000
(*m*) $r/t = 200$ and $r/l = 0.6$																
$\phi_k=30$:																
30	+13.41	0	−0.550	−0.1215	− 7.494	0	−0.177	+0.0459	−0.2086	0	−0.0732	−0.0024	− 53.95	0	− 8.721	+0.2589
20	− 2.95	+4.174	−2.106	−0.1221	− 0.317	−2.725	+0.756	+0.0512	−0.2355	−0.1233	−0.0394	−0.0016	− 4.01	−20.04	− 1.828	+0.4042
10	−19.88	−3.751	−2.624	−0.0867	+ 8.756	+0.171	+1.682	+0.0443	+0.3019	−0.1613	+0.0532	−0.0001	+ 66.27	+ 0.71	+ 5.559	+0.6931
0	+61.86	0	+0.500	0	−20.84	0	+0.866	0	+3.003	+0.6000	0	0	−166.4	0	0	+1.000

TABLE 14.2 (ASCE Table 2A) (*continued*)

ϕ	VERTICAL EDGE LOAD T_x (1)	S (2)	T_ϕ (3)	M_ϕ (4)	HORIZONTAL EDGE LOAD T_x (5)	S (6)	T_ϕ (7)	M_ϕ (8)	SHEAR EDGE LOAD T_x (9)	S (10)	T_ϕ (11)	M_ϕ (12)	EDGE MOMENT LOAD T_x (13)	S (14)	T_ϕ (15)	M_ϕ (16)
(*m*) (Continued)																
$\phi_k=35$:																
35	+17.48	0	+1.581	−0.0576	− 9.794	0	−1.155	+0.0231	+0.0614	0	−0.0363	−0.0020	− 48.68	0	−11.31	+0.0433
30	+13.76	+4.442	+0.942	−0.0648	− 8.004	−2.520	−0.794	+0.0279	+0.0019	+0.0113	−0.0381	−0.0020	− 40.77	−12.62	− 9.488	+0.0867
20	−10.37	+6.000	−2.570	−0.0989	+ 4.010	−3.973	+1.313	+0.0531	−0.2759	−0.0663	−0.0294	−0.0012	+ 17.11	−21.22	+ 1.522	+0.3710
10	−22.33	−5.090	−3.136	−0.0827	+11.70	+1.270	+2.252	+0.0523	+0.1839	−0.1613	+0.0454	−0.0000	+ 66.22	+ 5.83	+ 7.338	+0.7029
0	+68.85	0	+0.574	0	−30.32	0	+0.819	0	+3.204	+0.6000	0	0	−180.9	0	0	+1.000
$\phi_k=40$:																
40	+14.46	0	+2.530	−0.0052	− 8.900	0	−1.788	−0.0029	+0.2096	0	+0.0030	−0.0012	− 29.22	0	−10.53	−0.0850
30	+ 6.70	+6.504	+0.519	−0.0397	− 4.840	−4.143	−0.527	+0.0213	−0.0157	+0.0719	−0.0224	−0.0013	− 20.21	−14.50	− 6.191	+0.0578
20	−13.24	+4.992	−3.187	−0.0946	+ 6.608	−3.889	+2.003	+0.0630	−0.3733	−0.0458	−0.0384	−0.0010	+ 19.37	−16.37	+ 3.594	+0.3909
10	−19.32	−5.810	−3.114	−0.0812	+12.10	+2.231	+2.601	+0.0616	+0.1569	−0.1797	+0.0373	−0.0001	+ 56.09	+ 7.56	+ 7.285	+0.7139
0	+65.23	0	+0.643	0	−35.15	0	+0.766	0	+3.297	+0.6000	0	0	−169.2	0	0	+1.000
$\phi_k=45$:																
45	+ 8.60	0	+2.397	+0.0217	− 5.729	0	−1.858	−0.0205	+0.2347	0	+0.0286	−0.0004	− 8.38	0	− 7.554	−0.1241
40	+ 8.15	+2.319	+2.068	+0.0133	− 5.622	−1.562	−1.636	−0.0140	+0.1924	+0.0605	+0.0199	−0.0005	− 10.38	− 2.49	− 7.198	−0.0988
30	+ 2.68	+5.666	−0.311	−0.0418	− 2.980	−4.165	+0.041	+0.0293	−0.1065	+0.0944	−0.0306	−0.0011	− 15.36	−10.11	− 3.686	+0.0939
20	−11.68	+3.500	−3.237	−0.0948	+ 6.764	−3.438	+2.416	+0.0759	−0.4030	−0.0596	−0.0476	−0.0011	+ 12.42	−13.03	+ 3.667	+0.4113
10	−16.02	−5.641	−2.752	−0.0772	+12.12	+2.709	+2.752	+0.0695	+0.1906	−0.1895	+0.0361	−0.0001	+ 50.54	+ 6.80	+ 6.668	+0.7160
0	+58.53	0	+0.707	0	−38.04	0	+0.707	0	+3.268	+0.6000	0	0	−158.3	0	0	+1.000
$\phi_k=50$:																
50	+ 3.09	0	+1.709	+0.0284	− 1.816	0	−1.467	−0.0273	+0.1854	0	+0.0382	+0.0002	+ 7.18	0	− 3.972	−0.1088
40	+ 4.42	+1.981	+1.179	+0.0063	− 3.693	−1.374	−1.120	−0.0084	+0.1051	+0.0872	+0.0118	−0.0003	− 5.77	+ 1.40	− 4.634	−0.0572
30	+ 2.57	+4.287	−0.638	−0.0476	− 3.650	−3.722	+0.325	+0.0406	−0.1481	+0.0826	−0.0419	−0.0012	− 20.51	− 6.83	− 3.347	+0.1206
20	− 9.53	+2.775	−2.920	−0.0908	+ 6.372	−3.425	+2.581	+0.0862	−0.3819	−0.0757	−0.0497	−0.0012	+ 7.78	−12.83	+ 3.041	+0.4149
20	−14.34	−5.142	−2.383	−0.0706	+13.11	+2.955	+2.877	+0.0757	+0.2199	−0.1883	+0.0384	−0.0001	+ 51.00	+ 5.85	+ 6.371	+0.7134
0	+53.46	0	+0.766	0	−41.80	0	+0.643	0	+3.218	+0.6000	0	0	−156.6	0	0	+1.000

Source: Ref. [1].

TABLE 14.3 (ASCE Table 2B) Symmetrical Edge Loads on Simply Supported Cylindrical Shells; Displacements of Edge at $\phi = 0$

	Vertical Edge Load			Horizontal Edge Load			Shear Edge Load			Moment Edge Load		
	Vertical Displacement Δ_V—(Downward Positive)—											
	$V_L \frac{l^4}{r^3 t E} \times$ Col. (1) $\times \sin \frac{\pi x}{l}$			$H_L \frac{l^4}{r^3 t E} \times$ Col. (4) $\times \sin \frac{\pi x}{l}$			$S_L \frac{l^4}{r^3 t E} \times$ Col. (7) $\times \sin \frac{\pi x}{l}$			$M_L \times \frac{l^4}{r^4 t E} \times$ Col. (10) $\times \sin \frac{\pi x}{l}$		
	Horizontal Displacement Δ_H—(Inward Positive)—											
	$V_L \frac{l^4}{r^3 t E} \times$ Col. (2) $\times \sin \frac{\pi x}{l}$			$H_L \frac{l^4}{r^3 t E} \times$ Col. (5) $\times \sin \frac{\pi x}{l}$			$S_L \frac{l^4}{r^3 t E} \times$ Col. (8) $\times \sin \frac{\pi x}{l}$			$M_L \times \frac{l^4}{r^4 t E} \times$ Col. (11) $\times \sin \frac{\pi x}{l}$		
	Rotation θ— $V_L \frac{r^2}{E I} \times$ Col. (3) $\times \sin \frac{\pi x}{l}$			Rotation θ— $H_L \frac{r^2}{E I} \times$ Col. (6) $\times \sin \frac{\pi x}{l}$			Rotation θ— $S_L \frac{r^2}{E I} \times$ Col. (9) $\times \sin \frac{\pi x}{l}$			Rotation θ— $M_L \times \frac{r}{E I} \times$ Col. (12) $\times \sin \frac{\pi x}{l}$		
ϕ_k	Δ_V (1)	Δ_H (2)	θ (3)	Δ_V (4)	Δ_H (5)	θ (6)	Δ_V (7)	Δ_H (8)	θ (9)	Δ_V (10)	Δ_H (11)	θ (12)
	(a) $r/t = 100$											
$r/l = 0.1$:												
30	12.53	0.1445	0.1108	−0.1445	−0.0585	−0.04556	0.3401	0.000480	0.000363	−1.330	−0.5469	−0.5144
35	6.265	0.2508	0.1420	−0.2508	−0.1231	−0.07108	0.2178	0.001006	0.000544	−1.704	−0.8531	−0.5964
40	3.611	0.4009	0.1752	−0.4009	−0.2325	−0.1039	0.1443	0.002009	0.000822	−2.102	−1.247	−0.6749
45	2.591	0.5987	0.2088	−0.5981	−0.4040	−0.1444	0.1084	0.003771	0.001223	−2.505	−1.734	−0.7542
50	2.043	0.8434	0.2410	−0.8434	−0.6558	−0.1925	0.0792	0.006516	0.001776	−2.892	−2.312	−0.8290
$r/l = 0.2$:												
30	17.38	2.031	0.0982	−2.031	−0.8742	−0.04288	0.6950	0.00726	0.000353	−18.86	−8.233	−0.4893
35	12.87	3.538	0.1259	−3.538	−1.794	−0.06530	0.4733	0.02065	0.000678	−24.18	−12.54	−0.5564
40	12.51	5.544	0.1524	−5.544	−3.282	−0.09256	0.3715	0.04590	0.001145	−29.25	−17.77	−0.6159
45	13.47	7.956	0.1748	−7.956	−5.447	−0.1233	0.3402	0.08862	0.001757	−33.56	−23.67	−0.6651
50	14.54	10.50	0.1899	−10.50	−8.270	−0.1546	0.3610	0.1526	0.002478	−36.46	−29.70	−0.6986
$r/l = 0.3$:												
30	35.24	9.474	0.0891	−9.474	−4.010	−0.03866	1.187	0.06357	0.000536	−86.64	−37.58	−0.4499
35	37.31	15.65	0.1092	−15.65	−7.890	−0.05606	0.9583	0.1474	0.000923	−106.2	−54.50	−0.4965
40	42.38	22.99	0.1245	−22.99	−13.59	−0.07595	0.9561	0.2931	0.001413	−121.0	−73.83	−0.5288
45	46.06	30.12	0.1317	−30.12	−20.69	−0.09384	1.073	0.5020	0.001927	−127.9	−91.21	−0.5434
50	45.75	35.11	0.1285	−35.11	−27.93	−0.1070	1.209	0.7422	0.002334	−124.9	−104.0	−0.5402
$r/l = 0.4$:												
30	75.06	26.10	0.07760	−26.10	−11.06	−0.03483	1.940	0.2256	0.000632	−239.2	−104.0	−0.4049
35	87.57	40.75	0.09046	−40.75	−20.25	−0.04716	1.924	0.5080	0.000994	−277.7	−145.2	−0.4320
40	96.68	55.00	0.09596	−55.00	−32.68	−0.05921	2.189	0.9067	0.001384	−293.8	−181.4	−0.4417
45	96.10	64.69	0.09256	−64.69	−44.95	−0.06746	2.500	1.372	0.001641	−284.1	−207.1	−0.4348
50	85.15	67.32	0.08360	−67.32	−54.74	−0.07216	2.591	1.731	0.001732	−255.7	−220.8	−0.4228

TABLE 14.3 (ASCE Table 2B) (*continued*)

$r/l = 0.5$:												
30	142.2	54.14	0.06634	− 54.14	− 22.98	−0.02900	3.218	0.5995	0.000644	−497.6	−216.6	−0.3602
35	163.2	78.77	0.07239	− 78.77	− 39.97	−0.03811	3.588	1.202	0.000945	−542.9	−285.9	−0.3706
40	167.7	97.43	0.07124	− 97.43	− 58.47	−0.04484	4.141	1.938	0.001160	−535.2	−336.5	−0.3668
45	152.4	104.9	0.06505	−104.9	− 74.31	−0.04883	4.418	2.566	0.001242	−487.9	−366.2	−0.3568
50	127.5	103.9	0.05747	−103.9	− 87.21	−0.05166	4.246	2.945	0.001199	−431.0	−387.5	−0.3493
$r/l = 0.6$:												
30	238.5	88.47	0.05559	− 88.47	− 39.78	−0.02434	5.112	1.218	0.000620	−864.6	−378.6	−0.3247
35	257.7	126.7	0.05714	−126.7	− 64.74	−0.03049	5.976	2.242	0.000834	−888.7	−474.2	−0.3185
40	245.4	144.9	0.05366	−144.9	− 88.10	−0.03396	6.629	3.250	0.000932	−834.6	−528.2	−0.3171
45	212.3	149.5	0.04783	−149.5	−108.3	−0.03700	6.601	3.942	0.000918	−743.9	−575.4	−0.3042
50	175.8	143.7	0.04304	−147.9	−127.9	−0.03967	6.678	4.324	0.000865	−669.4	−618.7	−0.3086

(*b*) $r/t = 200$

$r/l = 0.1$:												
30	13.59	0.5598	0.10563	− 0.5598	− 0.3112	−0.04555	0.3446	0.001423	0.000252	− 5.070	− 2.186	−0.5143
35	7.732	0.9838	0.13813	− 0.9838	− 0.4916	−0.07100	0.2228	0.003352	0.000438	− 6.630	− 3.408	−0.5958
40	5.711	1.576	0.1719	− 1.576	− 0.9269	−0.1036	0.1579	0.007139	0.000718	− 8.250	− 4.974	−0.6749
45	5.186	2.354	0.2050	− 2.354	− 1.462	−0.1434	0.1237	0.01391	0.001117	− 9.842	− 6.883	−0.7503
50	5.265	3.292	0.2352	− 3.292	− 2.576	−0.1895	0.1083	0.02506	0.001649	− 11.29	− 9.097	−0.8195
$r/l = 0.2$:												
30	32.02	8.267	0.0986	− 8.267	− 3.492	−0.04257	0.7716	0.03561	0.000385	− 75.70	− 32.70	−0.4866
35	33.96	14.07	0.1245	− 14.07	− 7.090	−0.06423	0.6124	0.08739	0.000700	− 95.68	− 49.33	−0.5497
40	39.51	21.38	0.1468	− 21.38	− 12.62	−0.08898	0.6071	0.1815	0.001123	− 112.7	− 68.34	−0.5990
45	44.27	28.96	0.1600	− 28.96	− 19.84	−0.1132	0.6944	0.3237	0.001601	− 122.9	− 86.91	−0.6273
50	45.16	34.63	0.1600	− 34.63	− 27.44	−0.1321	0.8010	0.4950	0.002014	− 122.8	− 101.0	−0.6299
$r/l = 0.3$:												
30	110.9	37.12	0.08756	− 37.12	− 15.70	−0.03796	1.641	0.2500	0.000532	− 340.4	− 147.6	−0.4443
35	123.0	58.70	0.10342	− 58.70	− 29.63	−0.05150	1.782	0.5520	0.000868	− 402.2	− 200.2	−0.4792
40	136.4	78.62	0.10899	− 78.62	− 46.73	−0.06707	2.154	0.9824	0.001193	− 423.7	− 260.8	−0.4884
45	131.1	89.18	0.10274	− 89.18	− 62.16	−0.07481	2.432	1.410	0.001376	− 399.4	− 290.9	−0.4749
50	111.5	89.27	0.09013	− 89.27	− 73.18	−0.07807	2.437	1.695	0.001382	− 350.4	− 303.5	−0.4550
$r/l = 0.4$:												
30	247.6	98.82	0.07423	− 98.82	− 41.89	−0.03237	3.525	0.9471	0.000586	− 912.9	− 397.8	−0.3936
35	283.3	139.7	0.07945	−139.7	− 70.98	−0.04180	4.354	1.701	0.000837	− 976.3	− 513.6	−0.4011
40	273.8	161.7	0.07471	−161.7	− 97.64	−0.04717	4.977	2.518	0.000964	− 918.0	− 579.7	−0.3884
45	232.0	163.1	0.06551	−163.1	−117.2	−0.04975	4.925	3.016	0.000945	− 805.0	− 611.3	−0.3732
50	187.9	157.7	0.05768	−157.7	−135.5	−0.05270	4.456	3.239	0.000875	− 708.8	− 647.5	−0.3670
$r/l = 0.5$:												
30	473.4	194.4	0.06058	−194.4	− 82.77	−0.02666	6.899	2.116	0.000570	−1817.	− 800.0	−0.3421
35	486.4	244.3	0.05905	−244.3	−125.5	−0.03167	8.309	3.527	0.000704	−1772.	− 950.2	−0.3343
40	424.2	256.6	0.05240	−256.6	−158.6	−0.03414	8.519	4.483	0.000707	−1572.	−1024.	−0.3210
45	349.1	254.2	0.04653	−254.2	−189.1	−0.03671	7.792	4.924	0.000653	−1396.	−1101.	−0.3156
50	295.1	260.4	0.04398	−260.4	−234.5	−0.04173	6.701	5.075	0.000616	−1319.	−1252.	−0.3226
$r/l = 0.6$:												
30	756.0	315.3	0.04838	−315.3	−135.1	−0.02155	11.82	4.979	0.000510	−3009.	−1341.	−0.2961
35	706.1	360.6	0.04430	−360.6	−188.1	−0.02429	13.15	5.790	0.000555	−2756.	−1511.	−0.2847
40	591.7	367.3	0.03930	−367.3	−232.8	−0.02637	12.46	6.714	0.000519	−2445.	−1640.	−0.2778
45	498.8	374.8	0.03615	−374.8	−287.2	−0.02931	11.18	7.265	0.000486	−2249.	−1823.	−0.2782
50	431.9	389.3	0.03392	−389.3	−357.3	−0.03284	10.21	7.983	0.000481	−2110.	−2043.	−0.2807

Source: Ref. [1].

independently. If, however, these tables are to be applied to long shells without symmetrical loadings, all eight equations will have to be solved simultaneously.

Now, for the beam, $\bar{\delta}_{ij}$ and $\bar{\delta}_{i0}$ can be obtained from simple beam theory (refer to Chapter 1). Generally, edge beams have slender cross sections and hence are considered to be incapable of resisting any horizontal loads $\bar{X}_2$ or torsional moments $\bar{X}_4$. Thus we shall here derive formulas which can readily be used for finding stresses and deflections in edge beams under the action of loads $\bar{X}_1$ and $\bar{X}_3$ only. These derivations will also be useful when solving problems by the Schorer method or when writing an independent computer program.

14.3 DEFLECTIONS AND STRESSES IN EDGE BEAMS

We have seen that for solving problems by the ASCE tables or by Schorer theory (Chapter 13), we need to know the vertical and longitudinal deflections of the edge beam under sinusoidal vertical loads and longitudinal shear loads. For ready use, we shall derive these expressions under two typical loadings: (1) a vertical load, and (2) a longitudinal shear load, given, respectively, by

$$\bar{X}_1 = X_1 \sin \frac{\pi x}{L} \tag{14.2a}$$

and

$$\bar{X}_3 = X_3 \cos \frac{\pi x}{L} \tag{14.3a}$$

Vertical Load on Edge Beam

Figure 14.4 shows a beam, simply supported at $x = 0$ and L, and subjected to a vertical load $X_1 \sin(\pi x/L)$. We assume d as the depth, A as the cross-sectional area, and I as the moment of inertia about the major axis of the cross section of the beam. Deformation w is the downward vertical deflection, and u is the displacement along the x-axis. The differential equation for the deflection w is then (refer to Chapter 1)

$$EIw'''' = -X_1 \sin \frac{\pi x}{L}$$

We note that since there is only a vertical load acting on the beam, the axial force $T(x)$ on any section is zero. Thus we have

$$T(x) = 0 \tag{14.4a}$$

Now integrating the foregoing differential equation for w twice, we have

$$EIw'' = X_1 \frac{L^2}{\pi^2} \sin \frac{\pi x}{L} + c_1 x + c_2$$

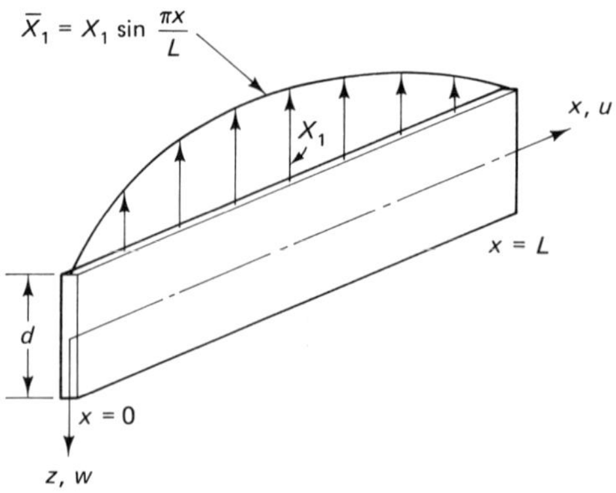

Figure 14.4 Simply supported edge beam subjected to a sinusoidally varying vertical load at its top face.

Noting that $M(x) = -EIw''$ is zero at $x = 0$ and $x = L$, we obtain the two constants c_1 and c_2 as zero. Hence we have

$$EIw'' = -M(x) = X_1 \frac{L^2}{\pi^2} \sin \frac{\pi x}{L} \tag{14.4b}$$

Integrating twice again with respect to x, we have

$$EIw = -X_1 \frac{L^4}{\pi^4} \sin \frac{\pi x}{L} + c_3 x + c_4$$

Again, constants c_3 and c_4 are zero, since $w = 0$ at $x = 0$ and $x = L$. Hence we find

$$EIw = -X_1 \frac{L^4}{\pi^4} \sin \frac{\pi x}{L} \tag{14.2b}$$

To obtain the longitudinal displacement at the top of the beam, we proceed as follows. Moment $M(x)$ is given by

$$M(x) = -EIw'' = -X_1 \frac{L^2}{\pi^2} \sin \frac{\pi x}{L} \tag{14.4b}$$

This moment gives a longitudinal stress σ_x at the top fiber of the beam (tensile positive) of

$$\sigma_x = -\frac{M(x)}{I}\frac{d}{2} = \frac{X_1}{I}\frac{L^2}{\pi^2}\frac{d}{2} \sin \frac{\pi x}{L}$$

Hence the strain at this extreme fiber is

$$\varepsilon_x = \frac{X_1}{EI}\frac{L^2}{\pi^2}\frac{d}{2} \sin \frac{\pi x}{L}$$

Due to symmetry, we have $u = 0$ at $x = L/2$. Hence displacement u for any other value of x at the top of the beam is given by

$$u = \int_{L/2}^{x} \varepsilon_x \, dx$$

$$= \int_{L/2}^{x} \frac{X_1}{EI} \frac{L^2}{\pi^2} \frac{d}{2} \sin \frac{\pi x}{L} \, dx$$

$$u = -\frac{X_1}{EI} \frac{L^3}{\pi^3} \frac{d}{2} \cos \frac{\pi x}{L} \tag{14.2c}$$

Longitudinal Shear on Edge Beam

Next we consider the longitudinal shear load applied at the top fiber of the beam, as shown in Fig. 14.5. Due to this self-equilibrating load, at any section $x = constant$, the axial tension force $T(x)$ in the beam is given by

$$T(x) = -\int_{0}^{x} X_3 \cos \frac{\pi x}{L} \, dx = -X_3 \frac{L}{\pi} \sin \frac{\pi x}{L} \tag{14.5a}$$

Transferring this force to the centroidal axis of the beam, we have an equivalent bending moment (sagging positive) at that section equal to

$$M(x) = -T(x) \frac{d}{2} = X_3 \frac{L}{\pi} \frac{d}{2} \sin \frac{\pi x}{L} \tag{14.5b}$$

Using the relation; $EIw'' = -M(x)$, we obtain

$$EIw'' = -X_3 \frac{L}{\pi} \frac{d}{2} \sin \frac{\pi x}{L}$$

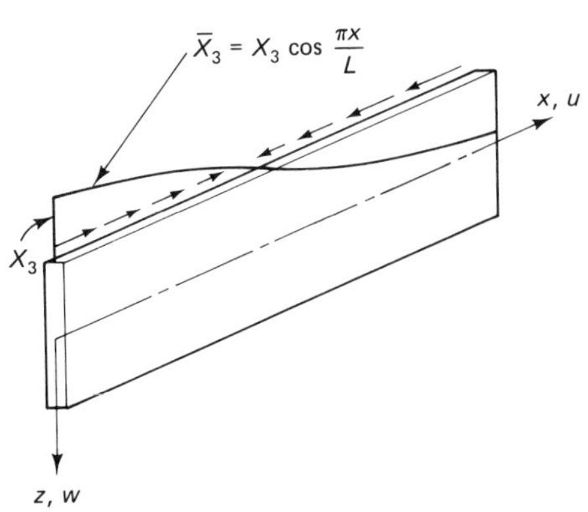

Figure 14.5 Simply supported edge beam subjected to a cosine varying longitudinal shear.

or

$$w = +\frac{X_3}{EI}\frac{L^3}{\pi^3}\frac{d}{2}\sin\frac{\pi x}{L} \tag{14.3b}$$

Now, due to both $T(x)$ and $M(x)$, the strain ε_x at the top fiber is found as

$$\begin{aligned}\varepsilon_x &= \frac{T(x)}{AE} - \frac{M(x)}{EI}\frac{d}{2}\\ &= \left(-\frac{X_3}{AE}\frac{L}{\pi} - \frac{X_3}{EI}\frac{L}{\pi}\frac{d^2}{4}\right)\sin\frac{\pi x}{L}\end{aligned}$$

which gives

$$u = \int_{L/2}^{x} \varepsilon_x\, dx = +\frac{X_3}{E}\frac{L^2}{\pi^2}\left(\frac{1}{A} + \frac{d^2}{4I}\right)\cos\frac{\pi x}{L} \tag{14.3c}$$

As expected, for $x > L/2$, u is negative for the loading shown in Fig. 14.5. Also, w is positive for all x. We shall summarize the results just obtained, for ready reference.

Edge Beam Stresses and Deformations

1. Vertically upward load:

$$\overline{X}_1 = X_1 \sin\frac{\pi x}{L} \tag{14.2a}$$

$$Ew = -\frac{X_1}{I}\frac{L^4}{\pi^4}\sin\frac{\pi x}{L} \tag{14.2b}$$

$$Eu = -\frac{X_1}{I}\frac{L^3}{\pi^3}\frac{d}{2}\cos\frac{\pi x}{L} \tag{14.2c}$$

$$T(x) = 0 \tag{14.4a}$$

$$M(x) = -X_1\frac{L^2}{\pi^2}\sin\frac{\pi x}{L} \tag{14.4b}$$

2. Longitudinal shear load:

$$\overline{X}_3 = X_3 \cos\frac{\pi x}{L} \tag{14.3a}$$

$$Ew = \frac{X_3}{I}\frac{L^3}{\pi^3}\frac{d}{2}\sin\frac{\pi x}{L} \tag{14.3b}$$

$$Eu = X_3\frac{L^2}{\pi^2}\left(\frac{1}{A} + \frac{d^2}{4I}\right)\cos\frac{\pi x}{L} \tag{14.3c}$$

$$T(x) = -X_3 \frac{L}{\pi} \sin \frac{\pi x}{L} \tag{14.5a}$$

$$M(x) = X_3 \frac{L}{\pi} \frac{d}{2} \sin \frac{\pi x}{L} \tag{14.5b}$$

$T(x)$ and $M(x)$ are assumed positive when tensile and sagging, respectively. Similar steps as above can be followed for obtaining corresponding formulas for the third harmonic $n = 3$. We shall next solve an example illustrating the use of the ASCE tables.

14.4 SOLUTION BY ASCE TABLES

Let us consider the problem of a single barrel shell with edge beams, as shown in Fig. 14.6. The span of the shell is 80 ft, $\phi_k = 30°$, and $a = 24$ ft. Hence the ratio $a/L = 0.3$, and assuming a thickness of 3 in. for the shell, we get $a/t = 96$. Since

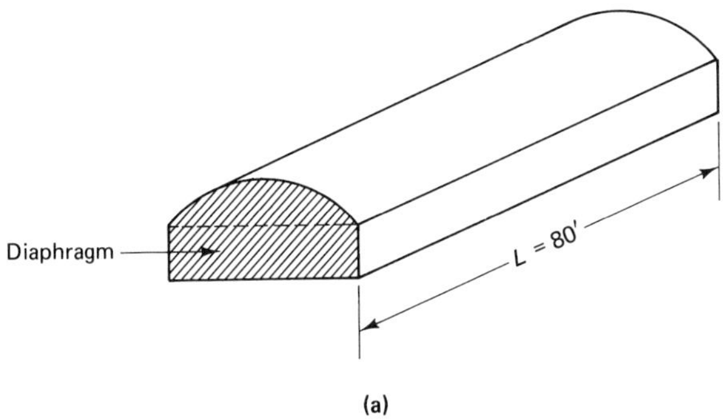

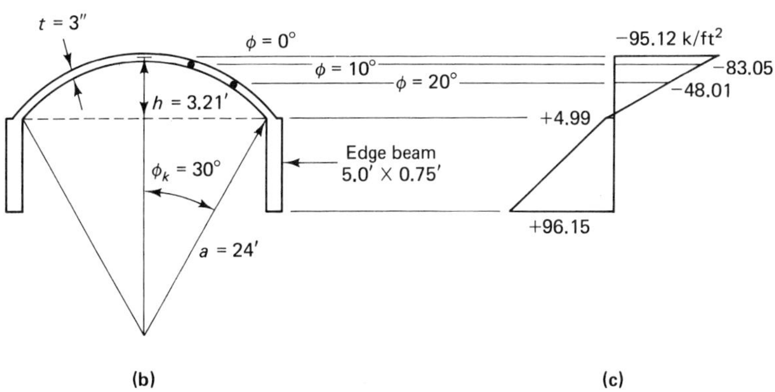

Figure 14.6 Example single barrel shell with edge beams. (a) Longitudinal configuration, (b) Cross section. (c) Variation of longitudinal stress over the depth.

this a/t ratio is very close to 100, we may use Table 2A with $a/t = 100$ and $a/L = 0.3$, without further interpolation. The rise of the shell is 3.21 ft, and assuming the depth of edge beams as 5.0 ft, the ratio of span to depth of cross section is $80/8.21 \approx 10$, which is a reasonable value. The assumed loading on the shell is summarized as follows:

Shell, 3 in. thick	= 37.5 lb/ft^2
Waterproofing	= 10.0
Live load	= 15.0
Total load	= 62.5 lb/ft^2

For simplicity, the total load (including live load) is assumed to be uniformly distributed over the surface area of the shell. We have to also consider the self-weight of the edge beams, which is $0.75(5.0)(150.0) = 562.5$ lb/ft for each beam. We shall assume that the edge beams are slender and thus cannot resist any horizontal or torsional loads. We shall now perform a step-by-step analysis of this single barrel shell with edge beams under the given loading.

Step I: Obtain Membrane Stress Resultants of the Shell Primary Structure

The uniformly distributed load $p = 62.5$ lb/ft^2 can be expanded into a Fourier series given by Eq. (10.6) in Chapter 10. Table 1B of the ASCE manual (Table 14.1) gives membrane forces under dead load for the first harmonic $n = 1$. In these tables, p_d represents the ordinate at $x = L/2$ of the first harmonic of the loading. Hence for a load constant along the span, we take $p_d = (4/\pi)(62.5) = 79.6$ lb/ft^2. Substituting this value into Table 1B, we obtain values for the stress resultants as indicated in Table 14.4. We see in this table that at the valley, we have $N_\phi = -1654.41 \sin(\pi x/L)$, which gives a horizontal component of $1654.41 \sin(\pi x/L) \cos(30°) = 1432.75 \sin(\pi x/L)$. Since the edge beam cannot resist such a horizontal force, this force must be applied back on the shell in the opposite direction. The

TABLE 14.4 Membrane Stress Resultants in the Example Shell for the First Harmonic of Loading

ϕ[a] (deg)	$\phi_k - \phi$	$N_x/\sin(\pi x/L)$ (lb/ft)	$N_{x\phi}/\cos(\pi x/L)$ (lb/ft)	$N_\phi/\sin(\pi x/L)$ (lb/ft)	$M_\phi/\sin(\pi x/L)$ (ft-lb/ft)
(crown)					
30	0	−4300.52	0.00	−1910.40	0.0
20	10	−4236.84	−703.66	−1881.36	0.0
10	20	−4041.56	−1386.95	−1795.20	0.0
0	30	−3723.16	−2026.93	−1654.41	0.0
(valley)					

[a]Angle ϕ is measured from the valley to the crown.

vertical component of N_ϕ is 827.2 sin $(\pi x/L)$, and this force should be applied in the opposite direction on the edge beam. The same can be said for the nonzero value of $N_{x\phi}$ at the edge. From Table 1B we also obtain at the valley (i.e., for $\phi = 0$),

$$\frac{E\Delta_v}{\sin(\pi x/L)} = 1{,}428{,}302.00$$

Also, we have

$$u = \int_{L/2}^{x} \varepsilon_x \, dx = \int_{L/2}^{x} \frac{N_x}{Et} \, dx$$

If $N_x = N_{x_1} \sin(\pi x/L)$, we obtain

$$Eu = -\frac{N_{x_1}}{t}\frac{L}{\pi}\cos\frac{\pi x}{L} \tag{14.6}$$

Substituting $N_{x_1} = -3723.16$ at $\phi = 0$ (from Table 14.4), we obtain at the edge

$$Eu = 379{,}237.96 \cos\frac{\pi x}{L}$$

Step II: Apply Corrective Edge Loads on the Shell

We apply the edge load $\overline{X}_2 = 1432.75 \sin(\pi x/L)$ back on the shell, and using Table 2A of the ASCE Manual, we obtain the stress resultants indicated in Table 14.5. From ASCE Table 2B, we also obtain, at $\phi = 0$,

$$E\Delta_v = -(1.60875 \times 10^8)\sin\frac{\pi x}{L}$$

Also, using Eq. (14.6), we have at $\phi = 0$,

$$Eu = \frac{10{,}516.39}{0.25}\left(\frac{80}{\pi}\right)\cos\frac{\pi x}{L} = 1{,}071{,}188.6 \cos\frac{\pi x}{L}$$

TABLE 14.5 Stress Resultants under the Edge Load $\overline{X}_2 = 1432.75 \sin(\pi x/L)$

ϕ (deg)	$N_x/\sin(\pi x/L)$ (lb/ft)	$N_{x\phi}/\cos(\pi x/L)$ (lb/ft)	$N_\phi/\sin(\pi x/L)$ (lb/ft)	$M_\phi/\sin(\pi x/L)$ (ft-lb/ft)
30	−4,237.76	0.0	1,371.29	3,779.02
20	−149.64	−458.96	1,397.22	3,376.71
10	−4,683.50	16.72	1,381.03	2,138.81
0	−10,516.39	0.0	1,240.76	0.0

From the results of these Steps I and II, we can make the following summary:

$$\frac{E\bar{\delta}_{10}^{S}}{\sin(\pi x/L)} = 1{,}428{,}302.0 - (1.60875 \times 10^8) = -1.594467 \times 10^8 \text{ lb/ft}$$

and

$$\frac{E\bar{\delta}_{30}^{S}}{\cos(\pi x/L)} = -379{,}237.96 - 1{,}071{,}188.6 = -1{,}450{,}426.6 \text{ lb/ft}$$

It may be noted in the above that if $u = u_1 \cos(\pi x/L)$ (where u_1 is a positive constant), then u is positive at $x = 0$; i.e., at $x = 0$, u is directed toward the midspan. This corresponds to a negative $\bar{\delta}_3$, since our assumed positive direction of $\bar{\delta}_3 = \delta_3 \cos(\pi x/L)$ for the shell is directed away from midspan at $x = L$ (Fig. 14.1). Hence for the shell we have $\bar{\delta}_3^S = -u$, $\bar{\delta}_{30}^S = -u$, etc.

Step III: Obtain Stress Resultants on the Primary Beam Structure

Computing the total vertical load on the primary structure of the beam, we have

First harmonic of the Fourier series expansion for the beam self-weight of (562.5) lb/ft	$= 716.2 \sin \dfrac{\pi x}{L}$ (acting downward)
Opposite action to the vertical component of N_ϕ at the edge (from Step I)	$= 827.2 \sin \dfrac{\pi x}{L}$ (acting downward)
Total load on beam primary structure	$= 1543.4 \sin \dfrac{\pi x}{L}$ (acting downward)

Results for this loading can be obtained by substituting $X_1 = -1543.4$ into Eqs. (14.2a–c) and (14.4a,b). We have, for the beam, a moment of inertia I equal to $\frac{1}{12}(0.75)(5^3) = 7.8125$ ft^4 and an area A of $0.75(5) = 3.75$ ft^2. Hence, from Eqs. (14.2a–c) and (14.4a,b), we find

$$\frac{Ew}{\sin(\pi x/L)} = +\frac{1543.4}{7.8125}\left(\frac{80^4}{\pi^4}\right) = +83{,}070{,}901.3 \text{ lb/ft}$$

$$\frac{Eu}{\cos(\pi x/L)} = \frac{1543.4}{7.8125}\left(\frac{80}{\pi}\right)^3 (2.5) = +8{,}155{,}466.7 \text{ lb/ft}$$

$$T(x) = 0$$

and

$$\frac{M(x)}{\sin(\pi x/L)} = 1{,}000{,}826.3 \text{ ft-lb}$$

We also have to apply on the beam the opposite of $N_{x\phi} = -2026.93 \cos(\pi x/L)$, acting at the shell edge (Table 14.4). This force has a positive value at $x = L$ and a negative value at $x = 0$. Hence, on the edge of the shell, the directions of these shears will be as shown in Fig. 14.7. On the edge beam, we apply opposite shears, as shown. Stresses and displacements in the beam are then found by substituting $X_3 = -2026.93$ into Eqs. (14.3a–c) and (14.5a,b). We thus obtain

$$\frac{Ew}{\sin(\pi x/L)} = -\frac{2026.93}{7.8125}\left(\frac{80^3}{\pi^3}\right)(2.5) = -10{,}710{,}483.4$$

$$\frac{Eu}{\cos(\pi x/L)} = -2026.93\left(\frac{80^2}{\pi^2}\right)\left[\frac{1}{3.75} + \frac{5^2}{4(7.8125)}\right] = -1{,}401{,}999.0$$

We also find

$$T(x) = 51{,}615.35 \sin\frac{\pi x}{L}$$

$$M(x) = -129{,}038.37 \sin\frac{\pi x}{L}$$

Adding the results for the two loadings, we obtain for the beam

$$\frac{E\bar{\delta}_{10}^{B}}{\sin(\pi x/L)} = -83{,}070{,}901.3 + 10{,}710{,}483.4 = -72{,}360{,}417.9 \text{ lb/ft}$$

$$\frac{E\bar{\delta}_{30}^{B}}{\cos(\pi x/L)} = 8{,}155{,}466.7 - 1{,}401{,}999.0 = 6{,}753{,}467.7 \text{ lb/ft}$$

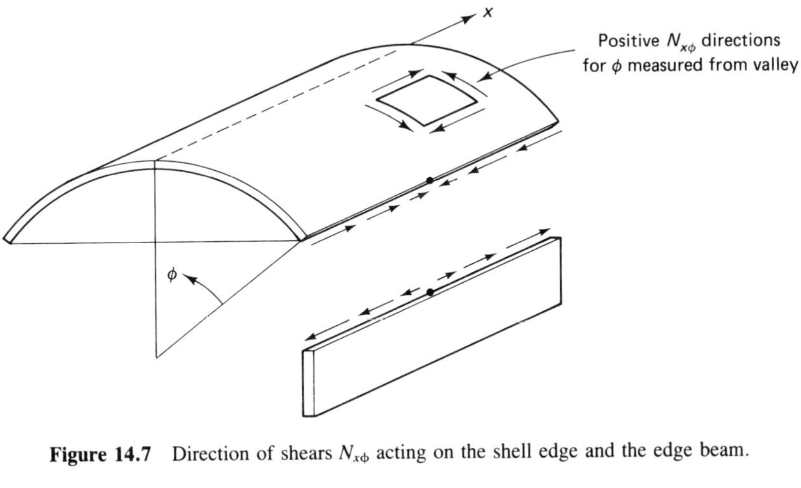

Figure 14.7 Direction of shears $N_{x\phi}$ acting on the shell edge and the edge beam.

and

$$T(x) = 51{,}615.35 \sin \frac{\pi x}{L} \text{ lb}$$

$$M(x) = (1{,}000{,}826.3 - 129{,}038.37) \sin \frac{\pi x}{L} = 871{,}787.93 \sin \frac{\pi x}{L} \text{ ft-lb}$$

For the beam it may be noted that when $u = u_1 \cos(\pi x/L)$ (where u_1 is positive), we find u as positive at $x = 0$; i.e., at $x = 0$, u is directed toward the midspan. Correspondingly, $\bar{\delta}_3$ will be positive, since our assumed positive direction of $\bar{\delta}_3 = \delta_3 \cos(\pi x/L)$ for the beam is directed toward the midspan at $x = 0$ (Fig. 14.3). Hence for the beam we have $\bar{\delta}_3^B = u$, $\bar{\delta}_{30}^B = u$, etc.

Step IV: Determine Flexibility Coefficients and Solve for Redundants

From Table 2B of the ASCE manual, we obtain for $\bar{X}_1 = \sin(\pi x/L)$, i.e., for $X_1 = 1$,

$$\frac{E\bar{\delta}_{11}^S}{\sin(\pi x/L)} = \frac{80^4}{24^3(0.25)}(35.24) = 417{,}659.26$$

and from Table 2A we find

$$N_x(\phi = 0) = \left(\frac{80}{24}\right)^2 (12.43) \sin \frac{\pi x}{L} = 138.114 \sin \frac{\pi x}{L}$$

Hence from Eq. (14.6), substituting for N_x, we have

$$Eu = -\frac{138.114}{0.25}\left(\frac{80}{\pi}\right) \cos \frac{\pi x}{L} = -14{,}068.15 \cos \frac{\pi x}{L}$$

or

$$E\bar{\delta}_{31}^S = 14{,}068.15 \cos \frac{\pi x}{L}$$

Similarly, under the edge load $\bar{X}_3 = (1) \cos(\pi x/L)$, we use the ASCE Table to find

$$\frac{E\bar{\delta}_{13}^S}{\sin(\pi x/L)} = \frac{80^4}{24^3}\left(\frac{1}{0.25}\right)(1.187) = 14{,}068.15$$

and

$$N_x(\phi = 0) = 1.102\left(\frac{80}{24}\right)^2 \sin \frac{\pi x}{L} = 12.244 \sin \frac{\pi x}{L}$$

Hence

$$Eu = -\frac{12.244}{0.25}\left(\frac{80}{\pi}\right)\cos\frac{\pi x}{L} = -1247.16\cos\frac{\pi x}{L}$$

or

$$E\bar{\delta}_{33}^{S} = 1247.16\cos\frac{\pi x}{L}$$

Now, for the beam, substituting $X_1 = 1$ into Eqs. (14.2b,c), we obtain

$$E\bar{\delta}_{11}^{B} = -Ew = 53{,}823.31\sin\frac{\pi x}{L}$$

$$E\bar{\delta}_{31}^{B} = Eu = -5284.09\cos\frac{\pi x}{L}$$

Also, for $X_3 = 1$, we obtain from Eqs. (14.3b,c),

$$E\bar{\delta}_{13}^{B} = -Ew = -5284.09\sin\frac{\pi x}{L}$$

$$E\bar{\delta}_{33}^{B} = Eu = 691.69\cos\frac{\pi x}{L}$$

The final values of $\bar{\delta}_{i0}$ and $\bar{\delta}_{ij}$ are obtained by adding respective values for the shell and the beam. Hence we get

$$E\bar{\delta}_{10} = (-1.594467 \times 10^8 - 72{,}360{,}417.9)\sin\frac{\pi x}{L} = -2.3180712 \times 10^8 \sin\frac{\pi x}{L}$$

$$E\bar{\delta}_{30} = (-1{,}450{,}426.6 + 6{,}753{,}467.7)\cos\frac{\pi x}{L} = 5{,}303{,}041.1\cos\frac{\pi x}{L}$$

$$E\bar{\delta}_{11} = (417{,}659.26 + 53{,}823.31)\sin\frac{\pi x}{L} = 471{,}482.57\sin\frac{\pi x}{L}$$

$$E\bar{\delta}_{13} = (14{,}068.15 - 5284.09)\sin\frac{\pi x}{L} = 8784.06\sin\frac{\pi x}{L}$$

$$E\bar{\delta}_{31} = (14{,}068.15 - 5284.09)\cos\frac{\pi x}{L} = 8784.06\cos\frac{\pi x}{L}$$

$$E\bar{\delta}_{33} = (1247.16 + 691.69)\cos\frac{\pi x}{L} = 1938.85\cos\frac{\pi x}{L}$$

It may be noted that for both the shell and the beam, coefficients $\bar{\delta}_{13}$ and $\bar{\delta}_{31}$ are equal, so that the reciprocity theorem will be satisfied. Dropping sine and cosine

TABLE 14.6 Stress Resultants under Edge Load $\bar{X}_1 = 592.64 \sin(\pi x/L)$

ϕ (deg)	$N_x/\sin(\pi x/L)$ (lb/ft)	$N_{x\phi}/\cos(\pi x/L)$ (lb/ft)	$N_\phi/\sin(\pi x/L)$ (lb/ft)	$M_\phi/\sin(\pi x/L)$ (ft-lb/ft)
30	−31,008.24	0.0	−1,992.46	−3,938.45
20	−23,494.88	−4,703.59	−1,559.83	−3,338.22
10	6,933.89	−6,499.29	−496.04	−1,813.48
0	81,850.17	0.0	296.32	0.0

terms, we obtain the following equations for X_1 and X_3, by substitution into Eqs. (14.1a).

$$471{,}482.57X_1 + 8784.06X_3 - 2.3180712 \times 10^8 = 0$$

$$8784.06X_1 + 1938.85X_3 + 5.3030411 \times 10^6 = 0$$

Solving these simultaneous equations, we get

$$X_1 = 592.64 \text{ lb/ft} \qquad \text{and} \qquad X_3 = -5420.12 \text{ lb/ft}$$

Step V: Obtain Final Stress Resultants

Using ASCE Table 2A, we obtain, for various values of ϕ, the stress resultants under edge loads $\bar{X}_1 = 592.64 \sin(\pi x/L)$ and $\bar{X}_3 = -5420.12 \cos(\pi x/L)$, as given in Tables 14.6 and 14.7, respectively. The final stress resultants in the shell are obtained by addition of values in Tables 14.4 to 14.7. These resultants are given in Table 14.8. The numerical quantities in the tables give the values of N_x, N_ϕ, and M_ϕ at midspan and the value of $N_{x\phi}$ at the supports.

Final values of tension and moment in the edge beam can also be obtained. Substituting $X_1 = 592.64$ into Eqs. (14.4a,b), we obtain

$$T(x) = 0 \qquad \text{and} \qquad M(x) = -384{,}300.71 \sin\frac{\pi x}{L}$$

TABLE 14.7 Stress Resultants under Edge Load $\bar{X}_3 = -5420.12 \cos(\pi x/L)$

ϕ (deg)	$N_x/\sin(\pi x/L)$ (lb/ft)	$N_{x\phi}/\cos(\pi x/L)$ (lb/ft)	$N_\phi/\sin(\pi x/L)$ (lb/ft)	$M_\phi/\sin(\pi x/L)$ (ft-lb/ft)
30	15,766.53	0.0	323.04	312.20
20	7,118.42	2,121.07	125.20	208.13
10	−19,578.68	1,355.03	−228.19	39.02
0	−66,366.36	−5,420.12	0.0	0.0

TABLE 14.8 Final Stress Resultants in Example Shell

ϕ (deg)	$N_x/\sin(\pi x/L)$ (lb/ft)	$N_{x\phi}/\cos(\pi x/L)$ (lb/ft)	$N_\phi/\sin(\pi x/L)$ (lb/ft)	$M_\phi/\sin(\pi x/L)$ (ft-lb/ft)
30	−23,779.99	0.0	−2,208.53	152.77
20	−20,762.94	−3,745.14	−1,918.77	246.62
10	−12,002.85	−6,514.49	−1,138.40	364.35
0	1,244.26	−7,447.05	−117.33	0.0

Also, substituting $X_3 = -5420.12$ into Eqs. (14.5a,b), we have

$$T(x) = 138{,}022.22 \sin \frac{\pi x}{L} \quad \text{and} \quad M(x) = -345{,}055.56 \sin \frac{\pi x}{L}$$

Adding these contributions to the values obtained in Step III, we obtain the final tension and moment in the beam as

$$T(x) = (51{,}615.35 + 138{,}022.22) \sin \frac{\pi x}{L} = 189{,}637.57 \sin \frac{\pi x}{L} \text{ lb}$$

$$M(x) = (871{,}787.93 - 384{,}300.71 - 345{,}055.56) \sin \frac{\pi x}{L}$$

$$= 142{,}431.66 \sin \frac{\pi x}{L} \text{ ft-lb}$$

At midspan, $x = L/2$, we can calculate stresses at the top and bottom fibers of the beam as

$$\sigma^B_{x,t} = \frac{189{,}637.57}{3.75} - \frac{142{,}431.66}{7.8125}(2.5)$$

$$= 50{,}570.02 - 45{,}578.13 = 4991.89 \text{ lb/ft}^2$$

and

$$\sigma^B_{x,b} = 50{,}570.02 + 45{,}578.13 = 96{,}148.15 \text{ lb/ft}^2$$

It may be noticed that from Table 14.8, the longitudinal stress at the edge of the shell, for $x = L/2$, is obtained as $(1244.26/0.25) = 4977.04$ lb/ft^2. This value matches well with the stress at the top of the beam, as expected. The variation of σ_x over the depth of the shell cross section, at $x = L/2$, is as shown in Fig. 14.6(c). This completes the analysis for this barrel shell structure. Reinforcement design details for this shell are considered in Chapter 15.

Other problems involving a single-barrel shell or an internal unit of a multibarrel shell, with or without edge beams, can be solved in a similar manner using the ASCE tables. From the example above, however, it is evident that such an analysis still involves lengthy calculations, although the procedure is straightforward. The calculations become even more tedious if the ASCE tables have to be interpolated for a given shell configuration. Even with the additional PCA tables [6], one may still need to use interpolation for a given shell. Hence it is desirable to consider the analysis of barrel shells using an alternate theory or using a computer program.

14.5 ANALYSIS OF CYLINDRICAL SHELLS AND FOLDED PLATES BY BEAM THEORY

It is known that cylindrical shells with large (L/a) values, spanning between diaphragm supports, tend to behave as simply supported beams. Folded plates, which are composed of interconnected plates, can also be considered to span as beams

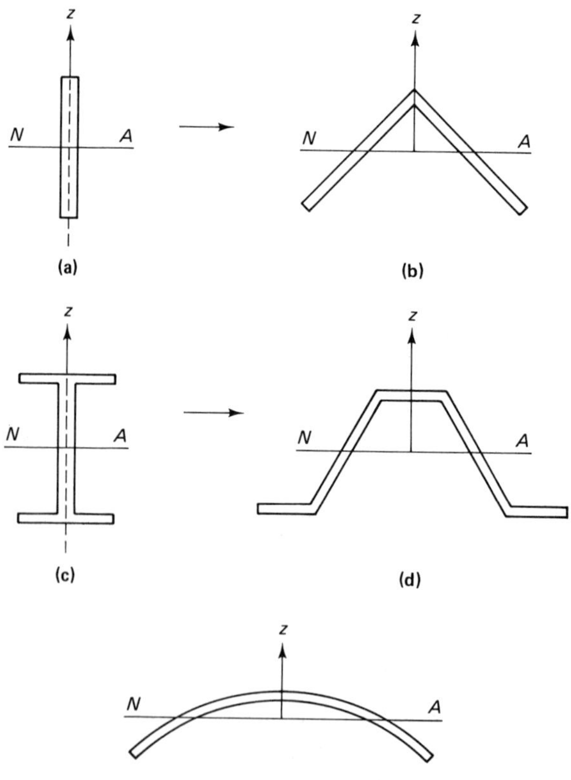

Figure 14.8 Symmetrical folded plate sections. (a) Normal rectangular beam. (b) Folded plate formed by splitting the rectangular section. (c) Normal I-section. (d) Trough section formed by splitting the I-section. (e) Open cylindrical section.

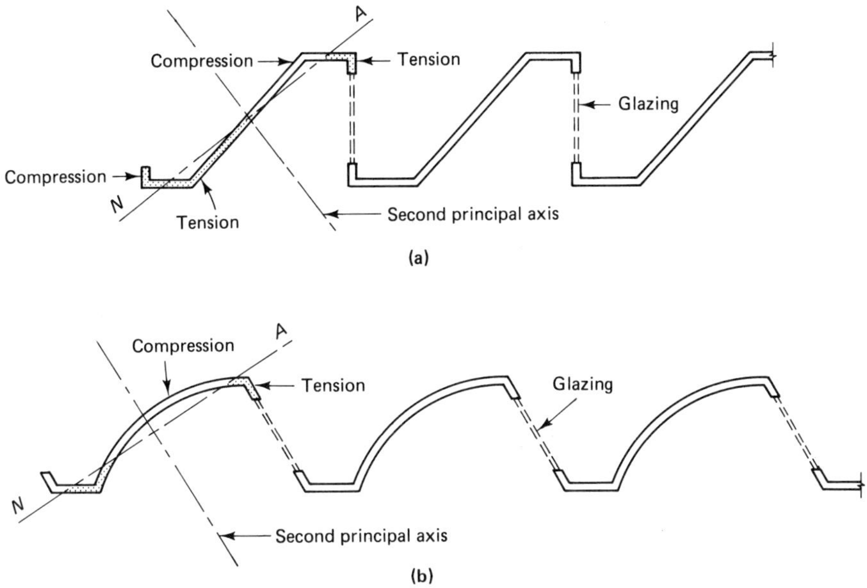

Figure 14.9 Unsymmetrical sections. (a) Northlight folded plate. (b) Northlight shell.

between diaphragm supports.* Figure 14.8(a) shows the rectangular cross section of a simple beam. If it is slit along the dashed line and spread apart, we get the cross section of an inverted V-shaped folded plate [Fig. 14.8(b)]. If the I-beam in Fig. 14.8(c) is slit in the center of the web and spread apart, we get the cross section of a trough-shaped folded plate [Fig. 14.8(d)]. If this cross section is curved, it gives the section of a cylindrical shell [Fig. 14.8(e)]. In a similar manner, we can obtain the cross section of a northlight folded plate or shell, as shown in Fig. 14.9. This leads us to consider that the longitudinal stresses in a folded plate or cylindrical shell could be obtained by using simple formulas of beam theory, e.g., $\sigma_x = (Mz/I)$, where M is the bending moment at any section $x = constant$ and I is the moment of inertia of the cross section about the neutral axis. For the symmetrical sections shown in Fig. 14.8(b), (d), and (e), the neutral axis is horizontal. However, for the northlight shells of Fig. 14.9 (a) and (b), with unsymmetrical Z-type sections, the neutral axis is inclined to the horizontal. Thus for such shells, longitudinal stresses must be calculated by resolving loads in the two principal directions. In a folded plate or a shell, there are also the transverse bending moments, shears, and in-plane forces (M_ϕ, Q_ϕ, and N_ϕ) which are generally not considered for simple beams. These have to be found by a secondary analysis called "arch analysis," in the case of shells or folded plates. The analysis of shells by this simple combined procedure is called "beam theory" analysis [8].

*For studying the various applications of folded plates and alternate methods of their analysis, the reader is referred to Reference [7].

Whenever it can be applied, the beam theory has certain advantages over other procedures of shell analysis. It is simple and gives the designer a physical insight in the behavior of the shell. It can be applied with equal ease to shells of variable thickness or to shells of noncircular cross sections (such as parabolic or catenary sections), as well as to northlight shells, and V-shaped, trough-shaped, or northlight-type folded plates.

In beam theory, we make all the assumptions of Schorer theory (Chapter 13), such as M_x, Q_x, $M_{x\phi}$, ν, ε_ϕ, and $\gamma_{x\phi}$ are zero. Deformations of the cross section (i.e., its arching or flattening) are not considered when we derive formulas for beams in strength of materials. For shells or folded plates, such deformations of the cross section in the transverse direction do take place, and they can change the resultants M_ϕ and N_ϕ obtained from simple beam theory. For single shells, such transverse deformations are not restrained, and hence beam theory would be applicable for L/a values larger than, say, 5 or more. With edge beams, the transverse vertical deformations are restrained to some extent, and we can apply beam theory for L/a greater than about 3. For an internal shell of a multibarrel shell structure, the horizontal deformation at the valley is completely restrained due to symmetry. Hence, in such cases, beam theory gives good results even for shells with $L/a \approx 1.67$ [4].

14.6 BEAM THEORY EQUATIONS

Beam Analysis

We shall consider a simply supported circular cylindrical shell, of span L, carrying a load p uniformly distributed over its surface area. The cross section of the shell is shown in Fig. 14.10. We shall measure ϕ from the crown.* Treating the shell as a beam of curved cross section, we can write the load q per unit length, along the beam span, as

$$q = 2ap\phi_k \tag{14.7}$$

From simple beam theory, the shear force and bending moment in the beam, at a distance x from the support, are found to be

$$V(x) = q\left(\frac{L}{2} - x\right) \tag{14.8a}$$

*This conforms to our original coordinate system. The ASCE tables measure ϕ from the edge, and hence only in Sections 14.2 to 14.4 did we consider ϕ as measured from the edge.

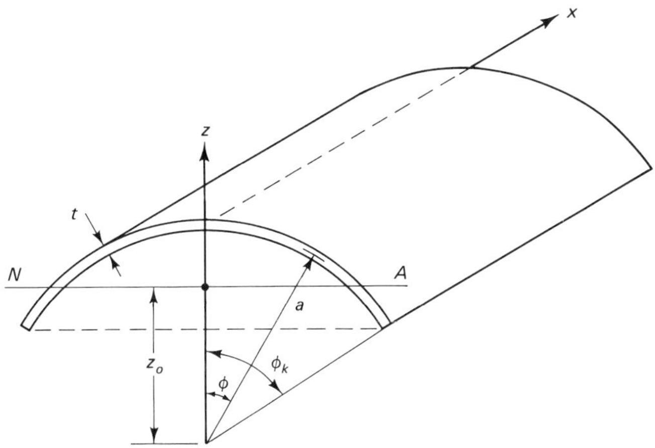

Figure 14.10 Configuration of simply supported open circular cylindrical shell.

$$M(x) = \frac{qL^2}{2}\left(\frac{x}{L} - \frac{x^2}{L^2}\right) \tag{14.8b}$$

Since the z-axis is an axis of symmetry, it is also a principal axis of the cross section. The second principal axis is the horizontal neutral axis (N.A.), about which the moment of area of the cross section is zero. If I is the moment of inertia of the cross section about the N.A., the longitudinal stress σ_x in the shell is given by

$$\sigma_x = -\frac{M(x)}{I}z \tag{14.8c}$$

where z is measured upward from the N.A. and σ_x is positive in tension. The location of the N.A. and the value of I can be calculated from the formulas of strength of materials. For a thin circular cross section, these quantities can readily be obtained from the following formulas derived by Chinn [9]:

$$z_o = \frac{a \sin \phi_k}{\phi_k} \tag{14.9a}$$

$$I = a^3 t\left[\phi_k + \sin \phi_k \left(\cos \phi_k - \frac{2 \sin \phi_k}{\phi_k}\right)\right] \tag{14.9b}$$

Values for I can also be readily calculated from Table 14.9 for known values of ϕ_k [4].

Having obtained the longitudinal stresses by normal beam theory, the stress resultants in the transverse direction are obtained by "arch analysis."

TABLE 14.9 Moment of Inertia for a Thin Circular Cross Section

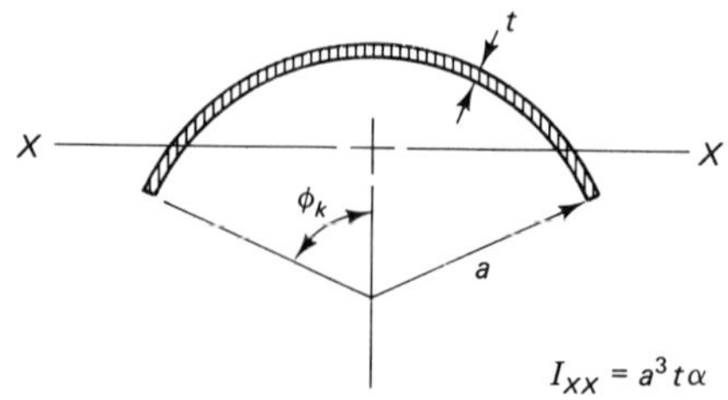

ϕ_k (deg)	α	ϕ_k (deg)	α
22.5	0.00041	37.5	0.00502
25.0	0.00068	40.0	0.00687
27.5	0.00110	45.0	0.01216
30.0	0.00168	50.0	0.02017
32.5	0.00249	55.0	0.03174
35.0	0.00358	60.0	0.04782

Source: Ref. [4].

Arch Analysis

If we consider a section *M-M* of the shell at a distance z from the N.A., as in Fig. 14.11(a), the longitudinal shear flow (and its complementary transverse shear) at x can be obtained from strength of materials. Since two edges of the shell are exposed in section *M-M*, we have

$$2N_{x\phi} = \frac{V(x)Q}{I}$$

or

$$N_{x\phi} = \frac{V(x)Q}{2I} \tag{14.10}$$

where $Q = A\bar{z}$ is the moment of area about the N.A. of the portion of the cross section above section *M-M*. Now we consider a slice of the shell between sections x and $x + dx$ [Fig. 14.11(b)]. Assuming that the edges $\phi = \phi_k$ are free, we note that this slice is in equilibrium under the external load q and the shears $N_{x\phi}$ and $N_{x\phi} + (\partial N_{x\phi}/\partial x)\,dx$, acting on the two faces of the slice. The difference in the two shears is called the "specific shear" and is obtained by using Eq. (14.10) as

$$\frac{\partial N_{x\phi}}{\partial x}\,dx = \frac{dV(x)}{dx}\,dx\,\frac{Q}{2I}$$

If dx is taken as unity, we have

$$\frac{\partial N_{x\phi}}{\partial x} = \text{specific shear} = q_s = \frac{dV(x)}{dx}\frac{Q}{2I} \tag{14.11a}$$

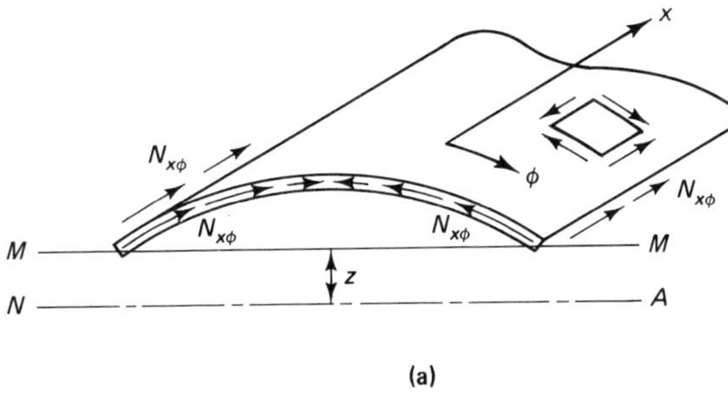

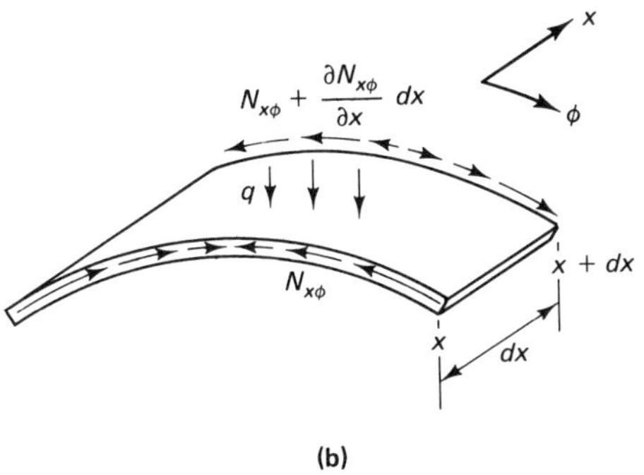

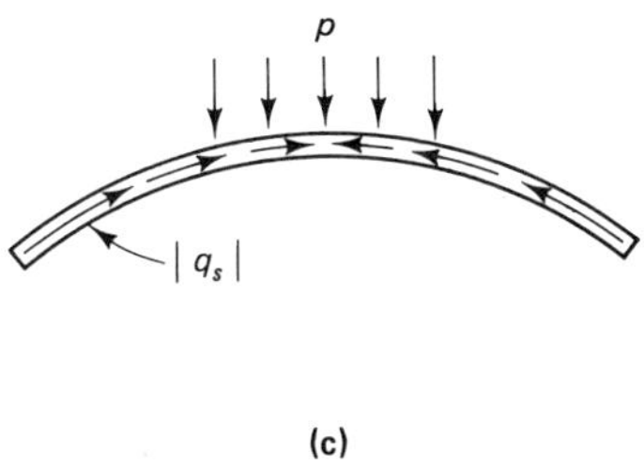

Figure 14.11 Arch analysis of a circular cylindrical cross section. (a) Shear flow on edge $x = constant$. (b) Element of shell. (c) Loads p and q_s on shell "arch."

Substituting Eq. (14.8a) into Eq. (14.11a), we have

$$q_s = -\frac{qQ}{2I} \tag{14.11b}$$

This specific shear is constant in x, for a shell of constant cross section which is subjected to a uniform load q. The negative sign in Eq. (14.11b) indicates that q_s is acting "upward"; i.e., it is directed from the valley to the crown. We can consider that a shell slice of unit length, taken at any section x along the span, is in equilibrium under the action of (1) the external load p, and (2) the specific shear load q_s. These loads are shown on the shell arch in Fig. 14.11(c). The values of q_s can be determined for different values of ϕ, from the valley to the crown, by calculating the values of Q at these points. For a circular cross section, Q at any point can be obtained from the following formula given by Chinn [9]:

$$Q = 2a^2t\left(\sin\phi - \frac{\phi}{\phi_k}\sin\phi_k\right) \tag{14.12}$$

Once the loads q and q_s are known, the internal stress resultants M_ϕ, Q_ϕ, and N_ϕ in the arch (i.e., in the shell itself) can be calculated from simple statics, starting from one edge, since the arch is statically determinate when the edges are free.

For an internal shell of a multibarrel shell structure, there will be a horizontal thrust and a moment acting at the valley of the arch. These are unknown quantities, and the arch is now statically indeterminate. The solution can be obtained by satisfying the conditions of zero horizontal displacement and rotation at the valley, by using any method for the structural analysis of statically indeterminate arches. Generally, the *elastic center method* is the simplest to use in such cases, and Chinn [9] has given formulas for simplifying the calculations for circular arches. Alternatively, it may be easier to divide the arch into a series of straight elements and analyze these under the loads q and q_s using a simple plane-frame computer program.

Having obtained σ_x from "beam analysis," $N_{x\phi}$ from Eq. (14.10), and M_ϕ, Q_ϕ, N_ϕ from "arch analysis," the analysis of the shell by beam theory is complete. Generally, for design, $N_{x\phi}$ values will be required at the support and all the other values will be required at midspan. We summarize below the steps involved in the analysis of shells (or folded plates) by beam theory.

Summary of Steps in Beam Theory Analysis—Load Uniformly Distributed over the Surface Area

Beam Analysis

Step I. Calculate the load per unit length, q, along the span, for the given distributed loading p. For a circular cross section, use Eq. (14.7).

Step II. Find the neutral axis location z_o, and then find the moment of inertia of the cross section, I, about the N.A. For a circular cross section, use Eqs. (14.9a,b) or Table 14.9.

Step III. Find the variation over depth of the maximum longitudinal stresses, σ_x at $x = L/2$, using Eqs. (14.8b,c).

Step IV. Find Q for different values of ϕ along the cross section, and also solve for corresponding $N_{x\phi}$ values at the support $x = 0$, from Eq. (14.10). For determining Q values of a circular cross section, Eq. (14.12) may be used.

Arch Analysis

Step V. Find the specific shear q_s at various points along the cross section, using Eq. (14.11b). For uniformly distributed loading, $V(x)$ varies linearly with x, and hence $dV(x)/dx$ in Eq. (14.11a) is equal to $[V(x = 0)/(L/2)]$. Comparing Eqs. (14.10) and (14.11a,b), we see that the absolute value of q_s is then simply found by dividing the value of $N_{x\phi}$ at $x = 0$, as obtained in Step IV, by $L/2$.

Step VI. Under loads p and q_s, analyze the arch by the elastic center method, or by a plane frame program, in order to obtain M_ϕ, Q_ϕ, and N_ϕ.

The procedure above can be completed in tabular form. To illustrate the procedure, we shall first solve the simple problem of a folded plate, using beam theory.

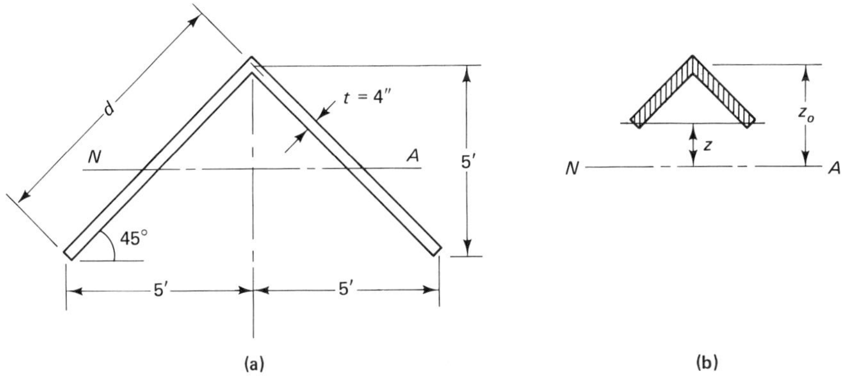

Figure 14.12 Example folded plate. (a) Configuration of cross section. (b) Area used in calculating Q.

14.7 FOLDED PLATE ANALYSIS USING BEAM THEORY

Figure 14.12(a) shows a single bay inverted V-shaped folded plate which spans 50 ft. We assume the following loading on this folded plate:

$$\begin{aligned} \text{Dead load} &= 50 \text{ lb/ft}^2 \\ \text{Finish} &= 15 \\ \text{Live load} &= 15 \\ \hline \text{Total load} &= 80 \text{ lb/ft}^2 \text{ (over the surface area)} \end{aligned}$$

Following the previously outlined steps, we have:

Step I. We calculate the load per unit length, q, along the beam span, as

$$q = 2(5)(\sqrt{2})(80) = 1131.37 \text{ lb/ft}$$

Step II. The neutral axis of the cross section will be at the midheight of the folded plate, i.e., at 2.5 ft from either the top or bottom edge. The moment of inertia of the cross section about the N.A. is found as

$$I = 2\left(\frac{td^3}{12}\right) \sin^2 \alpha$$

where d is the inclined depth of the plates and α is the angle of inclination of the plates. Substituting in the appropriate values gives

$$I = 2\left[\frac{(4/12)(5\sqrt{2})^3}{12}\right] \sin^2 45° = 9.821 \text{ ft}^4$$

Step III. The maximum beam moment is found from simple beam theory as

$$M_{\max} = M\left(x = \frac{L}{2}\right) = \frac{1131.37(50)^2}{8} = 353{,}553.1 \text{ ft-lb}$$

This gives a maximum longitudinal stress of

$$\sigma_x\left(x = \frac{L}{2}\right) = \pm\frac{353{,}553.1(2.5)}{9.821} = \pm 89{,}999.3 \text{ lb/ft}^2$$

or

$$\sigma_x = 625 \text{ psi}$$

Step IV. To find Q for various values of z, we consider a portion of the folded plate above the section z = *constant* [Fig. 14.12(b)]. The moment of this area about the N.A. is given by

$$Q = A\bar{z} = 2\left[(z_o - z)\frac{1}{\sin\alpha}t\right]\left[(z_o - z)\left(\frac{1}{2}\right) + z\right]$$

$$Q = (z_o^2 - z^2)\frac{t}{\sin\alpha} = (6.25 - z^2)(0.471)$$

Then $N_{x\phi}$ at the support is given by Eq. (14.10) as

$$N_{x\phi} = \frac{(1131.37)(25)}{2(9.821)}(6.25 - z^2)(0.471)$$

$$= 678.23(6.25 - z^2) \text{ lb/ft}$$

As expected, $N_{x\phi}$ varies parabolically over the depth, being zero at the top and bottom of the section, and having a maximum value of 4238.94 lb/ft at the neutral axis.

Step V. The specific shear q_s is given by

$$|q_s| = \frac{(N_{x\phi})_{\max}}{L/2}$$

$$= \frac{678.23}{25}(6.25 - z^2) = 27.13(6.25 - z^2) \text{ lb/ft}$$

This shear also varies parabolically over the depth and has values of zero at the top and bottom of the section.

Step VI. Considering an "arch" of 1-ft width in the x-direction, we can show the loads acting on it, as in Fig. 14.13(a). Values of N_ϕ, M_ϕ, and Q_ϕ, at a

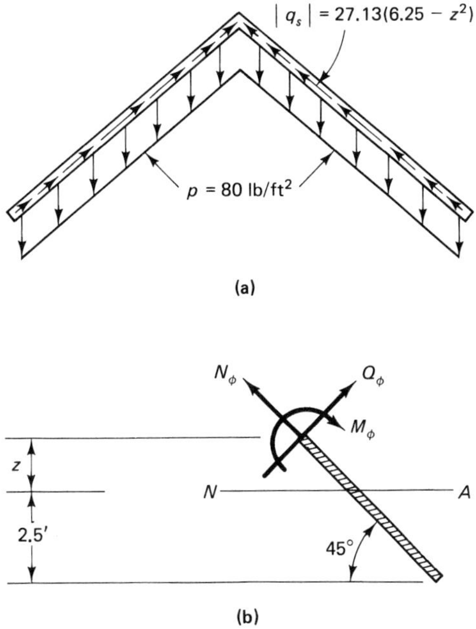

Figure 14.13 Loads acting on an "arch" of unit width. (a) External loading for analysis. (b) Internal resultants at section z from the neutral axis.

distance z from the N.A., can be obtained from simple statics. Thus the force N_ϕ can be found as

$$N_\phi = \int_{-2.5}^{z} -27.13(6.25 - z^2)\frac{dz}{\sin 45°} + 80.0 \sin 45° \left(\frac{z + 2.5}{\sin 45°}\right)$$

or

$$N_\phi = 12.789\, z^3 - 159.798\, z - 200$$

Also, we find the transverse shear Q_ϕ, from simple statics, as

$$Q_\phi = 80.0 \cos 45° \left(\frac{z + 2.5}{\sin 45°}\right) = (80.0)\, z + 200$$

and also the moment M_ϕ as

$$M_\phi = -80.0 \cos 45° \left(\frac{z + 2.5}{\sin 45°}\right)^2 \left(\frac{1}{2}\right)$$

or

$$M_\phi = -56.57(z + 2.5)^2$$

Maximum absolute values of Q_ϕ and M_ϕ occur at the crown, and these are 400.0 lb, and -1414.25 ft-lb, respectively. The maximum absolute value of N_ϕ occurs slightly below the crown at $z = 2.04$ ft, and is equal to -417.41 lb. These values are given per foot length along the span.

It may be noted that in this case, since the specific shears have a resultant that passes through the crown, the transverse bending moment at the crown is just equal to the moment of the load p about the crown on the "cantilevered" plates. Hence the thickness of the plate required will not be less than that of a cantilever slab of 5-ft span subjected to the given loading. In the case of a shell, however, the specific shears will not pass through the crown [Fig. 14.11(c)] and will thus cause a reduction in the cantilever moment due to p. Hence M_ϕ at the crown will be much less than the cantilever moment of the external load. We can therefore have a large section width and still maintain a small thickness for curved shells.

If the folded plate in this example were an internal bay of a multibarrel folded plate [Fig. 14.14(a)], we would then only have to consider an additional redundant horizontal force H_o and moment M_o at the valley [Fig. 14.14(b)], and solve the resulting statically indeterminate "arch," in order to obtain M_ϕ, Q_ϕ, and N_ϕ in Step VI. The earlier steps would remain unchanged.

Similar procedures can be followed step by step for the analysis of cylindrical shells also. For shells of unsymmetrical cross sections, such as northlight shells or northlight folded plates (Fig. 14.9), we have to first locate the principal axes of the cross section. These principal axes will be inclined to the vertical and horizontal directions, as shown. Beam action should then be considered for bending in both the principal planes under the components of loads in those planes. The specific shear at any point is also found as the addition of specific shears due to bending in both planes. The solution becomes somewhat complicated but can be compacted in tabular form [10].

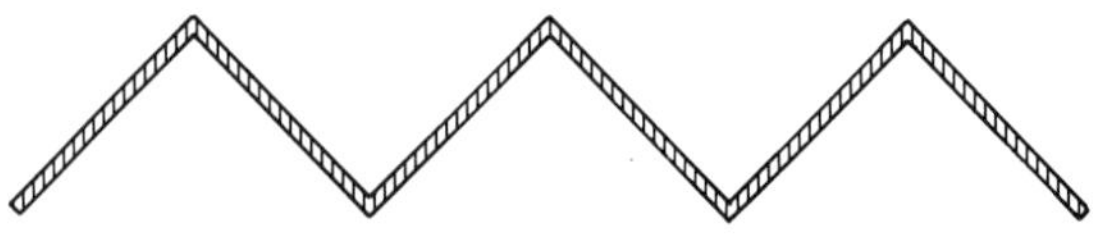

(a)

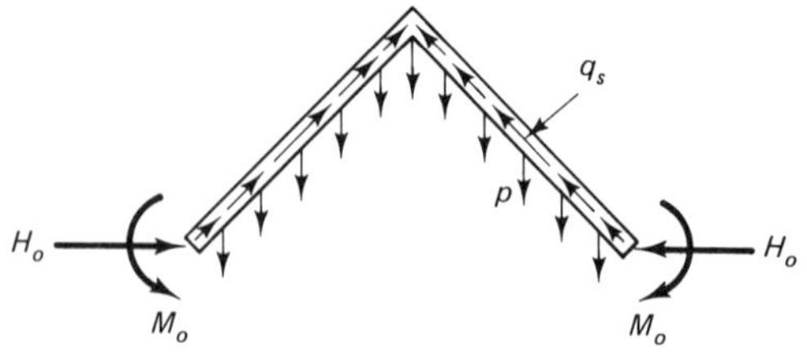

(b)

Figure 14.14 Multibarrel folded plate. (a) Typical section. (b) Loads acting on section for "arch analysis."

14.8 SHELL ANALYSIS USING THE TABLES OF PARME AND CONNER

In their discussion of the paper by Chinn [9], Parme and Conner [4] have given ready tables for finding the stress resultants in an internal shell of a multibarrel circular shell without edge beams, under a uniform dead load p_d and a live load p_u. These tables are here reproduced in Table 14.10 and have been prepared on the basis of beam theory. They give stress resultants directly for any practical value of ϕ_k. Hence these tables are of great help to the designer for dimensioning internal shells with $L/a > 1.67$ (the lower limit for applicability of beam theory for such shells).

Under vertical loads, the cross section of a "long" shell tends to flatten out under membrane stresses. Such flattening is not considered in beam theory, and hence beam theory may give erroneous values for M_ϕ and N_ϕ. This error is more predominant for live loading, which is constant over the plan width of the cross section. For dead loading, the equivalent load over the plan width of the cross section increases toward the edges and thus has a tendency to reduce the flattening. Hence the error due to flattening is not significant under dead load [4]. Parme and Conner [4] have presented a table giving the edge horizontal displacement under membrane response to live load. They have also given another table of corrections to be applied to the M_ϕ and N_ϕ values obtained by beam theory (Table 14.10) for live load, under a unit horizontal edge displacement. When live/snow loads form a significant part of the total load, these two tables of Reference [4] may be used to obtain corrected values of N_ϕ and M_ϕ. For many cases, however, live/snow loads are generally much smaller than dead loads, and hence such a correction may not change the total values of the stress resultants significantly. Thus Table 14.10 is extremely useful for performing a rapid analysis and design of cylindrical shells without edge beams.

14.9 SHELL ANALYSIS USING OTHER TABLES

We had seen from the example in Section 14.4 that even when using the ASCE tables, the solution of a problem involves considerable numerical work, which may easily present an opportunity for errors. Considering the importance to a practicing design engineer for confidently designing shell structures, many authors have given ready tables presenting values of internal stress resultants in circular cylindrical shells for various values of the shell parameters and boundary conditions. All the tables have been prepared using computer programs for the particular type of shell, with varying parameters.

Gibson [2] has given a series of tables wherein the angle ϕ_k is assumed to be 40°, the thickness 3 in., and the loading 56 lb/ft^2 plus edge beam weight. These tables consider single shells with edge beams and multibarrel (inner) shells with edge beams. The depth of the edge beam is kept to about $\frac{1}{20}$ of the span and its

TABLE 14.10 Parme and Conner's Results for Stress Resultants in an Internal Unit of a Multibarrel Shell without Edge Beams, Subjected to Vertical Live and Dead Loads

$$T_x = \frac{l^2}{r}\,[p_u \text{ Col. (1)} + p_d \text{ Col. (5)}]$$
$$T_\phi = r\,[p_u \text{ Col. (2)} + p_d \text{ Col. (6)}]$$
$$S = -l\,[p_u \text{ Col. (3)} + p_d \text{ Col. (7)}]$$
$$M_\phi = r^2\,[p_u \text{ Col. (4)} + p_d \text{ Col. (8)}]$$

		Uniform transverse load				Dead weight load			
		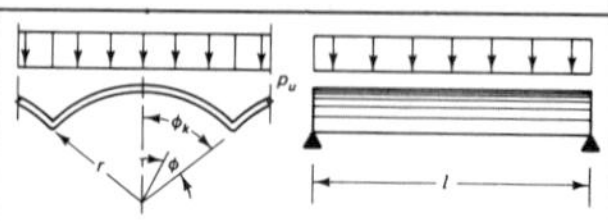				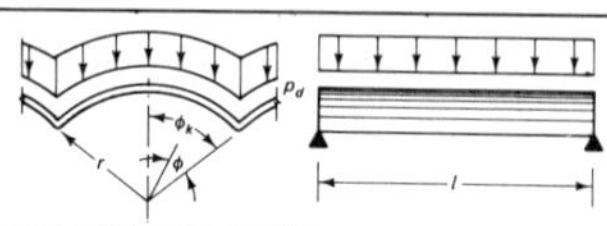			
ϕ_k, deg	ϕ	T_x (1)	T_ϕ (2)	S (3)	M_ϕ (4)	T_x (5)	T_ϕ (6)	S (7)	M_ϕ (8)
	ϕ_k	−6.010	−1.411	0.000	−0.00292	−6.167	−1.433	0.000	−0.00309
	$0.75\phi_k$	−4.875	−1.189	2.211	−0.00112	−5.003	−1.205	2.269	−0.00118
22.5	$0.50\phi_k$	−1.482	−0.614	3.533	0.00232	−1.521	−0.615	3.626	0.00245
	$0.25\phi_k$	4.137	0.049	3.084	0.00235	4.245	0.065	3.165	0.00249
	0	11.927	0.361	0.000	−0.00662	12.239	0.384	0.000	−0.00702
	ϕ_k	−4.855	−1.402	0.000	−0.00353	−5.012	−1.430	0.000	−0.00378
	$0.75\phi_k$	−3.937	−1.182	1.985	−0.00135	−4.064	−1.202	2.049	−0.00145
25.0	$0.50\phi_k$	−1.193	−0.612	3.170	0.00280	−1.232	−0.613	3.273	0.00300
	$0.25\phi_k$	3.342	0.044	2.765	0.00282	3.451	0.064	2.855	0.00304
	0	9.617	0.347	0.000	−0.00797	9.929	0.374	0.000	−0.00857
	ϕ_k	−4.000	−1.393	0.000	−0.00417	−4.158	−1.426	0.000	−0.00453
	$0.75\phi_k$	−3.242	−1.175	1.799	−0.00159	−3.370	−1.199	1.869	−0.00173
27.5	$0.50\phi_k$	−0.980	−0.609	2.871	0.00331	−1.018	−0.610	2.985	0.00360
	$0.25\phi_k$	2.755	0.038	2.503	0.00332	2.863	0.063	2.602	0.00363
	0	7.908	0.331	0.000	−0.00938	8.220	0.363	0.000	−0.01025
	ϕ_k	−3.350	−1.383	0.000	−0.00482	−3.508	−1.422	0.000	−0.00533
	$0.75\phi_k$	−2.714	−1.166	1.643	−0.00183	−2.842	−1.195	1.720	−0.00203
30.0	$0.50\phi_k$	−0.817	−0.606	2.622	0.00384	−0.856	−0.607	2.746	0.00424
	$0.25\phi_k$	2.308	0.032	2.284	0.00383	2.417	0.061	2.392	0.00426
	0	6.608	0.314	0.000	−0.01082	6.920	0.352	0.000	−0.01204
	ϕ_k	−2.844	−1.372	0.000	−0.00548	−3.002	−1.418	0.000	−0.00618
	$0.75\phi_k$	−2.303	−1.158	1.511	−0.00207	−2.431	−1.191	1.595	−0.00235
32.5	$0.50\phi_k$	−0.691	−0.603	2.410	0.00438	−0.729	−0.603	2.544	0.00492
	$0.25\phi_k$	1.960	0.026	2.098	0.00434	2.069	0.060	2.215	0.00492
	0	5.596	0.297	0.000	−0.01227	5.908	0.339	0.000	−0.01393
	ϕ_k	−2.442	−1.361	0.000	−0.00615	−2.601	−1.414	0.000	−0.00707
	$0.75\phi_k$	−1.977	−1.148	1.397	−0.00232	−2.105	−1.186	1.488	−0.00268
35.0	$0.50\phi_k$	−0.591	−0.599	2.227	0.00491	−0.629	−0.600	2.372	0.00565
	$0.25\phi_k$	1.684	0.019	1.938	0.00484	1.793	0.058	2.064	0.00561
	0	4.794	0.278	0.000	−0.01370	5.105	0.326	0.000	−0.01591
	ϕ_k	−2.118	−1.349	0.000	−0.00679	−2.278	−1.409	0.000	−0.00800
	$0.75\phi_k$	−1.714	−1.138	1.298	−0.00255	−1.842	−1.181	1.396	−0.00302
37.5	$0.50\phi_k$	−0.510	−0.596	2.069	0.00544	−0.548	−0.596	2.224	0.00640
	$0.25\phi_k$	1.461	0.012	1.798	0.00532	1.571	0.057	1.933	0.00632
	0	4.146	0.260	0.000	−0.01509	4.458	0.312	0.000	−0.01796
	ϕ_k	−1.853	−1.335	0.000	−0.00742	−2.013	−1.404	0.000	−0.00897
	$0.75\phi_k$	−1.498	−1.127	1.211	−0.00277	−1.627	−1.176	1.315	−0.00337

TABLE 14.10 *(continued)*

		Uniform transverse load				Dead weight load			
ϕ_k, deg	ϕ	T_x (1)	T_ϕ (2)	S (3)	M_ϕ (4)	T_x (5)	T_ϕ (6)	S (7)	M_ϕ (8)
40.0	$0.50\phi_k$	−0.443	−0.592	1.929	0.00595	−0.482	−0.592	2.095	0.00719
	$0.25\phi_k$	1.279	0.005	1.675	0.00578	1.389	0.055	1.819	0.00705
	0	3.616	0.241	0.000	−0.01641	3.928	0.297	0.000	−0.02006
	ϕ_k	−1.449	−1.307	0.000	−0.00853	−1.610	−1.393	0.000	−0.01096
	$0.75\phi_k$	−1.170	−1.104	1.065	−0.00316	−1.299	−1.165	1.183	−0.00408
45.0	$0.50\phi_k$	−0.343	−0.585	1.694	0.00688	−0.381	−0.583	1.882	0.00883
	$0.25\phi_k$	1.001	−0.011	1.468	0.00657	1.112	0.052	1.630	0.00854
	0	2.809	0.202	0.000	−0.01872	3.120	0.266	0.000	−0.02437
	ϕ_k	−1.160	−1.276	0.000	−0.00939	−1.322	−1.380	0.000	−0.01301
	$0.75\phi_k$	−0.935	−1.079	0.947	−0.00344	−1.065	−1.152	1.079	−0.00480
50.0	$0.50\phi_k$	−0.271	−0.578	1.504	0.00762	−0.308	−0.574	1.713	0.01054
	$0.25\phi_k$	0.802	−0.029	1.300	0.00714	0.914	0.049	1.481	0.01002
	0	2.232	0.164	0.000	−0.02042	2.543	0.234	0.000	−0.02871
	ϕ_k	−0.946	−1.242	0.000	−0.00989	−1.109	−1.367	0.000	−0.01506
	$0.75\phi_k$	−0.761	−1.053	0.849	−0.00358	−0.892	−1.139	0.995	−0.00549
55.0	$0.50\phi_k$	−0.217	−0.572	1.347	0.00807	−0.255	−0.563	1.578	0.01227
	$0.25\phi_k$	0.655	−0.048	1.161	0.00742	0.767	0.045	1.360	0.01144
	0	1.805	0.128	0.000	−0.02130	2.115	0.201	0.000	−0.03293
	ϕ_k	−0.783	−1.205	0.000	−0.00992	−0.947	−1.352	0.000	−0.01705
	$0.75\phi_k$	−0.629	−1.025	0.766	−0.00355	−0.761	−1.124	0.927	−0.00613
60.0	$0.50\phi_k$	−0.177	−0.566	1.213	0.00815	−0.214	−0.552	1.467	0.01398
	$0.25\phi_k$	0.543	−0.068	1.043	0.00734	0.656	0.042	1.261	0.01275
	0	1.481	0.095	0.000	−0.02118	1.790	0.167	0.000	−0.03688

Source: Ref. [4].

width is $\frac{1}{6}$ to $\frac{1}{8}$ of the depth. Tables are given for a wide range of shell spans and radii.

Wilby [5] has given graphs for finding internal stress resultants in circular cylindrical shells with constant thickness or with a varying thickness near the valley. Graphs are prepared by solving the governing equations for a set of boundary conditions and shell configurations on an analog computer. Tables are given in SI units and consider various span lengths for an external shell or an internal shell.

Wilby and Khwaja [3] have given extensive tables for internal stress resultants in northlight and barrel-type circular cylindrical shells. Values are given for external as well as internal shells. For northlight shells, two angles, 60° and 90°, are considered for the inclination of the glazing. The thickness is assumed to be 2.5 in., with some thickening near the edges. Donnel (Jenkins) theory was used in the analysis. Tables are given for various spans and widths, and are very easy to use for rapid design provided that the shell dimensions satisfy those in one of the tables.

For the design of V-shaped multibay folded plates, the Portland Cement Association [11] has published tables giving coefficients for transverse moments and longitudinal stresses. Only loads uniformly distributed over the plan area are considered. Due to continuity in the transverse direction, horizontal displacements

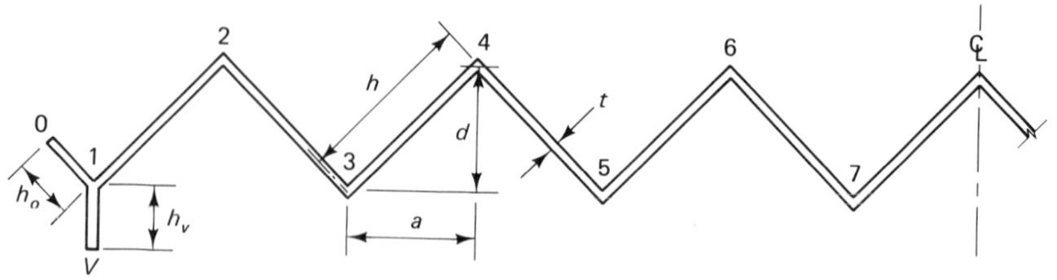

Figure 14.15 Multibay V-shaped folded plate, considered in the PCA tables.

at the internal valleys of the folded plate are zero. In a typical internal bay, the transverse moments at points 5, 6, 7, etc., in Fig. 14.15, are negative (hogging) with positive moments in between, as in a continuous beam. For the external bay, however, if the horizontal displacement is not restrained at the external valley (point 1), plate 1–2 deflects in the transverse direction as a cantilever and has a maximum transverse negative moment at point 2, as in a cantilever slab. This moment can be rather high, and it usually controls the thickness as well as the bay width. It is therefore desirable to control deflections of the end plate by providing a small stiffening arm 1–0 and an edge beam 1–V at the external edge (similar provisions can be made at external edges of barrel shells as well). The PCA tables give coefficients for various combinations of h_v/h, h_o/h, a/d, and L^2t/a^3, where L is the span length. Coefficients are given at external and internal node points 0, 1, 2 . . . , 5, etc., and are very useful for practical rapid design of such folded plate roofs. Similar but more exhaustive tables have been given by Varyani and Taneja [12]. Their tables consider northlight folded plates with different angles of glazing, V-shaped folded plates with or without glazing, and trough-shaped folded plates. These tables cover a wide range of parameters and are very useful in practice for quickly designing a given folded plate or for arriving at an economical and suitable shape for a given span and other requirements.

To simplify the complex task of analyzing barrel structures, we have seen that there are many aids available to a designer in the form of tables and graphs. These can be used for rapid preliminary design or as a check on more refined analyses by hand or computer. We shall briefly summarize some of these design aids below.

Summary of Design Aids for the Analysis of Circular Barrel Shells and Folded Plates

1. For the analysis of an internal unit of a multibarrel circular shell without edge beams, with L/a greater than at least 1.67, we can use Table 14.10 by Parme and Conner [4]. For large live loads, they have additional tables which give corrections for M_ϕ and N_ϕ. These corrections should be employed if live/snow

loads are large compared to the dead loads. These tables are very easy to use and are helpful for the rapid design of such shells.

2. For the analysis of internal or external shells, with or without edge beams, Gibson's tables [2] can be used for $\phi_k = 40°$ only.
3. For the analysis of internal or external shells, with or without edge beams, the tables by Wilby and Khwaja [3,5] are useful, and these consider various shell and beam dimensions. These tables are also useful for the analysis of northlight shells. The tables found in Reference [3] are especially easy to use for quick practical design of barrel shells.
4. The ASCE Manual and the PCA extension tables [1,6] are useful for the analysis of both long and short shells. However, these tables may still involve lengthy numerical calculations. Nevertheless, for short shells, such as silos, hangars, etc., these tables are useful, as few other tables are available for analyzing such shells.
5. For V-shaped multibay folded plates, the PCA tables [11] can be readily used. For all other common types of folded plates, the exhaustive tables found in Reference [12] are useful.

In addition to the foregoing aids, the designer can use a suitable computer program to rigorously analyze any particular shell without being required to compromise on shell dimensions or parameters to suit a particular table. We discuss next the analysis of barrel shells by using computer programs.

14.10 COMPUTER ANALYSIS OF CYLINDRICAL SHELLS AND FOLDED PLATES

In order to obtain a detailed understanding of the variation of stress resultants and deflections, it is often desirable to analyze a shell of given dimensions by a computer program. There are a number of available programs to choose from. If a design office wishes to develop a program for in-house use, it is best to develop a program that can be applied, even if somewhat approximately, to solve a wide variety of problems.

One can develop a program which uses Schorer's or Donnel's equations for the shell and equations such as Eqs. (14.2) to (14.5) for the edge beams, in order to solve for stresses in a circular cylindrical shell with or without edge beams. However, such a program would be applicable only to shells that have a circular cross section and which are simply supported at the ends $x = 0$ and $x = L$ (assuming that Fourier series are used in the x-direction, for separating the variables in the partial differential equation). Some tables [2,3,5,13] have been developed using such programs.

Finite Element programs, based on the stiffness method of analysis, would be more general. Such programs would be of two types:

(a) Programs that still use a Fourier sine, cosine, or any other series expansion in the x-direction, and hence reduce the problem to that of one-dimensional elements in the transverse direction.

(b) Two-dimensional finite element programs using plane or curved elements.

We shall discuss the applications and limitations of these two types of programs below.

Finite Element Analysis of Simply Supported Shells or Folded Plates

The analysis of simply supported shells or folded plates, of any cross-sectional shape and under any loading, can be conveniently carried out using the type-(a) programs. For this purpose the stiffness matrix derived by De Fries-Skene and Scordelis [14] is very useful. This stiffness matrix is developed for a rectangular plate element [Fig. 14.16(a)] and is based upon the plate bending and plane stress solutions given by Goldberg and Leve [15]. The element is simply supported at $x = 0$ and $x = L$ and has four possible displacements at each edge in the transverse direction.

The solution for the plate element can be obtained by using the plate bending equation for bending moment and shear loads and the plane stress equations for in-plane loads. In both cases, Fourier expansions are used in the x-direction, so that simple support conditions are satisfied at edges $x = 0$ and $x = L$. By assuming the other two edges to be fixed, and by applying one displacement (or load) on one edge at a time, stiffness coefficients for the plate can be obtained. The resulting (8×8) stiffness matrix is obtained in explicit form. Using such plate elements, which are viewed as straight lines in the transverse direction, one can closely idealize the cross section of any folded plate or of a cylindrical shell of circular or noncircular cross section. The stiffness matrix for the entire shell can be obtained by assembling the stiffness matrices for all plate elements, with proper transformation to a common set of axes (as per normal procedures of the stiffness method in structural analysis). Results can be obtained for any general loading, such as uniformly distributed along the span, triangular, partial loading, or prestress forces, by expanding the loads as Fourier series in x, and by inverting the structural stiffness matrix for each harmonic of the Fourier series. The final solution is obtained by taking the summation of solutions for a required number of harmonics.

Figures 14.16(b) and (c) show idealizations of circular barrel shells considered in Sections 13.4 and 14.4, using straight plate elements. For greater accuracy, smaller elements can be taken. However, dividing half of the shell into six or eight

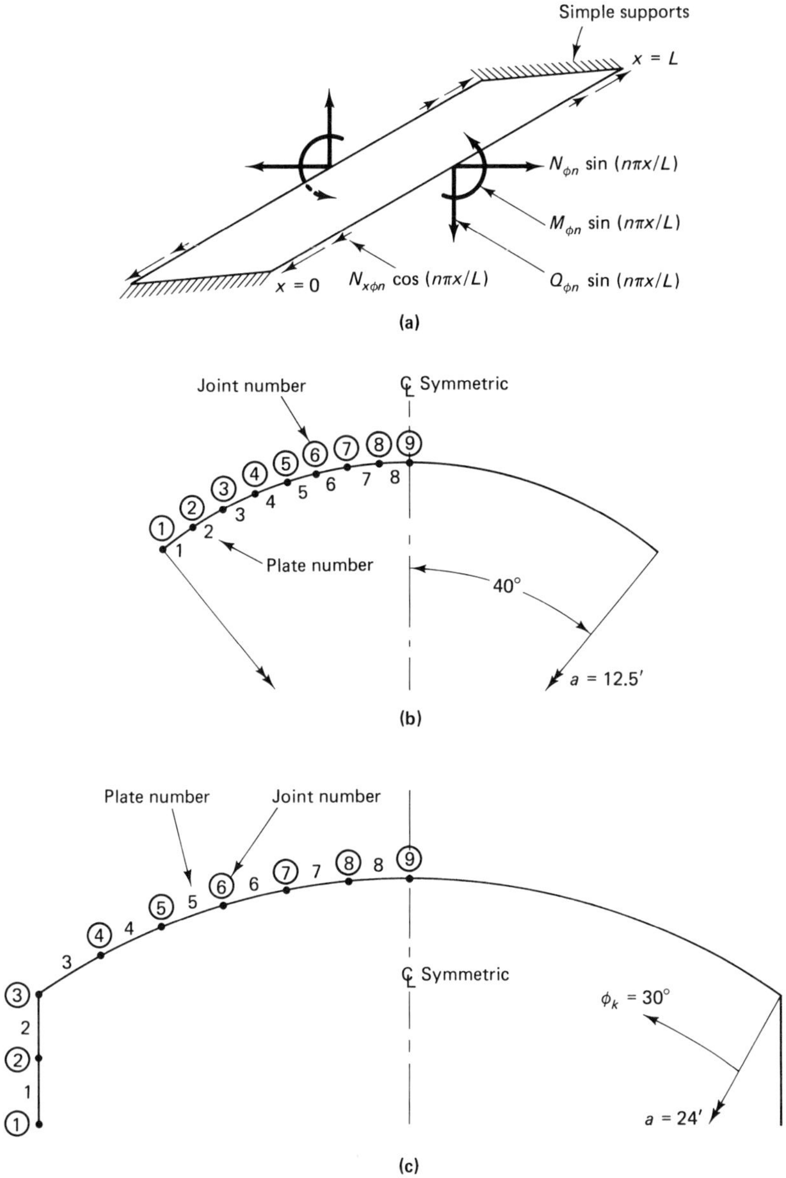

Figure 14.16 (a) Basic rectangular plate element used by De Fries-Skene and Scordelis (Goldberg and Leve) in a type-(a) Finite Element program. (b) Finite Element idealization of shell of Section 13.4. (c) Idealization of shell of Section 14.4

elements usually gives sufficient accuracy for any practical design. For the idealized shells of Fig. 14.16(b) and (c), the values of stress resultants obtained at midspan (and shears $N_{x\phi}$ obtained at the supports) with a computer program based on the stiffness matrix of Scordelis, are given in Tables 14.11 and 14.12 for the first harmonic $n = 1$. When using the computer program, one can easily calculate values which involve summation over more harmonics (such summation would be impractical by hand calculations). Generally, for a uniformly distributed load, Scordelis recommends summation over at last nine harmonics ($n = 1, 3, \ldots, 9$). This gives a maximum error of only 6.3% in M_ϕ, N_ϕ, etc. For partial or unsymmetrical loadings, a greater number of harmonics in the series has to be used for obtaining good convergence. The values in parentheses in Tables 14.11 and 14.12 have been calculated by taking a summation over nine harmonics. We can thus compare the results obtained by using one and nine harmonics in a stiffness analysis on the computer, with those results obtained earlier by hand calculations, based on the Schorer equations for the example of Section 13.4 (Table 13.6) and on the ASCE tables for the example of Section 14.4 (Table 14.8).

In the computer analysis by the stiffness method, the stress resultants were calculated by approximating the distributed loading as statically equivalent line loads at the joints. Hence values of N_ϕ and Q_ϕ at the edges, in Tables 14.11 and 14.12, are not zero but are equal to the respective components of this external load at the edge. At internal joints, since the shell is idealized into straight plate elements, forces N_ϕ and Q_ϕ for adjoining plates (which have different slopes) will be slightly

TABLE 14.11 Computer Analysis of Shell of Section 13.4 as an Equivalent Folded Plate[a]

ϕ (Measured from Crown) (deg)	M_ϕ (lb-ft/ft)	N_ϕ (lb/ft)	Q_ϕ (lb/ft)	$N_{x\phi}$ (At Support) (lb/ft)	N_x (lb/ft)
0	−749.61	−1,733.8	32.30	0.0	−16,934.75
(crown)	(−701.89)	(−1,472.3)	(28.05)	(0.0)	(−16,900.75)
10	−631.87	−1,520.45	116.96	2,599.0	−14,294.0
	(−590.98)	(−1,270.9)	(102.49)	(2,618.4)	(−14,058.5)
20	−346.49	−958.05	169.93	4,291.7	−5,627.0
	(−326.23)	(−767.6)	(145.42)	(4,656.4)	(−5,083.0)
30	−69.3	−296.07	122.87	3,940.5	11,113.5
	(−73.28)	(−225.35)	(102.29)	(4,851.0)	(11,226.5)
40	0.0	+26.41	+34.43	0.0	38,825.0
(valley)	(0.0)	(22.04)	(+28.75)	(0.0)	(36,515.0)

[a]Values outside and inside parentheses represent, respectively, results for summation over one and nine harmonics of the Fourier series.

TABLE 14.12 Computer Analysis of Shell of Section 14.4 as an Equivalent Folded Plate (Table in ASCE Sign Convention)[a,b]

ϕ (Measured from Valley) (deg)	N_x (lb/ft)	$N_{x\phi}$ (lb/ft)	N_ϕ (lb/ft)	M_ϕ (lb-ft/ft)
30 (crown)	−21,913.0 (−21,222.25)	0.0 (0.0)	−2,089.6 (−1,711.4)	+79.26 (+92.25)
20	−19,544.0 (−18,910.25)	−3,453.2 (−4,002.8)	−1,827.55 (−1,508.0)	+136.95 (+131.97)
10	−12,578.5 (−12,187.75)	−6,136.5 (−7,321.0)	−1,110.85 (−931.65)	+145.05 (+122.48)
0 (valley)	−1,560.25 (−1,589.55)	−7,335.0 (−8,758.6)	−90.0 (−79.18)	−350.9 (−299.43)

[a] $\sigma^B_{x,b}$ = 106,500.0 or (103,420.0) lb/ft^2 = stress at bottom of beam; $\sigma^B_{x,t}$ = −6241.0 or (−6358.2) lb/ft^2 = stress at top of beam.

[b] Values outside and inside parentheses represent, respectively, results for summation over one and nine harmonics of the Fourier series.

different. The average values of stress resultants N_ϕ and Q_ϕ, in adjoining plates, are presented in Tables 14.11 and 14.12 as the values acting at the joint.

Except for the values of M_ϕ in Table 14.12, all values for both examples agree reasonably well with those previously obtained by hand calculations. Whatever differences exist are not large enough to influence the design of a shell in practice. Even the values of M_ϕ in Table 13.6 or 14.8 are so small that they will most likely be resisted by the minimum steel provided. The designer should, however, keep in mind the negative moment generated at the junction of the shell and the beam in Table 14.12, even though simplified hand calculations (Table 14.8) give this moment as zero. This is because, in hand calculations, we had assumed that edge beams cannot resist any torsional moment. Hence M_ϕ at the junction of the shell and the beam was zero.

From Tables 14.11 and 14.12 it can also be seen that the differences in the results between summation over one or nine harmonics are not so large as to affect the practical design of the shell significantly. When running a computer analysis, one can, of course, use summation over a requisite number of harmonics for any given loading (nine for uniform loads), since this summation does not involve additional complexities.

Scordelis has also given a simplified stiffness matrix based on "beam theory," wherein the transverse stiffnesses are obtained assuming a beam in the transverse direction, rather than by considering a plate element. This approach has been further described by Chu and Pinjarkar [16]. Since this stiffness matrix does not consider plate theory (i.e., it neglects torsional moments $M_{x\phi}$, etc.), it fails to give good results for certain problems, such as that of a single inverted V-shaped folded plate with loading unsymmetric about the crown (say on one fold only), since statics cannot be satisfied. It is therefore desirable to use the more accurate "elasticity" stiffness matrix given by Scordelis and based on the plate solutions of Goldberg and Leve [15], which gives extremely good results. A wide variety of practical problems, including V- and trough-shaped folded plates, long and short cylindrical shells of varied cross section with constant or variable thickness, box girders with vertical or inclined webs, bridge decks, etc., for several different loadings, have been solved by the authors using this versatile program.

The main limitation of the Scordelis stiffness matrix is that it considers only simple supports at $x = 0$ and L. It cannot be used when the ends $x = 0$ or L are fixed, free, or continuous, although such conditions would not be commonly encountered in practice. It also cannot take into account a variation of thickness along the span (although it can account for thickness variation across the span by using a cross section made up of plates with different thicknesses). It can also not consider the presence of cutouts or transverse stiffeners.

Computer Analysis of Shells with Boundary Conditions Other Than "Simple Supports"

So far we have restricted our attention to the analysis of cylindrical shells which were simply supported on diaphragms at $x = 0$ and $x = L$. Although not very commonly constructed, we may also have to design shells which are cantilevers, propped cantilevers, or which are continuous over two or more spans. One example would be a simply supported barrel shell or folded plate roof which has a large overhang provided for architectural reasons or otherwise. Another example would be the vertical walls of a large underground reservoir. In this case, for a depth of water of 20 ft or more, such walls can be economically designed to have a trough or V-shaped folded plate cross section (in plan). This cross section would give a much larger lever arm for resisting bending moments (and hence a considerable reduction in reinforcement) for the same amount of concrete, as opposed to a wall of normal rectangular cross section. Such a wall has to be analyzed as a cantilever folded plate, fixed at the raft foundation, and subjected to a triangularly varying water load. If there is a roof slab, the wall becomes a propped cantilever folded plate.

The examples cited above cannot be analyzed by using a computer program based on the stiffness matrix of Scordelis. To solve problems with boundary conditions other than simple supports, we cannot use a Fourier sine or cosine series expansion in the x-direction, since these functions will no longer automatically satisfy the boundary conditions at the supports. In such cases we must expand the loads, stress resultants, and displacements as infinite series of "beam functions" in the longitudinal direction. The beam functions are orthogonal functions, like the sine and cosine functions, but are more complex, and are obtained from the vibration analyses of beams with corresponding boundary conditions (cantilever, propped cantilever, etc.). Because of the property of orthogonality of the beam functions, we can expand any function, such as loads, stress resultants, or displacements as infinite series of these functions in the x-direction. Using these series expansions, we can separate the variable x in the governing partial differential equations of the shell, while simultaneously satisfying boundary conditions in the longitudinal direction. This separation will lead to an ordinary differential equation in ϕ, which can be solved for each harmonic, while satisfying the boundary conditions at the transverse edges $\phi = constant$. Thus the procedure is, in principle, similar to the one adopted for simply supported shells (Chapter 13). However, the solution becomes very complicated, even for simple cases, if hand computations are to be performed [10]. A suitable computer program, similar to one using Schorer equations for simply supported shells, can be developed. In such a program the unknown constants of integration, like constants A, B, C, D in Chapter 13, would be obtained from the given boundary conditions at $\phi = \phi_k$ for each harmonic. However, development of a general stiffness matrix for plate elements with different boundary conditions at $x = 0$ and $x = L$, in a manner similar to Scordelis (Goldberg and Leve), would be a difficult task.

The programs above, based on infinite series expansion in the longitudinal direction, will have the same drawbacks as the type (a) programs for simply supported shells (i.e., they cannot take into account variation of thickness along the span, cutouts, presence of transverse stiffeners, etc.).

For "long" shells continuous over two or more spans, the ASCE Manual suggests a simple and approximate method for calculating the longitudinal stresses. In this procedure, the longitudinal stresses are first calculated for a simply supported shell of the same span length. Longitudinal stresses in the continuous shell are then obtained by multiplying the longitudinal stresses of the simply supported shell by a factor equal to the ratio of moments in continuous and simply supported beams of corresponding spans. Thus for a barrel shell, continuous over say two equal spans, we would first analyze the shell as if it were simply supported over a single span. The longitudinal stress at the continuous support of the actual shell is then obtained by multiplying the midspan longitudinal stress of the simply supported shell by a factor M_1/M_2. Here M_1 and M_2 are, respectively, (1) the moment at the continuous support of a corresponding two-span continuous beam and (2) the moment at midspan of a simply supported beam of the same span. Transverse stress resultants are assumed to be the same as in the simply supported shell. This method gives good results for practical design of long shells, which tend to behave as beams in the longitudinal direction. For short shells, the ASCE Manual gives equations for multiplication factors based on various cases of continuity. These factors have to be multiplied by the longitudinal stresses in a simply supported shell, in order to obtain corresponding values in the continuous shell.

Computer Analysis Using Two-Dimensional Finite Elements

As discussed, the computer programs of type (a), whether using infinite series of sine/cosine or of beam functions in the longitudinal direction, cannot be used for solving some problems which may arise in practice. As examples, we may have:

1. A roof shell with large cutouts for lighting or exhaust such that a detailed analysis of stresses around the cutouts may be needed.
2. A cantilever folded plate, such as for the wall of a water reservoir, discussed earlier, which may have a thickness increasing from its top to its bottom (i.e., shells having a variation of thickness along the x-direction).
3. Adjoining bays of northlight shells or folded plates, which are generally connected by mullions (struts) in which glazing is mounted. The structural contribution of the mullions gives continuity to adjoining shell bays, and helps reduce stresses in the otherwise, individual, Z-shaped, units.

Although these examples can always be solved approximately after carrying out a basic analysis of an idealized structure using type (a) programs, still, for a more rigorous solution, one should use a program with two-dimensional finite elements. Such a program would permit satisfying any boundary conditions at $x = 0$ and $x = L$, as well as on edges $\phi = constant$. By using a suitable finite element mesh, inclusion of cutouts, transverse stiffeners, variable thickness (along the span), etc., can be accommodated.

For such a finite element analysis, one can use a plane or curved rectangular element. A plane rectangular element program can be used for the analysis of folded plates or shells of any cross-sectional shape—not necessarily circular. Rockey and Evans [17] have given a stiffness matrix in explicit form for such an element. The size of the stiffness matrix of a single element becomes 24×24, considering six degrees of freedom (three displacements and three rotations) at each node of the element. (In the Scordelis stiffness matrix discussed earlier, the element stiffness matrix was only 8×8.) Hence even if a coarse mesh is used to idealize a given shell, the overall stiffness matrix for the entire shell will be very large and will require large computer memory and time.

Cantin and Clough [18] have given a stiffness matrix for a circular curved rectangular element. Thus this element can be used to idealize only circular cylindrical shells, and cannot be used for shells of other cross-sectional shapes or for folded plates. The size of the stiffness matrix for an individual element is again 24×24. A program using this element can easily be developed and can be used for analyzing circular barrels with any boundary conditions at the edges, as well as for analyzing shells with cutouts. However, again, the size of the stiffness matrix for a structure becomes quite large, especially if cutouts are present (since a finer mesh has to be used near such cutouts.)

There are, of course, a number of other finite element programs, using refined elements, which can be used for shell analysis. For this the reader is referred to some standard texts on finite element analysis [19].

REFERENCES

[1] "Design of Cylindrical Concrete Shell Roofs," *Manual of Engineering Practice, No. 31,* American Society of Civil Engineers, New York, 1952.

[2] Gibson, J. E., *Computer Analysis of Cylindrical Shells,* E. and F. N. Spon Ltd., London, 1961.

[3] Wilby, C. B., and Khwaja, I., *Concrete Shell Roofs,* Applied Science Publishers Ltd., Barking, Essex, England, 1977.

[4] Parme, A. L., and Conner, H. W., Discussion of the paper by Chinn (Reference [9]), *ACI j.,* p. 1584, Dec. 1959.

[5] Wilby, C. B., *Design Graphs for Concrete Shell Roofs,* Applied Science Publishers Ltd., Barking, Essex, England, 1980.

[6] "Coefficients for Design of Cylindrical Concrete Shell Roofs," Portland Cement Association, Skokie, Ill., 1959.

[7] Iffland, J. S. B., "Folded Plate Structures," *ASCE J. Struct. Div.,* Jan. 1979, pp. 111–122.

[8] Lundgren, H., *Cylindrical Shells,* Danish Technical Press, Copenhagen, 1960.

[9] Chinn, J., "Cylindrical Shell Analysis Simplified by the Beam Method," *ACI j.,* 55, May 1959.

[10] Ramaswamy, G. S., *Design and Construction of Concrete Shell Roofs,* Revised ed., R. E. Krieger, Malabar, Florida, 1984.

[11] "Direct Solution of Folded Plate Concrete Roofs," Advanced Engineering Bulletin, No. 3., Portland Cement Association, Skokie, Ill., 1960.

[12] Varyani, U. H., and Taneja, K. N., *Design Tables for Folded Plates,* Today and Tomorrow's Printers and Publishers, New Delhi, 1982.

[13] Gibson, J. E., *The Design of Shell Roofs,* E. and F. N. Spon Ltd., London, 1968.

[14] DeFries-Skene, A., and Scordelis, A. C., "Direct Stiffness Solution for Folded Plates," *ASCE J. Struct. Div.,* Aug. 1964, p. 3994.

[15] Goldberg, J. E., and Leve, H. L., "Theory of Prismatic Folded Plate Structures," International Association of Bridge and Structural Engineering, Zurich, No. 87, pp. 71–72.

[16] Chu, K. H., and Pinjarkar, S. G., "Multiple Folded Plate Structures," *ASCE J. Struct. Div.,* Apr. 1966, p. 4791.

[17] Rockey, K. C., and Evans, H. R., "A Finite Element Solution for Folded Plates," in *Space Structures,* ed. R.M. Davies, John Wiley & Sons, Inc., New York, 1967, p. 165.

[18] Cantin, G., and Clough, R. W., "A Curved Cylindrical Shell Finite Element," *AIAA J.,* Vol. 6., No. 6, 1968, p. 1057.

[19] Zienkiewicz, O. C., *The Finite Element Method,* 3rd ed., McGraw-Hill Book Company, New York, 1977.

PROBLEMS

14.1. Solve the shell problem of Section 14.4 using the ASCE tables, assuming the shell to be an internal shell of a multibarrel shell structure.

14.2. Figure P14.2 shows a shell roof over a bus stand or a cycle stand. Similar roofs can be seen on some platforms of the Washington, D.C., subway system. The shell spans 32 ft, and it is supported by diaphragm arches which cantilever out from columns. The shell cross section is as shown in the figure and has a radius of 8 ft. Assume a total load of 60 lb/ft^2 over the surface area of the shell. Obtain the longitudinal stresses and transverse moments, etc., by beam theory for this shell. Also obtain the shear loads from the shell on the diaphragm arches.

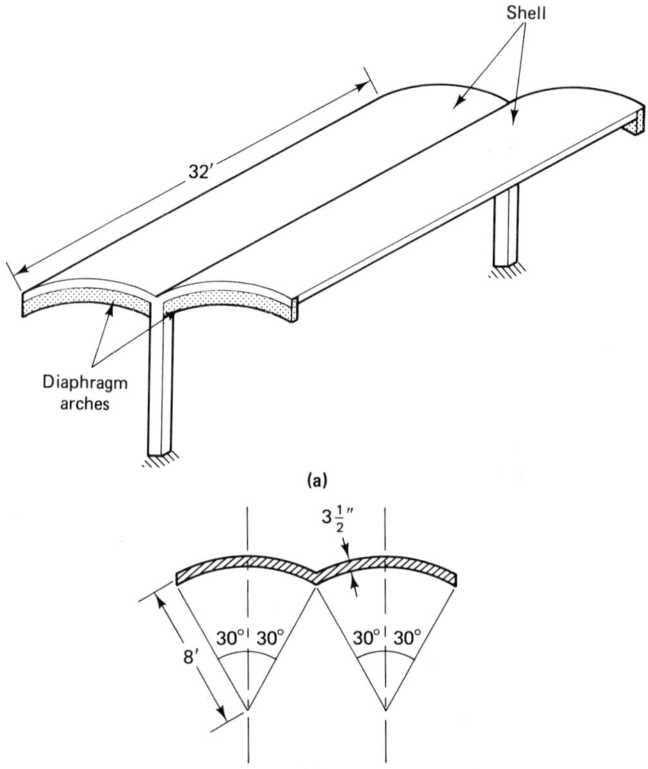

Figure P14.2 Barrel shell used as a platform cover. (a) Shell configuration. (b) Shell cross section.

14.3. Figure P14.3 shows an internal bay of a multibay folded plate roof structure, which spans 40 ft. Analyze the structure by beam theory for a total load of 75 lb/ft^2 of surface area. All plates may be assumed to be 4 in. thick.

14.4. Let the folded plate in Fig. P14.3 now be the cross section of a wall of a water tank. The height of the wall is 20 ft and the thicknesses may now be considered to be 6

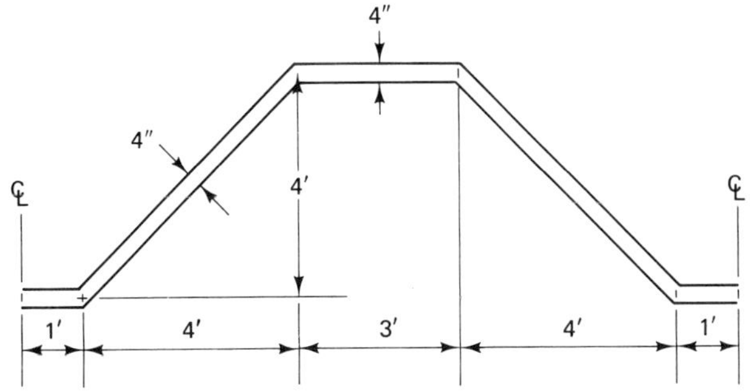

Figure P14.3 Internal unit of a multibay folded plate roof.

in. The wall is subjected to a water load with a 20-ft head. Analyze this cantilever folded plate wall using beam theory.

14.5. Figure P14.5 shows a multibay northlight folded plate structure which spans 45 ft. The thickness of all plates, except the verticals, is 4 in.; the small vertical plates (beams) are 8 in. thick. Analyze a typical unit for a total vertical load of 80 lb/ft^2 of surface area. Neglect any effects of continuity between the adjoining units through the truss action of the beams and the glazing verticals.

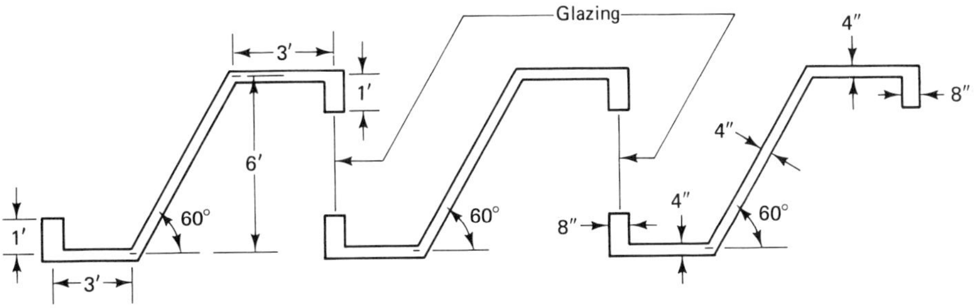

Figure P14.5 Multibay northlight folded plate.

14.6. Figure P14.6 shows a hat-shaped folded plate to be used as a roof unit. The unit spans 45 ft on simple supports and carries a vertical load of 70 lb/ft^2 over the surface area. The thickness of each plate is 4 in. Using beam theory, find M_ϕ, N_ϕ, and Q_ϕ in this roof unit.

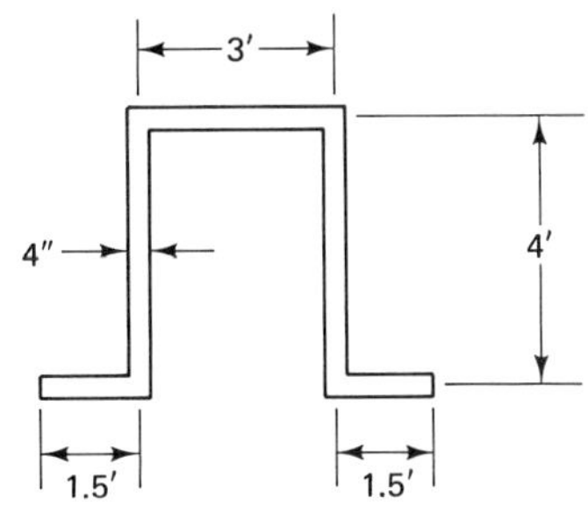

Figure P14.6 Hat-shaped folded plate used as a roof unit.

15

Design of Cylindrical Barrel Shells or Folded Plate Roofs

15.1 INTRODUCTION

Cylindrical barrel shells are curved in the transverse direction only, whereas individual plates of a folded plate are plane without any curvature at all. Since both of these types of "shells"* have no curvature in the direction of their spans, they can easily be constructed using straight-line forms, and they are the most popular types of shells for construction of roofs over large halls, industrial facilities, stores, aircraft hangars, and for large storage silos of fertilizer or cement plants—structures that require large column-free areas in one or both directions. For large spans, these shell roofs can prove to be more economical and more aesthetically pleasing than normal beam and slab construction.

In Chapters 12 to 14 we studied the analysis of barrel shells or folded plates of given dimensions, by solving the governing equations or by using approximate methods, tables, or computer programs. In practice, for a given plan layout, the structural engineer has first to decide on a suitable shell configuration: its profile, span length, bay widths, thicknesses, etc., which will satisfy the functional requirements of the client and the aesthetic demands of the architect. We shall briefly discuss this phase of the design process.

*In this chapter the term "shell" will generally refer to a barrel shell or a folded plate.

15.2 SELECTION OF SHELL CONFIGURATION AND DIMENSIONS

Although closer spacing of columns generally results in an economical structure, the requirements for spacing of columns, in both the longitudinal and transverse directions, is usually dictated by the client, depending on what size of column-free bays he can work with. The spacings of columns in the longitudinal direction thus determine the span length L of the shell roof to be designed. The loads from the shell are transmitted to the diaphragm beams, which span on columns in the transverse direction. Hence the width of each bay of a multibay shell need not have any relation to the column spacing in the transverse direction, since each span of the diaphragm beam can carry loads from one or more shell bays, or from any fractions thereof.

Knowing the span L of the shell for the roof of an industrial facility, store, etc., we would design the shell as a "long" shell, which behaves mainly as a beam in its span direction. Then, as for reinforced concrete beams, the depth of the cross section of the shell should be kept between $\frac{1}{8}$ to $\frac{1}{12}$ of its span. For barrel shells, since the compression flange is curved at the top, an increased depth of $\frac{1}{6}$ of its span may be used. For spans greater than about 80 ft, the depths become unduly large, and therefore, prestressing becomes economical. For prestressed shells, depths of about $\frac{1}{12}$ of the span, or less, can be used.

Now, before deciding on any other dimensions of the shell cross section, we must decide what shape to use—whether it be a cylindrical barrel shell or a folded plate—assuming, of course, that both configurations are architecturally acceptable.

15.3 BARREL SHELLS VERSUS FOLDED PLATES

Folded plates and "long" barrel shells both resist loads in a similar manner. Since individual plates of a folded plate have no curvature, they are easier to construct than the singly curved barrel shells. However, concrete and steel consumption in barrel shells is less than in a corresponding folded plate.

The transverse curvature of a barrel shell helps in reducing buckling tendencies under longitudinal compressive stresses. This is an advantage of the barrel shell in comparison to, say, a trough-shaped folded plate. The transverse moments in a barrel shell are generally much lower than those in a folded plate. As a simple example, consider the single-barrel shell solved in Section 13.4. This shell has a transverse bay width of 16.07 ft and carries a load of 62.5 lb/ft^2. The value of M_ϕ at the crown of the shell, which is also the maximum transverse moment, was found to be -958.02 ft-lb/ft (Table 13.6), or -701.89 ft-lb/ft by a more detailed computer analysis (Table 14.11). In a corresponding single-bay inverted V-shaped folded plate, with the same bay width and a slope of 40°, the transverse moment at the crown would just be the cantilever moment of the load acting on one arm of the folded plate. The inclined length of this arm is (8.035/cos 40°) = 10.49 ft. Hence the crown moment is equal to 62.5(10.49)(8.035/2) = 2633.97 ft-lb/ft. The slope of 40° for the folded plate will give a much greater depth of cross section (6.742 ft) than the 2.924 ft for the barrel shell. Even if an (impractical) slope, giving the same height of 2.924 ft at the crown, is assumed for the folded plate, the inclined length of the plate now becomes $(8.035^2 + 2.924^2)^{1/2} = 8.55$ ft, and the crown moment is equal to 62.5(8.55)(8.035/2) = 2146.85 ft-lb/ft. Here still, the transverse moment in the folded plate is much higher than that in the barrel shell. The reason for the higher transverse moment, as explained in Section 14.7, is that the specific shears in the folded plate act along the plates. When considering the freebody

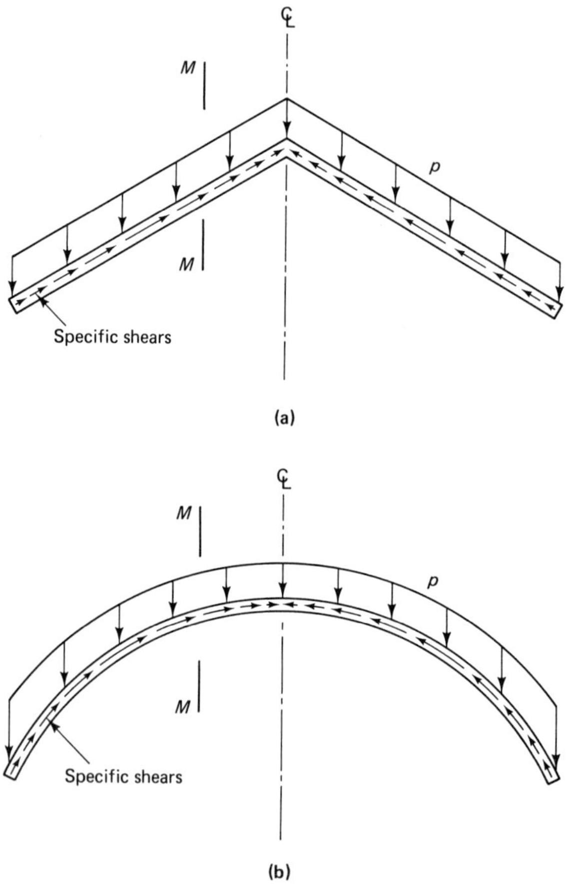

Figure 15.1 Specific shears in folded plates and barrel shells. (a) Specific shears act through the crown in a V-shaped folded plate and hence do not reduce the transverse moments. (b) In the barrel shell, specific shears cause a resisting moment which reduces the transverse moment due to the load.

of the folded plate to the left of any section *M-M* [Fig. 15.1(a)], we see that the transverse moment is caused by the cantilever moment of the loads. On the other hand, it can be seen that at a section *M-M* in the corresponding barrel shell [Fig. 15.1(b)], the cantilever moment of the loads is considerably reduced by the moment due to the specific shears.

Thus the thickness and transverse steel required for the folded plate above would be much greater than that required for the barrel shell. One, of course, would not design such a wide single-bay folded plate as that considered here. In multibay folded plates, the transverse moments are much less, as in a continuous slab, but they are still much greater than those in corresponding multibarrel shells.

Due to the reasons stated above, the thicknesses of most folded plates, for normal spans of 40 to 60 ft, are generally 3.5 to 4 in., whereas for the barrel shells they are 2 to 3 in. (3 in. being the most common). The choice of using a folded plate or a barrel shell cross section for a given span would thus depend on the relative savings in cost of concrete and steel compared to the added cost of preparing formwork for the curved barrels. These costs vary from country to country, as well as within a country, and the designer can make a decision after reviewing the local conditions.

We discuss next a few points that the designer should consider in arriving at the actual cross-sectional profile of a folded plate or a barrel shell, once he or she has selected one of the two alternatives.

15.4 PROPORTIONING THE CROSS SECTION OF A FOLDED PLATE

Knowing the span of a folded plate, the designer will usually choose either a V-shaped or a trough-shaped cross section, assuming that either one is architecturally acceptable. For V-shaped folded plates, the behavior is similar to a rectangular beam, and the depth of the section will be about $\frac{1}{8}$ to $\frac{1}{10}$ of the span. For trough shapes, the behavior is similar to a beam with an I-section, and hence the depth can be somewhat smaller, since I-sections give a higher moment of inertia for the same depth.

In most folded plates, the slope of the inclined plates is generally kept below 45° to avoid use of double formwork when concreting the plates. A slope between 35 and 45° is commonly used. It can be easily seen that, for similar slopes, the perimeter of a cross section of a trough-shaped folded plate is less than that of a V-shaped folded plate. For example, with a depth H and a slope of 45°, a V-shaped folded plate will cover a width $B = 2H$ and will have a perimeter of $2H/\sin 45° = 2.8H$. The perimeter-to-bay-width ratio in this case will then be 1.4. Now, if we consider a trough-shaped folded plate of the same depth, H, and assume that its top and bottom flanges have width H each, for a typical panel we have a perimeter of $2H + 2.8H = 4.8H$. The width covered in this case will be

$B = 2H + 2H = 4H$. The ratio of perimeter to width in this case is then 1.2. This is less than that for a V-shaped folded plate. Hence the concrete quantity for a trough-shaped cross section will be less than that for a V-shaped section. In addition, in the example above, since we assumed the same depth for both sections, the moment of inertia of the trough-shaped section will be much more than that for the V-shaped section. This will result in much lower longitudinal stresses and less required longitudinal steel reinforcement for the trough-shaped folded plate. Alternatively, for the same span and reinforcement, the required depth of a trough-shaped folded plate would be less than that for a V-shaped folded plate. Usually, the main longitudinal reinforcement is placed at the bottom, as in a reinforced concrete beam, and is calculated by dividing the bending moment by the lever arm. For trough-shaped folded plates, the main longitudinal reinforcement can be conveniently placed in the bottom flange, thus obtaining a maximum effective lever arm. This is opposed to the V-shaped folded plates, where, due to insufficient space, some main reinforcement has to be placed at upper levels in the inclined plates, thereby reducing the effective lever arm. Hence we conclude that trough-shaped folded plates are more economical in concrete and steel consumption than are V-shaped plates. The horizontal top and bottom flanges of the trough shapes should not be kept unduly large (generally, each is kept on the order of the depth H or less); otherwise, transverse moments may increase unnecessarily.

Thus, for a given span, the cross-sectional dimensions of a V- or trough-shaped folded plate can be designed. For spans of 40 to 60 ft, a thickness of 4 in. is commonly adopted for cast-in-place structures. For a larger span or smaller thickness, an intermediate transverse diaphragm connecting the various folds is beneficial in resisting buckling tendencies.

Figure 15.2(a) shows the variation of transverse moments obtained from a computer analysis for a typical trough-shaped folded plate spanning 60 ft. It is seen that the moments in internal panels vary as in a continuous slab, with negative (hogging) moments at the valley and crown, and positive moments in between. The end panel, however, behaves similar to a cantilever slab, with hogging moments over almost the entire end bay, unless there is a wall/column support along the external edge. The moments in the end bay usually control the thickness of the plate, since they are much larger than the moments in the interior panels. To minimize the transverse cantilever action of the end bay, a vertical upstand beam with a horizontal flange can be provided at the ends, as shown in Fig. 15.2(b). The variation of transverse moments in this case, for the same span and dimensions, are also shown. The vertical beam and the horizontal bottom flange at the end together restrain the horizontal and vertical deflections at the edge, and hence the moments in the end bay are now much less than in the previous case. The upstand leg is easily prone to instability, due to longitudinal compressive stresses at its top, and hence it should be connected to the next bay top flange by an intermediate diaphragm. Instead of the upstand beam, an inclined leg, or an inclined leg plus a downstand rib, as shown in Fig. 14.15, may be used.

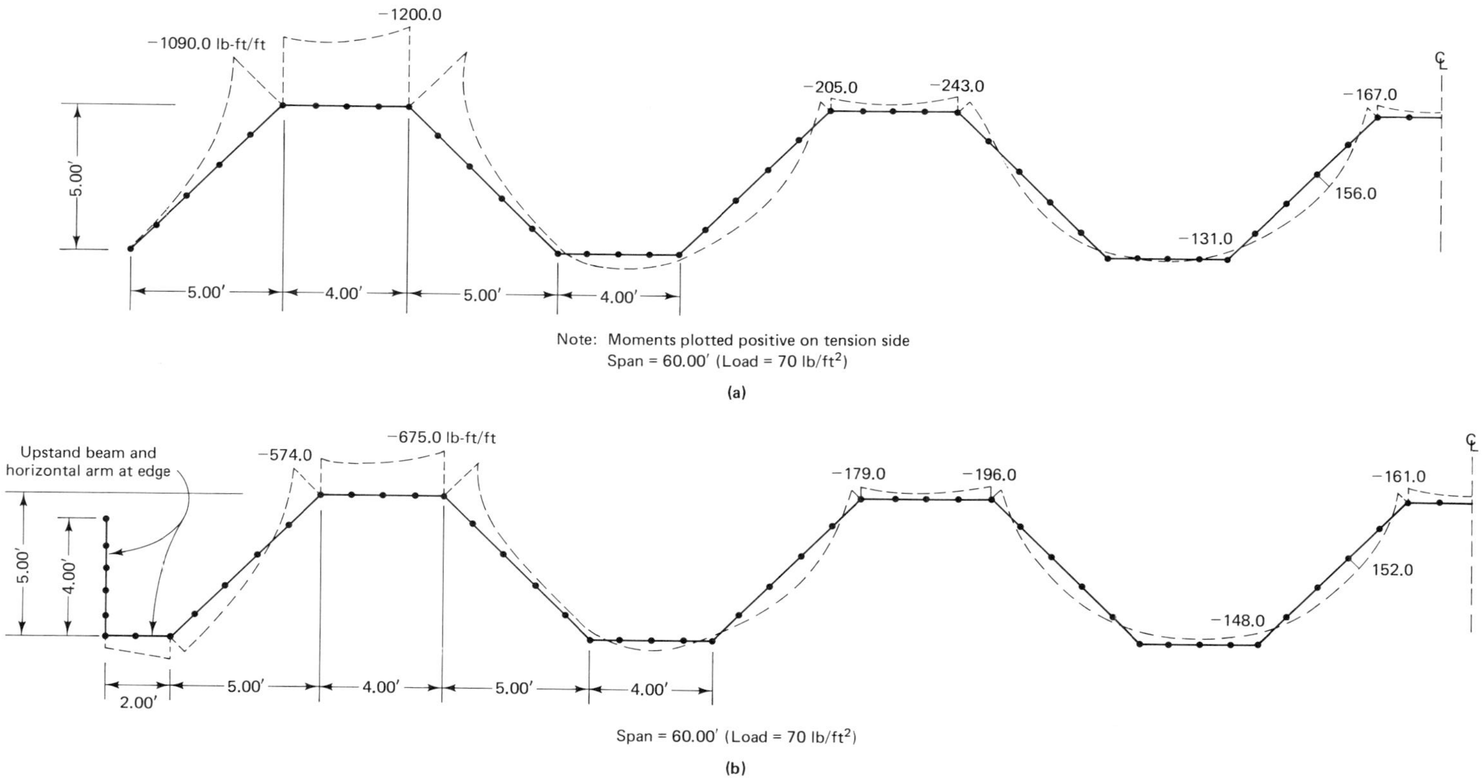

Figure 15.2 Variation of transverse moments for a typical trough-shaped folded plate that spans 60 ft. (a) Behavior of a roof without an upstand beam at its edge. (b) Reduction in end moments due to the presence of an "upstand beam."

15.5 PROPORTIONING THE DIMENSIONS OF A BARREL SHELL

For a given span L, the cross-sectional depth of a circular cylindrical barrel shell should also be kept between $\frac{1}{8}$ and $\frac{1}{12}$ of the span. Angle ϕ_k, the angle that the tangent at the valley makes with the horizontal, is kept between 20 and 45°. Angles between 30 and 40° are preferred. Angles greater than 45° are not used, so as to avoid double formwork when pouring concrete, and also because a larger perimeter is required to cover the same bay width. If the depth of the curved cross section of the shell itself is less than $L/8$ to $L/12$, the balance should be made up by the provision of sufficiently deep edge beams.

As an example, if we consider a single-bay "long" shell, L/a should be at least equal to 3 for beam action, where a is the radius of the shell. If $\phi_k = 35°$, the depth of the cross section of the shell is

$$H = a(1 - \cos 35°) = 0.1808a$$

This can be rewritten as

$$H = \frac{0.1808L}{L/a}$$

Now if we substitute $L/a = 3$, we have $H = 0.0603L$. This value is less than $L/12$, and hence the balance would be made up by providing edge beams.

Internal shells of a multibarrel shell exhibit beam action in the longitudinal direction, even for L/a as low as 1.67, since their horizontal displacements at the valleys are restrained due to symmetry. For such a shell, if $\phi_k = 40°$, we have

$$H = a(1 - \cos 40°) = 0.234a$$

or

$$H = \frac{0.234L}{L/a}$$

For $L/a = 1.67$, we then have

$$H = \frac{0.234L}{(1.67)}$$

$$H = 0.14L \simeq \frac{L}{7.1}$$

Hence, in this case, the depth of the shell itself appears to be sufficient, and no edge beams will be necessary. It can easily be checked that for L/a values between 1.67 and 3.0, and for angle ϕ_k varying between 30 and 40°, the bay width [which is equal to $2a(\sin \phi_k)$ or $2L \sin \phi_k/(L/a)$] varies between $\frac{1}{3}L$ and $\frac{3}{4}L$. Wilby and Khwaja [1] have given ready tables for finding stress resultants in internal and external bays of these shells, for several different spans ranging from 40 to 90 ft, with and without edge beams. For each span, values are given for at least three different bay widths: $\frac{1}{2}L$, $(L/2 + 2 \text{ ft})$, and $(L/2 - 2 \text{ ft})$. Angle ϕ_k in these tables generally ranges between 35 and 40°.

Based on the considerations discussed above, the designer should be able to determine the shell cross-sectional dimensions, such as a, ϕ_k, depth of edge beams, etc., for a given span. The thickness of the shell is generally kept between 2.5 and 3 in., 3 in. being more common to facilitate construction of cast-in-place shells. The edge beams are generally about 8 or 9 in. thick. As in folded plates, the transverse moments in an internal shell are generally hogging at the valley and crown, and are sagging in between. The values of these moments are very small, but for large bay widths, nominal thickening at the valleys may be necessary. As discussed for folded plates, unless the external edge of the end bay of the shell is supported on a wall (or beam and columns) along the span, an upstand edge beam or an inclined outstanding leg should be provided in order to reduce negative moments in the end bay.

15.6 DESIGN OF REINFORCEMENT

Having finalized the cross-sectional shape and profile, we can analyze the shell by any of the methods discussed in Chapters 13 and 14. We shall next consider the design of reinforcement in shell roofs.

The basis for design of the reinforcement will depend on the provisions in the applicable prevailing code of practice [2–4]. Provisions of such codes regarding minimum percentages of steel, type of steel (ordinary or high-strength bars), maximum spacing of bars, diameters, etc., have already been discussed briefly in Chapter 4, and they are also applicable to barrel shell or folded plate roofs. Once the analysis of the shell has been made to obtain the required internal stress resultants, a designer familiar with reinforced concrete design should have no difficulty in designing and detailing its reinforcement. Since the thickness of the shell will be small, the arrangement of reinforcement should be kept as simple as possible, so that placing of steel is facilitated.

We shall now discuss some aspects of the design of longitudinal, transverse, and shear reinforcement in barrel shells and folded plate structures.

Longitudinal Reinforcement

As in reinforced concrete beams, the longitudinal bending moment in the shell is assumed to be resisted by a couple formed from the compressive stresses in the concrete above the neutral axis and the tensile forcc in the reinforcement below. For an economical design, the total tensile force in the section below the neutral axis, obtained from an elastic analysis, is assumed to be resisted by the main steel reinforcement, which is placed at the bottom. Shells designed on this

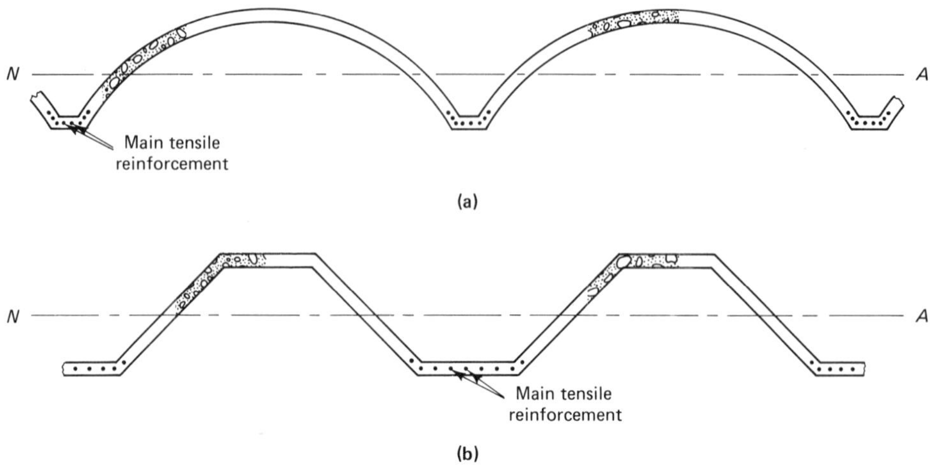

Figure 15.3 Tensile reinforcement arrangements for resisting longitudinal bending moments. (a) Typical placement for a multibarrel shell. (b) Typical placement for a trough-shaped folded plate.

basis have performed satisfactorily, and hence most codes permit this. This reinforcement is concentrated in the valleys of barrel shells without edge beams, at the bottom of edge beams in V-shaped folded plates where edge beams are provided, and in the bottom flanges of trough-shaped folded plates (Fig. 15.3). For resisting the hogging moment at the support of a continuous shell, this steel will, however, be placed near the top surface, above the neutral axis.

The size of the main tensile reinforcement should vary from No. 4 to No. 8 bars, as in a normal reinforced concrete beam. Some bars can be curtailed, as the stresses reduce near the supports. In the remaining tensile zone, as well as in the compression zone above the neutral axis, minimum steel for temperature and shrinkage must be provided. As per IASS (International Association of Shell and Spatial Structures) recommendations [2], this requirement amounts to a total of 0.6% of the shell cross section in the two orthogonal directions, or, say, 0.3% in the longitudinal direction alone. This reinforcement is provided in two layers, one layer on each face of the shell. The ACI code [3] specifies minimum steel of 0.35% in the remaining tensile zone and a lower percentage in the compression zone.

Transverse Reinforcement

In the transverse direction, reinforcement is required to resist the resultants M_ϕ and N_ϕ, the shears Q_ϕ generally being very small. The transverse section is thus designed as a section subjected to a bending moment plus direct compression or tension. In multibay barrels or folded plates, the bending moment is generally hogging at the valley and crown, and sagging in between. Thus steel is required at the top (outer face of the shell) at the valleys and the crowns, and at the bottom face in between. Due to a rapid change in the sign of the moment over a rather short distance, the ACI code [3], however, recommends placement of the steel on both top and bottom faces. Values of M_ϕ and N_ϕ are usually small, and hence minimum steel of say, 0.3%, provided for temperature and shrinkage, is sufficient for resisting the moments, except for large bay widths or for end bays.

The shell or plate reinforcement can be designed by simple elastic theory or by ultimate theory using proper load factors. The N_ϕ forces are usually small and could often be neglected when calculating the reinforcement. Bar sizes of No. 2 or No. 3 are generally sufficient in the transverse direction. Welded wire mesh may also be used for shells with shallow curvatures. When beam theory is used for the analysis (Chapter 14), one obtains M_ϕ and N_ϕ as constants along the span. Actually, however, these values will reduce near the supports. Hence for economical design, the spacing of transverse reinforcement could be increased in the quarter spans near the supports.

Shear Reinforcement

An element of the shell is subjected to in-plane longitudinal and transverse stress resultants N_x and N_ϕ, plus a shear $N_{x\phi}$, as shown in Fig. 15.4(a). N_x is compressive above the neutral axis and tensile below, whereas N_ϕ may be tensile or compressive. We must, therefore, calculate the principal tensile stress resultant and provide reinforcement to resist it. This resultant can be calculated by using the formulas derived in strength of materials for two-dimensional stresses. We obtain the maximum principal tension N_1 as

$$N_1 = \frac{N_x + N_\phi}{2} + \sqrt{\left(\frac{N_x - N_\phi}{2}\right)^2 + (N_{x\phi})^2} \tag{15.1a}$$

N_1 is inclined at an angle θ to the x-axis, which can be determined from

$$\tan 2\theta = \frac{2N_{x\phi}}{N_x - N_\phi} \tag{15.1b}$$

At the supports, N_x and N_ϕ are zero, and the element is in a state of pure shear, with only $N_{x\phi}$ acting. The principal tension has therefore the same numerical value as $N_{x\phi}$, and it is inclined at 45° to the x-axis at all points on the cross section. On the other hand, at midspan, $N_{x\phi}$ is zero, and N_x and N_ϕ themselves are in the directions of the principal stress resultants. Between $x = 0$ and $x = L/2$, the direction of the principal tension (below the neutral axis) changes from 45° with the x-axis, at the support, to 0° at the midspan, $x = L/2$. Thus the longitudinal steel should ideally be provided along the principal tensile stress trajectory, as shown in Fig. 15.4(b). This condition would not be very convenient in practice. Therefore, as for reinforced concrete beams, we provide main longitudinal steel along the x-axis, as discussed above, and also provide additional steel to resist the principal tension or "diagonal tension" near the supports. Since we are already providing

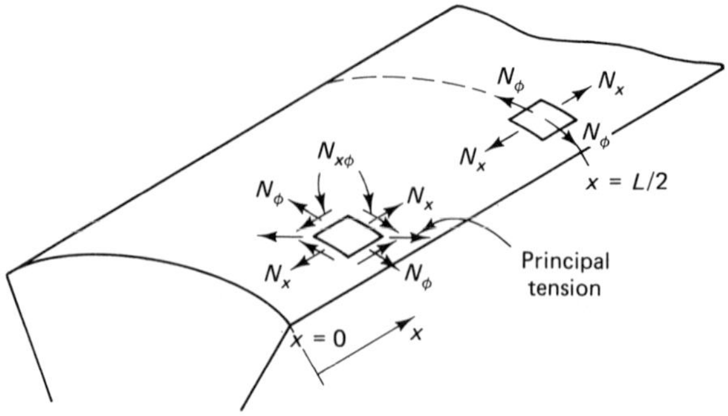

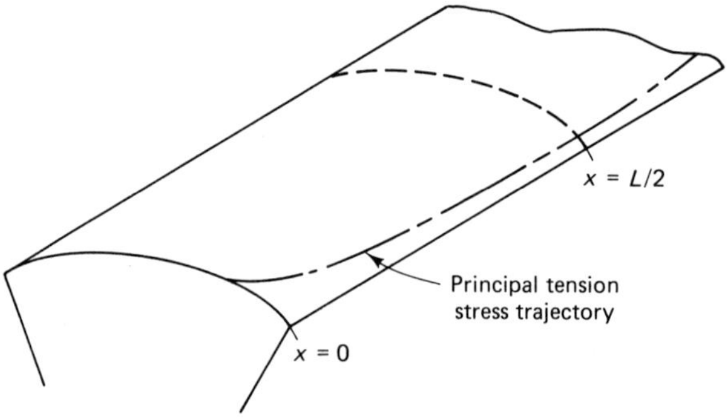

Figure 15.4 Stresses affecting shear reinforcement in a shell roof. (a) Resultants N_ϕ, N_x, and $N_{x\phi}$ acting on an element. (b) Stress trajectory, showing the direction of principal tension in the shell.

longitudinal steel for the N_x forces and transverse steel for N_ϕ (plus M_ϕ), as discussed previously, the additional steel is, strictly speaking, required only for the tension due to $N_{x\phi}$. However, it is common practice to calculate the principal tension due to $N_{x\phi}$, N_ϕ, and N_x near the supports, and to provide additional steel for resisting this tension, neglecting the transverse and longitudinal bars already provided. The additional steel is placed at an angle of 45° (in plan) with the x-axis, and is usually required in the region from support to quarter span [Fig. 15.5(a)].

As for reinforced concrete beams, we can also bend some of the main longitudinal bars, to resist part of the diagonal tension near the supports. However, this is not commonly done, since bending of bars along the curved barrel shell surface is difficult. (For folded plates, though, such bending is possible.) Also, the main bars would have a large diameter, whereas the shell thickness would be on the order of only 3 in.

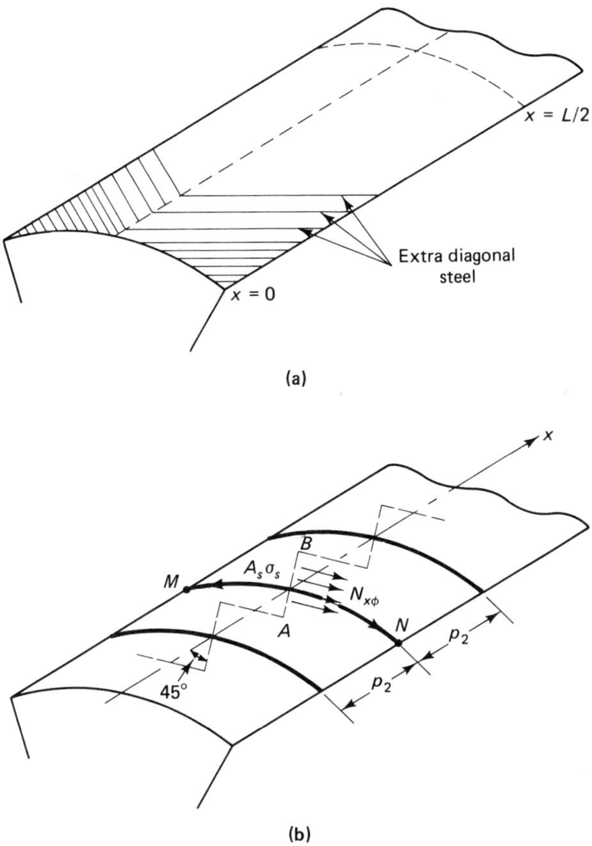

Figure 15.5 Shear reinforcement arrangements in shell roofs. (a) Additional reinforcement provided to resist the diagonal tension near supports. (b) Alternative arrangement, to reduce congestion of reinforcement, by including diagonal tension steel as a part of the transverse reinforcement.

Shear Reinforcement: Alternative Arrangement

With the provision of additional diagonal bars, there are now five layers of steel reinforcement near the supports: two transverse, two longitudinal, and one diagonal. This causes heavy congestion in a 2- to 4-in.-thick shell or folded plate. An alternative approach, which reduces this congestion, is to delete diagonal steel altogether and design the transverse reinforcement to resist the diagonal tension. The transverse reinforcement near the supports will now resist the effects of M_ϕ, N_ϕ, as well as those of $N_{x\phi}$.

A simplified procedure that one can follow for designing reinforcement to resist the diagonal tension due to $N_{x\phi}$ is similar to that followed for the design of stirrups in reinforced concrete beams. Spacing of reinforcement for resisting M_ϕ and N_ϕ is obtained independently, as discussed earlier for transverse reinforcement. If this spacing is p_1, and p_2 is the spacing of transverse reinforcement required for resisting the diagonal tension, then the final spacing, p, can be approximately obtained from the following equation (assuming identical bar sizes):

$$\frac{1}{p} = \frac{1}{p_1} + \frac{1}{p_2} \tag{15.2}$$

For finding p_2, let us proceed according to the derivation of the equation for the pitch of stirrups in reinforced concrete beams.

Figure 15.5(b) shows the transverse steel placed at a pitch p_2. A typical bar, MN, resists the diagonal tension acting on a portion AB which is inclined at 45° to both the x-axis and the bar. Length $\overline{AB}$ is equal to $p_2 \cos 45°$. Now, at the supports, the diagonal tension is numerically equal to $N_{x\phi}$. If σ_s is the allowable stress in the reinforcement MN, we must have

$$\sigma_s A_s \cos 45° = (N_{x\phi})(\overline{AB}) = N_{x\phi}(p_2) \cos 45°$$

or

$$p_2 = \frac{\sigma_s A_s}{N_{x\phi}} \tag{15.3}$$

where A_s is the cross-sectional area of the transverse reinforcement MN. Generally, the transverse reinforcement will be in two layers, one on each face. Then A_s will be equal to the area of two bars.

Using Eqs. (15.2) and (15.3), we can find the required spacing of transverse reinforcement at various distances x from the support. The effect of $N_{x\phi}$ will generally be very small beyond the quarter-span. Hence between quarter-span and midspan, spacing p_1, required for resisting M_ϕ and N_ϕ, will be sufficient. Between quarter-span and the support, this spacing should be reduced. Providing resistance for diagonal tension in this manner, without an additional layer of diagonal steel, simplifies the placing of reinforcement and is desirable in a thin shell. The authors have designed many barrel shells and folded plates in this manner, without adding any diagonal bars. It should be noted that Eq. (15.2) is an approximate equation which gives a spacing that is conservative compared to that obtained by calculating principal tensile stresses under forces N_x, N_ϕ, and $N_{x\phi}$.

We have thus seen that the design of main longitudinal and shear reinforcement for a shell roof is very similar to the design of such reinforcements in reinforced concrete beams. In fact, we have been discussing reinforcement for "long" shells, which behave essentially as beams in the longitudinal direction. The only additional feature for reinforcement in a shell is the consideration of transverse reinforcement.

15.7 ILLUSTRATIVE EXAMPLES FOR REINFORCEMENT DESIGN

We shall now illustrate the design of reinforcement for the two simple shells analyzed in Sections 13.4 and 14.4. We shall obtain the required areas of steel by simple elastic theory, although one may use the load factor method with equal ease. We shall also assume grade 40 steel, with an allowable tensile stress of 20,000 psi, since IASS recommendations [2] discourage the use of high-strength bars. In the authors' opinions, stresses in the high-strength reinforcement are generally restrained in simple shell structures, due to other code provisions of maximum bar spacing, etc., and hence high-strength bars have also been used.

Design of the Single-Barrel Shell of Section 13.4

We had obtained, in Table 13.6, stress resultants for a single-barrel shell spanning 42 ft. More accurate values obtained by a computer analysis were given in Table 14.11 for the first harmonic. Values after summation over nine harmonics were also given in parentheses in Table 14.11. We shall use these values for the design of reinforcement.

Longitudinal steel. Values of N_x at midspan, obtained from Table 14.11, are here plotted over the depth of the shell cross section, as shown in Fig. 15.6. It is seen that N_x varies almost linearly from top to bottom. The maximum compressive stress at the top is 16,900/(12)(3) = 469.44 psi, which is acceptable for a 3000-psi strength concrete. Generally, these stresses are kept lower than those in reinforced concrete beams, to avoid buckling of the thin shell.

If h_1 and h_2 denote the distances of the neutral axis from the top and bottom of the shell, respectively (Fig. 15.6), then assuming a linear variation of N_x for simplicity, we have

$$h_1 = 0.754 + \frac{5.08}{(5.08 + 11.23)} \times (1.675 - 0.754) = 1.041 \text{ ft}$$

and

$$h_2 = 2.924 - 1.041 = 1.883 \text{ ft}$$

Angle ϕ_o at the level of the neutral axis is then obtained from the equation

$$12.5(1 - \cos \phi_o) = 1.041$$

or

$$\phi_o = 23.55°$$

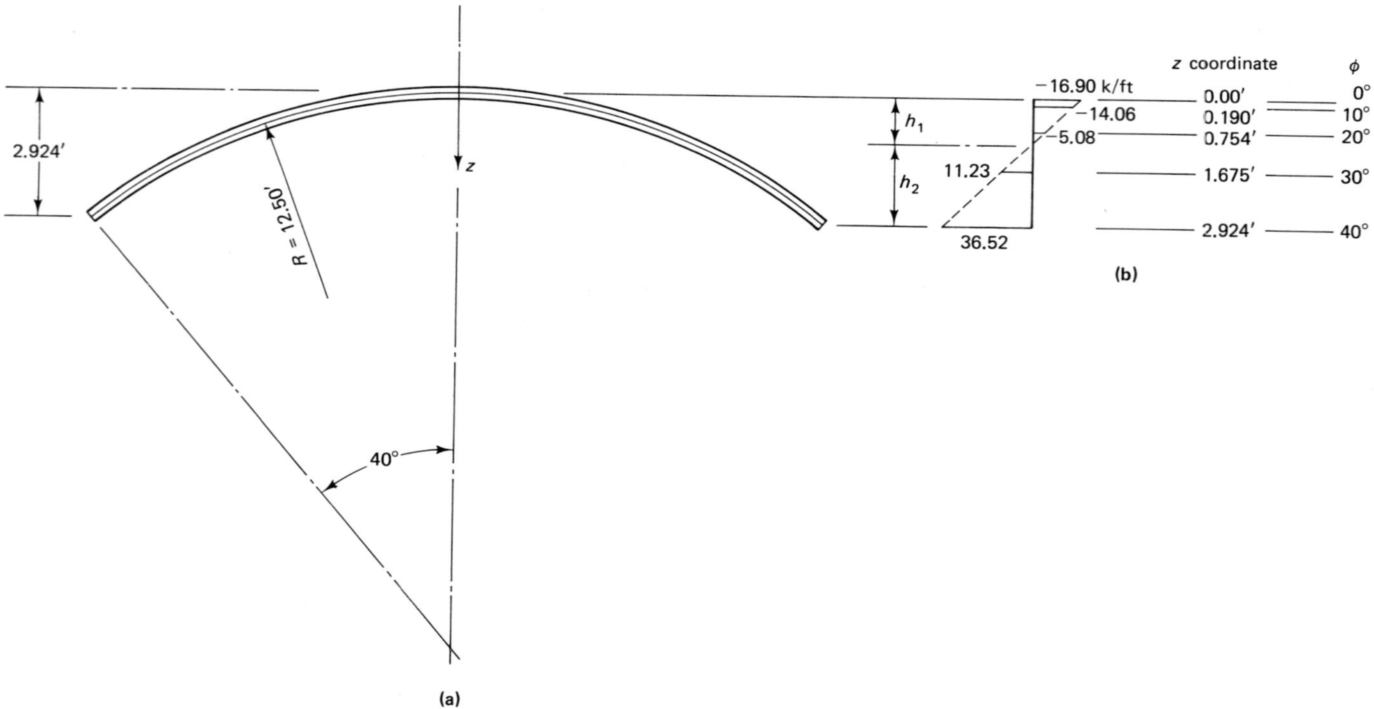

Figure 15.6 Stress resultant N_x obtained for the example single-barrel shell of Section 13.4. (a) Configuration of barrel. (b) Variation of N_x plotted versus coordinates z and ϕ.

Hence the curved length, L_1, of the shell (half portion) below the neutral axis, which is in tension, is given by

$$L_1 = 12.5(40° - 23.55°) \times \frac{\pi}{180} = 3.59 \text{ ft}$$

Thus the total tensile force is equal to (36.52/2)(3.59)(2) = 131.11 kips. The total area of main longitudinal steel, using grade 40 rebar, is then

$$A_s = \frac{131.11}{20} = 6.56 \text{ in.}^2$$

We can provide six No. 7 bars at the bottom of each half of the shell to accommodate this requirement. These bars can be in six rows, one bar in each layer, as shown in Fig. 15.7(a). With the transverse steel in two layers, the clear cover will be a little more than $\frac{1}{2}$ in. over the transverse bars, at the bottom of the shell. If required, the shell may be thickened in the bottom region.

The bending moment is resisted by a couple formed from the compressive force in the concrete and the tensile force in the steel. In computing the total tensile force above, the lever arm between the compressive and tensile forces was approximately $\frac{2}{3}h_1 + \frac{2}{3}h_2$, as seen from Fig. 15.6(b). For a symmetrical folded plate, which has the neutral axis at midheight, this would become two-thirds of the depth of the section. However, since we are concentrating the tension steel at the bottom of the shell, say with its center at about 6 in. from the bottom face, the actual effective lever arm will be $\frac{2}{3}h_1 + (h_2 - 0.5)$. For economy, one may use this new lever arm, and thus reduce the longitudinal steel in proportion to the ratio of the old to the new lever arm. However, we shall neglect such a refinement in this example.

From among the six No. 7 bars on each side, we can easily curtail two bars at quarter-span and continue the rest to the support. In the remaining tension zone, the ACI code [3] recommends a minimum steel of 0.35% of the cross-sectional area. This works out to be

$$\frac{0.35(12)(3)}{100} = 0.126 \text{ in.}^2/\text{ft}$$

Thus we can provide an area of 0.063 in.^{2}/ft on each face. The spacing of the bars should not exceed five times the shell thickness, i.e., 15 in. We can, therefore, provide No. 3 bars at 15 in. center-to-center on each face in the remaining tension zone below the neutral axis. In the compression zone, although we could reduce this steel to 0.3%, we shall keep the same steel throughout.

Transverse steel. We can now proceed to design the transverse reinforcement for this shell. Assuming a cover of $\frac{3}{4}$ in., we can take the effective depth as approximately 2 in. From Table 14.11 we note that M_ϕ is hogging throughout,

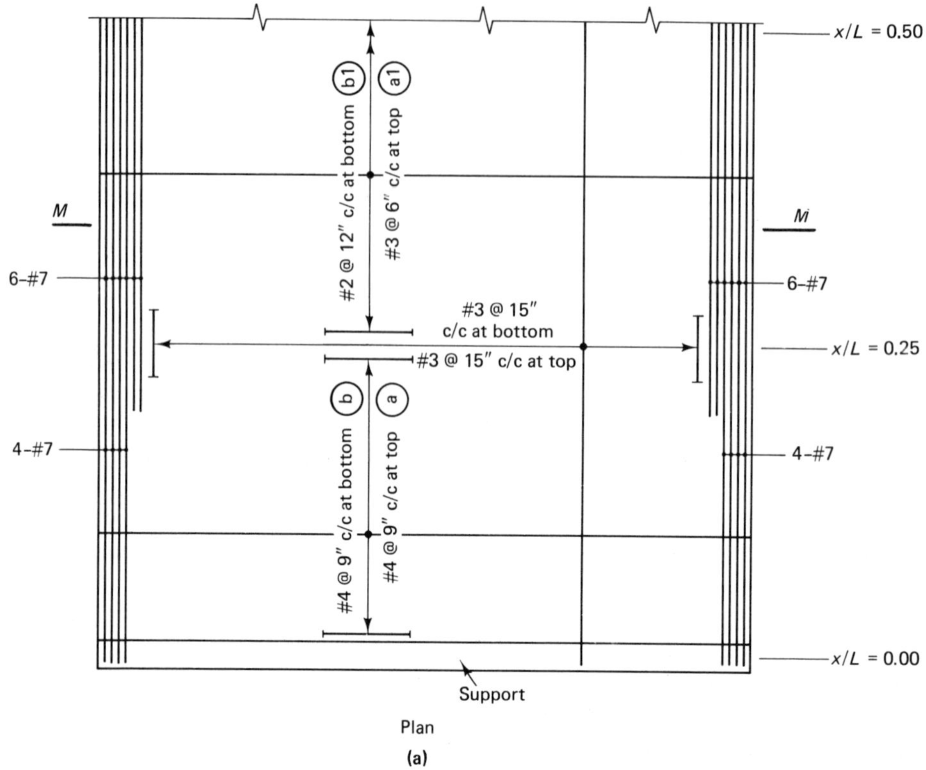

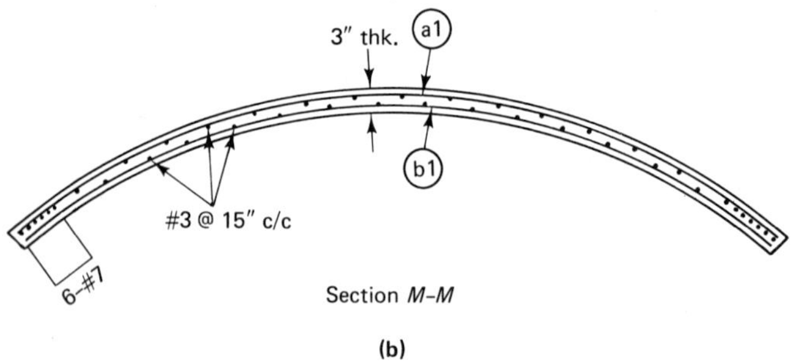

Figure 15.7 Reinforcement layout for the example shell of Section 13.4. (a) Plan layout. (b) Section view.

due to the cantilever action of the single barrel, and its maximum value is seen to be

$$\text{maximum } M_\phi = 701.89 \text{ ft-lb/ft}$$

The corresponding value of N_ϕ is

$$N_\phi = -1472.3 \text{ lb/ft}$$

Transferring this compressive force to the tension steel, we obtain a statically equivalent effective moment M_ϕ as

$$\text{effective } M_\phi \simeq 701.89 + \frac{1472.3(2 - 1.5)}{12} = 763.24 \text{ ft-lb/ft}$$

Hence the required steel area will be

$$A_s \simeq \frac{763.24(12)}{20{,}000(0.87)(2)} - \frac{1472.3}{20{,}000} = 0.19 \text{ in.}^2/\text{ft}$$

We can provide No. 3 bars at 6 in. center-to-center at the top face of the shell. Although there is no reversal of moment in this shell, we can also provide nominal steel at the bottom of No. 2 bars at 12 in. center-to-center.

Steel for diagonal tension. As discussed previously, we need to design for the diagonal tension in the shell. From Table 14.11 we find that at the supports, the maximum value of $N_{x\phi}$ is 4851 lb/ft at $\phi = 30°$, which is near the neutral axis. Hence the diagonal tension stress is $4851/(12)(3) = 134.75$ psi.

Assuming that this stress is resisted by transverse steel of No. 4 bars in the lower layer only, we obtain, from Eq. (15.3),

$$p_2 = \frac{20{,}000(0.2)}{4851} \times 12 = 9.89 \text{ in.}$$

Also, as calculated above, for resisting M_ϕ and N_ϕ, the area of transverse steel required at the top is 0.19 in.^{2}/ft, which can be satisfied by No. 4 bars at 12 in. center-to-center. Thus we can provide No. 4 bars at 9 in. center-to-center in both top and bottom layers near the support, in order to accommodate both the diagonal tension and transverse stresses.

As we move away from the support, at $x/L = 0.2$ and 0.3, the computer analysis (with summation over nine terms) gives maximum values (not listed in Table 14.11) of $N_{x\phi}$ as, respectively, 3606 and 2620 lb/ft. Similar values could also be obtained approximately by multiplying the maximum values at the support in Table 13.6 (or in Table 14.11 with $n = 1$) by $\cos(\pi x/L)$. (Of course, these values would not be as accurate as those obtained in the computer analysis.) At $x/L = 0.2$, the diagonal tension stress is thus a little over 100 psi [i.e., (3606/12)/(3)]. At $x/L = 0.25$, the stress will be much less than 100 psi, which the concrete itself can resist, and only nominal steel will be required at the bottom. Thus, from $x/L = 0.25$ to $x/L = 0.5$, we can provide the same transverse steel as that provided at midspan; i.e., No. 3 bars at 6 in. center-to-center at the top and No. 2 bars at 12 in. center-to-center at the bottom.

We did not actually calculate any values for the principal tension stress above. The value for this stress can easily be checked at $x/L = 0.25$ for $\phi = 30°$. Using linear interpolation, for simplicity, we have

$$N_{x\phi} = \frac{3606 + 2620}{2} = 3113.0 \text{ lb/ft}$$

Also, from Table 14.11, we find at $x/L = 0.5$,

$$N_\phi = -225.35 \text{ lb/ft}$$

and at $x/L = 0.25$,

$$N_x = 11{,}226.5(0.75) = 8419.88 \text{ lb/ft}$$

We have assumed here that at $x/L = 0.25$, N_x is 75% of its value at midspan (similar to the bending moment in a simply supported beam with uniform loading). We also assume N_ϕ is constant throughout the span. Substituting these values into Eq. (15.1a), we have

$$N_1 = \frac{8419.88 - 225.35}{2} + \sqrt{\left(\frac{8419.88 + 225.35}{2}\right)^2 + (3113.0)^2}$$

$$= 9424.16 \text{ lb/ft}$$

This resultant will give a tensile stress greater than 100 psi. N_1 makes an angle θ with the x-axis, which is given by

$$\tan 2\theta = \frac{2N_{x\phi}}{N_x - N_\phi} = \frac{2(3113.0)}{8419.88 + 225.35} = 0.72$$

or

$$\theta = 17.88°$$

Thus N_1 acts almost along the longitudinal direction. We have provided longitudinal steel at the bottom to take the full tensile force due to N_x, as in a "cracked section" of a reinforced concrete beam. Hence it would be incorrect to provide steel again for the principal tension, which includes the effect of N_x also. Thus our earlier procedure of providing steel for diagonal tension due to $N_{x\phi}$ is satisfactory, as for reinforced concrete beams.

The arrangement of reinforcement is as shown in Fig. 15.7. All longitudinal steel should be well anchored in the transverse diaphragms by L-bends. This also helps in resisting small moments, M_x, which develop at the junction of the shell and the diaphragm.

Design of the Example Barrel Shell of Section 14.4

The single shell with edge beams of Section 14.4 can be designed in a manner similar to the preceding example. For stress resultants we shall use the values in parentheses (involving a summation over nine terms of the Fourier series) found in Table 14.12 obtained from a computer analysis.

Longitudinal steel. From the N_x values in Table 14.12 it is seen that the shell is entirely in compression, the neutral axis lying slightly below the bottom of the shell. The longitudinal stresses at the top and bottom of the edge beams are -6358.2 and 103,420.0 lb/ft^2. Assuming a nearly linear variation in this stress, as also seen from Fig. 14.6(c), the height h_2 of the tensile zone from the bottom of the beam is

$$h_2 = \frac{103{,}420.0}{103{,}420.0 + 6358.2} \times 5.0 = 4.71 \text{ ft}$$

Thus we can calculate the total tensile force in the edge beam as

$$\frac{103.42}{2}(4.71)(0.75) = 182.67 \text{ kips}$$

Hence the required steel can be calculated as

$$A_s = \frac{182.67}{20} = 9.133 \text{ in.}^2$$

We can provide 12 No. 8 bars in the tension zone of each beam in six layers (two bars in each layer). Four of these bars could be curtailed at the quarter-span. In the remaining tension zone of the beam, we shall provide No. 4 bars at 12 in. center-to-center on each face. Now, looking at the compression region of the shell, from Table 14.12 we can calculate the maximum compressive stress in the concrete as

$$\frac{21{,}222.25}{12(3)} = 589.5 \text{ psi}$$

This stress is low for a 3,000 psi concrete strength. Hence, in the shell itself, we can provide nominal steel of No. 3 bars at 15 in. center-to-center on each face.

Transverse steel. For the design of transverse steel, we notice from Table 14.12 that the values for M_ϕ and N_ϕ are very small. Thus, we can provide No. 3 bars at 12 in. center-to-center on the top and bottom faces. This steel will, of course, have to be increased from quarter-span to the support, in order to cater for diagonal tension, as seen below.

Steel for diagonal tension. To design for diagonal tension in this shell, from Table 14.12, we find the maximum value of $N_{x\phi}$ as 8758.6 lb/ft, which occurs almost at the neutral axis, as expected. Assuming that this resultant is resisted by No. 4 bars in two layers, we obtain from Eq. (15.3),

$$p_2 = \frac{20{,}000(0.2)(2)}{8758.6} \times 12 = 10.96 \text{ in.}$$

The spacing p_1 of No. 4 bars required to resist the moment $M_\phi = 299.43$ ft-lb/ft (causing tension on the outer face) at this location is obtained as follows.* Neglecting the small force N_ϕ, we have

$$A_s = \frac{299.43(12)}{20{,}000(0.87)(2)} = 0.103 \text{ in}^2/\text{ft}$$

Hence we find

$$p_1 = \frac{0.2}{0.103} \times 12 = 23.30 \text{ in.}$$

Thus the pitch required for bars on the outer face will be given by

$$\frac{1}{p} = \frac{1}{23.30} + \frac{1}{10.96}$$

or

$$p = 7.45 \text{ in.}$$

We can use an average spacing for No. 4 bars of 8 in. center-to-center on both faces, since the top bars will resist the moment M_ϕ plus some diagonal tension, and the bottom bars will resist the remaining diagonal tension. We can continue this steel up to the quarter-span, where principal tension there becomes almost parallel to the x-axis. Between $x/L = 0.25$ and 0.5, we shall use No. 3 bars at 12 in. center-to-center on the top and bottom faces, as stated earlier. In the edge beam, a computer analysis yields values of $N_{x\phi}$ somewhat lower than considered above (not listed in Table 14.12). However, we can use the same steel for stirrups in the beam, i.e., No. 4 bars at 8 in. center-to-center up to the quarter-span, and No. 3 bars at 12 in. center-to-center between quarter-span and midspan. Figure 15.8 shows a cross section of the shell with its reinforcement.

We can see from the preceding examples that omitting diagonal steel for resisting principal tensions simplifies the arrangement of reinforcement in thin shells considerably, and this is very helpful at the construction site. Similar procedures can be used for designing the reinforcement of multibarrel shells or folded plates.

15.8 DESIGN OF "SHORT" CYLINDRICAL SHELLS

So far we have discussed mainly the design of "long" circular cylindrical shells or folded plates used for roof structures. These shells behave as beams of curved cross section in the longitudinal direction. For silos that store fertilizer or grains, the structure has to have a very large bay width, as well as a large height. At the top

*The value of M_ϕ given in Table 14.12 is at the midspan ($x/L = 0.5$). We assume it to be the same throughout the span. At the support itself, M_ϕ will theoretically be zero.

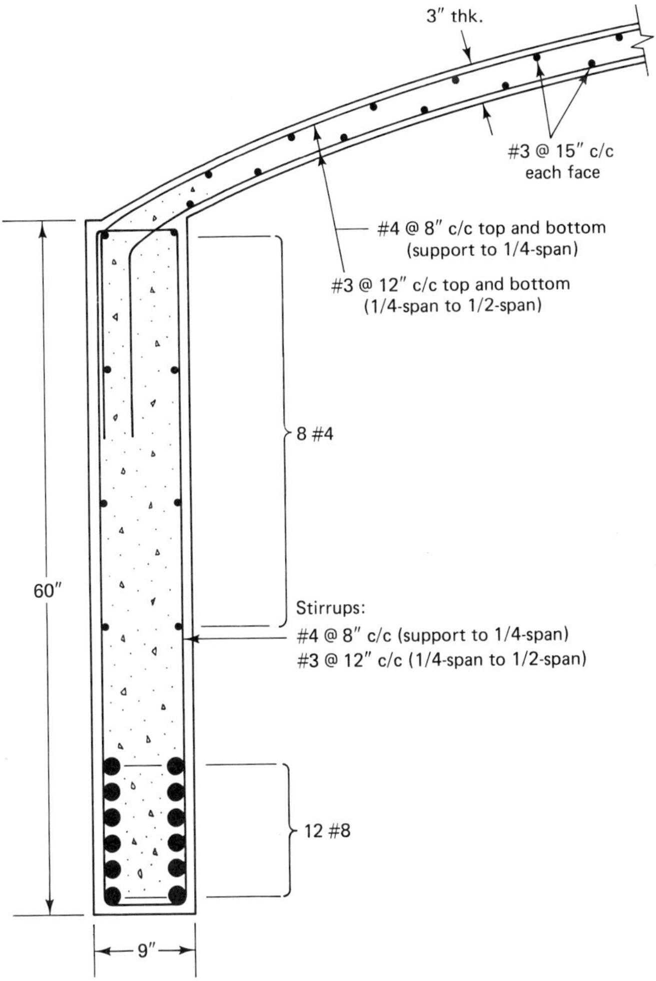

Figure 15.8 Reinforcement detail for the cross section of the example shell of Section14.4.

of the shell, a conveyor may be required for transporting the material. These structures consist of shells that have parabolic or catenary cross-sectional profiles, so that greater heights can be obtained in the same bay width, as compared to circular profiles. The shells span on arch ribs, which are typically at 20- to 30-ft centers. The shells transfer the loads to the arches, which are designed as two-pinned or fixed arches. A typical structure, of bay width 120 ft and height 75 ft, with a parabolic profile, is shown in Fig. 15.9. Similar structures may also be used for aircraft hangars.

In these structures, the shells do not behave as beams of curved cross section in the longitudinal direction. The longitudinal stress no longer varies (near) linearly

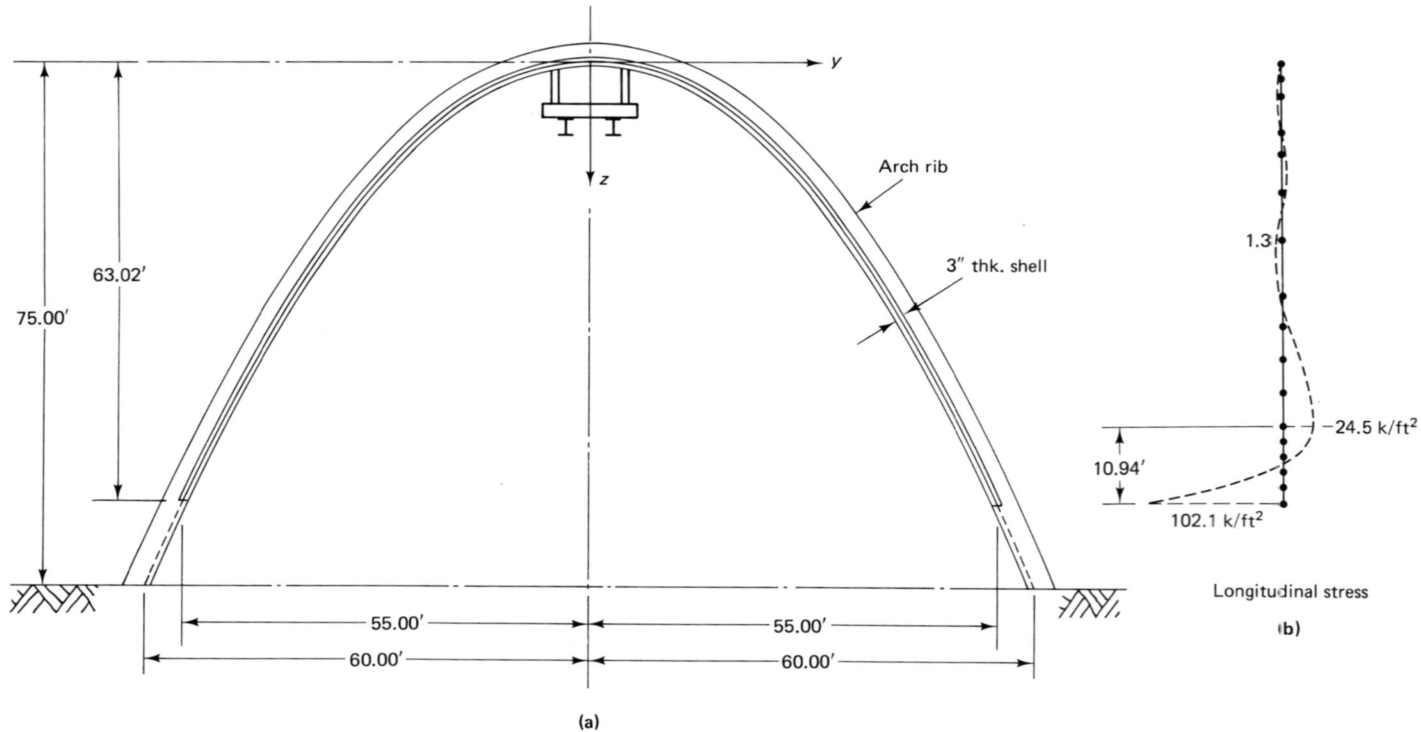

Figure 15.9 Silo or hangar structure with a parabolic profile. (a) Dimensions of the cross section. (b) Variation in the longitudinal stress at midspan under a load of 70 lb/ft^2 over the surface area.

from compression at the top to tension at the bottom. Figure 15.9(b) shows the variation in the longitudinal stress at the midspan of the shell shown in Fig. 15.9(a), assuming its span as 25 ft center-to-center over the arch ribs. The loading assumed was 70 lb/ft^2 over the surface area (the conveyor loading was not considered separately, for simplicity), and the midsurface profile of the shell was designed as a parabola given by the equation $y^2 = 48.0z$. This equation gives $z = 75$ ft for $y = 60$ ft. The shell is assumed to stop at $y = 55$ ft (i.e., at about 12 ft above the ground). The analysis was carried out with a computer program using the Scordelis plate element (discussed in Chapter 14), by idealizing the shell cross section as a number of straight plate elements, each simply supported at the arch ribs.

Figure 15.9(b) shows that only a small portion (about 11 ft) at the bottom of the shell acts as a beam spanning between the arch ribs. The remaining upper portion of the shell has negligible longitudinal stresses and transfers the loads to the lower beam portion through N_ϕ forces (i.e., by arch action only). The values of M_ϕ and N_ϕ are also very small (not plotted here), so that a 3-in.-thick shell with nominal reinforcement is sufficient, except in the bottom beam portion. Thus it is more economical to span the distance between the ribs as a shell than simply as a slab, which would require much greater thickness and reinforcement for spanning 25 ft.

For performing an approximate analysis of this shell, we can calculate N_ϕ and $N_{x\phi}$ by membrane analysis. Their nonzero values at the lower edges can be applied in the opposite direction as loads on the lower portion of the shell, considered as a simple (inclined) beam spanning between the arch ribs. Resultant N_ϕ will give loads in the plane of the beam, while $N_{x\phi}$ forces will give an axial force plus a bending moment in the beam. The portion of the shell near the bottom edge may be thickened to accommodate reinforcement.

Sometimes the shell and the arch ribs are combined into a single form whose cross section ϕ = *constant* is a continuous corrugated or trough shape. In that case the action of the structure is no longer that of shells spanning on arch ribs, but rather that of a continuous series of arches, each spanning a transverse distance of 120 ft.

15.9 DESIGN OF DIAPHRAGMS

In a barrel shell or folded plate roof structure, the shell transfers the loads to diaphragm supports by means of the internal shear forces $N_{x\phi}$. The diaphragms are then designed as simple transverse beams spanning between the columns and which carry the loads $N_{x\phi}$ from the shell. In a multibay shell, the columns may be spaced at distances longer than the shell bay width. In such a case, the diaphragm carries loads from two or more bays, and transfers these to the columns, thus spanning as a simply supported or continuous beam.

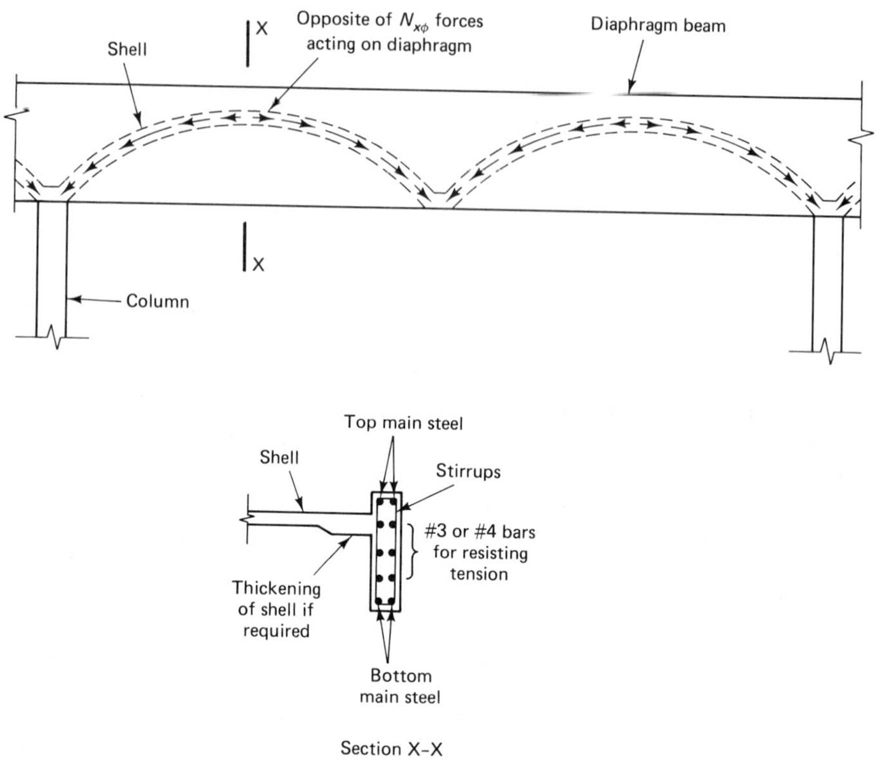

Figure 15.10 Multibarrel shell supported on a transverse diaphragm that spans as a simple rectangular beam.

A simple type of diaphragm would be a beam of rectangular cross section, as shown in Fig. 15.10. The height of the beam will depend on its span between the columns and will not be less than the depth of the shell itself. $N_{x\phi}$ forces from the shell are applied, in the opposite direction, as loads on the beam. Knowing $N_{x\phi}$ values at the node points ($\phi = 0°$, $10°$, $20°$, etc.) of the shell, we can multiply them by the sum of each one-half shell segment width on either side of the node, to obtain the inclined loads at the node points. These loads can then be resolved into vertical and horizontal components acting on the beam. The horizontal components are self-equilibrating, being symmetrical for a single-bay shell or for an internal bay of a multibarrel shell. For an external bay, the loads on the two halves may not be symmetrical; however, the resulting unbalanced force is equilibrated by a similar but opposite unbalanced force, acting on the other external bay (assuming a symmetrical configuration). Thus, obviously, there is no net horizontal load on the column (unless compatibility of the horizontal displacement of the beam with that of the column is considered). The horizontal loads thus cause only axial tension forces in the diaphragm beam, for which reinforcement throughout the depth of the beam is provided. The total of the vertical loads on the beam should add up to be one-half the load on the shell. These loads will give rise to bending moments and shears in the beam, for which appropriate bottom (and top, if the beam is continuous over columns) steel and stirrups are provided.

Design of a Diaphragm Beam for the Example Shell of Section 13.4

For the single barrel shell of Fig. 13.4, values of the resultant $N_{x\phi}$ at the support were summarized in Table 13.6. Also, values obtained by a computer analysis, with summation over nine harmonics, were given in parentheses in Table 14.11. We shall use the values of $N_{x\phi}$ found in Table 13.6 and obtained for the first harmonic. Opposite forces to $N_{x\phi}$ will act as loads on the diaphragms. The vertical and horizontal components of these loads are calculated, as shown in Table 15.1. In the table, the width of each chord is calculated as $10°(\pi/180)(12.5) = 2.182$ ft. The chord widths contributing to each node are as indicated.

Static check. We can perform a check on the statics of our analysis by comparing the load acting on the shell with that calculated from the vertical components of $N_{x\phi}$ acting on the diaphragm. The total vertical load on one-quarter of the shell is obtained from Table 15.1 as 8854.18 lb. The vertical load assumed on the shell, for a chord width s, is ps. The first term of the Fourier series of this load expanded over $x = 0$ to $x = L$ is given by $(4ps/\pi)\sin(\pi x/L)$. Hence, the total load on the chord, when considering the first term in the Fourier series, is given by

$$W = \int_0^L \frac{4(ps)}{\pi} \sin \frac{\pi x}{L}\, dx = \frac{8}{\pi^2}(ps)L = \frac{8}{\pi^2} \times \text{(actual total load)}$$

TABLE 15.1 Vertical and Horizontal Components of the Loads $N_{x\phi}$ That Act on the Diaphragm Beam of Section 13.4

Node	ϕ (deg)	$N_{x\phi}$ (lb/ft) (From Table 13.6)	Chord Width, s (ft)	Nodal Force, $F = N_{x\phi}\, s$ (lb)	Horizontal Component, $H = F \cos \phi$ (lb)	Vertical Component, $V = F \sin \phi$ (lb)
0	0	0.00	2.182	0.00	0.00	0.00
1	10	2692.64	2.182	5875.34	5786.08	1020.24
2	20	4465.80	2.182	9744.38	9156.72	3332.77
3	30	4125.73	2.182	9002.34	7796.26	4501.17
4 (Valley)	40	0.00	1.091	0.00	0.00	0.00
				Total Vertical Force	=	8854.18

where p is the distributed loading on the shell. In the present case, $L = 42$ ft and $p = 62.5$ lb/ft^2, and hence the load acting on one-quarter of the area of the shell is

$$W = \frac{8}{\pi^2}(62.5)\left(\frac{\pi}{180^\circ}(40^\circ)(12.5)\right)\left(\frac{42}{2}\right) = 9284.0 \text{ lb}$$

The percent error between this load and the total load of 8854.18 lb indicated in Table 15.1 is 4.63%, which is small. The simplified calculations made by us in Table 15.1 assume a linear variation of $N_{x\phi}$ from node to node. More refined calculations, using Simpson's trapezoidal rule or by calculating $N_{x\phi}$ values at closer node points, can be made.

The Fourier series for the load (ps) over the strip of width s and uniformly distributed over the span L is

$$ps = \sum_{n=1,3,5,\ldots}^{\infty} \frac{4ps}{n\pi} \sin\frac{n\pi x}{L}$$

Integrating the above, the Fourier series for the total load W will be

$$W = \int_0^L ps\,dx = \sum_{n=1,3,5,\ldots}^{\infty} \left(\frac{8}{n^2\pi^2}\right)(ps)L$$

This series has a $1/n^2$ type of convergence. Now $N_{x\phi}$ can be approximately assumed to vary linearly from $x = 0$ to L, as in a beam. Its even extension in the interval $(-L, L)$ will, therefore, have discontinuous derivatives at $x = -L$, 0, and L. Hence the Fourier series of $N_{x\phi}$ will also have a $1/n^2$ type of convergence. Thus, for a correct design, the values of $N_{x\phi}$ obtained in Table 13.6 for the first harmonic should be increased by multiplying by a factor $\pi^2/8 = 1.234$. For the longitudinal stress, this factor will be $\pi^3/32 = 0.969$. (Refer also to discussions of Examples 10.4 and 12.4.) Indeed, in Tables 14.11 and 14.12, we generally see an increase and decrease, respectively, in the values of $N_{x\phi}$ and N_x when we take a summation over nine terms of the series, as opposed to taking only one term. Since the summation of $1/n^2$ for $n = 1$ is 1, whereas for the first nine terms ($n = 1, 3, 5, \ldots, 9$) it is 1.183, the values of $N_{x\phi}$ within parentheses in Tables 14.11 and 14.12 should be about 18% higher than those outside the parentheses. The reader can verify that the agreement is reasonably good in Table 14.12, but it is not so good in Table 14.11. Of course, we cannot expect the agreement to be exact, because, in a detailed computer analysis, the values of $N_{x\phi}$ for each harmonic will also depend on the interaction with other stress resultants and displacements.

Design of diaphragm beam. We can now proceed to design the details of the diaphragm beam. Since we have only a one-bay shell, the span of the beam will simply be the bay width, 16.07 ft. (The effective span will be somewhat less, depending on the column width.) Since the span is small, the depth of the beam can be chosen as just that of the shell, or say, 39 in. Its width can be taken as 8

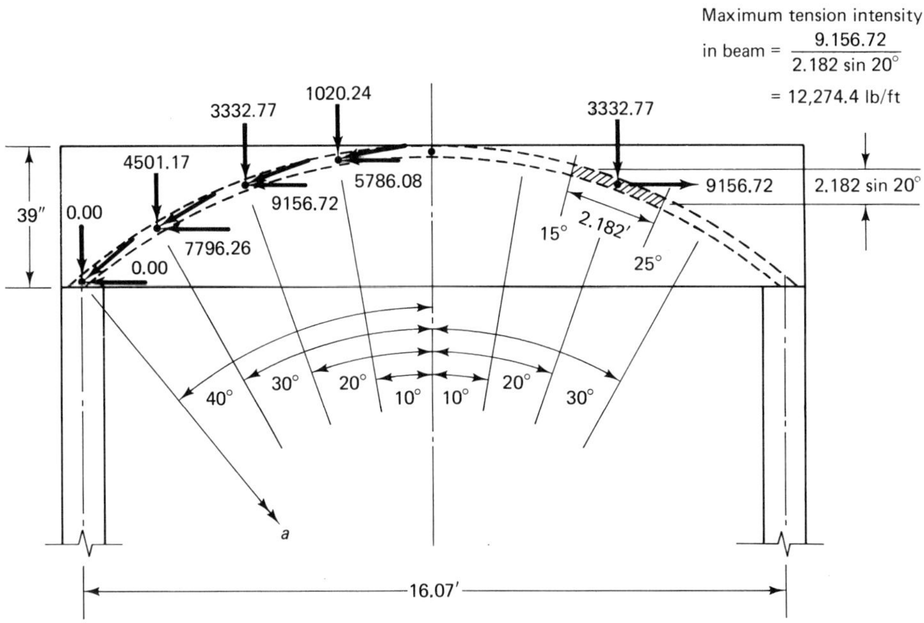

Figure 15.11 Loads acting on diaphragm beam for use in design.

in. The loads acting on the beam are shown in Fig. 15.11. We see that the maximum horizontal force of 9156.72 lb acts over a vertical height (length) of 2.182 sin 20° = 0.746 ft, as shown in Fig. 15.11. Hence the maximum intensity of axial tension in the beam is 9156.72/0.746 = 12,274.4 lb/ft. This force can be multiplied by $\pi^2/8$ in light of the discussion above. Hence we can calculate the required area of tension steel as

$$A_s = \frac{12{,}274.4(\pi^2/8)}{20{,}000}$$

$$= 0.757 \text{ in.}^2/\text{ft}$$

We can, therefore, simply provide No. 5 bars at 9 in. center-to-center on each face. These bars should be provided in the form of horizontal stirrups, so as to provide proper anchorage for the tensile forces at the beam ends.

The vertical loads on the diaphragm beam are symmetrical about the midspan. The maximum midspan bending moment on the beam is thus calculated (see Fig. 15.11) as

$$M_x = 1020.24(8.035 - 12.5 \sin 10°) + 3332.77(8.035 - 12.5 \sin 20°)$$
$$+ 4501.17(8.035 - 12.5 \sin 30°)$$
$$= 26{,}548.06 \text{ ft-lb}$$

When multiplied by $\pi^2/8$, we have $M_x = 32{,}752.35$ ft-lb. To this moment we must add the bending moment due to the self-weight of the beam, which is

$$\text{beam self-weight} = \left(\frac{150}{144}\right)(39)(8) = 325 \text{ lb/ft}$$

Hence the total moment at midspan is

$$M_x = 32{,}752.35 + 325\left(\frac{16.07^2}{8}\right) = 43{,}243.5 \text{ ft-lb}$$

For resisting this moment, two No. 6 bars at the bottom will be more than sufficient.

The maximum shear force in the beam is calculated as

$$Q_x = (1020.24 + 3332.77 + 4501.17)\left(\frac{\pi^2}{8}\right) + 325 \times 8.035$$

$$= 13{,}534.78 \text{ lb}$$

This force is very small for an 8-in. by 39-in. beam, and hence nominal two-legged stirrups of No. 3 bars at 12 in. center-to-center can be provided. The beam reinforcement detail is shown in Fig. 15.12.

The beam-type diaphragm considered so far has the disadvantage that it hides the smooth profile of the shell, which architects would like to be seen in the elevation of the building. For this, as well as other reasons, the shell is usually cantilevered a few feet beyond the diaphragms. Internal beam-type diaphragms in a multispan building also do not permit passing of service ducts, which otherwise could be conveniently accommodated just below the roof. For this reason the diaphragms may be designed as arches along the shell cross section. We have already seen such arched diaphragms for the shell in Fig. 15.9(a). The arches for barrel roofs should preferably have a tie beam connecting the columns. These roofs are also sometimes designed without tie beams, as single-bay or multibay arch portals. The arches can easily be designed to resist the shell loads, by using suitable tables or by using a simple plane-frame computer program for analysis. The Portland Cement Association [5] has also given ready tables for analyzing multibay circular arch portals.

For folded plates, a convenient way of avoiding the above-mentioned disadvantages of simple beam diaphragms is to provide a diaphragm beam below the folded plate, as shown in Fig. 15.13. Shears $N_{x\phi}$ from the shell are transmitted to the beam along the profile of the folded plate at the ends. Since these forces act along the inclined webs, they cause only compression in the web (and a small tension in the top flange), without any bending moment. Hence a little thickening of the plates at the ends, with nominal reinforcement, is sufficient to transmit these loads. For an end bay, the forces in the two halves of the top flange may not balance each other, but the resulting bending moments induced in the webs will be small and can be designed for. At the bottom of the plate, the diaphragm beam will be subjected to concentrated vertical and horizontal loads, for which it can readily be designed.

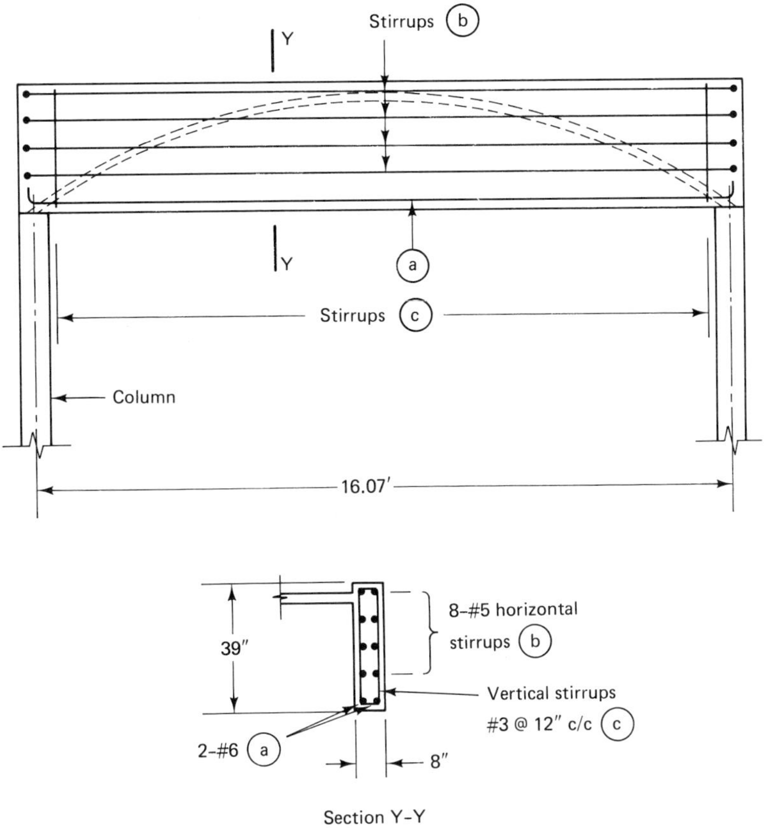

Figure 15.12 Diaphragm beam details for the single-barrel shell roof.

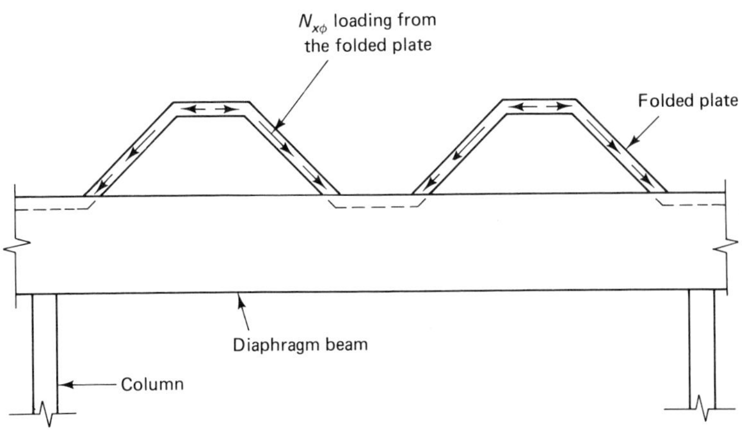

Figure 15.13 Folded plate spanning over a transverse diaphragm beam.

Such diaphragms for the folded plate permit free passage of ducts, etc., in the roof portion of a factory; however, they reduce some headroom along the column lines, since they are below the folded plate. For the barrel shell, a circular arch will not be a funicular curve for the $N_{x\phi}$ loads from the shell. Such an arch will therefore develop substantial bending moments and shear forces, and hence there is no advantage in providing a diaphragm beam below the shell.

15.10 EPILOGUE

Although, so far in this text, we have covered several topics of practical interest related to the analysis and design of shell structures, many aspects remain to be discussed. Topics such as stability of shells, their postelastic behavior, formwork and construction techniques of precast or cast-in-place shells, details of computer analysis techniques or programs, etc., have not been discussed. There are excellent references that discuss such topics in detail. IASS recommendations [2] also give a topic-wise bibliography of literature on shell structures.

As stated in the preface, the aim of this book has been to introduce the reader to a basic understanding of the procedures for the analysis and design of shell structures. It is hoped that by this time the reader has gained this understanding and now has sufficient insight into the behavior of different shells. He or she can use this knowledge to design shell structures with greater confidence and to take up further detailed study of topics of interest.

REFERENCES

[1] Wilby, C. B., and Khwaja, I., *Concrete Shell Roofs,* Applied Science Publishers Ltd., Barking, Essex, England, 1977.

[2] "Recommendations for Reinforced Concrete Shells and Folded Plates," International Association for Shell and Spatial Structures, Madrid, 1979.

[3] "Building Code Requirements for Reinforced Concrete—ACI 318-84," ACI Committee 318, American Concrete Institute, Detroit, 1984.

[4] "Criteria for the Design of Reinforced Concrete Shell Structures and Folded Plates," Indian Standard 2210-1962, Indian Standards Institution, New Delhi, 1963.

[5] "Design Constants for Circular Arch Bents," Advanced Engineering Bulletin 7, Portland Cement Association, Skokie, Ill., 1963.

PROBLEMS

15.1. Design the reinforcement for the folded plate analyzed in Section 14.7. Also design a suitable diaphragm support for it.

15.2. Design the reinforcement for the shell roof over a bus stand, given in Problem 14.2. Design the diaphragm supports for this shell, assuming them as arches along the profile of the shell cross section and cantilevering out from central columns.

15.3. Design the reinforcement for the multibay folded plate roof of Problem 14.3. Also, design the diaphragm as a beam below the folded plate, assuming that the columns are spaced at every two bays.

15.4. For the multibarrel shell of Problem 13.3, design the reinforcement using **(a)** diagonal steel, and **(b)** transverse steel, to resist the principal tensions. Design the diaphragm support of this shell assuming columns are spaced at every two bays.

15.5. For the single-barrel shell example in Section 15.7 (see also Section 13.4), design the diaphragm support assuming that **(a)** it is a tied arch, and **(b)** it has no tie but forms an arched gable with the column supports. Assume suitable sizes for the columns.

15.6. The silo shell shown in Fig. 15.9 is simply supported over a span of 25 ft and carries a load of 70 lb/ft^2 uniformly distibuted over its surface area. Obtain the membrane stress resultants in the shell using the equations derived in Section 12.4.

Now apply the nonzero values of N_ϕ and $N_{x\phi}$ at the edge (i.e., at $y = 55$ ft) back on the shell in the opposite direction, thus making the edges free. Assume that the lower 11 ft of the shell acts as an edge beam between the supports and resists these applied edge loads, as discussed in Section 15.8. Obtain the final stress resultants in the shell, and compare the approximate values of N_x thus obtained with those from the computer analysis given in Fig. 15.9(b).

Index